EXOTIC STATES OF NUCLEAR MATTER

Edited by

Umberto Lombardo
Università di Catania & INFN–LNS, Italy

Marcello Baldo
Fiorella Burgio
Hans-Josef Schulze
INFN, Sezione di Catania, Italy

EXOTIC STATES OF NUCLEAR MATTER

Proceedings of the
International Symposium
EXOCT07

Catania University, Italy   11 – 15 June 2007

World Scientific

NEW JERSEY • LONDON • SINGAPORE • BEIJING • SHANGHAI • HONG KONG • TAIPEI • CHENNAI

Published by

World Scientific Publishing Co. Pte. Ltd.

5 Toh Tuck Link, Singapore 596224

USA office: 27 Warren Street, Suite 401-402, Hackensack, NJ 07601

UK office: 57 Shelton Street, Covent Garden, London WC2H 9HE

British Library Cataloguing-in-Publication Data
A catalogue record for this book is available from the British Library.

EXOTIC STATES OF NUCLEAR MATTER AND NEUTRON STARS
Proceedings of the International Symposium on EXOCT07

ISBN-13 978-981-279-703-2
ISBN-10 981-279-703-3

Printed in Singapore by World Scientific Printers

PREFACE

The *International Symposium on Exotic States of Nuclear Matter* (EXOCT 2007) was held at the University of Catania, Italy during the period 11-15 June, 2007. The present volume contains the contents of most of the talks in chronological order within their individual sections and the poster presentations delivered during the conference.

The program of the symposium was devoted to nuclear matter theory and aspects of nuclear astrophysics, with emphasis on the link to the physics of heavy ion reactions at intermediate energy and nuclei far from the stability line.

Supernovae, protoneutron stars, and neutron stars are astrophysical compact objects which contain baryonic matter under very different physical conditions. From the core, where densities among the highest in the universe are present, to the surface, characterized by sub-saturation average density, baryonic matter is expected to undergo a set of phase transitions and changes of structure. Astrophysical observational data are of invaluable relevance for the development of our understanding of the behavior of nuclear matter in different conditions of density and temperature. Since in heavy ion collisions at intermediate energy nuclear matter is compressed and then expands, a link with the astrophysics of compact objects can be naturally established. Despite the different physical conditions, especially at the level of the dynamical time scale, the connection between the two fields has grown rapidly in the last few years, with a very fruitful cross-fertilizing process which is expected to continue in the future. The link with the physics of nuclei at the dripline is also promising for understanding the neutron star surface. Therefore, we thought that it was time to review the status of these exciting and challenging developments in astrophysics and nuclear physics.

The symposium was a unique opportunity to discuss the many aspects of nuclear matter under extreme conditions and the corresponding possible exotic states like hyperonic matter, kaon condensates, and quark matter, which can appear both in astrophysical compact objects and in heavy ion collision experiments. To this purpose, leading experts from astrophysics, nuclear physics, and elementary particle physics have delivered reviews and specialized seminars, which have highlighted the links among the different fields and the role of the underlying fundamental processes. Prospects in future astrophysical observations and heavy ion experiments, with present and planned apparatus, have been strongly emphasized and the proceedings volume will be surely a reference book for all the researchers working in this wide research area.

We thank the Department of Physics and Astronomy of Catania University for hosting the symposium, the National Institute of Nuclear Physics (INFN), the Faculty of Sciences, and the Provincia Regionale di Catania for their financial support. We acknowledge in particular the European Community project "Asia-Europe Link in Nuclear Physics and Astrophysics" [CN/ASIA-LINK/008(94791)] for its financial support and for the wide participation of members and young researchers involved in the project.

The symposium could not have been organized without the essential help of the INFN administrative staff members Anna Linda Magrí and Cettina Lombardo. Their continuous assistance throughout the conference period was greatly appreciated by all participants. To them go our most grateful acknowledgements.

Marcello Baldo
Fiorella Burgio
Umberto Lombardo (Chairman)
Hans-Josef Schulze

Catania, Italy
November 2007

ORGANIZING COMMITTEE

Marcello Baldo	– INFN, Sezione di Catania
Fiorella Burgio	– INFN, Sezione di Catania
Umberto Lombardo (Chairman)	– Università di Catania & INFN-LNS
Hans-Josef Schulze	– INFN, Sezione di Catania

INTERNATIONAL ADVISORY COMMITTEE

David Blaschke	– Wroclaw University, Poland
Ignazio Bombaci	– University of Pisa, Italy
John Clark	– Washington University, St. Louis, USA
Pawel Haensel	– CAMK, Warsaw, Poland
Morten Hjorth-Jensen	– Oslo University, Norway
Bennett Link	– Montana State University, Bozeman, USA
Zhongyu Ma	– CIAE Beijing, China
Jérôme Margueron	– IPN Orsay, France
Dany Page	– UNAM Mexico City, Mexico
Arturo Polls	– Barcelona University, Spain
Sanjay Reddy	– LANL, USA
Stefan Rosswog	– International University Bremen, Germany
Hiroyuki Sagawa	– Aizu University, Japan
Nicolae Sandulescu	– Bucharest University, Romania
Toshitaka Tatsumi	– Kyoto University, Japan
Dmitry Yakovlev	– IOFFE Institute, St. Petersburg, Russia
Wei Zuo	– IMP Lanzhou, China
Mikhail Zverev	– Kurchatov Institute, Moscow, Russia

CONTENTS

Part G Nuclear Superfluidity 359

Part H Poster Session 413

PART A

Theory of Nuclear Matter EOS and Symmetry Energy

CONSTRAINING THE NUCLEAR EQUATION OF STATE FROM ASTROPHYSICS AND HEAVY ION REACTIONS

CHRISTIAN FUCHS

Institute of Theoretical Physics, University of Tübingen, Germany
E-mail: christian.fuchs@uni-tuebingen.de

The quest for the nuclear equation of state (EoS) at high densities and/or extreme isospin is one of the longstanding problems of nuclear physics. In the last years substantial progress has been made to constrain the EoS both, from the astrophysical side and from accelerator based experiments. Heavy ion experiments support a soft EoS at moderate densities while recent neutron star observations require a "stiff" high density behavior. Ab initio calculations for the nuclear many-body problem make predictions for the density and isospin dependence of the EoS far away from the saturation point. Both, the constraints from astrophysics and accelerator based experiments are shown to be in agreement with the predictions from many-body theory.

Keywords: Nuclear equation of state; Neutron stars; Heavy ion reactions.

1. Introduction

The isospin dependence of the nuclear forces which at present is only little constrained by data will be explored by the forthcoming radioactive beam facilities at FAIR/GSI, SPIRAL2/GANIL and RIA. Since the knowledge of the nuclear equation-of-state (EoS) at supra-normal densities and extreme isospin is essential for our understanding of the nuclear forces as well as for astrophysical purposes, the determination of the EoS was already one of the primary goals when first relativistic heavy ion beams started to operate. A major result of the SIS100 program at the GSI is the observation of a soft EoS for symmetric matter in the explored density range up to 2-3 times saturation density. These accelerator based experiments are complemented by astrophysical observations. The recently observed most massive neutron star with $2.1 \pm 0.2 \left(^{+0.4}_{-0.5}\right) \mathrm{M}_\odot$ excludes exotic phases of high density matter and requires a relatively stiff EoS. Contrary to a naive expectation, the astrophysical observations do, however, not stand in contradiction with those from heavy ion reactions. Moreover, we are in the fortunate situation that ab initio calculations of the nuclear many-body problem predict a density and isospin behavior of the EoS which is in agreement with both observations.

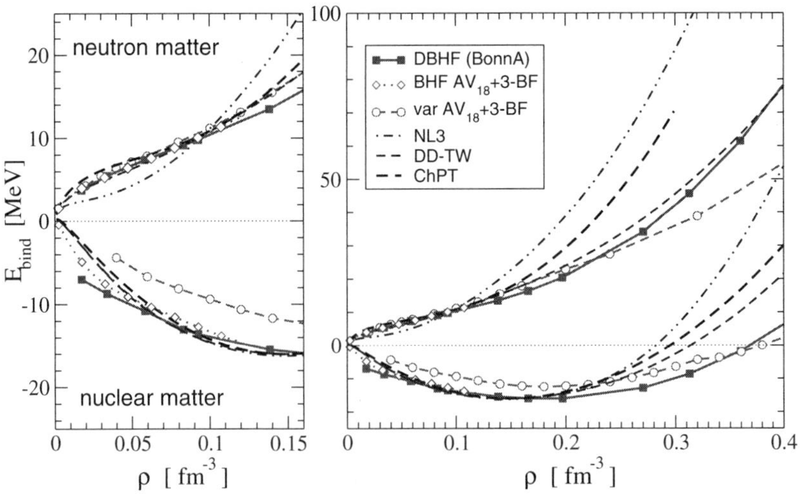

Fig. 1. EoS in nuclear matter and neutron matter. BHF/DBHF and variational calculations are compared to phenomenological density functionals (NL3, DD-TW) and ChPT+corr.. The left panel zooms the low density range. The Figure is taken from Ref. 6.

2. The EoS from Ab Initio Calculations

In *ab initio* calculations based on many-body techniques one derives the EoS from first principles, i.e. treating short-range and many-body correlations explicitly. This allows to make prediction for the high density behavior, at least in a range where hadrons are still the relevant degrees of freedom. A typical example for a successful many-body approach is Brueckner theory (for a recent review see Ref. 1). In the following we consider non-relativistic Brueckner and variational calculations[2] as well as relativistic Brueckner calculations.[3–5] It is a well known fact that non-relativistic approaches require the inclusion of - in net repulsive - three-body forces in order to obtain reasonable saturation properties. In relativistic treatments part of such diagrams, e.g. virtual excitations of nucleon-antinucleon pairs are already effectively included.

Fig. 1 compares now the predictions for nuclear and neutron matter from microscopic many-body calculations – DBHF[4] and the 'best' variational calculation with 3-BFs and boost corrections[2] – to phenomenological approaches (NL3 and DD-TW from Ref. 7) and an approach based on chiral pion-nucleon dynamics[8] (ChPT+corr.). As expected the phenomenological functionals agree well at and below saturation density where they are constrained by finite nuclei, but start to deviate substantially at supra-normal densities. In neutron matter the situation is even worse since the isospin dependence of the phenomenological functionals is less constrained. The predictive power of such density functionals at supra-normal densities is restricted. *Ab initio* calculations predict throughout a soft EoS in the density range relevant for heavy ion reactions at intermediate and low energies, i.e. up to about 3 ρ_0. Since the *nn* scattering length is large, neutron matter at subnu-

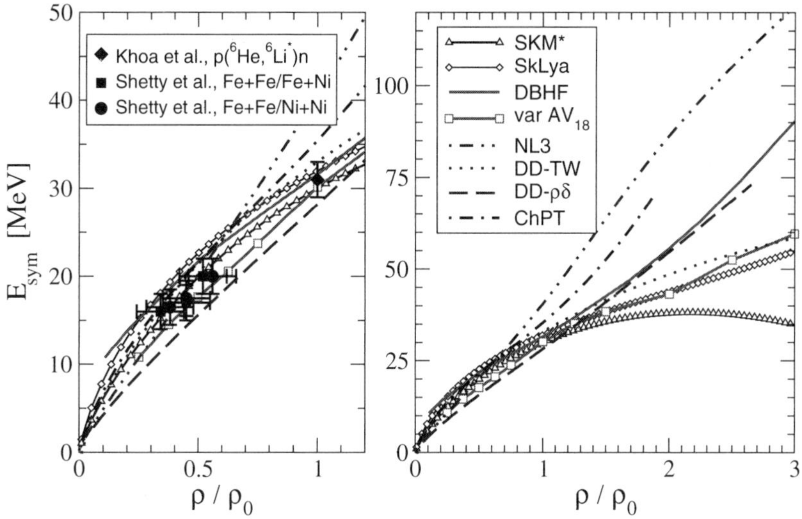

Fig. 2. Symmetry energy as a function of density as predicted by different models. The left panel shows the low density region while the right panel displays the high density range. Data are taken from Refs. 10 and 11.

clear densities is less model dependent. The microscopic calculations (BHF/DBHF, variational) agree well and results are consistent with 'exact' Quantum-Monte-Carlo calculations.[9]

Fig. 2 compares the symmetry energy predicted from the DBHF and variational calculations to that of the empirical density functionals already shown in Fig. 1. In addition the relativistic DD-$\rho\delta$ RMF functional[12] is included. Two Skyrme functionals, SkM* and the more recent Skyrme-Lyon force SkLya represent non-relativistic models. The left panel zooms the low density region while the right panel shows the high density behavior of $E_{\rm sym}$.

The low density part of the symmetry energy is in the meantime relatively well constraint by data. Recent NSCL-MSU heavy ion data in combination with transport calculations are consistent with a value of $E_{\rm sym} \approx 31$ MeV at ρ_0 and rule out extremely "stiff" and "soft" density dependences of the symmetry energy.[13] The same value has been extracted[10] from low energy elastic and (p,n) charge exchange reactions on isobaric analog states p($^6He,^6Li^*$)n measured at the HMI. At sub-normal densities recent data points have been extracted from the isoscaling behavior of fragment formation in low-energy heavy ion reactions with the corresponding experiments carried out at Texas A&M and NSCL-MSU.[11]

However, theoretical extrapolations to supra-normal densities diverge dramatically. This is crucial since the high density behavior of $E_{\rm sym}$ is essential for the structure and the stability of neutron stars. The microscopic models show a density dependence which can still be considered as *asy-stiff*. DBHF[4] is thereby stiffer than the variational results of Ref. 2. The density dependence is generally more complex than in RMF theory, in particular at high densities where $E_{\rm sym}$ shows a non-linear

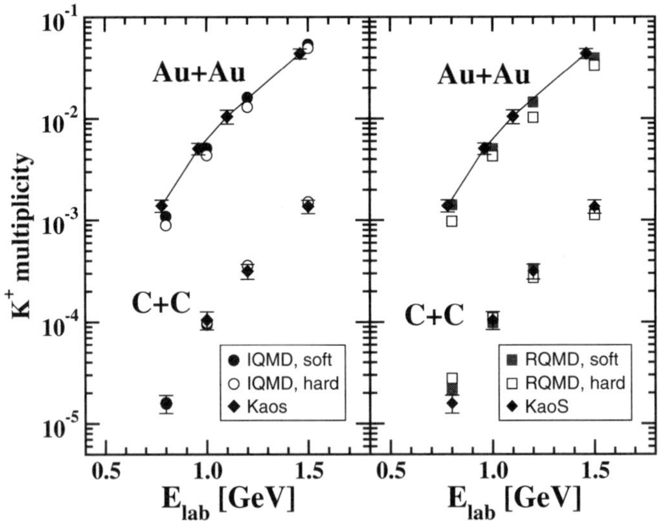

Fig. 3. Excitation function of the K^+ multiplicities in $Au + Au$ and $C + C$ reactions. RQMD[14] and IQMD[15] with in-medium kaon potential and using a hard/soft nuclear EoS are compared to data from the KaoS Collaboration.[16] The figure is taken from Ref. 6.

and more pronounced increase. Fig. 2 clearly demonstrates the necessity to better constrain the symmetry energy at supra-normal densities with the help of heavy ion reactions.

3. Constraints from Heavy Ion Reactions

Experimental data which put constraints on the symmetry energy have already been shown in Fig. 2. The problem of multi-fragmentation data from low and intermediate energy reactions is that they are restricted to sub-normal densities up to maximally saturation density. However, from low energetic isospin diffusion measurements at least the slope of the symmetry around saturation density could be extracted.[17] This puts already an important constraint on the models when extrapolated to higher densities. It is important to notice that the slopes predicted by the ab initio approaches (variational, DBHF) shown in Fig. 2 are consistent with the empirical values. Further going attempts to derive the symmetry energy at supra-normal densities from particle production in relativistic heavy ion reactions[12,18,19] have so far not yet led to firm conclusions since the corresponding signals are too small, e.g. the isospin dependence of kaon production.[20]

Firm conclusions could only be drawn on the symmetric part of the nuclear bulk properties. To explore supra-normal densities one has to increase the bombarding energy up to relativistic energies. This was one of the major motivation of the SIS100 project at the GSI where - according to transport calculation - densities between $1 \div 3\ \rho_0$ are reached at bombarding energies between $0.1 \div 2$ AGeV. Sensitive observables are the collective nucleon flow and subthreshold K^+ meson

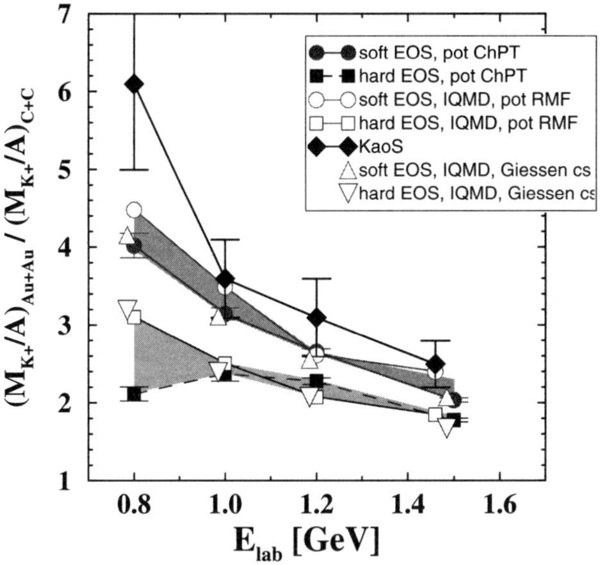

Fig. 4. Excitation function of the ratio R of K^+ multiplicities obtained in inclusive Au+Au over C+C reactions. RQMD[14] and IQMD[15] calculations are compared to KaoS data.[16] Figure is taken from Ref. 21.

production. In contrast to the flow signal which can be biased by surface effects and the momentum dependence of the optical potential, K^+ mesons turned out to an excellent probe for the high density phase of the reactions. At subthreshold energies the necessary energy has to be provided by multiple scattering processes which are highly collective effects. This ensures that the majority of the K^+ mesons is indeed produced at supra-normal densities. In the following I will concentrate on the kaon observable.

Subthreshold particles are rare probes. However, within the last decade the KaoS Collaboration has performed systematic high statistics measurements of the K^+ production far below threshold.[16,22] Based on this data situation, in Ref. 14 the question if valuable information on the nuclear EoS can be extracted has been revisited and it has been shown that subthreshold K^+ production provides indeed a suitable and reliable tool for this purpose. In subsequent investigations the stability of the EoS dependence has been proven.[15,21]

Excitation functions from KaoS[16,23] are shown in Fig. 3 and compared to RQMD[14,21] and IQMD[15] calculations. In both cases a soft (K=200 MeV) and a hard (K=380 MeV) EoS have been used within the transport approaches. Skyrme type forces supplemented by an empirical momentum dependence have been used. As expected the EoS dependence is more pronounced in the heavy Au+Au system while the light C+C system serves as a calibration. The effects become even more evident when the ratio R of the kaon multiplicities obtained in Au+Au over

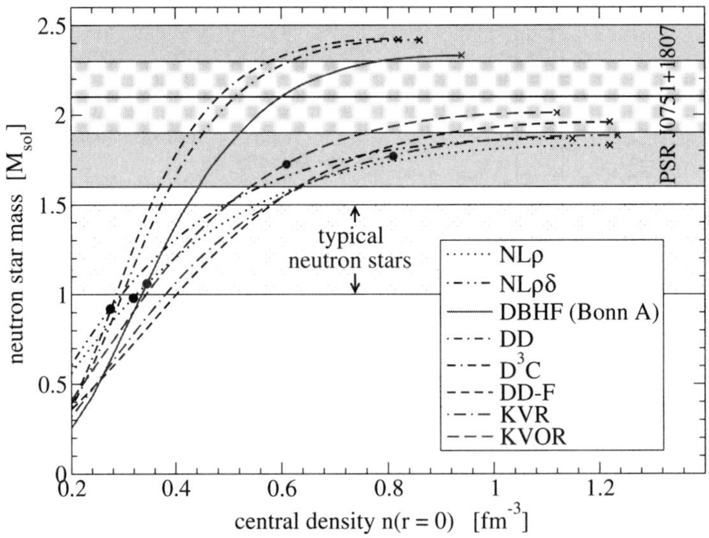

Fig. 5. Mass versus central density for compact star configurations obtained for various relativistic hadronic EsoS. Crosses denote the maximum mass configurations, filled dots mark the critical mass and central density values where the DU cooling process becomes possible. According to the DU constraint, it should not occur in "typical NSs" for which masses are expected from population synthesis[28] to lie in the lower grey horizontal band. The dark and light grey horizontal bands around 2.1 $M_\odot$ denote the 1σ and 2σ confidence levels, respectively, for the mass measurement of PSR J0751+1807.[29] Figure is taken from Ref. 30.

C+C reactions (normalized to the corresponding mass numbers) is built.[14,16] Such a ratio has the advantage that possible uncertainties which might still exist in the theoretical calculations cancel out to large extent. Comparing the ratio shown in Fig. 4 to the experimental data from KaoS,[16] where the increase of R is even more pronounced, strongly favors a soft equation of state. This result is in agreement with the conclusion drawn from the alternative flow observable.[24–27]

4. Constraints from Neutron Stars

Measurements of "extreme" values, like large masses or radii, huge luminosities etc. as provided by compact stars offer good opportunities to gain deeper insight into the physics of matter under extreme conditions. There has been substantial progress in recent time from the astrophysical side.

The most spectacular observation was probably the recent measurement[29] on PSR J0751+1807, a millisecond pulsar in a binary system with a helium white dwarf secondary, which implies a pulsar mass of $2.1 \pm 0.2\,(^{+0.4}_{-0.5})$ M$_\odot$ with 1σ (2σ) confidence. Therefore, a reliable EoS has to describe neutron star (NS) masses of at least 1.9 M$_\odot$ (1σ) in a strong, or 1.6 M$_\odot$ (2σ) in a weak interpretation. This condition limits the softness of the EoS in neutron star (NS) matter. One might therefore be worried about an apparent contradiction between the constraints derived from neutron stars and those from heavy ion reactions. While heavy ion reactions favor a

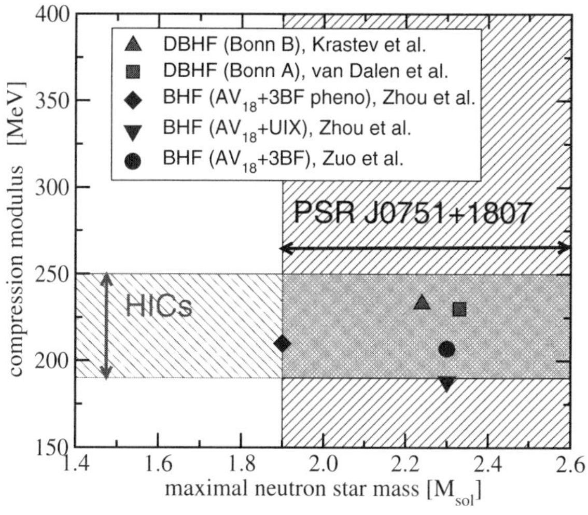

Fig. 6. Combination of the constraints on the EoS derived from the maximal neutron star mass criterium and the heavy ion collisions constraining the compression modulus. Values of various microscopic BHF and DBHF many-body calculations are show.

soft EoS, PSR J0751+1807 requires a stiff EoS. The corresponding constraints are, however, complementary rather than contradictory. Intermediate energy heavy-ion reactions, e.g. subthreshold kaon production, constrains the EoS at densities up to $2 \div 3\ \rho_0$ while the maximum NS mass is more sensitive to the high density behavior of the EoS. Combining the two constraints implies that the EoS should be *soft at moderate densities and stiff at high densities*. Such a behavior is predicted by microscopic many-body calculations (see Fig. 6). DBHF, BHF or variational calculations, typically, lead to maximum NS masses between $2.1 \div 2.3\ M_\odot$ and are therefore in accordance with PSR J0751+1807, as can be seen from Fig. 5 and Fig. 6 which combines the results from heavy ion collisions and the maximal mass constraint.

There exist several other constraints on the nuclear EoS which can be derived from observations of compact stars, see e.g. Refs. 30,31. Among these, the most promising one is the Direct Urca (DU) process which is essentially driven by the proton fraction inside the NS.[32] DU processes, e.g. the neutron β-decay $n \rightarrow p + e^- + \bar{\nu}_e$, are very efficient regarding their neutrino production, even in super-fluid NM and cool NSs too fast to be in accordance with data from thermally observable NSs. Therefore, one can suppose that no DU processes should occur below the upper mass limit for "typical" NSs, i.e. $M_{DU} \geq 1.5\ M_\odot$ ($1.35\ M_\odot$ in a weak interpretation). These limits come from a population synthesis of young, nearby NSs[28] and masses of NS binaries.[29] While the present DBHF EoS leads to too fast neutrino cooling this behavior can be avoided if a phase transition to quark matter is assumed.[33] Thus a quark phase is not ruled out by the maximum NS mass. However, corresponding quark EoSs have to be almost as stiff as typical hadronic EoSs.[33]

5. Summary

Heavy ion reactions provide in the meantime reliable constraints on the ispspin dependence of the nuclear EoS at sub-normal densities up to saturation density and for the symmetric part up to - as an conservative estimate - two times saturation density. These are complemented by astrophysical constraints at high densities. The present situation is in fair agreement with the predictions from nuclear many-body theory.

References

1. M. Baldo and C. Maieron, *J. Phys. G* **34**, R243 (2007).
2. A. Akmal, V. R. Pandharipande, D. G. Ravenhall, *Phys. Rev. C* **58**, 1804 (1998).
3. T. Gross-Boelting, C. Fuchs and A. Faessler, *Nucl. Phys. A* **648**, 105 (1999).
4. E. van Dalen, C. Fuchs and A. Faessler, *Nucl. Phys. A* **744**, 227 (2004); *Phys. Rev. C* **72**, 065803 (2005); *Phys. Rev. Lett.* **95**, 022302 (2005); *Eur. Phys. J. A* **31**, 29 (2007).
5. P. G. Krastev and F. Sammarrunca, *Phys. Rev. C* **74**, 025808 (2006).
6. C. Fuchs and H. H. Wolter, *Euro. Phys. J. A* **30**, 5 (2006).
7. S. Typel and H. H. Wolter, *Nucl. Phys. A* **656**, 331 (1999).
8. P. Finelli, N. Kaiser, D. Vretenar and W. Weise, *Nucl. Phys. A* **735**, 449 (2004).
9. J. Carlson *et al.*, *Phys. Rev. C* **68**, 025802 (2003).
10. D. T. Khoa, W. von Oertzen, H. G. Bohlen and S. Ohkubo, *J. Phys. G* **33**, R111 (2007); D. T. Khoa *et al.*, *Nucl. Phys. A* **759**, 3 (2005).
11. D. V. Shetty, S. J. Yennello and G. A. Souliotis, arXiv:0704.0471 [nucl-ex] (2007).
12. V. Baran, M. Colonna, V. Greco and M. Di Toro, *Phys. Rep.* **410**, 335 (2005).
13. D. V. Shetty, S. J. Yennello and G. A. Souliotis, *Phys. Rev. C* **75**, 034602 (2007).
14. C. Fuchs, A. Faessler, E. Zabrodin and Y. E. Zheng, *Phys. Rev. Lett.* **86**, 1974 (2001).
15. Ch. Hartnack, H. Oeschler and J. Aichelin, *Phys. Rev. Lett.* **96**, 012302 (2006).
16. C. Sturm *et al.* [KaoS Collaboration], *Phys. Rev. Lett.* **86**, 39 (2001).
17. L. W. Chen, C. M. Ko and B. A. Li, *Phys. Rev. Lett.* **94**, 032701 (2005).
18. G. Ferini *et al.*, *Phys. Rev. Lett.* **97**, 202301 (2006).
19. T. Gaitanos *et al.*, *Nucl. Phys. A* **732**, 24 (2004).
20. X. Lopez *et al.* [FOPI Collaboration], *Phys. Rev. C* **75**, 011901 (2007).
21. C. Fuchs, *Prog. Part. Nucl. Phys.* **56**, 1 (2006).
22. A. Schmah *et al.* [KaoS Collaboraton], *Phys. Rev. C* **71**, 064907 (2005).
23. F. Laue *et al.* [KaoS Collaboration], *Phys. Rev. Lett.* **82**, 1640 (1999).
24. P. Danielewicz, *Nucl. Phys. A* **673**, 275 (2000).
25. T. Gaitanos*et al.*, *Eur. Phys. J. A* **12**, 421 (2001); C. Fuchs and T. Gaitanos, *Nucl. Phys. A* **714**, 643 (2003).
26. A. Andronic *et al.* [FOPI Collaboration], *Phys. Rev. C* **64**, 041604 (2001); *Phys. Rev. C* **67**, 034907 (2003).
27. G. Stoicea *et al.* [FOPI Collaboration], *Phys. Rev. Lett.* **92**, 072303 (2004).
28. S. Popov, H. Grigorian, R. Turolla and D. Blaschke, *Astron. Astrophys.* **448**, 327 (2006).
29. D. J. Nice *et al.*, *Astrophys. J.* **634**, 1242 (2005).
30. T. Klähn *et al.*, *Phys. Rev. C* **74**, 035802 (2006).
31. A. W. Steiner, M. Prakash, J. M. Lattimer and P. J. Ellis, *Phys. Rep.* **411**, 325 (2005).
32. J. M. Lattimer, C. J. Pethick, M. Prakash and P. Haensel, *Phys. Rev. Lett.* **66**, 2701 (1991).
33. T. Klähn *et al.*, nucl-th/0609067 (2006).

IN-MEDIUM HADRONIC INTERACTIONS AND THE NUCLEAR EQUATION OF STATE

F. SAMMARRUCA

Physics Department, University of Idaho,
Moscow, IDAHO 83844-0903, U.S.A.
E-mail: fsammarr@uidaho.edu

Microscopic studies of nuclear matter under diverse conditions of density and asymmetry are of great contemporary interest. Concerning terrestrial applications, they relate to future experimental facilities that will make it possible to study systems with extreme neutron-to-proton ratio. In this talk, I will review recent efforts of my group aimed at exploring nuclear interactions in the medium through the nuclear equation of state. The approach we take is microscopic and relativistic, with the predicted EoS properties derived from realistic nucleon-nucleon potentials. I will also discuss work in progress. Most recently, we completed a DBHF calculation of the Λ hyperon binding energy in nuclear matter.

Keywords: Nuclear matter; Equation of state; In-medium hadronic interactions.

1. Introduction

The properties of hadronic interactions in a dense environment is a problem of fundamental relevance in nuclear physics. Such properties are typically expressed in terms of the nuclear equation of state (EoS), a relation between energy and density (and possibly other thermodynamic quantities) in infinite nuclear matter. The infinite geometry of nuclear matter eliminates surface effects and makes this system a clean "laboratory" for developing and testing models of the nuclear force in the medium.

The project I review in this talk is broad-scoped. We have examined several EoS-sensitive systems/phenomena on a consistent footing with the purpose of gaining a broad overview of various aspects of the EoS. We hope this will help identify patterns and common problems.

Our approach is microscopic, with the starting point being realistic free-space interactions. In particular, we apply the Bonn B[1] potential in the Dirac-Brueckner-Hartree-Fock (DBHF) approach to asymmetric nuclear matter as done earlier by the Oslo group.[2] The details of the calculations have been described previously.[3] As it has been known for a long time, the DBHF approximation allows a more realistic description of the saturation properties of symmetric nuclear matter as compared with the conventional Brueckner scheme. The leading relativistic effect characteristic of the DBHF model turns out to be a very efficient saturation mechanism. We recall that such correction has been shown to simulate a many-body effect (the "Z-diagrams") through mixing of positive- and negative-energy Dirac spinors by the scalar interaction.[4]

In what follows, I will review some of our recent results and discuss on-going work. I stress again that these efforts belong to the broader context of learning more about the behavior of the nuclear force in the medium using the EoS of infinite matter (under diverse conditions of isospin and spin asymmetry) as an exploratory tool. I also emphasize the importance of fully exploiting empirical information, which is becoming more available through collisions of neutron-rich nuclei.

2. Isospin-Asymmetric Nuclear Matter

2.1. *Seeking laboratory constraints to the symmetry potential*

Reactions induced by neutron-rich nuclei can probe the isospin dependence of the EoS. Through careful analyses of heavy-ion collision (HIC) dynamics one can identify observables that are sensitive to the asymmetric part of the EOS. Among those, for instance, is the neutron-to-proton ratio in semiperipheral collisions of asymmetric nuclei at Fermi energies.[5]

In transport models of heavy-ion collisions, particles drift in the presence of an average potential while undergoing two-body collisions. Isospin-dependent dynamics is typically included through the *symmetry potential* and isospin-dependent effective cross sections. Effective cross sections (ECS) will not be discussed in this talk. However, it is worth mentioning that they play an outstanding role for the determination of quantities such as the nucleon mean-free path in nuclear matter, the nuclear transparency function, and, eventually, the size of exotic nuclei.

The symmetry potential is closely related to the single-neutron/proton potentials in asymmetric matter, which we obtain self-consistently with the effective interaction. Those are shown in Fig. 1 as a function of the asymmetry parameter $\alpha = \frac{\rho_n - \rho_p}{\rho}$. The approximate linear behavior, consistent with a quadratic dependence on α of the *average* potential energy, is apparent. Clearly, the isospin splitting of the single-particle potential will be effective in separating the collision dynamics of neutrons and protons. The symmetry potential is shown in Fig. 2 where it is compared with empirical information on the isovector part of the optical potential (the early Lane potential).[6] The decreasing strength with increasing energy is in agreement with optical potential data.

2.2. *Summary of EOS results and comparison with recent constraints*

In Table 1 we summarize the main properties of the symmetric and asymmetric parts of our EoS. Those include saturation energy and density, incompressibility K, the skewness parameter K', the symmetry energy, and the symmetry pressure, L. The EoS for symmetric and neutron matter and the symmetry energy are displayed in Figs. 3-4.

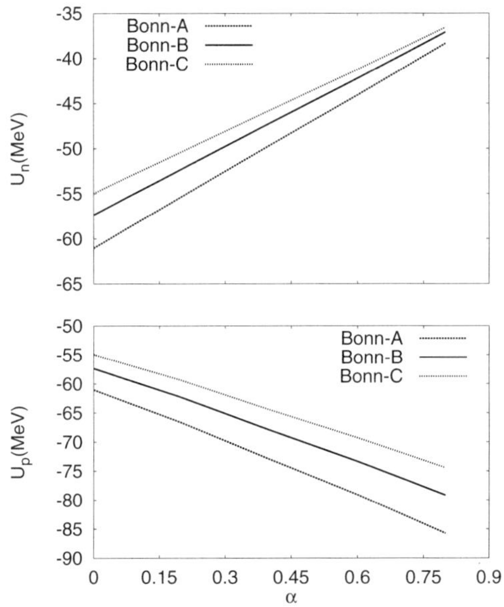

Fig. 1. The single-neutron (upper panel) and single-proton (lower panel) potentials as a function of the asymmetry parameter for fixed average density ($k_F = 1.4fm^{-1}$) and momentum ($k = k_F$).

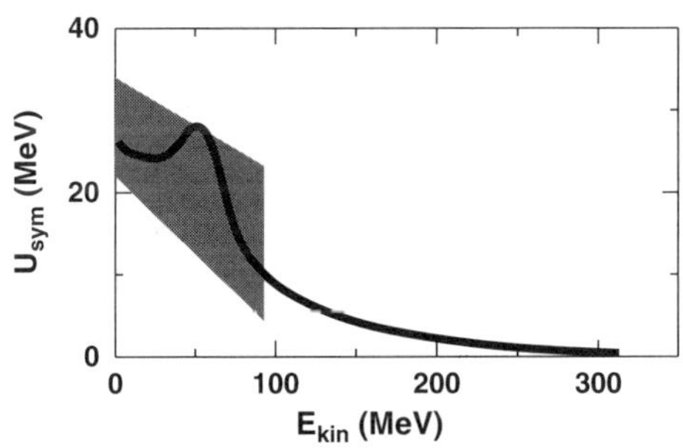

Fig. 2. The symmetry potential as a function of the nucleon kinetic energy at $k_F = 1.4fm^{-1}$. The shaded area represents empirical information from optical potential data.

A recent analysis to constrain the EoS using compact star phenomenology and HIC data can be found in Ref. 7. While the saturation energy is not dramatically different between models, the incompressibility values spread over a wider range. Major model dependence is found for the K' parameter, where a negative value indicates a very stiff EoS at high density. That is the case for models with parameters fitted to the properties of finite nuclei, whereas flow data require a soft EoS at the higher densities and thus a larger K'. The L parameter also spreads considerably, unlike the symmetry energy which tends to be similar in most models. For L, a combination of experimental information on neutron skin thickness in nuclei and

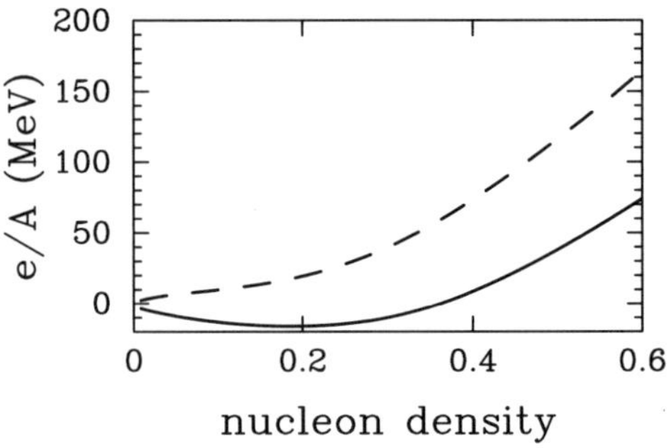

Fig. 3. Energy/particle in symmetric (solid) and neutron (dashed) matter, as a function of density (in fm^{-3}).

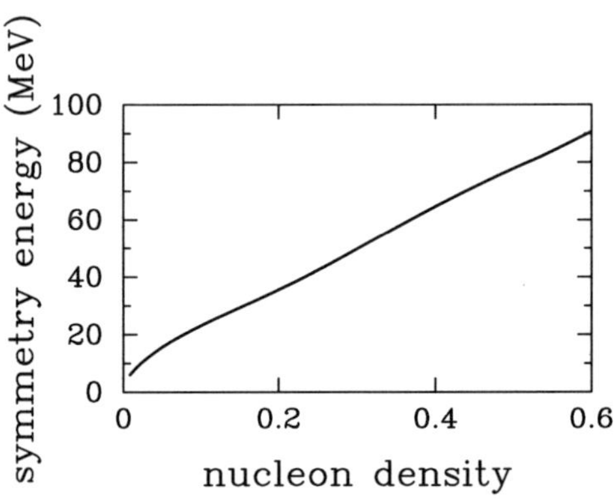

Fig. 4. Symmetry energy as a function of density (in fm^{-3}).

isospin diffusion data sets the constraint 62 MeV $< L <$ 107 MeV.

Overall, our EoS parameters compare reasonably well with most of those constraints. They also compare well with those from other DBHF calculations reported in Ref. 7.

Table 1. An overview of our predicted properties for the EOS of symmetric matter and neutron matter.

Parameter	Predicted Value
e_s	-16.14 MeV
ρ_s	0.185 fm^{-3}
K	259 MeV
K'	506 MeV
$e_{sym}(\rho_s)$	33.7 MeV
$L(\rho_s)$	70.1 MeV

The parameters in Table 1 are defined by an expansion of the energy written as

$$e = e_s + \frac{K}{18}\epsilon^2 - \frac{K'}{162}\epsilon^3 + ... + \alpha^2(e_{sym} + \frac{L}{3}\epsilon + ...), \tag{1}$$

with $\epsilon = (\rho - \rho_s)/\rho_s$, and ρ_s is the saturation density.

3. Spin-Polarized Neutron Matter

In this section we move on to a another issue presently discussed in the literature with regard to exotic states of nuclear/neutron matter, namely the aspect of spin asymmetry and (possible) spin instabilities.

The problem of spin-polarized neutron/nuclear matter has a long history. Extensive work on the this topic has been done by Vidaña, Polls, Ramos, Bombaci, Müther, and more (see bibliography of Ref. 8 for a more complete list). The major driving force behind these efforts is the search for ferromagnetic instabilities, namely the existence of a polarized state with lower energy than the unpolarized, which naturally would lead to a spontaneous transition. Presently, conclusions differ widely concerning the existence of such transition and its onset density.

A coupled self-consistency scheme similar to the one described in Ref. 3 was developed to calculate the EoS of polarized neutron matter. The details are described in Ref. 8. As done previously for the case of isospin asymmetry, the single-particle potential (for upward and downward polarized neutrons), is obtained self-consistently with the effective interaction. Schematically

$$U_i = \int G_{ij} + \int G_{ii} \tag{2}$$

with $i, j{=}u, d$. The nearly linear dependence of the single-particle potentials on the spin-asymmetry parameter[8]

$$U_{u/d}(\rho, \beta) \approx U_0(\rho, 0) \pm U_s(\rho)\beta \tag{3}$$

with β the spin-asymmetry parameter, is reminescent of the analogous case for isospin asymmetry and may be suggestive of a possible way to seek constraints on U_s, the "spin Lane potential", similarly to what was discussed above for the isovector optical potential. Namely, one can write, for a nucleus,

$$U \approx U_0 + U_\sigma(\mathbf{s} \cdot \mathbf{\Sigma})/A \tag{4}$$

with $\mathbf{s}$ and $\mathbf{\Sigma}$ the spins of the projectile nucleon and the target nucleus, respectively, and extract an obvious relation with the previous equation. (In practice, the situation for an actual scattering experiment on a polarized nucleus would require a more complicated parametrization than the one above, as normally a spin-unsaturated nucleus is also isospin-unsaturated.)

As already implied by the linear dependence of the single-particle potential displayed in Eq. (3), the dependence of the average energy on the asymmetry parameter is approximately quadratic,[8]

$$e(\rho, \beta) \approx e(\rho, 0) + S(\rho)\beta^2, \tag{5}$$

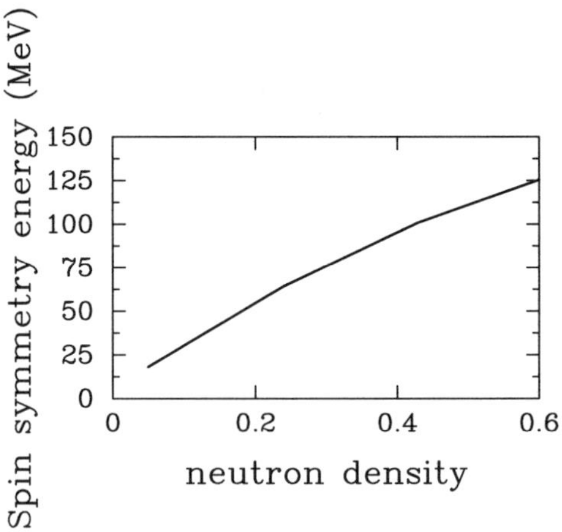

Fig. 5. Spin symmetry energy as a function of the neutron density (in fm^{-3}).

where $S(\rho)$ is the spin symmetry energy, shown in Fig. 5. The spin symmetry energy can be related to the magnetic susceptibility through

$$\chi = \frac{\rho\mu^2}{2S(\rho)}. \tag{6}$$

The rise of $S(\rho)$ with density shows a tendency to slow down, a mechanism that we attribute to increased repulsion in large even-singlet waves (which contribute only to the energy of the unpolarized state). This could be interpreted as a precursor of spin instability. In Table 2 we show predicted values of the ratio χ_F/χ, where χ_F is the susceptibility of a free Fermi gas.

Concerning the possibility of laboratory constraints which may help shed light on these issues, magnetic properties are of course closely related to the strength of the effective interaction in the spin-spin channel, which suggests to look into the G_0 Landau parameter. With simple arguments, the latter can be related to the susceptibility ratio and the effective mass as

$$\frac{\chi}{\chi_F} = \frac{m^*/m}{1+G_0}. \tag{7}$$

Thus, $G_0 \leq -1$ signifies spin instability. (Notice the analogy with the formally similar relation between the incompressibility ratio K/K_F and the parameter F_0, where $F_0 \leq -1$ signifies that nuclear matter is unstable.) At this time, no reliable constraints on G_0 are available, due to the fact that spin collective modes have not been observed with sufficient strength.

In closing this section, we note that we found similar considerations concerning the trend of the magnetic susceptibility to hold for symmetric nuclear matter as well as for neutron matter.

Table 2. Ratio χ_F/χ at different densities.

ρ/ρ_0	χ_F/χ
0.5383	2.1455
0.8886	2.3022
1.3651	2.4950
1.9873	2.6411
2.7741	2.6824
3.7457	2.6548

4. Work in Progress: Non-Nucleonic Degrees of Freedom

There are important motivations for considering strange baryons in nuclear matter. The presence of hyperons in stellar matter tends to soften the EoS, with the consequence that the predicted neutron star maximum masses become considerably smaller. With recent constraints allowing maximum masses larger than previously accepted limits, accurate calculations which include strangeness become important and timely. On the other hand, remaining within terrestrial nuclear physics, studies of hyperon energies in nuclear matter naturally complements our knowledge of finite hypernuclei.

The nucleon and the Λ potentials in nuclear matter are the solution of a coupled self-consistency problem, which reads, schematically

$$U_N = \int_{k<k_F^N} G_{NN} + \int_{k<k_F^\Lambda} G_{N\Lambda}, \qquad (8)$$

$$U_\Lambda = \int_{k<k_F^N} G_{\Lambda N} + \int_{k<k_F^\Lambda} G_{\Lambda\Lambda}.$$

To confront the simplest possible scenario, one may consider the case of symmetric nuclear matter at some Fermi momentum k_F^N in the presence of a "Λ impurity", namely $k_F^\Lambda \approx 0$. Under these conditions, the problem stated above simplifies considerably. Such calculation was done in Ref. 9 within the Brueckner scheme.

We have done a similar calculation but made use of the latest nucleon-hyperon (NY) potential of Ref. 10, which was provided by the Jülich group. In a first approach, we have taken the single-nucleon potential from a separate calculation of symmetric matter. (Notice that the Λ potential is quite insensitive to the choice of U_N, as reported in Ref. 9 and as we have observed as well.) The parameters of the Λ potential, on the other hand, are calculated self-consistently with the $G_{N\Lambda}$ interaction, which is the solution of the Bethe-Goldstone equation with one-boson exchange nucleon-hyperon potentials. In the Brueckner calculation, angle-averaged Pauli blocking and dispersive effects are included. Once the single-particle potential is obtained, the value of $-U_\Lambda(p)$ at p=0 provides the Λ binding energy in nuclear matter, B_Λ.

As shown and discussed extensively in Ref. 10, there are several remarkable differences between this model and the older NY Jülich potential,[11] and those seem to have a large impact on nuclear matter results. The main new feature of this model is a microscopic model of correlated $\pi\pi$ and $K\bar{K}$ exchange to constrain both the σ and ρ contributions.[10] With the new model, we obtain considerably more attraction than Reuber *et al.*,[9] about 49 MeV at k_F^N=1.35 fm^{-1} for B_Λ. We have also incorporated the DBHF effect in this calculation (which amounts to involving the Λ single-particle Dirac wave function in the self-consistent calculation through the effective mass) and find a moderate reduction of B_Λ by 3-4 MeV. A detailed report of this project is forthcoming.

The natural extension of this preliminary calculation will be a DBHF self-consistent calculation of U_N, U_Λ, and U_Σ for diverse Λ and Σ concentrations.

5. Summary and Conclusions

I have presented a summary of recent results from my group as well as on-going work. Our scopes are broad and involve several aspects of nuclear matter, the common denominator being the behavior of the nuclear force, including its isospin and spin dependence, in the medium. I have stressed the importance of seeking and exploiting laboratory constraints. In the future, coherent effort from theory, experiment, and observations will be the key to improving our knowledge of nuclear matter and its exotic states.

Acknowledgments

Support from the U.S. Department of Energy under Grant No. DE-FG02-03ER41270 is acknowledged. I am grateful to Johann Haidenbauer for providing the nucleon-hyperon potential code and for useful communications.

References

1. R. Machleidt, *Adv. Nucl. Phys.* **19**, 189 (1989).
2. G. Bao, L. Engvik, M. Hjorth-Jensen, E. Osnes and E. Østgaard, *Nucl. Phys. A* **575**, 707 (1994).
3. D. Alonso and F. Sammarruca, *Phys. Rev. C* **67**, 054301 (2003).
4. G. E. Brown, W. Weise, G. Baym and J. Speth, *Comments Nucl. Part. Phys.* **17**, 39 (1987).
5. V. Baran, M. Colonna, M. Di Toro, M. Zielinska-Pfabe and H. H. Wolter, *Phys. Rev. C* **72**, 064620 (2005).
6. A. M. Lane, *Nucl. Phys.* **35**, 676 (1962).
7. T. Klähn *et al.*, *Phys. Rev. C* **74**, 0635802 (2006).
8. F. Sammarruca and P. Krastev, *Phys. Rev. C* **75**, 034315 (2007).
9. A. Reuber, K. Holinde and J. Speth, *Nucl. Phys. A* **570**, 543 (1994).
10. J. Haidenbauer and Ulf-G. Meissner, *Phys. Rev. C* **72**, 044005 (2005).
11. B. Holzenkamp, K. Holinde and J. Speth, *Nucl. Phys. A* **500**, 485 (1989).

EOS AND SINGLE-PARTICLE PROPERTIES OF ISOSPIN-ASYMMETRIC NUCLEAR MATTER WITHIN THE BRUECKNER THEORY

W. ZUO,[1,*] U. LOMBARDO[2,3] and H.-J. SCHULZE[4]

[1] *Institute of Modern Physics, Chinese Academy of Sciences, Lanzhou 730000, China*
**E-mail: zuowei@impcas.ac.cn*
[2] *INFN-LNS, Via S. Sofia 62, I-95123 Catania, Italy*
[3] *Physics Department, Catania University, Via S. Sofia 64, I-95123 Catania, Italy*
[4] *INFN Sezione di Catania, Via S. Sofia 64, I-95123 Catania, Italy*

We investigate the equation of state (EOS) and single-particle (s.p.) properties of asymmetric nuclear matter (including the neutron and proton s.p. potentials and effective masses, symmetry potential, neutron/proton effective mass splitting) by using the Brueckner approach extended to include microscopic three-body forces. We have concentrated on the isospin-dependence of the properties of asymmetric nuclear matter and the TBF effects. We have discussed especially the TBF-induced rearrangement effect on the s.p. properties and their isospin behavior in neutron-rich nuclear medium.

Keywords: Asymmetric nuclear matter; Symmetry energy; Brueckner theory; Three-body force; Symmetry potential; TBF-induced rearrangement effect.

1. Introduction

One of the most important aims of heavy ion collisions (HIC) is to extract reliable information on the equation of state (EOS) and single-particle (s.p.) properties of isospin asymmetric nuclear matter (ANM). The properties of ANM at low densities around normal nuclear matter density, especially their isospin dependence, play a crucial role in predicting the properties of neutron-rich nuclei far from the nuclear stability line, such as the radius, the neutron-skin thickness and the density distribution. Besides its general interest in nuclear physics and heavy ion physics, the properties of ANM at high densities are closely related to many astrophysical problems and are expected to be extremely important for understanding many observational phenomena in astrophysics. For example, the high-density behavior of symmetry energy determines the proton fraction in β-stable (n,p,e,μ) neutron star matter, and thus is decisive for the cooling mechanism of neutron stars. The EOS of ANM is a fundamental input of the stellar structure model and its stiffness at high densities determines the predicted maximum mass of (n,p,e,μ) neutron stars. Microscopically the properties of ANM have been extensively investigated by adopting various theoretical approaches. In the present paper, we shall report our research on the properties of ANM within the Brueckner-Hartree-Fock (BHF)framework.

2. Theoretical Approaches

Our investigation is based on the Brueckner-Bethe-Goldstone (BBG) theory for ANM.[1,2] The extension of the BBG scheme to include three-body forces can be found in Ref. 3,4. The starting point of the BBG scheme is the Brueckner reaction G matrix, which satisfies the Bethe-Goldstone (BG) equation. In solving the BG equation for the G-matrix, we adopt the continuous choice for the s.p. potential since it provides a much faster convergency of the hole-line expansion than the gap choice.[5] Under the continuous choice, the s.p. potential describes physically at the BHF level the nuclear mean field felt by a nucleon in nuclear medium.

The realistic nucleon-nucleon (NN) interaction v_{NN} in the present calculation contains two parts, i.e., the Argonne V_{18} (AV_{18}) two-body interaction plus the contribution of the microscopic TBF based on the meson-exchange current approach.[4] In our BHF calculation, the TBF contribution has been included by reducing the TBF to an equivalently effective two-body interaction according to the standard scheme as described in Ref. 4. In r-space, the equivalent two-body force $V_3^{\rm eff}$ reads:

$$\begin{aligned}\langle \vec{r}_1{}'\vec{r}_2{}'|V_3^{\rm eff}|\vec{r}_1\vec{r}_2\rangle = \frac{1}{4}{\rm Tr}\sum_n \int {\rm d}\vec{r}_3{\rm d}\vec{r}_3{}'\phi_n^*(\vec{r}_3{}')(1-\eta(r'_{13}))(1-\eta(r'_{23})) \qquad (1)\\ \times W_3(\vec{r}_1{}'\vec{r}_2{}'\vec{r}_3{}'|\vec{r}_1\vec{r}_2\vec{r}_3)\phi_n(\vec{r}_3)(1-\eta(r_{13}))(1-\eta(r_{23})).\end{aligned}$$

It is worth stressing that the effective force $V_3^{\rm eff}$ depends strongly on density. It is the density dependence of the $V_3^{\rm eff}$ that induces the TBF rearrangement contribution to the s.p. properties in nuclear medium within the BHF framework. The TBF rearrangement contribution to the s.p. potential can be calculated as follows:[6]

$$U^{\rm tbf}(k) \approx \frac{1}{2}\sum_{k_1k_2} n_{k_1}n_{k_2}\left\langle k_1k_2\left|\frac{\delta V_3^{\rm eff}}{\delta n_k}\right|k_1k_2\right\rangle_A . \qquad (2)$$

3. Results and Discussions

3.1. *TBF effect and relativistic effect on the EOS of SNM*

In the left panel of Fig. 1 is reported the calculated energy per nucleon of symmetric nuclear matter (SNM) vs. density in several cases. As expected, the TBF gives a repulsive contribution to the nuclear EOS. At low densities, the TBF effect turns out to be fairly small. As the density increases, the TBF repulsion increases rapidly and makes the EOS at high densities much stiffer as compared to the case of not including the TBF. The repulsive contribution from the TBF turns out to improve remarkably the predicted saturation properties of SNM. To discuss the connection between the relativistic effect in the Dirac BHF (DBHF) approach and the TBF effect, we report in the right panel of Fig. 1 the contribution of the $2\sigma-(N\bar{N})$ TBF (i.e., the Z-diagram contribution) to the EOS of SNM (solid curve) in comparison with the relativistic correction from the DBHF approach (symbols).[7] It is clear from the comparison that the most important relativistic correction to the EOS of nuclear matter in the DBHF approach can be reproduced quantitatively by the

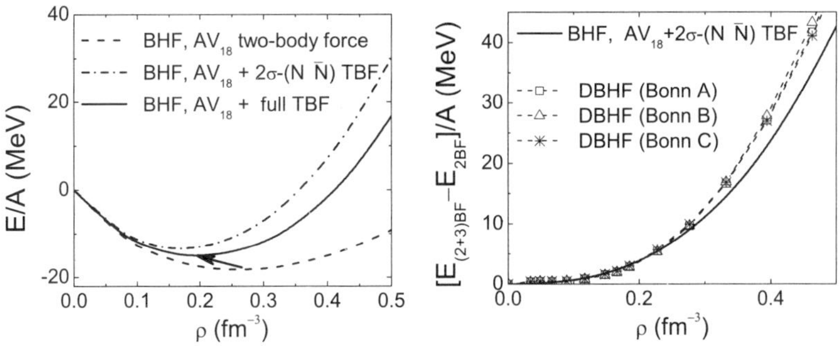

Fig. 1. Left panel: the energy per nucleon of SNM vs. density. Solid curve: adopting the AV_{18} plus the full TBF; Dot-dashed curve: adopting the AV_{18} plus the $2\sigma - (N\bar{N})$ component of the TBF; Dashed curve: using the pure AV_{18} two-body interaction. Right panel: TBF contribution to the nuclear EOS in comparison with the relativistic correction taken from Ref. 7.

$2\sigma - N\bar{N}$ component of the microscopic TBF. However, by comparing the solid and dot-dashed curves in the left panel we can see that the TBF components from the other elementary processes becomes important at high densities and can not be completely neglected even around the saturation density as discussed in Ref. 3.

3.2. *Isospin dependence of the EOS of ANM*

To see clearly the isospin dependence of the EOS of ANM, we report in Fig. 2 the isovector part of the EOS as a function of β^2, where β is the isospin asymmetry of ANM. The left and middle panels show the results for cold nuclear matter at several typical values of density in the two cases of including and not including the TBF contribution, respectively. In the right panel is given the results at finite-temperature ($T = 20$MeV). We find that the EOS of ANM fulfills satisfactorily a quadratic dependence on asymmetry parameter β in the whole asymmetry range $0 \leq \beta \leq 1$ in both cases with and without the TBF contribution.[1,8] Accordingly, the EOS of ANM may be expressed as: $E_A(\rho, T, \beta) = E_A(\rho, T, 0) + E_{sym}(\rho, T)\beta^2$, where $E_{sym}(\rho, T)$ is the symmetry energy. From the right panel we notice that the

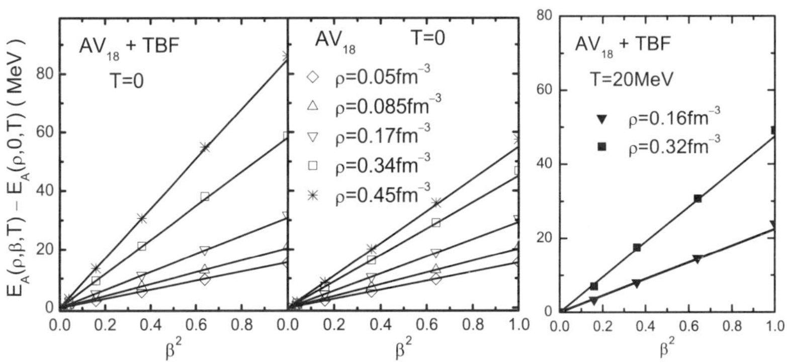

Fig. 2. Difference between the energy per nucleon of ANM and that of SNM.

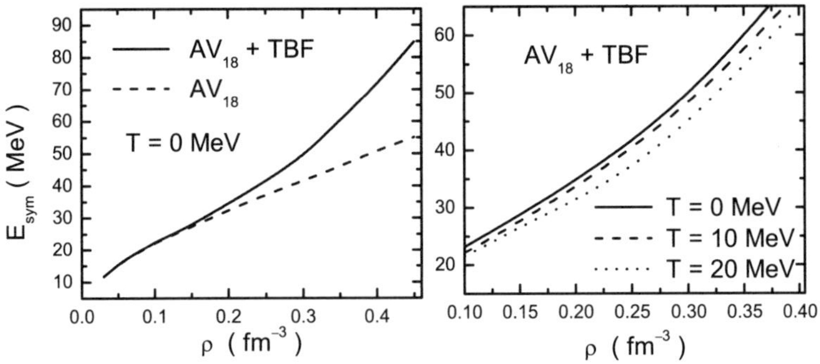

Fig. 3. TBF effect (left panel) and thermal effect (right panel) on symmetry energy.

linear dependence of the EOS of ANM on β^2 can be extended to finite-temperature case. The above result provides an microscopic support for the empirical β^2-law extracted from the nuclear mass table and extended its validity up to the highest isospin asymmetry. The β^2-law for ANM is also obtained recently by Gad *et al.*[9] within the self-consistent Green function approach. The β^2-law of the EOS of ANM leads to two important consequences. First, it indicates that the EOS of ANM at any isospin asymmetry is determined completely by the EOS of SNM and the symmetry energy. Second, the above β^2-law implies that the difference of the neutron and proton chemical potentials in β-stable neutron star matter is determined by the symmetry energy in an explicit way: $\mu_n - \mu_p = 4\beta E_{sym}$ and thus the symmetry energy plays a crucial role in predicting the composition of neutron stars.

3.3. *TBF effect and thermal effect on symmetry energy*

In Fig. 3 is reported the predicted symmetry energy vs. density. In the left panel is shown the TBF effect on symmetry energy. And the right panel displays the thermal effect on symmetry energy. In the case of not including the TBF, the density dependence of symmetry energy follows approximately $E_{sym} \propto \rho^{0.6}$ in the whole density region of $\rho \leq 0.5\text{fm}^{-3}$. The TBF effect is reasonably small at low densities around and below the saturation density. The TBF provides a repulsive contribution to symmetry energy at high densities and its repulsion increases rapidly as a function of density. Consequently, inclusion of the TBF makes the stiffness of the symmetry energy at high densities become remarkably different from that at low densities, i.e., it changes the predicted density dependence of symmetry energy at high densities from asy-soft to asy-stiff. As for the thermal effect, from the right panel we see that the symmetry energy decreases as the temperature increases.

3.4. *TBF rearrangement effect on nucleon s.p. potential in SNM*

Within the BHF framework, the TBF is expected to affect the s.p. potential in two different ways: first, it influences the s.p. potential directly via its modification of the G-matrix; second, it may induce a strong rearrangement contribution. It has

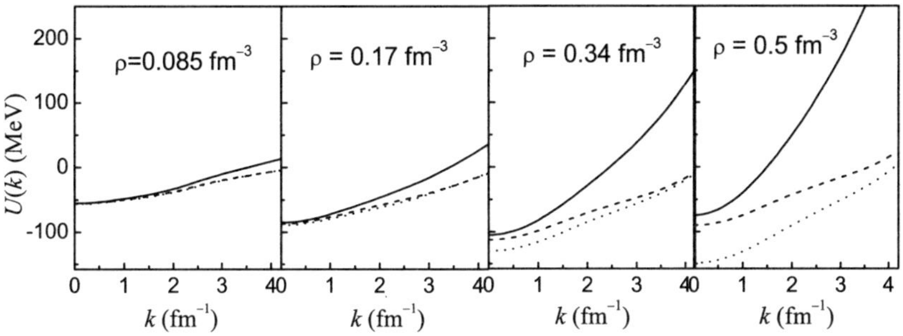

Fig. 4. The predicted s.p. potentials for SNM at several densities in three cases: without the TBF contribution (dotted curves); including the TBF effect only via the G-matrix (dashed curves); including the TBF effect via both the G-matrix and the rearrangement contribution (solid curves).

been shown in Ref. 10, the BHF s.p. potential predicted without including TBFs is too attractive and its momentum dependence is too weak at large densities and high momenta for describing the high-energy elliptic flow data. To discuss the TBF effect, we compare in Fig. 4 the s.p. potentials obtained in three different cases. It is seen that the s.p. potential without the TBF is most attractive. At low densities, the TBF effects are reasonably small. The TBF effects become significant rapidly as the density increases. The TBF effect via the G-matrix alone provides an moderate repulsion at high densities which is more pronounced at lower momenta and weakens the momentum dependence of the s.p. potential. The TBF-induced rearrangement effect provides an additional repulsive and strongly momentum-dependent contribution at high densities and high momenta. The repulsion due to the TBF rearrangement is shown to be much stronger at high momenta as compared to that via the G-matrix. The rearrangement repulsion induced by the TBF reduces largely the attraction and enhances strongly the momentum-dependence of the s.p. potential at large densities and high momenta. It improves substantially the agreement between the microscopic BHF s.p. potential and the parametrized one of Ref. 10.

3.5. *Isospin dependence of neutron and proton s.p. potentials*

The s.p. potentials felt by protons and neutrons in ANM are basic inputs of transport models for HIC and play an important role in connecting the experimental observables in HIC and the nuclear EOS.[11] In neutron-rich nuclear matter, it is shown[1,2] that the proton s.p. potential becomes more attractive and the neutron one becomes more repulsive as the neutron excess increases. The isospin dependence of the neutron and proton s.p. potentials in neutron-rich matter may be described by the symmetry potential defined as: $U_{sym} = (U_n - U_p)/2\beta$. In Ref. 12 we have shown that the symmetry potential predicted in the BHF framework is almost independent on β, indicating a linear dependence of the neutron and proton s.p. potentials on β and providing an microscopic support for the Lane potential. We have also found that the symmetry potentials obtained by the BHF approach and the DBHF approach of Refs. 13,14 display an overall agreement. Whereas the phe-

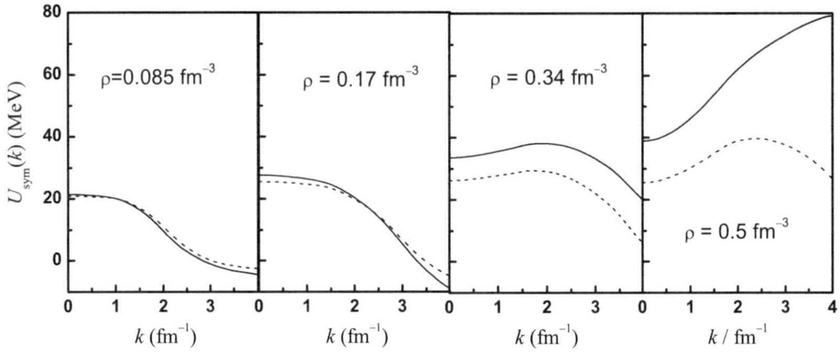

Fig. 5. Symmetry potential vs. momentum. Dashed curves: without including the TBF rearrangement contribution; solid curves: including the TBF rearrangement contribution.

nomenological parametrizations of symmetry potential[15] adopted in the dynamical simulations of HIC show a remarkably different behavior as a function of density and momentum from our microscopic $U_{\rm sym}$. In Fig. 5, we show the TBF rearrangement effect on symmetry potential. It is seen that the predicted symmetry potential depends strongly on both density and momentum. At low densities around and below $\rho_0 = 0.17{\rm fm}^{-3}$, the TBF rearrangement effect turns out to be almost negligible and the $U_{\rm sym}$ above the Fermi surface decreases rapidly as a function of momentum. As the density increases, the TBF rearrangement effect becomes pronounced. One can see that the TBF rearrangement enhances considerably the $U_{\rm sym}$ at high densities, i.e., it enhances the repulsion (attraction) of the $U_{\rm sym}$ on neutrons (protons).

3.6. *Neutron-proton effective mass splitting in neutron-rich matter*

In neutron-rich nuclear matter, the neutron-proton (n-p) effective mass splitting measures the difference between the isospin-behaviors of the neutron and proton effective masses.

In Fig. 6, we show the TBF rearrangement effect on the n-p effective mass splitting. In the microscopic BHF framework the predicted neutron effective mass is seen to be larger than the proton one in neutron-rich matter, i.e., $m_n^* \geq m_p^*$, in both cases with and without including the TBF rearrangement contribution.[6] It is found[2,12] that inclusion of the effect of ground state correlations (EBHF approach) leads to

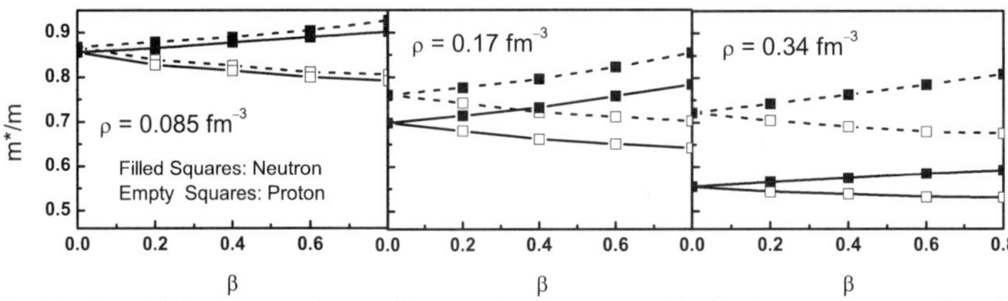

Fig. 6. Neutron (filled squares) and Proton (empty squares) effective masses vs. β obtained in the two cases of including (solid) and not including (dashed) the TBF rearrangement contribution.

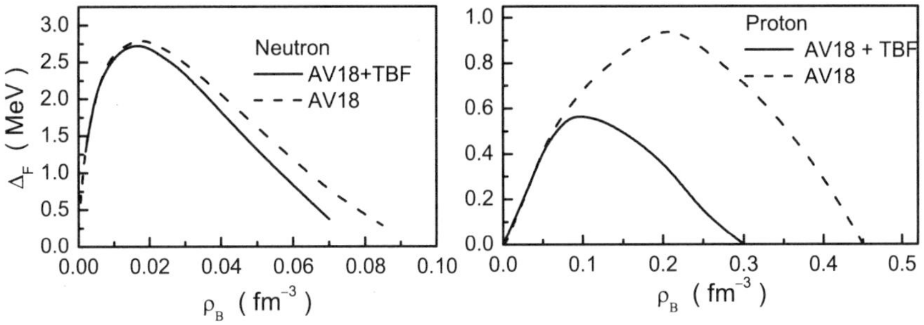

Fig. 7. Neutron $^1S_0$ pairing gaps (left panel) and proton $^1S_0$ pairing gaps (right panel) vs. total nucleon density in neutron star matter for the two cases of including and not including the TBF.

an overall enhancement of the neutron and proton effective masses, but does not affect their isospin splitting. In the relativistic DBHF approach, it has been shown in Refs. 16 that the splitting of the nonrelativistic-type effective masses is consistent with the predictions of the BHF approach. At low densities the TBF rearrangement effect is fairly small. At high densities, the TBF-induced rearrangement contribution reduces remarkably not only the neutron and proton effective masses but also the magnitude of their isospin splitting in dense neutron-rich matter. In Ref. 12, we have found that within the BHF framework, the isospin splitting of the neutron and proton effective masses is determined essentially by their k-masses and is controlled to a large extent by the nature of the NN interaction.

3.7. *TBF effect on nucleon superfluidity in neutron star matter*

Nucleon superfluidity is expected to be significant for the cooling processes via neutrino emission, the properties of rotating dynamics, the post-glitch timing observation, the possible vertex pinning of neutron stars. In Fig. 7 we show the TBF effect on the neutron and proton $^1S_0$ pairing gaps in β-stable neutron star matter. It is seen that the neutron superfluidity phase in the $^1S_0$ channel can only occur in the low-density region and the TBF reduces only slightly the neutron $^1S_0$ pairing gap. However, the proton $^1S_0$ superfluid phase extends to much higher densities with a smaller peak gap value as compared to the neutron one and the TBF weakens strongly the proton superfluidity in the $^1S_0$ channel. The strong weakening of the proton $^1S_0$ superfluidity due to the TBF may has important implication for modeling the neutron-star cooling scenario.[17] We have also found the TBF enhances remarkably the $^3PF_2$ neutron superfluidity in neutron star matter.

4. Summary

In summary, we report our recent investigation on the properties of ANM within the Brueckner theory. We show that the microscopic TBF provides a repulsive contribution to the EOS of nuclear matter and improves remarkably the predicted saturation properties. We find that the main relativistic correction to the nuclear EOS in the

DBHF approach can be reproduced quantitatively by the $2\sigma - N\bar{N}$ component of the TBF. However, the TBF components from the other elementary processes become significant at high densities. The empirical parabolic law for the EOS of ANM can be extended to the highest asymmetry and to finite-temperature case in both cases of including and not including the TBF. The TBF leads to a strong stiffening of the density dependence of symmetry energy at high densities. The neutron-proton effective mass splitting turns out to be $m_n^* \geq m_p^*$ in neutron-rich matter. The TBF induces a strongly repulsive and momentum-dependent rearrangement contribution to s.p. potential in dense nuclear medium. The TBF rearrangement effect reduces considerably the attraction and enhances strongly the momentum-dependent of the BHF s.p. potential at high densities and high momenta, and reduces remarkably the magnitude of the *n-p* effective mass splitting in dense neutron-rich matter.

Acknowledgments

The work is supported in part by the National Natural Science Foundation of China (10575119,10775061), the Knowledge Innovation Project(KJCX3-SYW-N2) of Chinese Academy of Sciences, the Major State Basic Research Developing Program of China under No. 2007CB815004, the CAS/SAFEA International Partnership Program for Creative Research Teams (CXTD-J2005-1), and the Asia-Link project [CN/ASIA-LINK/008(94791)] of the European Commission.

References

1. I. Bombaci and U. Lombardo, *Phys. Rev. C* **44**, 1892 (1991).
2. W. Zuo, I. Bombaci and U. Lombardo, *Phys. Rev. C* **60**, 024605 (1999).
3. W. Zuo, A. Lejeune, U. Lombardo and J.-F. Mathiot, *Nucl. Phys. A* **706**, 418 (2002).
4. P. Grangé, A. Lejeune, M. Martzolff and J.-F.Mathiot, *Phys. Rev. C* **40**, 1040 (1989).
5. H. Q. Song, M. Baldo, G. Giansiracusa *et al.*, *Phys. Rev. Lett.* **81**, 1584 (1998).
6. W. Zuo, U. Lombardo, H.-J. Schulze *et al.*, *Phys. Rev. C* **74**, 014317 (2006).
7. R. Brockmann and R. Machleidt, *Phys. Rev. C* **42**, 1965 (1990).
8. W. Zuo, A. Lejeune, U. Lombardo and J.-F. Mathiot, *Eur. Phys. J. A* **14**, 469 (2002).
9. Kh. Gad and Kh. S. A. Hassaneen, *Nucl. Phys. A* **793**, 67 (2007).
10. P. Danielewicz, *Nucl. Phys. A* **673**, 375 (2000).
11. B. A. Li, *Phys. Rev. Lett.* **88**, 192702 (2002); L. W. Chen, C. M. Ko and Bao-An Li, *Phys. Rev. Lett.* **94**, 032701 (2005).
12. W. Zuo, L. G. Cao, B. A. Li *et al.*, *Phys. Rev. C* **72**, 014005 (2005).
13. E. N. E. van Dalen, C. Fuchs and A. Faessler, *Nucl. Phys. A* **744**, 227 (2004).
14. F. Sammarruca, W. Barredo and P. Krastev, *Phys. Rev. C* **71**, 064306 (2005).
15. C. B. Das, S. Das Gupta, C. Gale and B. A. Li, *Phys. Rev. C* **67**, 034611 (2003).
16. E. N. E. van Dalen, C. Fuchs and A. Faessler, *Phys. Rev. Lett.* **95**, 022302 (2005); Z. Y. Ma, J. Rong, B. Q. Chen *et al.*, *Phys. Lett. B* **604**, 170 (2004).
17. A. D. Kaminker, M. E. Gusakov, D. G. Yakovlev *et al.*, *Mon. Not. R. Astron. Soc.* **365**, 1300 (2006).

THERMODYNAMICS OF CORRELATED NUCLEAR MATTER

A. POLLS* and A. RAMOS

Departament d'Estructura i Constituents de la Matèria,
Universitat de Barcelona, Avda. Diagonal 647,
E-08028 Barcelona, Spain
**E-mail: artur@ecm.ub.es*

A. RIOS

National Superconducting Cyclotron Laboratory,
East Lansing, 48823-1 Michigan, USA

H. MÜTHER

Institut für Theoretische Physik, Universität Tübingen,
D-72076 Tübingen, Germany

The Self-Consistent Green's Function's (SCGF) method at the level of the ladder approximation is used to calculate the free-energy of symmetric nuclear matter. The ladder approximation considers the propagation of particles and holes which translates into the incorporation of significant correlations in the wave function. These correlations are reflected in the shape of the single-particle spectral functions. An essential ingredient of the free energy is the entropy, which is calculated with the Luttinger-Ward formalism. In this approach, the finite width of the quasi-particle states is explicitly taken into account. It turns out that the entropy measures thermal effects and the effect of dynamical correlations, already present at zero temperature, is not so relevant for the calculation of the entropy. Preliminary results for the liquid-gas phase transition are also presented.

Keywords: Nuclear matter; Nuclear many-body problem; Temperature; Entropy.

1. Introduction

The nuclear equation of state (EOS) is a necessary tool in the understanding of some astrophysical scenarios as well as in the description of heavy ion collisions. Conversely, one also expects that the large amount of experimental information and observational data should be used to constraint the behavior of the nuclear EOS.[1] Ideally, if the EOS has been constructed microscopically starting from the bare nucleon-nucleon (NN) interaction and using a consistent many-body approach, one should be able to translate the requirements to the EOS in constraints for the NN interaction. In some situations, i.e. the thermal evolution of protoneutron stars or intermediate energy heavy ion collisions, one can reach temperatures of the order of tenths of MeV, as large as nuclear energy scales, and therefore for the description of these systems it is necessary to look for approaches that take into account the dynamical correlations induced by the strong NN interactions as well as the thermal effects. This is a very old problem which has been faced by the nuclear many-body community for a long time. However, in spite of the efforts and the enormous progress, the problem is not completely solved.

Some of the difficulties are linked to the interplay between the determination of the bare NN interaction and the many-body problem itself. In the last years there has been a substantial progress in both aspects.[2] Here we will assume that the bare NN interaction is well defined: all calculations reported here have been obtained with the CDBONN potential[3] and will concentrate in the many-body problem at finite temperature. It seems natural to require that the calculations should be performed in a thermodynamically (TD) consistent many-body scheme. However, the two approaches commonly used at zero temperature, namely the Brueckner-Hartree-Fock (BHF) approach and the Correlated Basis Function theory have theoretical difficulties in their generalization to finite temperature. It turns out that the Self-Consistent Green's function (SCGF) method, based on the perturbation expansion of the one-body propagator can be naturally extended to finite temperature, keeping the thermodynamical consistency of the approach.[4,5] The study of the propagation of a nucleon in the medium provides access to all single-particle properties and also allows one to calculate the energy per particle provided that the NN interaction is of two-body nature. This is very convenient in the sense that one does not need to know all the details of the wave function but only how the system reacts when one takes out or adds a particle to it. A microscopic calculation dealing with nuclear systems should take into account the modifications that the short range and tensor components of the NN interaction introduce in the wave function which can not be longer described in terms of a mean-field Slater determinant.

In the following, we investigate the thermodynamic properties of symmetric nuclear matter (SNM), an ideal, infinite system composed of the same amount of neutrons and protons interacting via a realistic interaction.

2. Single-Particle Properties

The single-particle propagator, in the grand-canonical ensemble, is defined according to

$$iG(\mathbf{k}t, \mathbf{k}'t') = Tr\left\{\hat{\rho} T[a_{\mathbf{k}}(t) a^{\dagger}_{\mathbf{k}'}(t')]\right\}, \tag{1}$$

where we have introduced the density matrix operator

$$\hat{\rho} = \frac{1}{Z} e^{-\beta(\hat{H} - \mu \hat{N})}, \tag{2}$$

and the partition function

$$Z = Tr\left\{e^{-\beta(\hat{H} - \mu \hat{N})}\right\}. \tag{3}$$

In these equations, β denotes the inverse temperature and μ is the chemical potential of the system. The symbol T stands for the time-ordering operator. Finally,

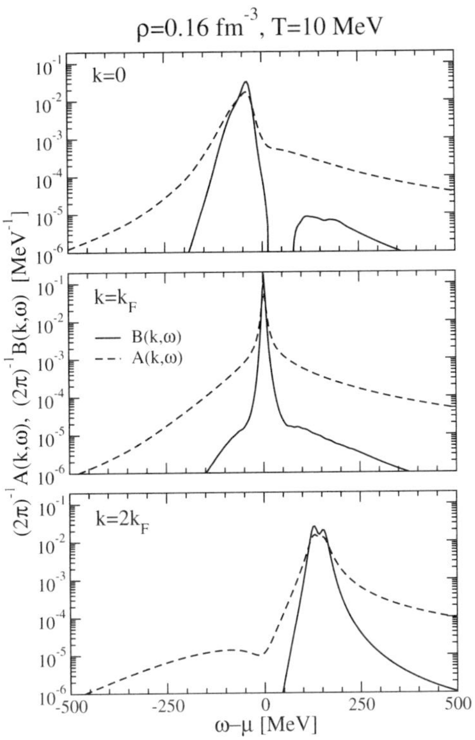

Fig. 1. Spectral functions $A(k,\omega)$ (dashed lines) and $B(k,\omega)$ (solid lines) at $\rho = 0.16$ fm^{-3} and $T = 10$ MeV for three different momenta $k = 0, k_F$ and $2k_F$.

the traces Tr are to be taken over all energy and particle number eigenstates of the system.

A Lehmann representation allows to write the propagator in terms of the single-particle spectral functions

$$G(k,\omega) = \frac{1}{2\pi}\int_{-\infty}^{\infty} d\omega' \frac{A^{>}(k,\omega')}{\omega - \omega' + i\eta} + \frac{1}{2\pi}\int_{-\infty}^{\infty} d\omega' \frac{A^{<}(k,\omega')}{\omega - \omega' - i\eta}, \tag{4}$$

which are defined as:

$$A^{<}(k,\omega) = 2\pi \sum_{nm} \frac{e^{-\beta(E_m - \mu N_m)}}{Z} \mid \langle \Psi_n \mid a_k \mid \Psi_m \rangle \mid^2 \delta\left[\omega - (E_m - E_n)\right]. \tag{5}$$

A similar definition holds for $A^{>}(k,\omega)$ with the use of a creation operator, $a_k^\dagger$. Note the presence in Eq. (5) of the thermal average on the initial states, which accounts for the thermal population of the excited states. This is different from the zero temperature case, where the ground state is considered to be the only possible initial state. A careful analysis of the definitions of $A^<$ and $A^>$ allows one to establish a simple relation between both functions:

$$A^{>}(k,\omega) = e^{\beta(\omega-\mu)} A^{<}(k,\omega). \tag{6}$$

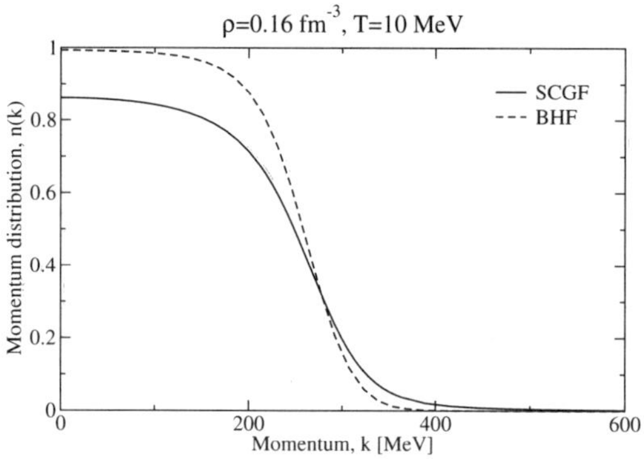

Fig. 2. Nucleon momentum distributions in the SCGF approach (full lines) and the BHF approach (dashed lines) at $\rho = 0.16$ fm^{-3} and $T = 10$ MeV.

In contrast with the $T = 0$ case, the energy domains of $A^<$ and $A^>$ are not separate and both functions are defined for all energies. The total spectral function $A(k,\omega)$ is defined as the sum of the two functions, $A^<$ and $A^>$, and the following relation holds: $A^<(k,\omega) = A(k,\omega)f(\omega)$, where $f(\omega) = (1 + e^{\beta(\omega-\mu)})^{-1}$ is the Fermi-Dirac distribution. The full propagator can be expressed in terms of the spectral function and therefore all the single-particle properties can be obtained from $A(k,\omega)$. The momentum distribution, for instance, is given by:

$$n(k,T) = \frac{1}{2\pi}\int_{-\infty}^{\infty} A^<(k,\omega)d\omega = \frac{1}{2\pi}\int_{-\infty}^{\infty} A(k,\omega)f(w)d\omega. \tag{7}$$

The single-particle Green's function can be obtained from the self-energy through a Dyson equation. The self-energy accounts for the interactions of a particle with the particles in the medium and is obtained from the renormalized NN interaction in the medium. This effective interaction is calculated in the ladder approximation, propagating fully dressed particles and holes in the intermediate states of the ladder diagrams. The ladder scheme is the minimal reliable approximation to treat short-range correlations. By construction, the modification of the single-particle properties affects the effective interaction among the nucleons in the medium and both things, the NN interaction in the medium and the single-particle propagator, should be determined in a self-consistent way. In this procedure, the chemical potential $\tilde{\mu}$ is determined at each iteration by inverting:

$$\rho = \nu \int \frac{d^3k}{(2\pi)^3} n(k,\tilde{\mu}), \tag{8}$$

where ν denotes the spin-isospin degeneracy of the system ($\nu = 4$ in symmetric nuclear matter). Once convergence is reached, one obtains the single-particle propagator from Dyson's equation and, from its imaginary part, the spectral function $A(k,\omega)$. At this point, one can calculate several micro- and macroscopic properties of the system. For instance, the momentum distribution given above or the energy

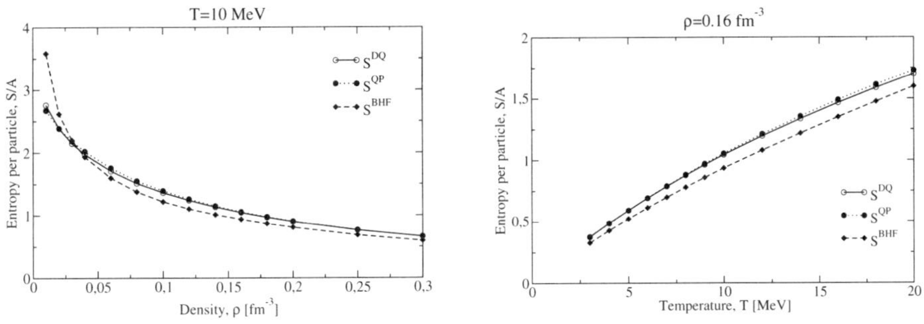

(a) (b)

Fig. 3. (a) Different approximations to the entropy as a function of density at a temperature T=10 MeV. (b) Different approximations to the entropy as a function of temperature at a density $\rho = 0.16$ fm−3.

per particle of the system, accessible from the Galitskii-Migdal-Koltun sum rule:

$$\frac{E}{N}(\rho, T) = \frac{\nu}{\rho} \int \frac{d^3k}{(2\pi)^3} \int_{-\infty}^{\infty} \frac{d\omega}{2\pi} \frac{1}{2} \left(\frac{k^2}{2m} + \omega \right) A(k, \omega) f(\omega). \tag{9}$$

The spectral functions (dashed lines) as a function of the energy for three characteristic momenta, $k = 0, k_f$ and $2k_f$ at $\rho = 0.16$ fm^{-3} and at a temperature of $T = 10$ MeV are reported in Fig. 1. The spectral functions show a peak, located at the so called quasi-particle energy. This peak gets narrower when the momentum is close to the Fermi surface. The presence of the high energy tails in the spectral function is an indication of the importance of the correlations. The momentum distribution at the same density and temperature is shown in Fig. 2, for both the standard BHF approach and the SCGF. The BHF $n(k)$ does not contain correlation effects and is very similar to a normal Fermi distribution whit single-particle energies defined by the BHF quasi-particle energies. The SCGF $n(k)$ contains besides thermal effects, important short-range and tensor correlations, which are reflected in the depletion of the occupation at low momentum and in a larger occupation than the BHF momentum distribution at large momenta.

3. Thermodynamical Properties of Nuclear Matter

For a complete thermodynamical description of the system one should compute the free energy, $F = E - TS$. Therefore one needs a suitable method to compute the entropy. It turns out that the knowledge of the single-particle propagator allows for the construction of the grand-potential, Ω, in the framework of the Luttinger-Ward formalism.[6,7] From the grand-potential, one can obtain the entropy through the thermodynamical relation $S = -\frac{\partial \Omega}{\partial T} |_{\mu}$. To this end, it is convenient to split this entropy in two terms, $S = S^{DQ} + S'$. The first one is the so-called dynamical quasi-particle (DQ) entropy:

$$S^{DQ} = \nu \int \frac{d^3k}{(2\pi)^3} \int_{-\infty}^{\infty} \frac{d\omega}{2\pi} \sigma(\omega) B(k, \omega), \tag{10}$$

formed by the convolution of a statistical factor, $\sigma(\omega) = -f(\omega)\ln f(\omega) - [1 - f(\omega)]\ln[1-f(\omega)]$, times a certain spectral function $B(k,\omega)$, closely related to the propagator and the single-particle spectral function, $A(k,\omega)$:

$$B(k,\omega) = A(k,\omega)\left[1 - \frac{\partial Re\Sigma(k,\omega)}{\partial\omega}\right] - 2\frac{\partial ReG(k,\omega)}{\partial\omega} Im\Sigma(k,\omega_+). \quad (11)$$

S^{DQ} takes into account the correlations of the dressed particles in the medium and includes finite width effects. The energy dependence of the spectral function $B(k,\omega)$ (solid lines) is also shown in Fig. 1, for three momenta $k = 0, k_F$ and $2k_F$, at $\rho = 0.16$ fm^{-3} and $T = 10$ MeV. The energy tails of $B(k,\omega)$ are shorter than for $A(k,\omega)$ and the strength is mostly concentrated in the quasi-particle peak. This is in agreement with the idea that the entropy is almost not affected by the width that correlations induce to the quasi-particle peak.

In the following, we shall make the assumption that the second term of the entropy, S', is negligible. This approximation leads to TD consistent results, supporting the assumption that the contribution of S' is small in the density and temperature range explored. The density dependence of the entropy per particle at $T = 10$ MeV within different approximations is shown in Fig. 3(a). The full dynamical quasi-particle entropy (S^{DQ}) is depicted in solid lines. A quasi-particle approximation to the entropy (S^{QP}), obtained by reducing the function $B(k,\omega)$ to a delta function with all the strength at the quasi-particle peak, is given by the dotted line. The fact that both fall on top of each other indicates that the entropy is scarcely influenced by the width of the B spectral function, which is induced mainly by short range and tensor correlations. In addition we also show the entropy obtained in the BHF approximations (S^{BHF}) with dashed lines. The expression for S^{BHF} is the same as for S^{QP} but using the BHF quasi-particle energies. Therefore, the difference between the S^{BHF} and S^{QP} comes from the shift in quasi-particle energies and chemical potentials produced by the propagation of holes in the SCGF method. The qualitative behavior is the same in all approximations. As expected, for a given temperature, the entropy decreases with density. In the second panel of Fig. 3 we show the temperature dependence of the entropy for a fixed density, $\rho = 0.16$ fm^{-3}, computed with the same approximations as discussed above. We find again a good agreement between the different approaches. In all cases, the entropy increases with temperature and approaches a linear dependence at low T characteristic of a normal Fermi fluid. The relatively small differences observed between the BHF and SCGF entropies are in agreement with the general conclusion that the entropy is a mesure of thermal effects, and not so much of the dynamical effects induced by NN interactions.[8]

4. Liquid-Gas Phase Transition

The density dependence of the free-energy per particle at $T = 10$ MeV is shown in Fig. 4(a). F^{SCGF} is more repulsive than F^{BHF}. As in both approaches the entropies are similar, one concludes that the difference is due to the repulsive effect that the

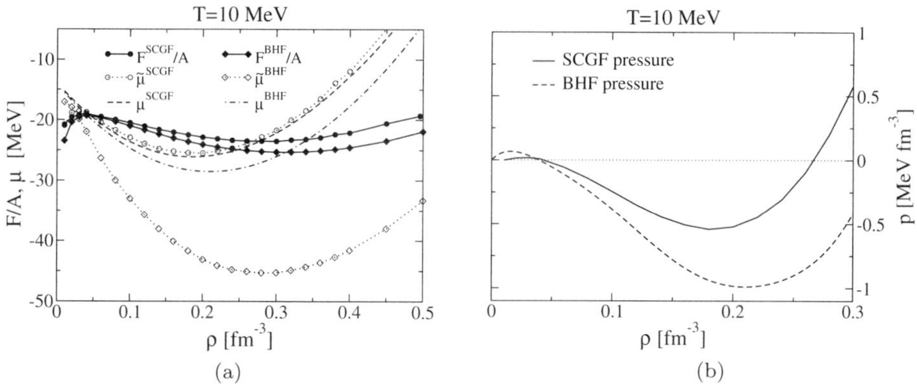

Fig. 4. (a) Free energies per particle (full lines) and microscopic chemical potentials $\tilde{\mu}$ (dotted lines) for the SCGF (circles) and BHF (diamonds) approaches. The macroscopic chemical potentials μ are also shown (dashed line for SCGF and dash-dotted line for BHF). (b) Pressure of symmetric nuclear matter within the SCGF (solid) and BHF (dashed) approaches.

propagation of holes has on the energy. This repulsion increases with density. In the same figure, we report the chemical potentials calculated either from a normalization condition, $\rho = \sum_k n(k, \tilde{\mu})$, (microscopic chemical potential) or from a derivative of the free energy, $\mu = \left(\frac{\partial F}{\partial N}\right)_T$ (macroscopic chemical potential). While the differences between $\tilde{\mu}$ and μ in the BHF approach can be larger than 10 MeV, the SCGF results fulfill TD consistency with an accuracy of the order of an MeV in the full density range.

The pressure as a function of the density for both SCGF and the BHF approaches at $T = 10$ MeV is shown in Fig. 4(b). The repulsive effect of holes in the free energy is translated in a larger pressure in the SCGF approach with respect to the BHF at intermediate densities. In both approaches, one observes a range of densities where the pressure decreases with density, signaling a mechanical instability which is usually associated to a liquid-gas phase transition. The range of densities and temperatures where this takes place defines the so-called spinodal region. This mechanical instability is associated to a liquid-gas phase transition. Detailed investigations of the phase diagram of nuclear matter have been carried out by means of heavy ion collisions. As an outcome of these experiments it has been possible to explore the liquid-gas phase transition which takes place at sub-saturation nuclear densities and for relatively moderate temperatures.[9,10]

The liquid-gas coexistence line is obtained by finding the gas and liquid densities such that the equations $\mu(\rho_g) = \mu(\rho_l)$ and $p(\rho_g) = p(\rho_l)$ are simultaneously satisfied. Finding the corresponding spinodal and coexistence regions for each temperature, one obtains the phase diagrams of Fig. 5 for the BHF (left panel) and the SCGF (right panel) approaches. The critical temperature for the liquid-gas phase transition of SNM corresponds to the maximum of the spinodal and coexistence lines. While for the SCGF scheme, $T_c \sim 18$ MeV, for the BHF it is substantially larger, close to

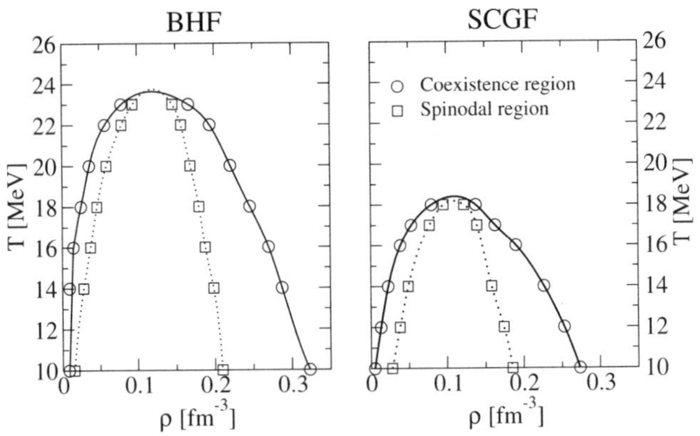

Fig. 5. Coexistence (circles) and spinodal (squares) regions for SNM within the BHF (left panel) and SCGF (right panel) approaches for the CDBONN interaction.

$T_c \sim 24$ MeV. The differences in the critical density for the two approaches seem to be smaller, $\rho_c \sim 0.1$ fm^{-3} for both cases. One could be tempted to say that the effect of the hole-hole propagation on the critical properties of the liquid-gas phase transition is rather large. However, before drawing any further conclusions, one should test the effect that the use of other realistic NN potentials, other partial waves (beyond the J=4 used here) or even the inclusion of three-body forces can have on the critical temperatures and densities. Note, however, that since we are exploring the region of sub-saturation densities, one expects the effects of the latter to be particularly small.

Acknowledgements

This work was partially supported by the NSF under Grant No. PHY-0555893, by the MEC (Spain) and FEDER under Grant No. FIS2005-03142, by the Generalitat de Catalunya under Grant No. 2005SGR-00343 and by the European Union under contract RII3-CT-2004-506078.

References

1. M. Baldo, *Nuclear Methods and the Nuclear Equation of State*, Int. Rev. of Nucl. Phys, Vol. 9 (World Scientific, Singapore, 1999).
2. H. Müther and A. Polls, *Prog. Part. Nucl. Phys.* **45**, 243 (2000).
3. R. Machleidt, F. Sammarruca and Y. Song, *Phys. Rev. C* **53**, R1483 (1996).
4. T. Frick and H. Müther, *Phys. Rev. C* **68**, 034310 (2003).
5. T. Frick, H. Müther, A. Rios, A. Polls and A. Ramos, *Phys. Rev. C* **71**, 014313 (2005).
6. G. M. Carneiro and C. J. Pethick, *Phys. Rev. B* **11**, 1106 (1975).
7. J. M. Luttinger and J. C. Ward, *Phys. Rev.* **118**, 1417 (1960).
8. A. Rios, A. Polls, A. Ramos and H. Müther, *Phys. Rev. C* **74**, 054317 (2006).
9. J. Pochodzalla *et al.*, *Phys. Rev. Lett.* **75**, 1040 (1995).
10. P. Danielewicz, R. Lacey and W. G. Lynch, *Science* **298**, 1592 (2002).

THE VALIDITY OF THE LOCV FORMALISM AND NEUTRON STAR PROPERTIES

H. R. MOSHFEGH,* M. M. MODARRES, A. RAJABI and E. MAFI

Department of Physics, University of Tehran,
Tehran, 1439955961, Iran
**E-mail: Hmoshfegh@ut.ac.ir*

We develop a method to make a very good estimate of the three-body cluster energy by using the state dependent correlation functions which have been produced by the lowest order constraint variational (LOCV) method. It is shown that the LOCV normalization constraint plays a major role in the convergence of the cluster expansion. Finally using the equation of state derived from LOCV formalism for neutron star matter, we calculate the neutron star properties such as its mass-radius relation and minimum mass.

Keywords: Equation of state; Cluster expansion; Neutron star.

1. Introduction

The method of lowest order constrained variational (LOCV) method has been reviewed in several papers.[1] In order to test the convergence of LOCV method for nuclear matter, we performed calculation beyond the lowest order (the two-body cluster term) and the three-body cluster energy was calculated with the state-averaged correlation functions.[2] The smallness of the normalization parameter (the convergence parameter) and the three-body cluster energy where indicated that at least up to the twice nuclear matter density, our expansion converges reasonably and it is a good approximation to stop after the two-body cluster terms.[2]

Recently we have investigated various properties of cold and hot asymmetrical nuclear matter using LOCV method based on cluster expansion theory.[3] This approach has been further generalized to include more sophisticated interactions such as AV_{14}, UV_{14}, AV_{18} and the new charge dependent $Reid93$ potentials.[1,4]

In this article we cheque the validity of LOCV by calculating the three-body cluster energy and using the state dependent correlation functions and effective interactions. It is conclude that the LOCV technique is good enough to calculate other properties of quantum fluids. In this respect, the properties of neutron star is calculated by using the equation of state which comes from LOCV method.The results are comparable with other many-body models.

2. The LOCV Formalism

We consider a trial many-body wave function of the form:[1,3]

$$\psi = F\varphi, \tag{1}$$

where φ is a Slater determinant of plane waves of N independent nucleons (ideal Fermi gas wave function), and F is a N-body correlation operator that can be given by the product of the two-body correlation operators (Jastrow form). Now using the above trial wave function, we construct a cluster expansion for the expectation value of our Hamiltonian.

$$E([F]) = \frac{1}{A}\frac{\langle\psi|H|\psi\rangle}{\langle\psi|\psi\rangle} = E_1 + E_2 + E_3 \cdots \geq E_0, \tag{2}$$

where E_0 is the through ground-state energy. $E_1 = \frac{1}{N}\sum\langle i|T_1|i\rangle$ is one-body kinetic energy and E_2 is two-body clusters energy and one can write it as:

$$E_2 = \frac{1}{2N}\sum\langle ij|W(12)|ij\rangle_a, \tag{3}$$

where the "effective interaction operator" is given by the following equation:

$$W(12) = [F^\dagger(12), [T_1 + T_2, F(12)]] + F^\dagger(12)V(12)F(12), \tag{4}$$

where T is the kinetic energy operator. In the lowest order, we truncate above series afterE_2. Typically the many-body Hamiltonian is expressed as:

$$H = \sum_{i=1}^{N} p_i^2/2m + \sum_{i<j} V(ij), \tag{5}$$

where $V(ij)$ is the two-nucleon interaction. Minimizing the two-body cluster energy under normalization constraint, $[\chi =< \psi|\psi > -1] = 0$, (that we impose on the channel tow-body correlation functions) we obtain a set of Euler-Lagrange equations. By solving these equations,[1,3] we can calculate correlation functions and consequently the non relativistic two-body energy E_2.

3. Three-body Cluster Energy

In general the three-body cluster term expression in the energy expectation value has the following form:[2]

$$E_3 = E_{3(2)} + E_{3(3)}, \tag{6}$$

$E_{3(2)}$ and $E_{3(3)}$ are defined as,

$$E_{3(2)} = -\frac{1}{N}\sum\langle ik|(F^\dagger(12)F(12) - 1)|ik\rangle_a\langle ij|W(12)|ij\rangle_a, \tag{7}$$

$$E_{3(3)} = -\frac{1}{N3!}\sum\langle ijk|W(123)|ijk\rangle_a, \tag{8}$$

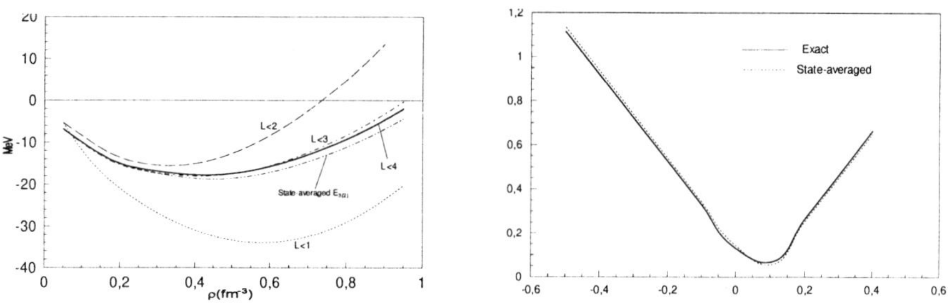

Fig. 1. Left: $E_{3(2)}$ versus density. Right: $|\frac{E_{3(2)}}{E_2}|$ versus convergence parameter.

where

$$W(ijk) = \frac{1}{2}[F_3^\dagger(123), [T_1 + T_2 + T_3, F_3(123)]] + adj.$$
$$+(F_3^\dagger(123)V(12)F_3(123) - W(12) + same \;\; for \;\; pairs \;\; 23 \;\; and \;\; 13). \qquad (9)$$

In Ref. 2 it has been shown that by define a state averaged two-body correlation function and effective interaction the contribution of E_3 is less than 1 MeV up to twice of nuclear matter saturation densities and main contribution comes from $E_{3(2)}$. In this work, by using the addition of angular momentum theorem, the partial wave expansion and various orthogonality relations, we calculate the explicit form of $E_{3(2)}$:

$$E_{3(2)} = -\frac{1}{N} \sum_{k_i k_j k_k, LSJT, L'S'J'T'} \int_0^\infty dr dr' (rr')^2 [1 - (-1)^{L+S+T}][1 - (-1)^{L'+S'+T'}]$$
$$\frac{(4\pi)^4}{2\Omega^2} |j_L(k_{ik} r)|^2 |j_{L'}(k_{ij} r')|^2 W^{L'+S'+T'}(r') \sum_{M_L M_{L'}} |Y_{LM_L}(\hat{k}_{ik})|^2 |Y_{L'M'_L}(\hat{k}_{ij})|^2$$
$$(2T+1)(2T'+1) \sum_{\sigma_i \sigma_j \sigma_k} \sum_{M_S M_{S'}} \sum_{M_J M_{J'}} |C^{SM_S}_{\frac{1}{2}\sigma_i \frac{1}{2}\sigma_k}|^2 |C^{S'M'_S}_{\frac{1}{2}\sigma_i \frac{1}{2}\sigma_j}|^2 |C^{JM_J}_{LM_L SM_S}|^2 |C^{J'M'_J}_{L'M'_L S'M'_S}|^2,$$

where $C^{\alpha m_\alpha}_{\beta m_\beta \gamma m_\gamma}$ are familiar Clebsch-Gordon coefficients, $Y_{LM_L}(\hat{k})$ are the spherical harmonics and $j_L(x)$ are spherical Bessel functions.

In the left panel of Fig. 1 we have plotted $E_{3(2)}$ versus density up to various summation on the relative angular momentum L i.e. 0,1,2 and 3 for Reid potential. The result shows that there is no much difference between summation up to $L = 2$ and $L = 3$. So we can conclude that the contribution of the $L > 3$ are negligible. In order to see the effect of normalization constraint, we have plotted the absolute ratio of state-dependent $E_{3(2)}$ to two-body cluster energy versus the convergence parameter χ at $\rho = 0.17 fm^{-3}$ for Reid potential. The dash line is the state-average results. In general, it is seen that three-body cluster energy is much smaller than E_2 and their ratio becomes very small near $\chi \simeq 0$. It is also found that minimum of above ratio is independent of density. The result shows that the state-average as well as LOCV approximation is working quite well.

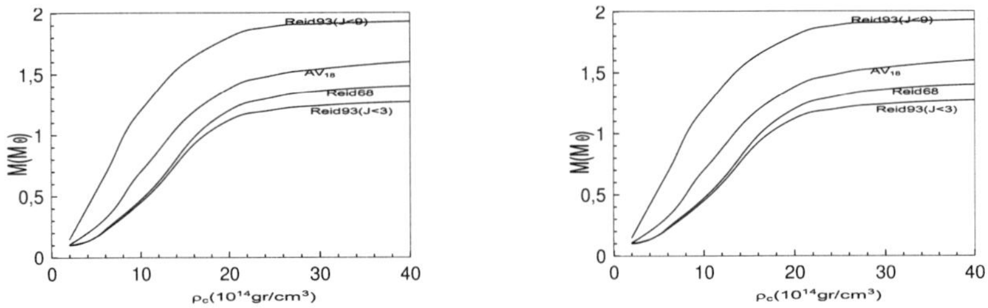

Fig. 2. Left: Gravitational mass of a neutron star versus central density. Right: Mass-radius relation for a neutron star.

4. Neutron Star Properties

The equation of state (EOS) of neutron star matter has a key role in determining the neutron star structure. We study the neutron star structure using EOS that comes from LOCV calculation employing the Reid potential. Using the EOS of neutron star matter, we can calculate the neutron star mass and radius as a function of central mass density, ρ_c, by numerical integrating the general relativistic equation of hydrostatic equilibrium, Tolman-Oppenheimer-Volkoff (TOV) equation:

$$\frac{dP}{dr} = -\frac{G}{r^2}[\rho_g(r) + P(r)/c^2]\frac{m(r) + 4\pi r^3 P(r)/c^2}{1 - \frac{2Gm(r)}{rc^2}}, \tag{10}$$

where $\rho_g = \rho[E(\rho)+mc^2]$ and $m(r) = \int_0^r 4\pi r'^2 \rho_g(r')dr'$. By selecting a central mass density under the boundary conditions $P(0) = P_c, m(0) = 0$, we integrate the TOV equation outwards to a radius $r = R$ at which P vanishes. This yields the neutron star radius R and mass $M = m(R)$. The calculated neutron star gravitational mass (in solar mass $M_\odot$ units) as a function of central mass density with $Reid68$, $Reid93(j < 3)$ and AV_{18} potentials is presented at the left panel of Fig. 2. In the right panel of this figure we have plotted gravitational mass versus radius. Our results show that at low densities neutron star masses exhibit a minimum ($\simeq 0.1M_\odot$) which is nearly independent of EOS and a maximum mass between $1.4M_\odot$ and $1.9M_\odot$ which is strongly dependent on the equation of state. It is seen that the given maximum mass for Reid equation of state shows a good consistency with the accurate observations of radio pulsars.[5]

We would like to thank the University of Tehran and the Institute for Research and Planning in Higher Education for supporting us.

References

1. M. Modarres and H. R. Moshfegh, *Prog. Theor. Phys.* **112**, 21 (2004).
2. H. R. Moshfegh and M. Modarres, *J. Phys. G* **24**, 821 (1998).
3. H. R. Moshfegh and M. Modarres, *Nucl. Phys. A* **792**, 201 (2007).
4. G. H. Bordbar and M. Modarres, *Phys. Rev. C* **57**, 7114 (1998).
5. S. E. Thorsett and D. Chakrabary, *Astrophys. J.* **512**, 288 (1999).

FERROMAGNETIC INSTABILITIES OF NEUTRON MATTER: MICROSCOPIC VERSUS PHENOMENOLOGICAL APPROACHES

ISAAC VIDAÑA

Departament d'Estructura i Constituents de la Matèria, Universitat de Barcelona, Avda. Diagonal 647, Barcelona E-08028, Spain

The magnetic susceptibility of spin-polarized neutron matter is calculated at both zero and finite temperature within (i) the microscopic framework of the Brueckner-Hartree-Fock formalism, and (ii) using the Skyrme and Gogny phenomenological forces. The microscopic results show no indication of a ferromagnetic transition at any density and temperature. In the case of the Skyrme interaction, it is shown that the critical density at which ferromagnetism takes place decreases with temperature. This unexpected feature is associated to an anomalous behavior of the entropy that becomes larger for the polarized phase than for the unpolarized one above a certain critical density. For the Gogny force, the results show two different behaviors: whereas the D1P parametrization exhibits a ferromagnetic transition at a density of $\rho \sim 1.31$ fm^{-3} whose onset increases with temperature, no sign of such a transition is found for D1 at any density and temperature, in agreement with the microscopic calculations.

Keywords: Neutron star matter; Ferromagnetic transition; Finite temperature.

1. Introduction

Since the suggestion of Pacini[1] and Gold[2] pulsars are believed to be rapidly rotating neutron stars with strong surface magnetic fields in the range $10^{12} - 10^{13}$ Gauss. Despite the great theoretical effort of the last forty years, there is still no general consensus regarding the mechanism to generate such strong magnetic fields in a neutron star. From the nuclear physics point of view one of the most interesting and stimulating mechanism which have been suggested is the possible existence of a phase transition to a ferromagnetic state at densities corresponding to the theoretically stable neutron stars and, therefore, of a ferromagnetic core in the liquid interior of such compact objects. Such a possibility has been considered since long ago by several authors within different theoretical approaches,[3–6] but the results are still contradictory. In this work, we compare the magnetic susceptibility of spin-polarized neutron matter at both zero and finite temperature obtained within (i) the microscopic framework of the Brueckner-Hartree-Fock formalism, and (ii) using the Skyrme and Gogny phenomenological forces. The paper is organized as follows: spin polarized neutron matter is presented in Section 2, the main results are shown in Section 3, and the conlcusions are given in Section 4.

2. Spin-Polarized Neutron Matter

Spin-polarized neutron matter is an infinite nuclear system composed of two different fermionic components: neutrons with spin up and neutrons with spin down having densities $\rho_\uparrow$ and $\rho_\downarrow$, respectively. The total density of the system is given by $\rho = \rho_\uparrow + \rho_\downarrow$, whereas the degree of polarization of the system can be characterized by the so-called spin polarization parameter $\Delta = (\rho_\uparrow - \rho_\downarrow)/(\rho_\uparrow + \rho_\downarrow)$. We are particularly interested in the magnetic susceptibility χ, which characterizes the response of the system to a magnetic field and gives a measure of the energy required to produce a net spin alignement in the direction of the field. The stability of the system against spin fluctuations are guarenteed if $\chi > 0$, a change of sign on χ would signal the onset of a ferromagnetic transition. The magnetic susceptibility can be written as

$$\chi = \frac{\mu^2 \rho}{\left(\frac{\partial^2 (F/A)}{\partial \Delta^2}\right)_{\Delta=0}}, \tag{1}$$

being F/A the total free energy per particle of the system.

Here we have followed two approaches in order to evaluate the free energy per particle and, therefore the magnetic susceptibility of spin polarized matter. The first one is a *microscopic approach* based on an extension of the Brueckner–Hartree–Fock approximation of the to (i) the case in which neutron matter is arbitrarily asymmetric in the spin degree of freedom, and (ii) to the case of finite temperature. This approach starts with the construction of the neutron-neutron G-matrix, which describes in an effective way the interaction between two neutrons with spin up and down in the presence of a surrounding medium. The single-particle potentials of neutrons with spin up and down, from which finally the total free energy per particle can be evaluated, are obtained in a selfconsistent way together with the G-matrix (see Refs. 4,5 for details). The calculations have been carried out using the Argonne V18 nucleon-nucleon force supplemented by a three-body force of the Urbana type in order to reproduce the saturation properties of symmetric nuclear matter.

The second approach is a *phenomenological* one based on the well known Skyrme and Gogny nucleon-nucleon forces. Here we have considered the Skyrme forces SLy4[7] and SkI3[8] and the Gogny forces D1[9] and D1P.[10] In this case the total free energy per particle is obtained in the Hartree–Fock approximation as the sum of the kinetic energy associated to the Fermi gas of polarized neutron matter and the expectation value of the Skyrme and Gogny interactions between the wave function describing two free Fermi seas corresponding to neutrons with two different spin orientations (see Ref. 6 for details).

3. Results

The ratio between the magnetic susceptibility of the free Fermi gas and the corresponding magnetic susceptibility of interacting neutron matter is shown in Fig. 1

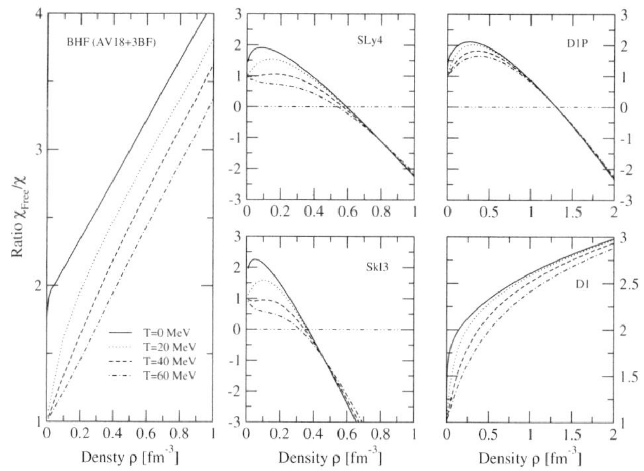

Fig. 1. Ratio between the magnetic susceptibility of the free Fermi gas and the corresponding magnetic susceptibility of interacting neutron matter as a function of density for several temperatures, for the Brueckner (left panel), Skyrme (middle panels) and Gogny (right panels) calculations.

as a function of density for several temperatures, for the Brueckner (left panel), Skyrme (middle panels) and Gogny (right panels) calculations. In the case of the Brueckner calculation and the Gogny one with the D1 parametrization, this ratio increases as the density increases at any temperature and no signal of a change of such a trend is expected at higher densities. This is an indication that a ferromagnetic transition, whose onset would be signaled by the density at which this ratio becomes zero, is not seen and not expected at larger densities either. On the other hand, for the Skyrme and the Gogny D1P interactions, the ratio become negative at $\rho \sim 0.60$ fm^{-3}, ~ 0.4 fm^{-3} and ~ 1.3 fm^{-3} for the SLy4, SkI3 and D1P interactions respectively. Contrary to what can be intuitively expected, the onset density for the magnetic instability decreases with temperature in the case of the Skyrme interaction. This unexpected behaviour is associated to an anomaly of the entropy: above a certain critical density, the entropy of the polarized phase turns out to be larger than that of the unpolarized one. We showed in Ref. 6 that this nonintuitive behaviour of the entropy is, in fact, a consequence of the dependence of the entropy on the effective mass of the neutrons with different spin component (See Ref. 6 for details).

In Fig. 2 we show the behaviour of the entropy per particle S/A as a funtion of the spin polarization at a fixed density for several temperatures. The entropy is symmetric with respect the nonpolarized state ($\Delta = 0$) and almost parabolic in Δ. Whereas the Brueckner and Gogny calculations show a maximum at zero polarization, in agreement with the intuitive idea that the polarized phase is more "ordered" than the nonpolarized one, the Skyrme results show the opposite behaviour: the entropy of the fully polarized system is larger than the nonpolarized one, giving the anomalous behaviour of the entropy with the polarization, we mentioned before.

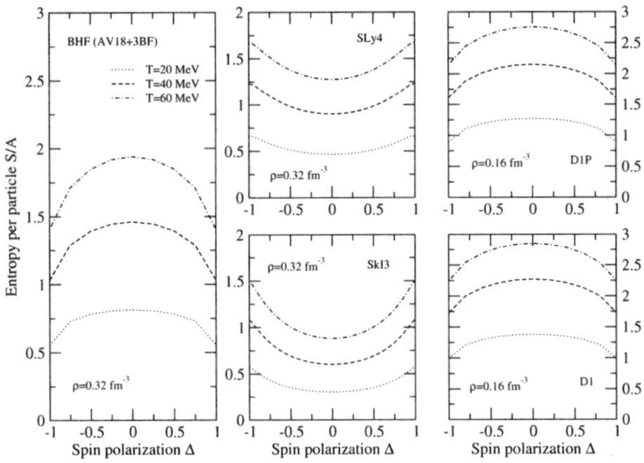

Fig. 2. Entropy per particle as a function of the spin polarization Δ at a fixed density and several temperatures, for the Brueckner (left panel), Skyrme (middle panels) and Gogny (right panels) calculations.

4. Conclusions

We have calculated the magnetic susceptibility of spin-polarized neutron matter at both zero and finite temperature within (i) the microscopic framework of the Brueckner-Hartree-Fock formalism, and (ii) using the Skyrme and Gogny phenomenological forces. Whereas the microscopic and the D1 parametrization of the Gogny force show no indication of a ferromagnetic transition at any density and temperature; the Skyrme and the Gogny D1P force predict such a transition at densities about 0.6 and 1.3 fm^{-3}, respectively. In addition we have found, in the case of the Skyrme interaction, that the critical density at which ferromagnetism takes place decreases with temperature. This unexpected feature is associated to an anomalous behavior of the entropy that becomes larger for the polarized phase than for the unpolarized one above a certain critical density.

References

1. F. Pacini, *Nature* **216**, 567 (1967).
2. T. Gold, *Nature* **218**, 731 (1968).
3. D. H. Brownel and J. Callaway, *Nuovo Cimento B* **60**, 169 (1969); M. J. Rice, *Phys. Lett. A* **29**, 637 (1969); E. Østgaard, *Nucl. Phys. A* **154**, 202 (1970); V. R. Pandharipande, V. K. Garde and J. K. Srivastava, *Phys. Lett. B* **38**, 485 (1972); A. Vidaurre, J. Navarro and J. Bernabeu, *Astron. Astrophys.* **135**, 361 (1984); S. Fantoni, A. Sarsa and K. E. Schmidt, *Phys. Rev. Lett.* **87**, 181101 (2001).
4. I. Vidaña, A. Polls and A. Ramos, *Phys. Rev. C* **65**, 035804 (2002).
5. I. Bombaci, A. Polls, A. Ramos, A. Rios and I. Vidaña, *Phys. Lett. B* **632**, 638 (2005).
6. A. Rios, A. Polls and I. Vidaña, *Phys. Rev. C* **71**, 055802 (2005); D. López-Val, A. Rios, A. Polls and I. Vidaña, *Phys. Rev. C* **74**, 068801 (2006).
7. E. Chabanat *et al.*, *Nucl. Phys. A* **627**, 710 (1997).
8. P.-G. Reinhard and H. Flocard, *Nucl. Phys. A* **684**, 467 (1995).
9. J. Dechargé and D. Gogny, *Phys. Rev. C* **21**, 1568 (1980).
10. M. Farine *et al.*, *J. Phys. G* **25**, 863 (1999).

SIGMA MESON AND NUCLEAR MATTER SATURATION

A. B. SANTRA
Nuclear Physics Division, Bhabha Atomic Research Centre, Mumbai 400085, India
E-mail: santra@barc.gov.in

U. LOMBARDO
Dipartimento di Fisica, Universita di Catania and INFN-LNS, Via S. Sofia 44, I-95123 Catania, Italy
E-mail: lombardo@lns.infn.it

We have found that it is possible to understand all the saturation observables of symmetric nuclear matter by incorporating in-medium modification of the parameters of sigma meson alone. In the framework of Brueckner-Bethe-Goldstone formalism with Bonn-B potential as two-body interaction, in-medium modification (density independent reduction) of σ-meson-nucleon coupling constant by about 3.5% and σ-meson mass by about 6.8% is enough to understand nuclear matter saturation observables.

Keywords: Nuclear matter equation of state; Medium modification of hadronic properties.

1. Introduction

It is known since the time of the formulation of quantum chromodynamics (QCD), the fundamental theory of strong interactions, that the basic degrees of freedom of strong interaction are quarks and gluons, not the mesons and nucleons, the usual hadorns. Hadrons are colorless composites of quarks and low-lying excitations in the QCD vacuum. The QCD vacuum, in which hadronic excitations are built, thus, is not the perturbative vacuum; and it does not share some of the symmetries of (classical) QCD lagrangian. For example, the chiral $SU(2)_L \times SU(2)_R$ symmetry of QCD lagrangian (in the massless limit of u and d quarks) is not manifest in the vacuum. The true QCD vacuum is realized from the perturbative vacuum through a phase transition,[1] known as the confinement-deconfinement or the chiral transitions. Low energy effective QCD models and lattice QCD calculations reveal that, in the limit $m_q \to 0$, the order parameter of chiral transition is the quark condensate, $\langle \bar{q}q \rangle$. We can identify particles corresponding to the fluctuation of the order parameter; the pions, almost massless pseudoscalar-isovector collective modes associated with the spontaneous breaking of chiral symmetry, and the σ meson, scalar-isoscalar collective mode, corresponds to phase and modulus fluctuation respectively. Though properties of pion and its interactions with nucleon are well understood, we have very little understanding about the σ meson. This is mainly because of the long standing problem of experimental identification of scalar mesons due to their large decay widths causing strong overlap between resonances and background. However, it appears that many experimental facts of hadron physics can be understood in a simple way, if we identify the σ-meson. Including a σ-pole with $m_\sigma < 1$ GeV, it is possible to reproduce[2] the $\pi - \pi$ phase shifts in I = 0, J = 0 channel in a model independent framework, respecting chiral symmetry, unitarity and crossing symmetry. Three-body decays of charm mesons, $D^+ \rightarrow \sigma\pi^+ \rightarrow \pi^-\pi^+\pi^+$, give strong evidence[3] of a scalar-isoscalar resonance with mass $m_\sigma = 478 \pm 17\,\mathrm{MeV}/c^2$, and width, $\Gamma_\sigma = 324 \pm 21\,\mathrm{MeV}/c^2$. The origin of the enhancement of $\Delta I = 1/2$ processes in kaon decay $K^0 \rightarrow \pi^+\pi^-$, $K^0 \rightarrow \pi^0\pi^0$ can be traced[4] to the sigma propagator in $K \rightarrow \sigma \rightarrow \pi\pi$, in the framework of effective description (NJL model) of low energy QCD, where σ-meson is treated as a dynamical degree of freedom. The intermediate range attraction of nucleon-nucleon interaction has been shown[5] to be generated by the exchange of scalar-isoscalar σ meson with $m_\sigma \sim 400 - 700$ MeV in the meson theoretical model of the nuclear force.

Traditional problem of nuclear matter calculations is to understand its ground state observables. In case of symmetric nuclear matter, the saturation density, ρ_0, the binding energy per nucleon, E/A, at the saturation point and the incompressibility, K, are respectively known to be, $\rho_0 = 0.17 \pm 0.02$ fm^{-3}, $E/A = -16\pm 1$ MeV and $K = 210\pm 30$ MeV. Calculations in the framework of Brueckner-Hartree-Fock (BHF) and Dirac-Brueckner-Hartree-Fock (DBHF) using microscopic or phenomenological two body NN interaction potential fail to reproduce the above data.

Hadron properties are expected to change inside the nuclear medium due to restoration of chiral symmetry. Theoretical studies toward understanding modification of hadron parameters in nuclear medium was started after pointing out by Brown and Rho[6] that masses of hadrons would scale in nuclear medium. Calculations[7,8] of EOS of symmetric nuclear matter using Bonn-B potential as two-body interaction in the framework of non-relativistic BHF and Dirac-Brueckner-Hartree-Fock (DBHF) and including the in-medium modification of the hadrons as given by Brown-Rho[6] do not reproduce the nuclear matter saturation observables completely.

Given the important role that the σ-meson plays in hadron physics, we explore the role of in-medium modification σ-meson spectral function on nuclear matter EOS. In section 2, we describe the method of generating in-mediun NN potential due to exchange of σ-meson. In section 3, we give the results of our calculation of EOS of symmetric nuclear matter including in-medium modification of sigma meson parameters, and summary of the presentation is given in section 4.

2. Modification of In-Medium NN Potential

Density dependent in-medium NN-potential due to σ-meson exchange can be written as,

$$\tilde{V}_{NN}^{\sigma}(q,\omega,\rho) = \int V^{\sigma}(q,\omega,\mu,g)\, S^{\sigma}(\mu,\rho)\, d\mu^2 , \tag{1}$$

where $S^{\sigma}(\mu,\rho)$ is density dependent normalized spectral function of σ-meson and $V^{\sigma}(q,\omega,\mu,g)$ is free NN-potential due to σ-exchange. The modification of of spectral function at finite density can be taken into account by allowing the mass and σ-NN coupling constant dependent on density in NN-potential with zero width σ meson exchange. Taking estimate of $S^{\sigma}(\mu,\rho)$ from Aouissat et al[9] we find, computing Eq. (1), the dependence of sigma meson mass (m_{σ}^{*}) and σ-NN coupling (g_{σ}^{*}) with density as,

$$m_{\sigma}^{*} = \frac{m_{\sigma}}{1+0.028\sqrt{\rho/\rho_0}} , \qquad g_{\sigma}^{*} = \frac{g_{\sigma}}{1+0.15\sqrt{\rho/\rho_0}} , \tag{2}$$

where ρ_0 is the saturation density, and m_{σ} and g_{σ} are respectively the mass of σ meson and σ-NN coupling constant as taken in the Bonn-B potential.

3. Equation of State of Symmetric Nuclear Matter

We calculate the equation of state using non-relativistic BHF method with Bonn-B as two body NN potential including modifications of its σ-meson parameters as given in Eq. (2). We get the saturation density, ρ_0, and energy per nucleon, E/A, from the EOS as, $\rho_0 = 0.13 fm^{-3}$, and $E/A = -6MeV$ which are very far from the observed saturation density and energy per particle. We then try to find out possible extent of modification of the parameters of sigma meson so that the EOS reproduces the saturation observables. We take the ansatz for in-medium σ-NN coupling and σ mass as,

$$g_{\sigma}^{*} = g_{\sigma}\left(1-\alpha\frac{\rho}{\rho_0}\right) , \qquad m_{\sigma}^{*} = m_{\sigma}\left(1-\beta\frac{\rho}{\rho_0}\right) . \tag{3}$$

Calculating the equation of state in non-relativistic BHF formalism with Bonn-B as two body NN potential including modifications of its σ-meson parameters as given in Eq. (3), we found that for no values of α and β the observed saturation density and the energy per particle could be reproduced by the EOS. We then make the in-medium mass of σ-meson as density independent, but keep the in-medium σ-NN coupling density dependent as,

$$g_{\sigma}^{*} = g_{\sigma}\left(1-\alpha\frac{\rho}{\rho_0}\right) , \qquad m_{\sigma}^{*} = m_{\sigma}(1-\gamma) . \tag{4}$$

Calculating the equation of state in non-relativistic BHF formalism with Bonn-B as two body NN potential including modifications of its σ-meson parameters as given in Eq. (4), we found that all the saturation observables can be reproduced for $\alpha = 0.068$ and $\gamma = 0.035$. The results are shown in Fig. 1. The saturation observables obtained from the EOS are $\rho_0 = 0.17\,\mathrm{fm}^{-3}$, $\mathrm{E/A} = -16\,\mathrm{MeV}$ and $\mathrm{K} = 230\,\mathrm{MeV}$.

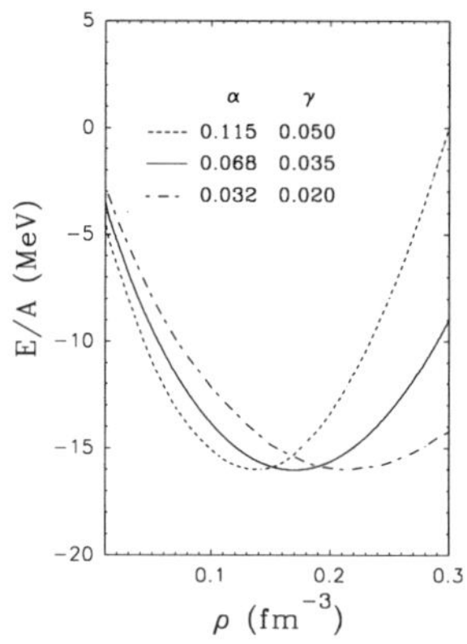

Fig. 1. Equation of state of symmetric nuclear matter with Bonn potential including modification of sigma meson parameters alone.

4. Summary

In hadron physics, σ-meson plays an important role. Identifying σ-meson as the particle corresponding to the fluctuation of the modulus of the order parameter of chiral transition, we expect that role of σ-meson will be important in nuclear matter too. We have estimated how much modification of the parameters of σ-meson alone will be required to understand all the saturation observables of nuclear matter simultaneously in the framework of non-relativistic BHF calculation with Bonn-B potential. We find that density independent reduction of σ-meson mass by a very small amount (3.5%), and, density dependent reduction of σ-NN coupling constant also by a very small amount (6.8% at the saturation density), is sufficient for understanding the nuclear matter saturation observables.

Acknowledgments

We thank Professor Ruprecht Machleidt for giving his codes of nuclear matter calculation and Bonn-potential, and for many discussions over email. ABS thanks the organizers of the symposium and INFN, Catania, for financial support to participate in this symposium.

References

1. F. Karsch, *Lecture Notes in Physics* **583**, 209 (2002).
2. K. Igi and K. Hikasa, *Phys. Rev. D* **59**, 034005 (1999).
3. E. M. Aitala et al., *Phys. Rev. Lett.* **86**, 770 (2001).
4. T. Morozumi, C. S. Lim and A. I. Sanda, *Phys. Rev. Lett.* **65**, 404 (1990).
5. R. Machleidt, *Adv. Nucl. Phys.* **19**, 1 (1989).
6. G. E. Brown and M. Rho, *Phys. Rev. Lett.* **66**, 2720 (1991).
7. A. B. Santra and U. Lombardo, *Phys. Rev. C* **62**, 018202 (2000).
8. R. Rapp, R. Machleidt, J. W. Durso and G. E. Brown, *Phys. Rev. Lett.* **82**, 1829 (1999).
9. Z. Aouissat, G. Chanfray, P. Schuck and J. Wambach, *Phys. Rev. C* **61**, 012202 (2000).

RAMIFICATIONS OF THE NUCLEAR SYMMETRY ENERGY FOR NEUTRON STARS, NUCLEI AND HEAVY-ION COLLISIONS

A. W. STEINER

Joint Institute for Nuclear Astrophysics, National Superconducting Cyclotron Laboratory, and the Department of Physics and Astronomy, Michigan State University, East Lansing, MI 48824-2320, USA

B.-A. LI

Department of Physics, Texas A&M University-Commerce, Commerce, TX 75429-3011, USA

M. PRAKASH

Department of Physics and Astronomy, Ohio University, Athens, OH 45701-2979, USA

The pervasive role of the nuclear symmetry energy in establishing some nuclear static and dynamical properties, and in governing some attributes of neutron star properties is highlighted.

Keywords: Symmetry energy; Nuclear matter; Nuclei; Neutron stars; Intermediate-energy heavy-ion collisions.

1. Introduction

The nuclear symmetry energy, $E_s = \frac{1}{2}\frac{\partial^2 E}{\partial \delta^2}$, where $\delta = (n_n - n_p)/(n = n_n + n_p)$ with n_n and n_p denoting the neutron and proton densities and $E \equiv E(n, \delta)$ is the energy per particle, measures the stiffness encountered in making a system of nucleons isospin-asymmetric. Figure 1 schematically shows how the symmetry energy connects several nuclear and astrophysical observables. The difference between the energy per baryon of pure neutron matter and that of symmetric nuclear matter (containing equal numbers of neutrons and protons) at any particular density is largely given by the density dependent symmetry energy. Below, we highlight some recent work that has shed light on some of the connections between the symmetry energy and data from terrestrial experiments and astrophysical observations.

2. The Skin Thicknesses of Heavy Nuclei

Traditionally, constraints on the nuclear symmetry energy have been derived from mass measurements of nuclei. For neutron star physics, the correlation that exists[2,3]

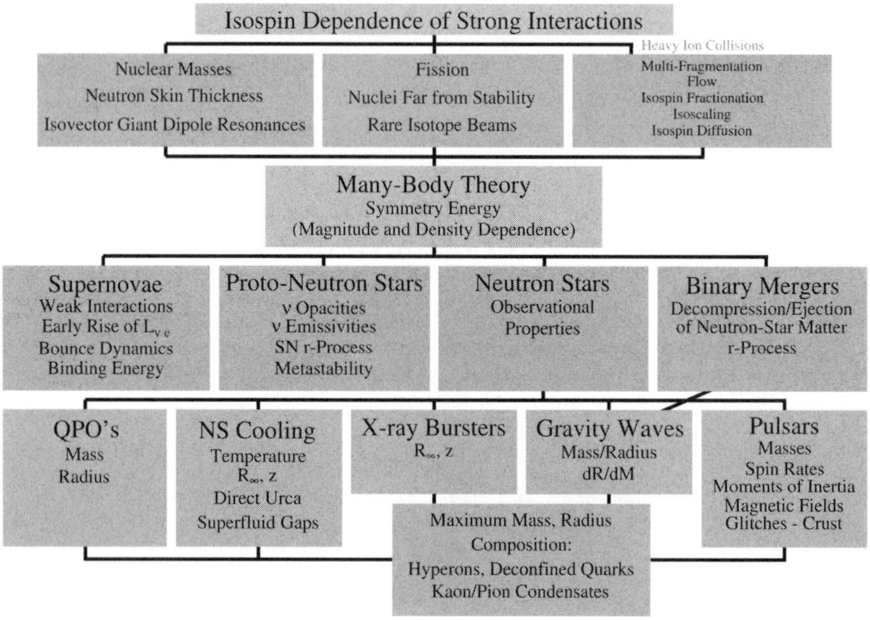

Fig. 1. The nuclear physics observables (top panels) and astrophysical observables (lower panels) are both connected to the nuclear symmetry energy, which is determined from nuclear many-body physics (adapted from Ref. 1).

between the neutron skin thicknesses in heavy nuclei, δR (the difference between the neutron and proton root-mean-square radii), and the pressure P of pure neutron matter at a density of $n \simeq 0.1$ fm^{-3} is particularly useful. Accurate measurements of δR can establish an empirical calibration point for the pressure of neutron star matter at subnuclear densities. The connection between the neutron skin thickness and the symmetry energy has been known from Bodmer's work in the 60's.[4] The Typel-Brown correlation between δR and $P(5n_0/8)$, which is closely related to the density derivative of the symmetry energy, demonstrates clearly the new information that could be obtained by accurate measurements of skin thicknesses in heavy nuclei. Figure 2 shows the correlation between the skin thickness δR of ^{208}Pb and the pressure of beta-equilibrated matter at 0.1 fm^{-3} for several potential models (based on the Skyrme interaction) and field-theoretical models.[1] The skin thickness of ^{208}Pb is scheduled to be measured at the Jefferson Lab in the summer of 2008 in the PREX experiment[5,6] and will likely provide a stringent constraint.

3. Intermediate-Energy Heavy-Ion Collisions

The possibility of determining the equation of state (EOS) of nucleonic matter from heavy-ion collisions has been discussed for almost 30 years. A number of heavy ion collision probes of the symmetry energy have been proposed including isospin frac-

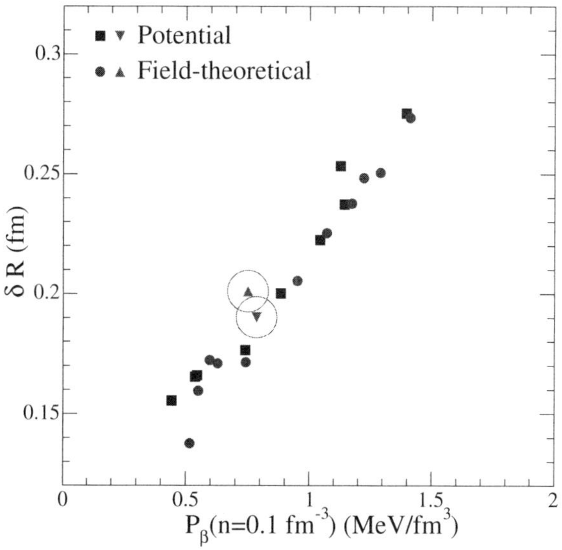

Fig. 2. The correlation between neutron skin thickness of ^{208}Pb and the pressure of neutron star matter for the models described in Ref. 1.

tionation,[7,8] isoscaling,[9,10] neutron-proton differential collective flow,[11] pion production,[12] isospin diffusion,[13] and neutron-proton correlation functions.[14] Determination of the EOS from heavy-ion data involves comparisons of transport model simulations[15–17] with experimental data given an input EOS. Transport model simulations track the evolution of the phase space distribution function (not necessarily that of an equilibrium distribution function at finite temperature) at momenta that usually exceed those found in the static initial configurations. As the driving force is the density functional derivative of the energy density (the in-medium cross sections control the collision integral), access to the cold EOS at high densities is afforded.

Isospin diffusion is caused by the exchange of neutrons and protons between nuclei in a heavy-ion collision (Fig. 3). This diffusion process, driven by the symmetry energy, moves the target and projectile nuclei toward isospin symmetry. To gain access to details of the diffusion process, fragment emission during and after the collision must be taken into account. This requirement is achieved by considering the ratio[18]

$$R_\delta = \frac{2\delta^{A+B} - \delta^{A+A} - \delta^{B+B}}{\delta^{A+A} - \delta^{B+B}}, \qquad (1)$$

where A and B denote nuclei with different isospin asymmetries and δ is the isospin asymmetry of the projectile-like fragment.

Recently, isospin diffusion has been exploited to constrain the symmetry energy from reactions involving ^{112}Sn and ^{124}Sn at the NSCL,[13,20] leading to $R_\delta \sim 0.46$. In conjunction with an isospin- and momentum-dependent transport model,

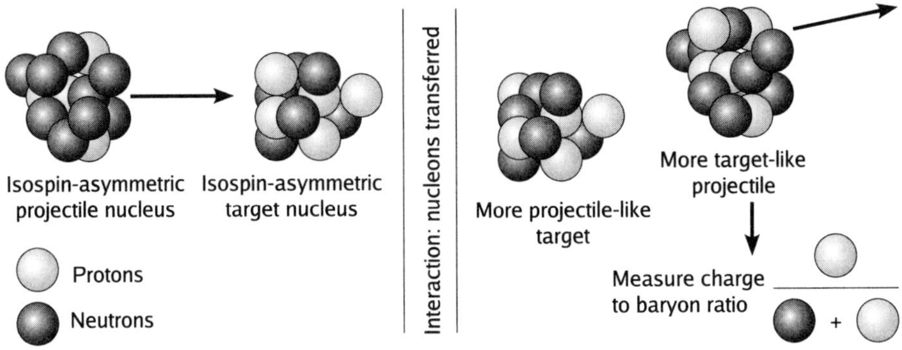

Fig. 3. Isospin diffusion process during a heavy-ion collision (from Ref. 19).

IBUU04,[21,22] the NSCL data on isospin diffusion can be used to constrain the symmetry energy. Writing the symmetry energy in terms of a kinetic part and a potential part as

$$E_s(n) = S_v(n/n_0)^\gamma \,, \tag{2}$$

where S_v is the symmetry energy at the nuclear equilibrium density, $n_0 = 0.16\ \mathrm{fm}^{-3}$, the constraint $0.69 < \gamma < 1.05$ has been found.[20,23] This constraint is consistent with the symmetry energy inherent in the EOS (computed using Monte Carlo simulations with input two- and three-body interactions which are matched to nucleon-nucleon scattering phase shifts and the energy levels of light nuclei) of Akmal, et al.[24] (APR). This constraint also rules out models with values of $\gamma > 1.05$ found in some field-theoretical models.

Because intermediate-energy heavy-ion collisions provide a constraint on the symmetry energy at the same densities as would be probed by a measurement of the neutron skin thickness of ^{208}Pb, the NSCL data also provides a restrictive range for its neutron skin thickness. This connection was used in Refs. 25 and 23 to show that the neutron skin thickness of ^{208}Pb should be at least greater than 0.15 fm in order to be consistent with the NSCL data, with values between 0.23 fm and 0.27 fm favored by the transport model simulations (Fig. 4).

4. Neutron Star Radii

Neutron star radii tend to probe the density dependence of the symmetry energy around the nuclear equilibrium density, $n_0 = 0.16\ \mathrm{fm}^{-3}$. Lattimer and Prakash[26] found that the radius R of a neutron star exhibits the power law correlation:

$$R \simeq C(n, M)\ [P(n)]^{0.23-0.26} \,, \tag{3}$$

where $P(n)$ is the total pressure inclusive of leptonic contributions evaluated at a density n in the range n_0 to $2n_0$, and $C(n, M)$ is a number that depends on

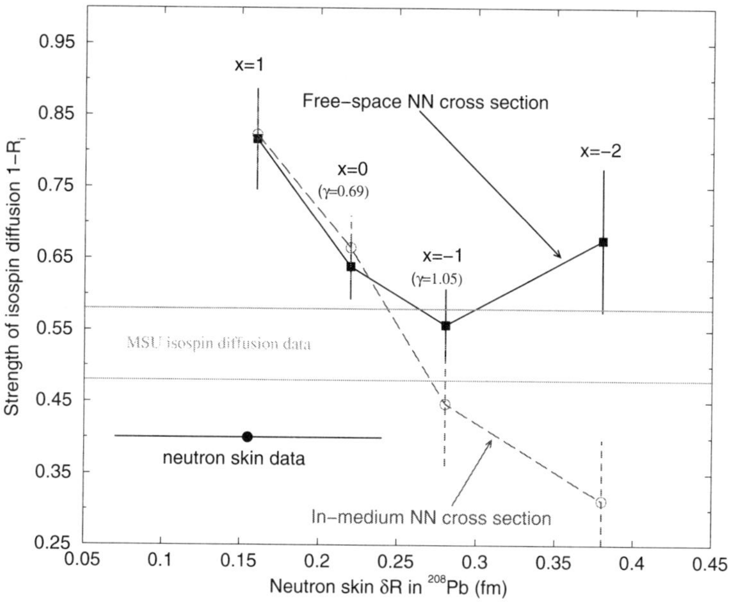

Fig. 4. The correlation between the neutron skin thickness and the amount of isospin diffusion in collisions between Sn isotopes as measured at the NSCL (from Ref. 23).

the density n at which the pressure is evaluated and on the stellar mass M. The left panel in Fig. 5 shows this correlation as $RP^{-\alpha}$ versus R for stars of mass 1.4 $M_\odot$. Neutron star radius measurements, especially those with uncertainties less than about 0.5 km, constrain the symmetry energy above the nuclear equilibrium density. These constraints will be much improved when simultaneous mass and radius measurements of the same object become available.

If there is no phase transition between n_0 and a few times n_0, the range in which neutron star radii are determined mainly by the symmetry energy, results from the isospin diffusion data at the NSCL can be used to constrain neutron star radii.[27] As only EOSs with symmetry energies between $x = 0$ and $x = -1$ (where x is a parameter designed to vary the density dependence of the symmetry energy without modifying the magnitude of the symmetry energy at n_0 or the isospin-symmetric part of the EOS) are consistent with the isospin diffusion data, this range of x values is representative of the possible variation in neutron star structure that is consistent with terrestrial data. Neutron star radii, while being strong functions of the symmetry energy, are also affected by contributions from the isospin-symmetric part of the EOS, especially at high densities. About 5% difference is representative of the radius uncertainty stemming from the symmetric part of the EOS. The conclusion of Ref. 27 is that only radii between 11.5 and 13.6 km (or radiation radii between 14.4 and 16.3 km) are consistent with the $x = 0$ and $x = -1$ EOSs (see the right panel in Fig. 5).

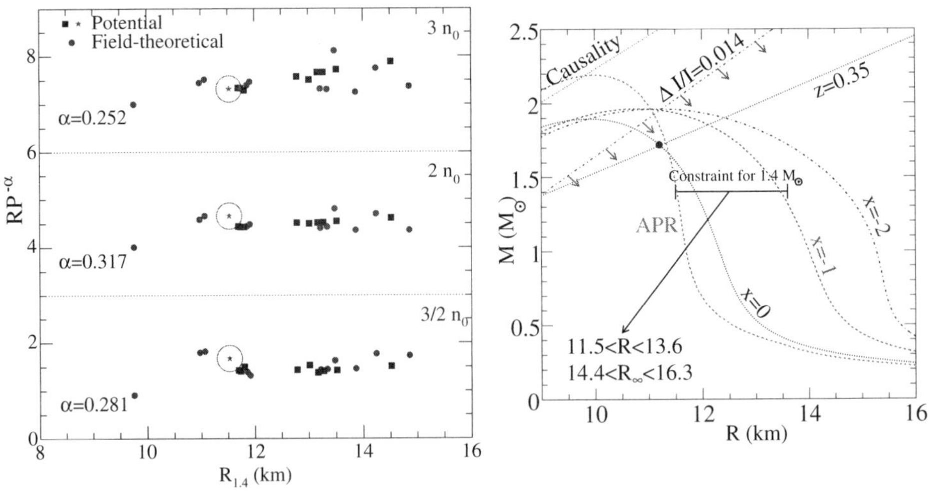

Fig. 5. *(Left)* The value $RP^{-\alpha}$ for several field-theoretical and potential models considered in Ref. 1. The values of α correspond to the density indicated in the upper right corner of each block. *(Right)* The constraint on the radius of 1.4 solar mass neutron stars as determined from isospin diffusion in heavy-ion collisions (from Ref. 27).

5. The Direct Urca Process

The long-term cooling of a neutron star is chiefly determined by its composition. Beta equilibrium and charge neutrality determine the proton fraction in neutron-star matter and thus the critical density for the onset of the direct Urca processes $n \to p + e + \bar{\nu}$ and $e + p \to n + \nu$, which cool the star more rapidly than the modified Urca processes in which an additional nucleon is present.[28] The direct Urca processes, however, require a sufficient amount of protons in matter (of order 10-14%).

Larger symmetry energies induce larger proton fractions (with matter being closer to isospin-symmetric) and smaller critical densities for the onset of the direct Urca processes. However, higher than quadratic terms in the energy $E(n, \delta)$ of isospin asymmetric matter can have an important role to play.[29] Recently, Steiner[30] has shown that quartic terms in $E(n, \delta)$ play an important role in determining the critical density for the direct Urca process. Such terms can be easily generated within the context of field-theoretical models. This is demonstrated in Fig. 6 for the Akmal, et al., (1998) EOS, for which the relative size of the quartic and quadratic terms is parametrized by η as described in Ref. 30. The mass and radius are virtually unchanged, whereas the threshold density for the direct Urca process changes by more than a factor of two.

Gusakov, et al.,[31] have investigated the cooling of neutron stars using the EOS of APR.[24] They found that, because of the direct Urca process, stars with masses

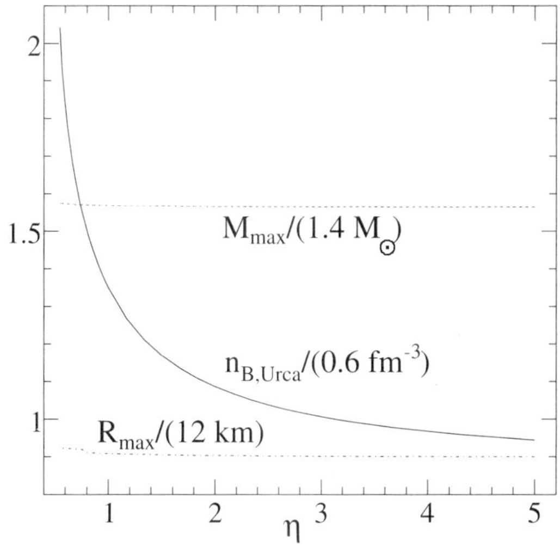

Fig. 6. The maximum mass, the radius of the maximum mass star, and the critical density for the direct Urca process as a function of η, which describes the strength of quartic terms in the symmetry energy (from Ref. 30).

larger than about 1.7 $M_{\odot}$ cool so rapidly as to be cooler than nearly all of the observed neutron stars. Our work offers a possible resolution: quartic terms can play a role at high density to turn off the direct Urca process thus making the computed cooling curves match the comparatively warm neutron stars.

6. Outlook

In addition to the role of the symmetry energy in the few areas highlighted here, its importance in controlling the cooling times of transient x-ray bursters, seismic activity of neutron star surfaces, ejection of baryons during binary mergers, etc., is only beginning to be appreciated. The PREX experiment, heavy-ion experiments, neutron star mass, radius and surface temperature measurements, observations of transient x-ray bursters etc., all hold keys to pin down the magnitude and density dependence of the symmetry energy in addition to delineating its pervasive role.

Acknowledgments

Research support for A. W. S. by the Joint Institute for Nuclear Astrophysics under the NSF-PFC grant PHY 02-16783, for B-A. L. in parts by the NSF under Grant No. PHY0652548 and the Research Corporation under Award No. 7123, and for M. P. by the DOE under Grant. DE-FG02-93ER40756 is gratefully acknowledged.

References

1. A. W. Steiner, M. Prakash, J. M. Lattimer and P. J. Ellis, *Phys. Rep.* **411**, 325 (2005).
2. B. A. Brown, *Phys. Rev. Lett.* **85**, 5296 (2000).
3. S. Typel and B. A. Brown, *Phys. Rev. C* **64**, 027302 (2001).
4. A. R. Bodmer, *Nucl. Phys.* **17**, 388 (1960).
5. R. Michaels, P. A. Souder and G. M. Urciuoli, *Jefferson Laboratory Proposal* **PR–00–003** (2000).
6. C. J. Horowitz, S. J. Pollock, P. A. Souder and R. Michaels, *Phys. Rev. C* **63**, 025501 (2001).
7. H. Muller and B. Serot, *Phys. Rev. C* **52**, 2072 (1995).
8. B. A. Li and C. M. Ko, *Nucl. Phys. A* **618**, 498 (1997).
9. M. B. Tsang, W. A. Friedman, C. K. Gelbke, W. G. Lynch, G. Verde and H. S. Xu, *Phys. Rev. Lett.* **86**, 5023 (2001).
10. A. Ono, P. Danielewicz, W. A. Friedman, W. G. Lynch and M. B. Tsang, *Phys. Rev. C* **68**, 051601(R) (2003).
11. B.-A. Li, *Phys. Rev. Lett.* **85**, 4221 (2000).
12. B.-A. Li, *Phys. Rev. Lett.* **88**, 192701 (2002).
13. M. B. Tsang, T. X. Liu, L. Shi, P. Danielewicz, C. K. Gelbke, X. D. Liu, W. G. Lynch, W. P. Tan, G. Verde, A. Wagner, H. S. Xu, W. A. Friedman, L. Beaulieu, B. Davin, R. T. de Souza, Y. Larochelle, T. Lefort, R. Yanez, V. E. Viola, Jr., R. J. Charity and S. L. G., *Phys. Rev. Lett.* **92**, 062701 (2004).
14. L.-W. Chen, V. Greco, C. M. Ko. and B.-A. Li, *Phys. Rev. Lett.* **90**, 162701 (2003).
15. B.-A. Li, C. M. Ko and W. Bauer, *Int. Jour. Mod. Phys. E* **7**, 147 (1998).
16. B.-A. Li and W. U. Schröder (eds.), *Isospin Physics in Heavy-Ion Collisions at Intermediate Energies* (Nova Science Publishers, Inc., New York, 2001).
17. P. Danielewicz, R. Lacey and W. G. Lynch, *Science* **298**, 1592 (2002).
18. F. Rami, Y. Leifels, B. de Schauenburg, A. Gobbi, B. Hong, J. P. Alard, A. Andronic, R. Averbeck, V. Barret, Z. Basrak, N. Bastid, I. Belyaev, A. Bendarag and G. Berek, *Phys. Rev. Lett.* **84**, 1120 (2000).
19. J. M. Lattimer and M. Prakash, *Phys. Rep.* **442**, 109 (2006).
20. L.-W. Chen, C. M. Ko and B.-A. Li, *Phys. Rev. Lett.* **94**, 032701 (2005).
21. B.-A. Li, C. B. Das, S. Das Gupta and C. Gale, *Phys. Rev. C* **69**, 011603(R) (2004).
22. B.-A. Li, C. B. Das, S. Das Gupta and C. Gale, *Nucl. Phys. A* **735**, 563 (2004).
23. B.-A. Li and L.-W. Chen, *Phys. Rev. C* **72**, 064611 (2005).
24. A. Akmal, V. R. Pandharipande and D. G. Ravenhall, *Phys. Rev. C* **58**, 1804 (1998).
25. A. W. Steiner and B.-A. Li, *Phys. Rev. C* **72**, 041601 (2005).
26. J. M. Lattimer and M. Prakash, *Astrophys. J.* **550**, 426 (2001).
27. B.-A. Li and A. W. Steiner, *Phys. Lett. B* **642**, 436 (2006).
28. J. M. Lattimer, C. J. Pethick, M. Prakash and P. Haensel, *Phys. Rev. Lett.* **66**, 2701 (1991).
29. H. Muether, M. Prakash and T. L. Ainsworth, *Phys. Lett. B* **199**, 469 (1988).
30. A. W. Steiner, *Phys. Rev. C* **74**, 045808 (2006).
31. M. E. Gusakov, A. D. Kaminker, D. G. Yakovlev and O. Y. Gnedin, *Mon. Not. R. Astron. Soc.* **363**, 555 (2005).

THE SYMMETRY ENERGY IN NUCLEI AND NUCLEAR MATTER

A. E. L. DIEPERINK

Kernfysisch Versneller Instituut,
NL-9747AA Groningen, The Netherlands
E-mail: dieperink@kvi.nl

We discuss to what extent information on ground-state properties of finite nuclei (energies and radii) can be used to obtain constraints on the symmetry energy in nuclear matter and its dependence on the density. The present approach is based upon a generalized Weizsäcker formula for ground-state energies. In particular effects from the Wigner energy and shell structure on the symmetry energy are investigated. Data on the neutron skin is used as an additional source of information.

Keywords: Nuclear symmetry energy; Liquid drop model; Shell effects; Neutron skin.

1. Introduction

The nuclear symmetry energy is an important ingredient in the description of properties of proto-neutron stars. The equation of state, the proton fraction and the pressure are strongly affected by the density dependence of the symmetry energy in nuclear matter. Conventionally the symmetry energy is expanded around the saturation density ρ_0 as

$$S(\rho) = a_4 + p_0 \frac{(\rho - \rho_0)}{\rho_0^2} + \Delta K \frac{(\rho - \rho_0)^2}{18\rho_0^2}. \tag{1}$$

Most microscopic calculations are either based upon realistic nucleon-nucleon interactions (using Brueckner or variational techniques) or mean-field models using parameters fitted to data of finite nuclei. In practice predictions for the symmetry energy vary substantially: *e.g.*, $a_4 \equiv S(\rho_0) = 28$–$36$ MeV, whereas predictions for the slope parameter, p_0, can vary by a factor three.

In practice, only the empirical value of $a_4 \sim 29$ MeV has been extracted with reasonable accuracy from finite nuclei by fitting ground-state energies using the Weizsäcker mass formula. Little information on the slope, p_0, is available.

Therefore the question arises whether one can obtain quantitative constraints from finite nuclei. Naturally one may distinguish two regions, those containing information on sub-saturation densities ($\rho < \rho_0$) and those on supra-normal densities ($\rho > \rho_0$). While the latter requires a (model-dependent) interpretation of results from heavy-ion reactions (diffusion of neutron-proton asymmetry), the former region can be addressed by analyzing static properties of nuclei. This will be the subject of the present contribution.

To make contact with finite systems, one uses the semi-empirical mass formula[1] which contains information on the average values of bulk and surface binding energies (for isospin symmetric systems) and (bulk) symmetry and Coulomb energies. In the past it has been realized that the symmetry and Coulomb energies not only have a bulk contribution but one from the surface as well. This leads to a generalized "liquid drop" description of finite nuclei and is akin to the treatment in nuclear collective models (*e.g.* describing properties of the giant dipole resonance) where it is equally essential to include a surface symmetry energy in addition to the bulk symmetry energy.[2] More recently, in connection with the semi-empirical mass formula, it has been stressed by Danielewicz[3] that in order to provide a consistent description of nuclei with neutron excess it is necessary to consider a *surface symmetry term* in addition to the bulk symmetry energy. The main purpose of this contribution is to show that in determining the surface symmetry energy several other corrections to the liquid-drop model (LDM) should be dealt with as well, in particular those due to shell structure and the effect of neutron-proton correlations (Wigner energy).

2. Extended Liquid Drop Model

The conventional semi-empirical liquid drop mass formula gives the binding energy of a nucleus as

$$B(N,Z) = a_{\rm v}A - a_{\rm surf}A^{2/3} - a_{\rm sym}\frac{(N-Z)^2}{A} - a_{\rm c}\frac{Z(Z-1)}{A^{1/3}} + a_{\rm p}\frac{\Delta(N,Z)}{A^{1/2}}, \qquad (2)$$

with N and Z the number of neutrons and protons, and $A = N + Z$. The terms on the right-hand side of Eq. (2) represent the bulk or volume, surface, symmetry, Coulomb and pairing energies, respectively, and $\Delta(N,Z)$ is a simple parametrization of pairing which is 1 for even-even, 0 for odd-mass and -1 for odd-odd nuclei.

2.1. *Surface symmetry energy*

It has been pointed out[3] that in Eq. (2) volume and surface terms are not separated in a consistent way: one needs also to separate the volume and surface contributions to the symmetry energy. To accomplish this is not trivial. In a rigorous derivation one first introduces the concept of surface tension; the latter can then be decomposed into isospin symmetric and asymmetric contributions.[3,4] In practice the result can be obtained in a more schematic way by decomposing the total particle asymmetry, $N-Z$, into volume(v) and surface(s) terms, $N-Z = N_{\rm v}-Z_{\rm v}+N_{\rm s}-Z_{\rm s}$, and requiring that the symmetry energy (quadratic in the asymmetry) scales with particle numbers as

$$B_{\rm sym}(N,Z) = S_{\rm v}\frac{(N_{\rm v}-Z_{\rm v})^2}{A} + S_{\rm s}\frac{(N_{\rm s}-Z_{\rm s})^2}{A^{2/3}}. \qquad (3)$$

Minimization of Eq. (3) under fixed $N-Z$ leads to a generalized formula for the binding energy of a nucleus in which the symmetry energy depends on two independent parameters, S_v, the volume symmetry energy, and the ratio $y \equiv S_v/S_s$

$$B(N,Z) = a_v A - a_{surf} A^{2/3} - \frac{S_v}{1+yA^{-1/3}} \frac{(N-Z)^2}{A} - a_c \frac{Z(Z-1)}{A^{1/3}} + a_p \frac{\Delta(N,Z)}{A^{1/2}}. \quad (4)$$

It should be noted that in other works (e.g. Steiner et *al.*[4]) the ratio y^{-1} rather than y appears; this difference originates from a different definition of surface symmetry energy terms of surface tension.

2.2. *Shell corrections*

It is well known that shell corrections to the LDM formula play an important role. Many methods have been proposed to deal with shell effects, *e.g.*, those developed by Möller and Nix.[5] Here we use a simple prescription[6] which is closely related to the ideas used in the interacting boson approximation (IBA) model.[7] This model describes collective degrees of freedom in nuclei away from closed shells, and suggests that the relevant physics ingredient is the number of valence particles (neutrons and/or protons) with respect to the nearest closed shells (taken here to be $N, Z = 2, 8, 20, 28, 50, 82, 126, 184$), where particles beyond mid-shell are counted as holes. In our present work we add to the LDM expression a two-parameter term

$$B_{shell}(n_n, n_p) = a_1 F_{max} + a_2 F_{max}^2, \quad (5)$$

where $F_{max} = (n_n + n_p)/2$ with n_i the number of valence neutrons or protons, of particle or hole character. This is equivalent to counting bosons in the neutron-proton IBA model where F_{max} is the maximum F spin.[7]

From the fit one finds that the linear term is repulsive whereas the quadratic term is attractive.

2.3. *Wigner energy*

Nuclei with $N = Z$ are in general more strongly bound as compared to the LDM formula; this effect can be incorporated by including an additional term (known as the Wigner energy) in mass formulas. The origin of the Wigner energy which has been discussed by several groups, *e.g.*, Satuła *et al.*[10] and Jänecke *et al.*,[8] can be understood microscopically as an effect from the overlap of neutron and proton wave functions which is maximal in $N = Z$ nuclei. The Wigner contribution to the binding energy is usually decomposed into two parts[5]

$$B_w(N,Z) = -W(A)|N-Z| - d(A)\delta_{N,Z}\pi_{np}, \quad (6)$$

where π_{np} equals 1 for odd-odd nuclei and 0 otherwise. Whereas the Wigner energy has a clear physical meaning for light nuclei it has been applied also to heavy nuclei.[8] For the purpose of the symmetry energy only the linear term is relevant; its effect can be taken into account by replacing $(N-Z)^2 = 4T_z^2$ by $4T(T+r)$ where r ia a shell dependent parameter.[8] For the determination of the parameters in the symmetry energy the choice of the value r is rather crucial, since r is of order $T/10$.

In practice it turns out that the actual value of r depends whether or not shell corrections are taken into account. This can be explained qualitatively: we can distinguish two different situations (i) valence neutrons and protons are both particle- or hole-like; in this case there is no net contribution from the shell correction to isovector separation energies if one replaces a neutron by a proton; (ii) valence neutrons are particles (holes) while the protons are holes (particles); in the latter case the shell correction can be expressed in terms of isospin since $n_n + n_p = \Omega - (N - Z) = \Omega - 2T$ where Ω is the degeneracy of the last neutron shell. One can easily see (using Eq.(5)) that in this case $B_{\text{shell}}(N, Z)$ can be expressed as $aT(T + b) + c$, i.e. of the form $a_{\text{sym}}T(T + r)$. In practice several mass regions are dominated by case (ii), e.g. the rare-earth region $[82 < N < 126, 50 < Z < 82]$ (where Jänecke reported a large r value), with $[82 < N < 104,\ 66 < Z < 82]$ and $[82 < Z < 104, 104 < N < 126]$. A study of the rare-earth region reveals that after inclusion of the shell effect the value of r falls in the range 1.0-1.5, as compared to 2.5-4.0 without. Therefore since we do include shell corrections explicitly we take $r = 1$.

2.4. *Coulomb energy*

The simple expression for the Coulomb energy $E_c = Z^2/(r_0 A^{1/3})$ can easily be improved by adding several corrections. First the partition of the proton number into surface and volume terms, leads to the modification of the type $\frac{3}{5}Z^2 \rightarrow \frac{3}{5}Z_v^2 + Z_s^2 + \frac{1}{2}Z_s Z_v$, where the last term corresponds to the Coulomb energy of a uniform shell of a sphere. After substitution of the expressions for Z_s, Z_v one has[3]

$$\Delta E_c = a_c \frac{Z(Z-1)}{A^{1/3}} \frac{N-Z}{6Z} \frac{1}{1 + yA^{1/3}}. \tag{7}$$

One may recognize this as a generalization of the charge radius $R_c = r_0 A^{1/3}$. In practice as noted in Ref. 12 a better fit to the charge radius is obtained with a more general form, namely $R_c = 1 + a\frac{N-Z}{A} + b(\frac{N-Z}{A})^2$, where a, b are parameters fitted to radii. Note that the isovector term in R_c is present in the neutron skin $\Delta R = R_n - R_p$ (see below). We also add an exchange term of the form $E_c^{\text{exch}} = \frac{a'(Z(Z-1))^{2/3}}{R_c}$. Inclusion of both effects leads to 10% improvement in fitting masses.

3. Fitting to Masses

We start with a fit to binding energies using the full expression for the extended LDM. In this way the bulk parameters and those for the shell corrections and pairing are fixed.

For the purpose of an accurate determination of the parameters of the symmetry energy it is more efficient to separate isospin asymmetric properties from the isospin symmetric ones. This can be achieved by considering differences of masses. In practice two methods can be used.

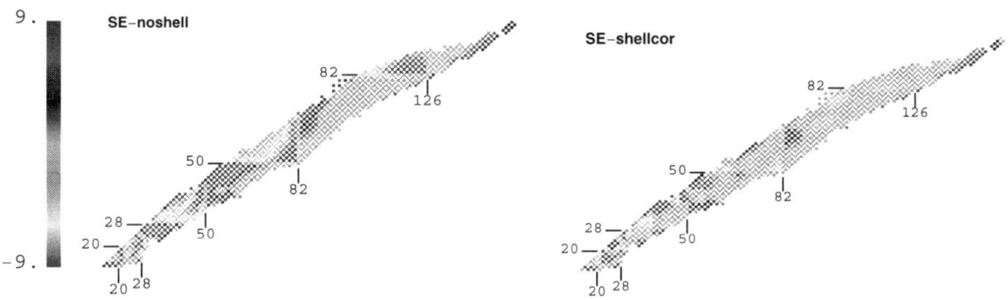

Fig. 1. (Color Online) Difference $a_{sym}(\exp) - a_{sym}(\text{fit})$ without (left) and with (right) the application of shell corrections.

3.1. *Fit to isobaric analogue states*

The excitation energies of isobaric analogue states (IAS) in nuclei can be used to determine the symmetry energy in a more direct way.[11] The differences in excitation energies between states with isospin T and T' in the same nucleus can be related to differences of masses in neighboring nuclei and a Coulomb correction, $\Delta_{T,T'}(A,T_z) = M(A,T',T_z) - M(A,T,T_z) + \Delta E_c(T'_z,T_z,A) - (T'-T)\Delta M$, with ΔM the mass difference between neutron and proton and ΔE_c the Coulomb displacement energy. If one assumes charge independence of the nuclear hamiltonian H the IAS energy can be expressed solely in terms of differences of symmetry and pairing energies $\Delta_{T,T'}(A,T_z) = E_{\text{sym}}(A,T') - E_{\text{sym}}(A,T) + E_{\text{pair}}(A,T') - E_{\text{pair}}(A,T)$. Hence $a_{\text{sym}} = A\Delta_{TT'}/(T'(T'+1) - T(T+1))$. On the other hand the set of observed IAS energies is much smaller than that of the full mass table; alternatively Jänecke[8] has made use of computed IAS energies obtained with parametrized Coulomb displacement energies.

3.2. *Fit to separation energies*

A closely related quantity is the isovector chemical potential $\mu_a = (\mu_n - \mu_p) = \frac{dE}{d\rho_n} - \frac{dE}{d\rho_p}$, which can be expressed as

$$\mu_a = \frac{1}{2}[B(N+1,Z) - B(N,Z+1) + B(N-1,Z) - B(N,Z-1)]. \tag{8}$$

Clearly from the LDM formula one obtains a direct relation with a_{sym}

$$\mu_a = 2\frac{N-Z}{A}a_{\text{sym}} + 2a_c[\frac{Z-1}{(A-1)^{1/3}} + \frac{Z+1}{(A+1)^{1/3}} + ...]. \tag{9}$$

The application of this method to extract the symmetry energy requires a rather accurate description of the Coulomb contribution. The Wigner contribution can be taken into account by replacing $N-Z$ by $N-Z+1$. As a result we deal with a two-parameter fit of $a_{\text{sym}} = S_v/(1+yA^{-1/3})$. Fig. 1 shows a plot of the difference of observed and fitted a_{sym} with and without application of shell corrections. Without the correction the shell structure shows up clearly as a regular triangular pattern.

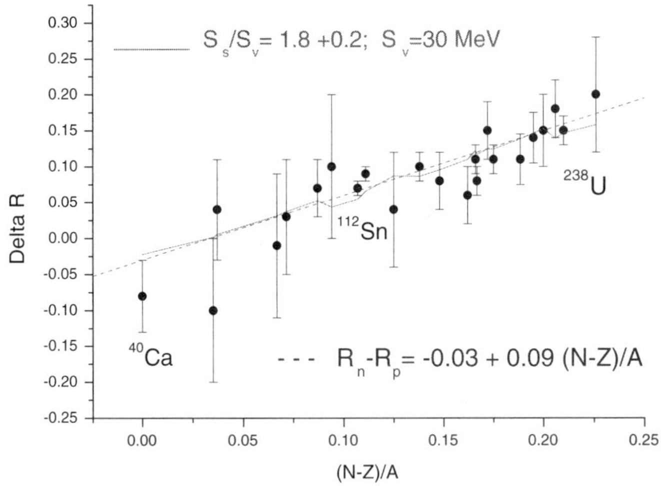

Fig. 2. The neutron-skin data ΔR from anti-protonic atoms for nuclei between ^{40}Ca and ^{238}. The full line is obtained from the LDM expression (10) with a volume symmetry energy $S_{\rm v} = 30$ MeV and a volume-to-surface ratio $y = 1.8$. The dashed line is the fit $\Delta R = -0.03 + 0.90(N - Z)/A$.[13]

Inclusion of the correction reduces the deviations for large A substantially; some remaining deviations, e.g. near $Z = 40, N = 50$, might be due to the fact that $Z = 40$ is not included as a magic number at present. The resulting best fit parameters S_v, y show some sensitivity to whether or not the correction for shell effects is applied.

4. Correlations

It has been noted by Danielewicz[3] and Steiner *et al.*[4] that in the fit to nuclear binding energies the parameters $S_{\rm v}$ and $S_{\rm s}$ are strongly correlated. In the correlation plot of $S_{\rm v}$ *versus* $y = S_{\rm v}/S_{\rm s}$ one obtains for the rms deviation a narrow valley described by the linear relation $S_{\rm v} = a + by$. This correlation can be understood qualitatively from the observation that in the two-parameter fitting function $S_{\rm v}/(1 + yA^{-1/3})$, in first approximation for heavy nuclei $A^{-1/3}$ can be replaced by its average value (weighted with $(N - Z)^2/A$), $\langle A^{-1/3}\rangle \approx 0.185$ for nuclei with $A > 20$. To further constrain the surface SE one needs supplementary information; this is provided by data on neutron skins.

5. Neutron Skin

The procedure leading to the result (4) also yields a direct relation between the skin ΔR and the symmetry energy parameters $(S_{\rm v}, S_{\rm s})$:

$$\frac{R_{\rm n} - R_{\rm p}}{R} = \frac{A(N_{\rm s} - Z_{\rm s})}{6NZ} = \frac{A}{6NZ}\,\frac{N - Z - a_c Z A^{2/3}(12S_{\rm v})^{-1}}{1 + y^{-1}A^{1/3}}, \tag{10}$$

which is valid for the difference of sharp-sphere radii.[3] We note that in the absence of the Coulomb contribution the neutron skin depends on the ratio y only.

A similar correlation between the neutron skin and (the derivative of) the symmetry energy in nuclear matter for, *e.g.*, ^{208}Pb was observed by Brown[15] and by Furnstahl[16] in mean-field models.

The extracted experimental information on ΔR is in general the result of a model-dependent analysis of nuclear reactions (elastic scattering of protons and neutrons, anti-protonic atoms, giant-resonance excitations). In order to minimize the model dependence, we have fitted data on radii obtained with one specific experimental tool only, namely from anti-protonic atoms,[13] available for targets between ^{40}Ca and ^{238}U; we do *not* use information on unstable light nuclei, which we believe to be more model dependent. Since the dependence of ΔR on S_{v} is weak, we may adopt the value $S_{\mathrm{v}} = 30$ MeV as it was obtained from ground-state energies, and determine the ratio y_{s} from the neutron-skin data. The resulting fit obtained by using Eq. (10) is shown in Fig. 2 and yields a value of $y = 1.8$. It is seen that in the limited mass region considered the simple two-parameter fit[13] $\Delta R = a + b(N-Z)/A = -0.03 + 0.90(N-Z)/A$ works equally well. That parametrization has, however, no obvious physical interpretation, the negative contribution from the a coefficient becomes unphysical for light $N = Z$ nuclei and it does not properly describe the Coulomb term for heavy systems.

At this point one might wonder whether one should take into account shell corrections to radii, similar to the corrections to ground-state energies. In practice one should expect two types of effects, namely collective ones related to the deformation, and single-particle effects from the filling of particular isolated sub-shells. The former could be included in a similar way as shell corrections to masses by adding terms linear in the number of valence nucleons. On the other hand treating sub-shell effects (expected to be of importance near closed-shell nuclei like the filling of the $h_{11/2}$ neutron shell in the Sn isotopes) would require a more microscopic approach, which is beyond the scope of the present paper.

It is clear from Fig. 10 that in practice in fitting ΔR over a larger mass region the symmetry energy is basically determined by the overall slope and that inclusion of shell effects even if they occur would not affect the result significantly.

6. Results

From the fit to separation energies we find $S_v = 31$ MeV, $y = 2.2 \pm 0.2$. Inclusion of skin data leads to $S_v = 30 \pm 1.0$MeV and $y = 1.8 \pm 0.2$, to be compared with the result from Danielewicz: $S_v = 31 \pm 1.2$MeV $y = 2.8 \pm 0.2$.

7. Asymmetric Nuclear Matter

Clearly the ratio $S_{\mathrm{v}}/S_{\mathrm{s}}$ is a measure for the density dependence of the symmetry energy in nuclear matter. In a macroscopic (liquid-drop) type of approach one may relate the ratio $S_{\mathrm{v}}/S_{\mathrm{s}}$ to $S(\rho_0)/S(\rho_1)$ at a selected subsaturation density ρ_1 by integrating across the surface (using the Local Density Approximation):[3]

$$\frac{S_{\rm v}}{S_{\rm s}} \approx \frac{3}{R\rho_0} \int \rho(r) \left(\frac{S(\rho_0)}{S(\rho)} - 1 \right) dr. \tag{11}$$

This result can be derived from the minimization of the second-order (in the asymmetry) energy functional $E = E_0 + \int S(\rho)\rho(\rho_{\rm a}/\rho)^2 dr$ under fixed particle numbers $A = \int \rho dr$ and $N - Z = \int \rho_{\rm a} dr$ with $\rho_{\rm a} \equiv \rho_{\rm n} - \rho_{\rm p}$. The application of Eq. (11) becomes even simpler if one assumes a power parametrization for the density dependence of the symmetry energy, $S(\rho) = S_{\rm v} \times (\rho/\rho_0)^\gamma$. Using the value of $y_{\rm s} = 1.8 \pm 0.3$ obtained from fitting neutron-skin data, we find $\gamma = 0.5 \pm 0.1$ although it is difficult to give a quantitative estimate of the error due to the Thomas-Fermi approximation. This result is consistent with $0.55 < \gamma < 0.79$ reported by Danielewicz.[17] It is also of interest to compare to the results from (relativistic) mean-field calculations. Recently Piekarewicz[14] reported that the use of two sets of parameters, which both describe properties of finite nuclei, can lead to a drastically different density dependence of the symmetry energy: $S_{\rm v} = 36.9$ (32.7) MeV and $\gamma = 0.98$ (0.64) for NL3 and FSUGold, respectively. There have been also attempts to constrain the value of γ using heavy-ion reactions (isospin diffusion); it was found[18] that $0.7 < \gamma < 1.1$.

References

1. A. Bohr and B. R. Mottelson, *Nuclear Structure II. Nuclear Deformations* (Benjamin, New York, 1975).
2. E. Lipparini and S. Stringari, *Phys. Rep.* **175**, 103 (1989).
3. P. Danielewicz, *Nucl. Phys.* **A727**, 233 (2003).
4. A. W. Steiner, M. Prakash, J. M. Lattimer and P. J. Ellis, *Phys. Rep.* **411**, 325 (2005).
5. P. Möller and R. Nix, *Nucl. Phys.* **A536**, 20 (1992).
6. A. E. L. Dieperink and P. van Isacker, *Eur. Phys. J.* **A32**, 11 (2007).
7. F. Iachello and A. Arima, *The Interacting Boson Model*, (Cambridge University Press, Cambridge, 1987).
8. J. Jänecke and T. W. O'Donnell, *Phys. Lett.* **B605**, 87 (2005).
9. W. Satuła, D. J. Dean, J. Gary, S. Mizutori and W. Nazarewicz, *Phys. Lett.* **B407**, 103 (1997).
10. W. Satuła and R. Wyss, *Phys. Rev. Lett.* **86**, 4488 (2001) 4488; *ibid.* **87**, 052504 (2001).
11. P. Danielewicz, nucl-th/0607030.
12. J. Duflo and A. P. Zuker, *Phys. Rev.* **C66**, 051304(R) (2002).
13. A. Trzcinska *et al.*, *Phys. Rev. Lett.* **87**, 082501 (2001).
14. J. Piekarewicz, *Phys. Rev.* **C73**, 044325 (2006).
15. B. A. Brown, *Phys. Rev. Lett.* **85**, 5296 (2000).
16. R. Furnstahl, *Nucl. Phys.* **A706**, 85 (2002).
17. P. Danielewicz, nucl-th/0411115.
18. B.-A. Li, L.-W. Chen, C. M. Ko and A. W. Steiner, nucl-th/06011028.

PROBING THE SYMMETRY ENERGY AT SUPRA-SATURATION DENSITIES

M. DI TORO,* M. COLONNA, G. FERINI, V. GRECO and J. RIZZO

Laboratori Nazionali del Sud, Via S. Sofia 62, I-95123 Catania, Italy and Physics-Astronomy Dept., University of Catania
** E-mail: ditoro@lns.infn.it*

V. BARAN

NIPNE-HH, Bucharest and Bucharest University, Romania

T. GAITANOS

Institut für Theoretische Physik, Universität Giessen, D-35392 Giessen, Germany

LIU BO

IHEP, Beijing, China

G. LALAZISSIS and V. PRASSA

Dept. of Theoretical Physics, Aristotle University, Thessaloniki Gr-54124, Greece

H. H. WOLTER

Dept. für Physik, Universität München, D-85748 Garching, Germany

We show that the phenomenology of isospin effects on heavy ion reactions at intermediate energies (few AGeV range) is extremely rich and can allow a "direct" study of the covariant structure of the isovector interaction in a high density hadron medium. We work within a relativistic transport frame, beyond a cascade picture, consistently derived from effective Lagrangians, where isospin effects are accounted for in the mean field and collision terms. Rather sensitive observables are proposed from collective flows ("differential" flows) and from pion/kaon production (π^-/π^+, K^0/K^+ yields). For the latter point relevant non-equilibrium effects are stressed. The possibility of the transition to a mixed hadron-quark phase, at high baryon and isospin density, is finally suggested. Some signatures could come from an expected "neutron trapping" effect.

Keywords: Relativistic heavy ion collisions; EOS at high baryon density.

1. Introduction

Recently the development of new heavy ion facilities (radioactive beams) has driven the interest on the dynamical behaviour of asymmetric matter, see the recent reviews Refs. 1,2. Here we focus our attention on relativistic heavy ion collisions, that provide a unique terrestrial opportunity to probe the in-medium nuclear interaction in high density and high momentum regions. An effective Lagrangian approach to the hadron interacting system is extended to the isospin degree of freedom: within the same frame equilibrium properties (EoS[3]) and transport dynamics can be consistently derived.

Within a covariant picture of the nuclear mean field, for the description of the symmetry energy at saturation (a_4 parameter of the Weizsäecker mass formula) (a) only the Lorentz vector ρ mesonic field, and (b) both, the vector ρ (repulsive) and scalar δ (attractive) effective fields[4,5] can be included. In the latter case the competition between scalar and vector fields leads to a stiffer symmetry term at high density.[2,4] We present here observable effects in the dynamics of heavy ion collisions. We focus our attention on collective isospin flows, in particular the elliptic ones, and on the isospin content of particle production, in particular kaons. We finally show that in the compression stage of isospin asymmetric collisions we can enter a mixed deconfined phase, if the EoS conditions for the existence of quark stars are met.

2. Relativistic Transport

The starting point is a simple phenomenological version of the Non-Linear (with respect to the iso-scalar, Lorentz scalar σ field) effective nucleon-boson field theory, the Quantum-Hadro-Dynamics.[3] According to this picture the presence of the hadronic medium leads to effective masses and momenta $M^* = M + \Sigma_s$, $k^{*\mu} = k^\mu - \Sigma^\mu$, with Σ_s, Σ^μ scalar and vector self-energies. For asymmetric matter the self-energies are different for protons and neutrons, depending on the isovector meson contributions. We will call the corresponding models as $NL\rho$ and $NL\rho\delta$, respectively, and just NL the case without isovector interactions. For the more general $NL\rho\delta$ case the self-energies of protons and neutrons read:

$$\begin{aligned} \Sigma_s(p,n) &= -f_\sigma \sigma(\rho_s) \pm f_\delta \rho_{s3}, \\ \Sigma^\mu(p,n) &= f_\omega j^\mu \mp f_\rho j_3^\mu, \end{aligned} \qquad (1)$$

(upper signs for neutrons), where $\rho_s = \rho_{sp} + \rho_{sn}$, $j^\alpha = j_p^\alpha + j_n^\alpha$, $\rho_{s3} = \rho_{sp} - \rho_{sn}$, $j_3^\alpha = j_p^\alpha - j_n^\alpha$ are the total and isospin scalar densities and currents and $f_{\sigma,\omega,\rho,\delta}$ are the coupling constants of the various mesonic fields. $\sigma(\rho_s)$ is the solution of the non linear equation for the σ field.[2,4]

For the description of heavy ion collisions we solve the covariant transport equation of the Boltzmann type within the Relativistic Landau Vlasov (RLV) method, using phase-space Gaussian test particles,[6] and applying a Monte-Carlo procedure for the hard hadron collisions. The collision term includes elastic and inelastic processes involving the production/absorption of the $\Delta(1232 MeV)$ and $N^*(1440 MeV)$ resonances as well as their decays into pion channels.[7]

3. Collective Flows

The flow observables can be seen respectively as the first and second coefficients of a Fourier expansion of the azimuthal distribution: $\frac{dN}{d\phi}(y, p_t) \approx 1 + 2V_1 cos(\phi) +$

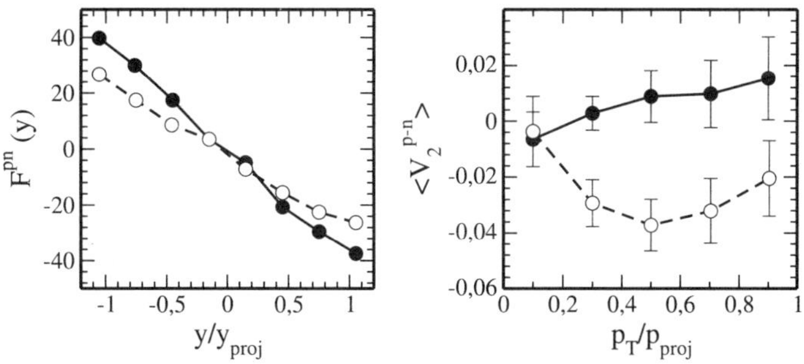

Fig. 1. Differential neutron-proton flows for the $^{132}Sn+^{124}Sn$ reaction at 1.5 AGeV ($b = 6$ fm) from the two different models for the isovector mean fields. Left: Transverse Flows. Right: Elliptic Flows. Full circles and solid line: $NL\rho\delta$. Open circles and dashed line: $NL\rho$.

$2V_2cos(2\phi)$ where $p_t = \sqrt{p_x^2+p_y^2}$ is the transverse momentum and y the rapidity along beam direction. The transverse flow can be expressed as: $V_1(y,p_t) = \langle \frac{p_x}{p_t} \rangle$. The sideward (transverse) flow is a deflection of forwards and backwards moving particles, within the reaction plane. The second coefficient of the expansion defines the elliptic flow given by $V_2(y,p_t) = \langle \frac{p_x^2-p_y^2}{p_t^2} \rangle$.

It measures the competition between in-plane and out-of-plane emissions. The sign of V_2 indicates the azimuthal anisotropy of emission: particles can be preferentially emitted either in the reaction plane ($V_2 > 0$) or out-of-plane ($squeeze - out$, $V_2 < 0$).[8] For the isospin effects the neutron-proton flow differences $V_{1,2}^{(n-p)}(y,p_t) \equiv V_{1,2}^n(y,p_t) - V_{1,2}^p(y,p_t)$ have been suggested as very useful probes of the isovector part of the EoS since they appear rather insensitive to the isoscalar potential and to the in medium nuclear cross sections,[9] In heavy-ion collisions around 1 AGeV with radioactive beams, differential flows will directly exploit the Lorentz nature of a scalar and a vector field. In Fig. 1 transverse and elliptic differential flows are shown for the $^{132}Sn+^{124}Sn$ reaction at 1.5 AGeV ($b = 6$ fm),[9] The effect of the different structure of the isovector channel is clear. Particularly evident is the splitting in the high p_t region of the elliptic flow. In the $(\rho+\delta)$ dynamics the high-p_t neutrons show a much larger $squeeze - out$. This is fully consistent with an early emission (more spectator shadowing) due to the larger ρ-field in the compression stage. We expect similar effects, even enhanced, from the measurements of differential flows for light isobars, like ^3H *vs.* ^3He.

4. Isospin Effects on Sub-Threshold Kaon Production at Intermediate Energies

Kaon production has been proven to be a reliable observable for the high density EoS in the isoscalar sector.[10,11] Here we show that the $K^{0,+}$ production (in particular the K^0/K^+ yield ratio) can be also used to probe the isovector part of the EoS.[12,13]

Using our RMF transport approach we analyze pion and kaon production in central $^{197}Au+^{197}Au$ collisions in the 0.8-1.8 AGeV beam energy range, comparing

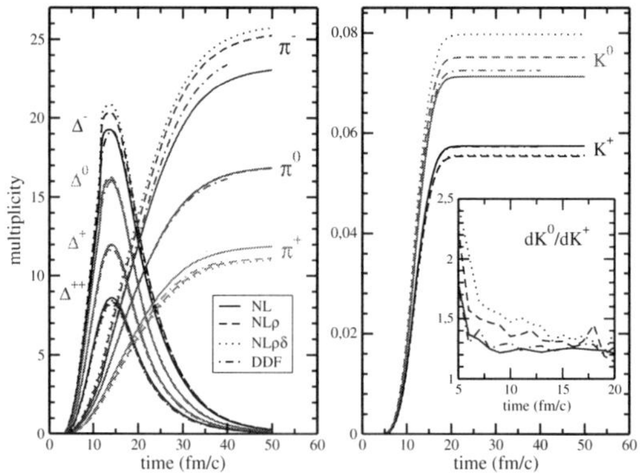

Fig. 2. Time evolution of the $\Delta^{\pm,0,++}$ resonances and pions $\pi^{\pm,0}$ (left), and kaons ($K^{+,0}$ (right) for a central ($b = 0$ fm impact parameter) Au+Au collision at 1 AGeV incident energy. Transport calculation using the $NL, NL\rho, NL\rho\delta$ and DDF models for the iso-vector part of the nuclear EoS are shown. The inset contains the differential K^0/K^+ ratio as a function of the kaon emission time.

models giving the same "soft" EoS for symmetric matter and with different effective field choices for E_{sym}. We will use three Lagrangians with constant nucleon-meson couplings (NL... type, see before) and one with density dependent couplings (DDF, see Ref. 5), recently suggested for better nucleonic properties of neutron stars.[14,15]

Fig. 2 reports the temporal evolution of $\Delta^{\pm,0,++}$ resonances, pions ($\pi^{\pm,0}$) and kaons ($K^{+,0}$) for central Au+Au collisions at 1 AGeV.

It is clear that, while the pion yield freezes out at times of the order of $50fm/c$, i.e. at the final stage of the reaction (and at low densities), kaon production occur within the very early (compression) stage, and the yield saturates at around $20fm/c$. From Fig. 2 we see that the pion results are weakly dependent on the isospin part of the nuclear mean field. However, a slight increase (decrease) in the π^- (π^+) multiplicity is observed when going from the NL (or DDF) to the $NL\rho$ and then to the $NL\rho\delta$ model, i.e. increasing the vector contribution f_ρ in the isovector channel. This trend is more pronounced for kaons, see the right panel, due to the high density selection of the source and the proximity to the production threshold. Consistently, as shown in the insert, larger effects are expected for early emitted kaons, reflecting the early N/Z of the system.

When isovector fields are included the symmetry potential energy in neutron-rich matter is repulsive for neutrons and attractive for protons. In a HIC this leads to a fast, pre-equilibrium, emission of neutrons. Such a *mean field* mechanism, often referred to as isospin fractionation,[1,2] is responsible for a reduction of the neutron to proton ratio during the high density phase, with direct consequences on particle production in inelastic NN collisions.

Threshold effects represent a more subtle point. The energy conservation in a hadron collision in general has to be formulated in terms of the canonical momenta, i.e. for a reaction $1+2 \to 3+4$ as $s_{in} = (k_1^\mu + k_2^\mu)^2 = (k_3^\mu + k_4^\mu)^2 = s_{out}$. Since hadrons are propagating with effective (kinetic) momenta and masses, an equivalent relation

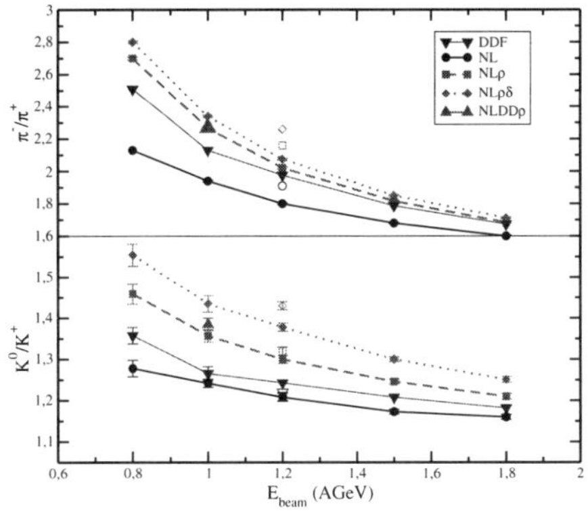

Fig. 3. π^-/π^+ (upper) and K^+/K^0 (lower) ratios as a function of the incident energy for the same reaction and models as in Fig. 2. In addition we present, for $E_{beam} = 1$ AGeV, $NL\rho$ results with a density dependent ρ-coupling (triangles). The *open* symbols at 1.2 AGeV show the corresponding results for a $^{132}Sn + ^{124}Sn$ collision, more neutron rich. Note the different scale for the π^-/π^+ ratios.

should be formulated starting from the effective in-medium quantities $k^{*\mu} = k^\mu - \Sigma^\mu$ and $m^* = m + \Sigma_s$, where Σ_s and Σ^μ are the scalar and vector self-energies, Eqs. (1). The self-energy contributions will influence the particle production at the level of thresholds as well as of the phase space available in the final channel.

Finally the beam energy dependence of the π^-/π^+ (left) and K^0/K^+ (right) ratios is shown in Fig. 4. At each energy we see an increase of the yield ratios with the models $NL \to DDF \to NL\rho \to NL\rho\delta$. The effect is larger for the K^0/K^+ compared to the π^-/π^+ ratio. This is due to the subthreshold production and to the fact that the isospin effect enters twice in the two-step production of kaons, see Ref. 12. Between the two extreme DDF and $NL\rho\delta$ isovector interaction models, the variations in the ratios are of the order of $14-16\%$ for kaons, while they reduce to about $8-10\%$ for pions. Interestingly the Iso-EoS effect for pions is increasing at lower energies, when approaching the production threshold.

We have to note that in a previous study of kaon production in excited nuclear matter the dependence of the K^0/K^+ yield ratio on the effective isovector interaction appears much larger (see Fig. 8 of Ref. 7). The point is that in the non-equilibrium case of a heavy ion collision the asymmetry of the source where kaons are produced is in fact reduced by the $n \to p$ "transformation", due to the favored $nn \to p\Delta^-$ processes. This effect is almost absent at equilibrium due to the inverse transitions, see Fig. 3 of Ref. 7. Moreover in infinite nuclear matter even the fast neutron emission is not present. This result clearly shows that chemical equilibrium models can lead to uncorrect results when used for transient states of an *open* system.

5. Testing Deconfinement at High Isospin Density

The hadronic matter is expected to undergo a phase transition into a deconfined phase of quarks and gluons at large densities and/or high temperatures. On very

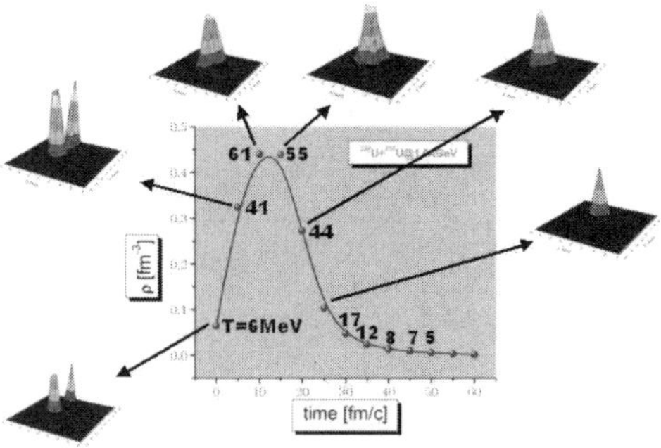

Fig. 4. $^{238}U+^{238}U$, 1 AGeV, semicentral. Correlation between density, temperature, momentum thermalization inside a cubic cell, 2.5 fm wide, in the center of mass of the system.

general grounds, the transition's critical densities are expected to depend on the isospin of the system, but no experimental tests of this dependence have been performed so far. Moreover, up to now, data on the phase transition have been extracted from ultrarelativistic collisions, when large temperatures but low baryon densities are reached. In order to check the possibility of observing some precursor signals of some new physics even in collisions of stable nuclei at intermediate energies we have performed some event simulations for the collision of very heavy, neutron-rich, elements. We have chosen the reaction $^{238}U+^{238}U$ (average proton fraction $Z/A = 0.39$) at 1 AGeV and semicentral impact parameter $b = 7\ fm$ just to increase the neutron excess in the interacting region. In Fig. 4 we report the evolution of momentum distribution and baryon density in a space cell located in the c.m. of the system. We see that after about 10 fm/c a nice local equilibration is achieved. We have a unique Fermi distribution and from a simple fit we can evaluate the local temperature (black numbers in MeV). We note that a rather exotic nuclear matter is formed in a transient time of the order of 10 fm/c, with baryon density around $3-4\rho_0$, temperature 50-60 MeV, energy density 500 MeV fm^{-3} and proton fraction between 0.35 and 0.40, likely inside the estimated mixed phase region.

Here we study the isospin dependence of the transition densities. Concerning the hadronic phase, we use the relativistic non-linear model of Glendenning-Moszkowski (in particular the "soft" $GM3$ choice),[16] where the isovector part is treated just with ρ meson coupling, and the iso-stiffer $NL\rho\delta$ interaction.[17] For the quark phase we consider the MIT bag model with various bag pressure constants. In particular we are interested in those parameter sets which would allow the existence of quark stars,[18] i.e. parameters sets for which the so-called Witten-Bodmer hypothesis is satisfied.[19,20] One of the aim of our work it to show that if quark stars are indeed possible, it is then very likely to find signals of the formation of a mixed quark-hadron phase in intermediate-energy heavy-ion experiments.[17]

The structure of the mixed phase is obtained by imposing the Gibbs conditions[21,22] for chemical potentials and pressure and by requiring the conservation of

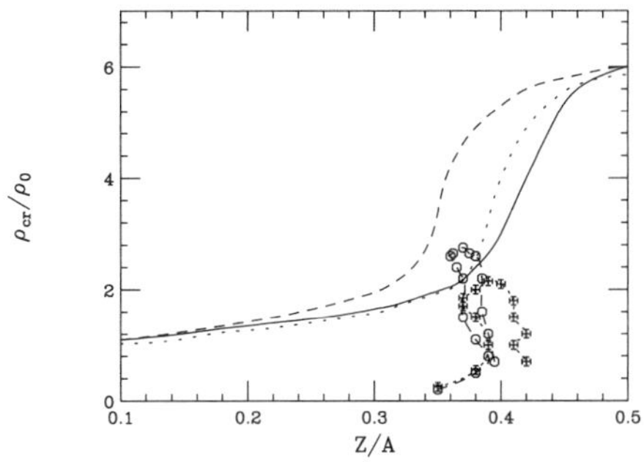

Fig. 5. Variation of the transition density with proton fraction for various hadronic *EoS* parameterizations. Dotted line: *GM*3 parametrization; dashed line: $NL\rho$ parametrization; solid line: $NL\rho\delta$ parametrization. For the quark *EoS*, the *MIT* bag model with $B^{1/4}$=150 MeV. The points represent the path followed in the interaction zone during a semi-central ^{132}Sn+^{132}Sn collision at 1 AGeV (circles) and at 300 AMeV (crosses).

the total baryon and isospin densities

$$\begin{aligned}
&\mu_B^{(H)} = \mu_B^{(Q)}, \quad \mu_3^{(H)} = \mu_3^{(Q)}, \\
&P^{(H)}(T, \mu_{B,3}^{(H)}) = P^{(Q)}(T, \mu_{B,3}^{(Q)}), \\
&\rho_B = (1-\chi)\rho_B^H + \chi\rho_B^Q, \\
&\rho_3 = (1-\chi)\rho_3^H + \chi\rho_3^Q,
\end{aligned} \tag{2}$$

where χ is the fraction of quark matter in the mixed phase.

In this way we get the *binodal* surface which gives the phase coexistence region in the (T, ρ_B, ρ_3) space. For a fixed value of the conserved charge ρ_3 we will study the boundaries of the mixed phase region in the (T, ρ_B) plane. In the hadronic phase the charge chemical potential is given by $\mu_3 = 2E_{sym}(\rho_B)\frac{\rho_3}{\rho_B}$. Thus, we expect critical densities rather sensitive to the isovector channel in the hadronic *EoS*.

In Fig. 5 we show the crossing density ρ_{cr} separating nuclear matter from the quark-nucleon mixed phase, as a function of the proton fraction Z/A. We can see the effect of the δ-coupling towards an *earlier* crossing due to the larger symmetry repulsion at high baryon densities. In the same figure we report the paths in the $(\rho, Z/A)$ plane followed in the c.m. region during the collision of the n-rich ^{132}Sn+^{132}Sn system, at different energies. At 300 AMeV we are just reaching the border of the mixed phase, and we are well inside it at 1 AGeV. Statistical We expect a *neutron trapping* effect, supported by statistical fluctuations as well as by a symmetry energy difference in the two phases. In fact while in the hadron phase we have a large neutron potential repulsion (in particular in the $NL\rho\delta$ case), in the quark phase we only have the much smaller kinetic contribution. Observables related to such neutron "trapping" could be an inversion in the trend of the formation of neutron rich fragments and/or of the π^-/π^+, K^0/K^+ yield ratios for reaction products coming from high density regions, i.e. with large transverse momenta. In general we would expect a modification of the rapidity distribution of the emitted "isospin", with an enhancement at mid-rapidity joint to large event by event fluctuations.

6. Perspectives

We have shown that collisions of n-rich heavy ions at intermediate energies can bring new information on the isovector part of the in-medium interaction at high baryon densities, qualitatively different from equilibrium EoS properties. We have presented quantitative results for differential collective flows and yields of charged pion and kaon ratios. Important non-equilibrium effects for particle production are stressed. Finally our study supports the possibility of observing precursor signals of the phase transition to a mixed hadron-quark matter at high baryon density in the collision, central or semi-central, of neutron-rich heavy ions in the energy range of a few GeV per nucleon.

Acknowledgements

We warmly thank A. Drago and A. Lavagno for the intense collaboration on the mixed hadron-quark phase transition at high baryon and isospin density.

References

1. B. A. Li and W. U. Schroeder (Eds.), *Isospin Physics in Heavy-Ion Collisions at Intermediate Energies*, (Nova Science, New York, 2001).
2. V. Baran, M. Colonna, V. Greco and M. Di Toro, *Phys. Rep.* **410**, 335 (2005).
3. B. D. Serot and J. D. Walecka, *Advances in Nuclear Physics* **16**, 1, eds. J. W. Negele and E. Vogt, (Plenum, N.Y., 1986).
4. B. Liu, V. Greco, V. Baran, M. Colonna and M. Di Toro, *Phys. Rev. C* **65**, 045201 (2002).
5. T. Gaitanos *et al.*, *Nucl. Phys. A* **732**, 24 (2004).
6. C. Fuchs and H. H. Wolter, *Nucl. Phys. A* **589**, 732 (1995).
7. G. Ferini, M. Colonna, T. Gaitanos and M. Di Toro, *Nucl. Phys. A* **762**, 147 (2005).
8. P. Danielewicz, *Nucl. Phys. A* **673**, 375 (2000).
9. V. Greco *et al.*, *Phys. Lett. B* **562**, 215 (2003).
10. C. Fuchs, *Prog. Part. Nucl. Phys.* **56**, 1 (2006).
11. C. Hartnack, H. Oeschler and J. Aichelin, *Phys. Rev. Lett.* **96**, 012302 (2006).
12. G. Ferini *et al.*, *Phys. Rev. Lett.* **97**, 202301 (2006).
13. V. Prassa *et al.*, *Nucl. Phys. A* **789**, 311 (2007).
14. T. Klähn *et al.*, *Phys. Rev. C* **74**, 035802 (2006).
15. B. Liu *et al.*, *Phys. Rev. C* **75**, 048801 (2007).
16. N. K. Glendenning and S. A. Moszkowski, *Phys. Rev. Lett.* **67**, 2414 (1991).
17. M. Di Toro, A. Drago, T. Gaitanos, V. Greco and A. Lavagno, *Nucl. Phys. A* **775**, 102 (2006).
18. A. Drago and A. Lavagno, *Phys. Lett. B* **511**, 229 (2001).
19. E. Witten, *Phys. Rev. D* **30**, 272 (1984).
20. A. R. Bodmer, *Phys. Rev. D* **4**, 1601 (1971).
21. L. D. Landau and L. Lifshitz, *Statistical Physics*, (Pergamon Press, Oxford, 1969).
22. N. K. Glendenning, *Phys. Rev. D* **46**, 1274 (1992).

INVESTIGATION OF LOW-DENSITY SYMMETRY ENERGY VIA NUCLEON AND FRAGMENT OBSERVABLES

HERMANN H. WOLTER

Physics Department, University of Munich, D-85748 Garching, Germany
E-mail: hermann.wolter@lmu.de

J. RIZZO, M. COLONNA, M. DI TORO and V. GRECO

Lab. Nazionali del Sud, INFN, I-95123 Catania, Italy

V. BARAN

Univ. of Bucharest and NIPNE-HH, Bucharest, Romania

M. ZIELINSKA-PFABE

Physics Dep., Smith College, Northampton, Mass., USA

With stochastic transport simulations we study in detail central and peripheral collisions at Fermi energies and suggest new observables, sensitive to the symmetry energy below normal density.

Keywords: Symmetry energy; Isospin transport coefficients; Neck fragmentation.

There has been much interest in recent years in the determination of the nuclear symmetry energy as a function of density, which is important for the structure of exotic nuclei as well as for astrophysical processes. Heavy ion collisions (HIC) present an attractive way to constrain the the existing models for this poorly-determined isovector equation-of-state (iso-EOS),[1] which can be investigated both at densities above and below normal density with relativistic energies and in the Fermi energy domain, respectively. However, observables, which are both sensitive to the iso-EOS and testable experimentally, still have to be identified clearly.[2,3]

In this report we discuss dissipative collisions at Fermi energies. Isospin dynamics at low and intermediate energies and its relation to the symmetry energy has, in fact, attracted much attention in recent years in experiment as well as in theory.[4–7] Here we focus our attention on pre-equilibrium emission in central collisions and on the charge equilibration dynamics in peripheral collisions, where we expect to see symmetry energy effects. The interesting feature at Fermi energies is the onset of collective flows due to compression and expansion of the interacting nuclear matter. The isospin transport takes place in regions with density and asymmetry variations,

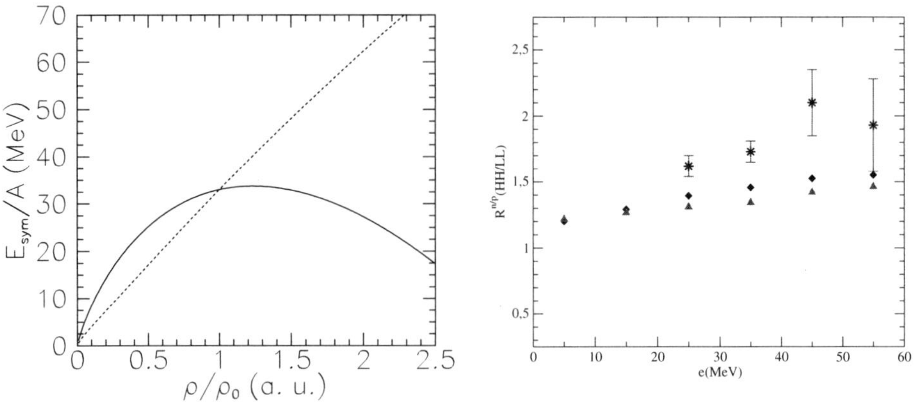

(b)

(a)

Fig. 1. (a) Density dependence of the symmetry energies used in the simulations presented here: asy-soft (solid) and asy-stiff (dashed). (b) Double ratio of emitted neutron over proton yield in central $^{124}Sn + ^{124}Sn$ (HH) over $^{112}Sn + ^{112}Sn$ (LL) collisions at 50 AMeV with asy-stiff (red triangles) and asy-soft (black diamonds) EOS as a function of nucleon kinetic energy. Data of Famiano et al.,[13] are given as (blue) stars.

and thus we expect to have contributions to the isospin current from charge and mass drift mechanisms.

We perform *ab initio* collision simulations using the microscopic Stochastic Mean Field (SMF) model. It is based on mean field transport theory with correlations from hard nucleon-nucleon collisions and with stochastic fluctuations acting on the mean phase-space trajectory.[8,9] Stochasticity is essential in order to allow the growth of dynamical instabilities with fragment production, and to obtain physical widths of distributions of observables. A detailed description of the procedure is given in Ref. 2 and in Refs. therein.

We have used a generalized form of effective interaction with momentum dependent terms in the isoscalar and the isovector channel,[10] which is an asymmetric extension of the Gale-Bertsch-DasGupta (GBD) force.[11] The parameters are chosen to give a soft equation of state for *symmetric* nuclear matter (compressibility modulus 215 MeV, isoscalar effective mass $m^*/m = 0.67$), which is held fixed. Here we want to test the sensitivity of isospin transport observables to two essentially different behaviors of the symmetry energy around saturation: *asy-soft* and *asy-stiff*.[2] In Fig. 1a we show the density dependence for these two typical choices. The *isoscalar* momentum dependence has been found to be important for the general dynamics, in particular the particle flow, of HIC.[12] To discuss its influence also on isospin dynamics we consider here interactions with (MD) and without (MI) momentum dependence. The *isovector* momentum dependence changes the proton/neutron effective masses and is still very controversial.[1,2] It is most effective at higher energies, but it effects are also evident in the Fermi energy range in pre-equilibrium emission.[10]

Isospin transport is closely connected to the value and the slope of the symmetry energy at a given density. In fact, the p/n currents can be expressed as

$$\mathbf{j}_{p/n} = D^{\rho}_{p/n}\nabla\rho - D^{\beta}_{p/n}\nabla\beta$$

with $D^{\rho}_{p/n}$ the mass (drift), and $D^{\beta}_{p/n}$ the isospin transport (diffusion) coefficients (asymmetry $\beta = (N-Z)/A$), which are directly given by the n, p chemical potentials.[4] Of special interest here is the difference of neutrons and protons currents (iso-vector current) for which the transport coefficients are proportional to[4]

$$D^{\rho}_{n} - D^{\rho}_{p} \propto 4\beta\frac{\partial E_{sym}}{\partial\rho},$$
$$D^{\beta}_{n} - D^{\beta}_{p} \propto 4\rho E_{sym}.$$

Referring back to Fig. 1a we see that isospin drift and diffusion behave very differently for the two iso-EOS's.

Preequilibrium nucleons and light clusters are emitted in the approach and overlap stages of a HIC. The ratio of neutron to proton yields, (resp. of isobaric light cluster yields) carries information on the isospin forces. To reduce effects of secondary emission double ratios between different reactions have been investigated for nucleons[13,14] and IMF's.[15] We show a result from our calulations for nucleons in the "gas" phase (defined by a density cut of $\rho/\rho_0 < 1/6$) in Fig. 1b. It is seen that the asy-soft EOS is more effective, since the symmetry energy is higher below normal density (see Fig. 1a).

However, the iso-EOS effect is not very large, and both results are considerably below the data,[13] which, unlike in our calculations, are taken with a transverse angular cut. Already the results for the single n/p yield ratios deviate strongly from the experiment, which are very much higher for low energy nucleons. Our result is also in contrast to other calculations,[13,14] which show a stronger iso-EOS effect. Obviously, both the experimental data and the calculations have to be understood better. We also mention, that the effect of different neutron/proton effective masses, i.e. of an isovector momentum dependence, is of considerable influence already in this energy range,[10] but does not resolve the discrepancies noted above. The calculations shown in Fig. 1b are taken for the choice $m^*_n > m^*_p$.

In peripheral collisions of nuclei with different asymmetries isospin is equilibrated through the neck, mainly due to isospin gradients (diffusion). The amount of isospin transport has been measured with the so-called imbalance (or isospin transport) ratio,[16] which is defined as

$$R^{\beta}_{P,T} = 2\frac{\beta^{M} - \beta^{eq}}{\beta^{HH} - \beta^{LL}},$$

with $\beta^{eq} = (\beta^{HH} + \beta^{LL})/2$. Instead of the asymmetry $\beta = (N-Z)/A$, one has also considered other isospin sensitive quantities, such as isoscaling coefficients.[19]

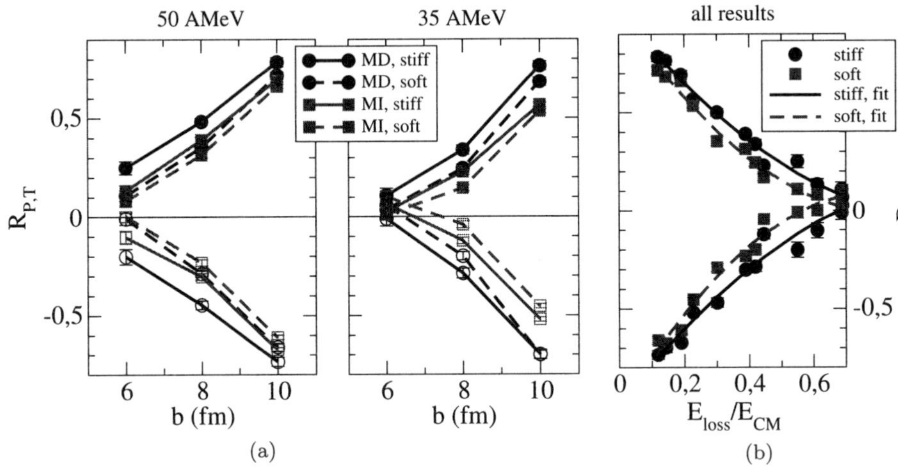

Fig. 2. (a) Imbalance ratios for $Sn + Sn$ collisions for incident energies of 50 (left) and 35 AMeV (right) as a function of the impact parameter for asy-stiff and asy-soft EOS (signatures see legend box), projectile rapidity (upper curves), target rapidity (lower curves). (b) Imbalance ratio for all results in part (a) but as a function of relative energy loss for asy-stiff (black dots) and asy-soft (red squares) EOS. Quadratic fit to all points for the asy-stiff (solid), resp. asy-soft (dashed) EOS.

The indices HH and LL refer to the symmetric reaction between the heavy (n-rich,^{124}Sn) and the light (n-poor,^{112}Sn) systems, while M refers to the mixed reaction, and P, T denote the PLF and TLF rapidity regions. Clearly, this ratio is ± 1 for complete transparency, resp. complete rebound, while it is zero for complete equilibration. Indeed, it can be shown, that the imbalance ratio depends on the magnitude of the symmetry energy and the interaction time. It is a very sensitive observable, magnifying small differences in asymmetry.

Results for our system as a function of impact parameter for different beam energies are shown in Fig. 2a. It is seen that the equilibration is larger (R smaller) for an asy-soft EOS (as expected from above), and for MI interactions and lower energy. The last two observation are fairly obvious, since they are due to longer interaction times (since the collision is also faster for the repulsive MD forces). It is therefore profitable to consider the imbalance ratio as a function of the interaction time, or an observable which is closely correlated to it. Such an observable is the total kinetic energy loss, which has been extensively investigated in dissipative collisions.[17] We thus show the imbalance ratios as a function of the relative energy loss per particle in Fig. 2b. Here all results for MD/MI interactions and 35/50 AMeV are collected and - to guide the eye - are fitted by a quadratic curve for the asy-stiff and asy-soft EOS's separately. It is now seen that the results presented in this way are *only* sensitive to the iso-EOS, albeit with some scatter. Such a representation should be useful to obtain a unifying picture of different experiments as well as calculations.

In peripheral collisions a third intermediate mass particle (IMF) can appear, which originates in the rupture of the neck (ternary event, neck fragmentation).[18]

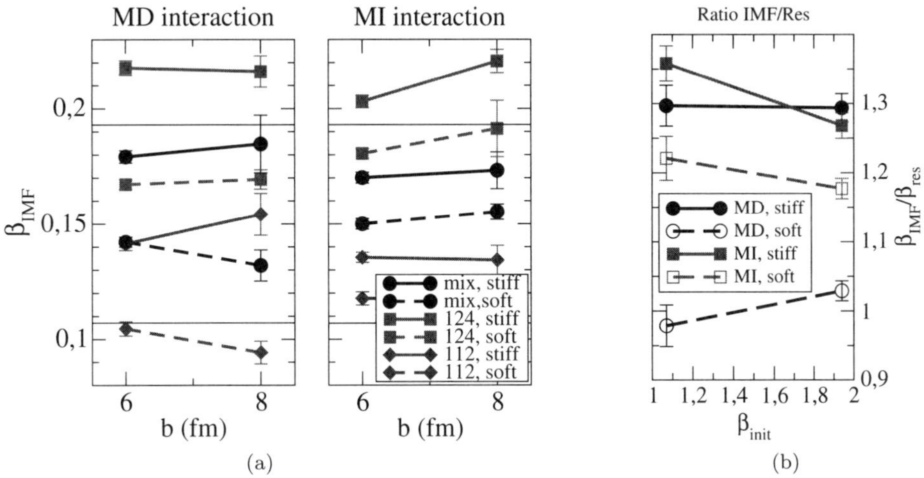

Fig. 3. (a) Asymmetries of IMF's in ternary $Sn+Sn$ reactions at 50 AMeV as a function of impact parameter for MD (left panel) and MI (right panel) interactions for mixed and symmetric $Sn+Sn$ collisions for asy-stiff and asy-soft EOS's (see legend box). Horizontal thin lines: asymmetries of ^{124}Sn and ^{112}Sn, respectively. (b) Ratios of asymmetries of IMF to residues for symmetric $Sn+Sn$ reactions at 50 AMeV as a function of the initial asymmetry for $b = 6fm$ for MD and MI interactions (see legend box).

This has been studied also experimentally in asymmetric collisions, in particular with respect to velocity correlations between the residues and the IMF and to the alignment of the IMF.[20] In this report the isospin content of the IMF is of special interest, since it is mainly influenced by isospin transport due to density gradients, i.e. isopin drift or migration, which according to the above is governed by the slope of the symmetry energy below normal density. The asymmetry of the IMF in ternary reactions in our systems is shown in Fig. 3a both for the symmetric and the mixed $Sn+Sn$ collisions.

The asymmetry of the IMF is larger, i.e. the IMF is more n-rich, for the stiff relative to the soft iso-EOS, since the former exhibits a larger isospin migration due to the larger slope of the symmetry energy below saturation. This is clearly the case for the symmetric reactions, but it is also true for the mixed reactions, where there is a competition with isospin diffusion, which depends on the value of the symmetry energy and it is larger for the soft iso-EOS. Our result then shows that the isospin migration is the dominating effect for the asymmetry of the neck fragments. It is also seen that the difference between stiff and soft iso-EOS is particularly large for MD interactions, which can be traced back to the fact that the IMF's originate from more compact configurations.

The sensitivity to the IMF asymmetry can be enhanced by taking the ratio relative to the asymmetry of the residue, which is shown (only for the more transparent symmetric reactions) in Fig. 3b, as a function of the initial asymmetry. The large, almost 30% effect for the more realistic MD interaction is noteworthy. Thus this quantitiy, which should also not be very sensitive to secondary evaporation, may constitute a promising observable to gain more information on the symmetry energy. One may even consider double ratios of this quantity in reactions of HH over LL *Sn* isotopes.

We have shown that there exist several observables which should be able to yield information on the ill-determined low-density symmetry energy. Here we have investigated ratios of pre-equilibrium particles in central collisions and isospin transport between the residues and to the neck in peripheral collisions. We suggest to study the imbalance ratio not only as a function of centrality, but also depending on the energy loss in the reaction. We have also identified the asymmetry of an IMF from the neck as a promising observable. Unfortunately, however, both the agreement of theoretical calculations with each other as well as the comparison with experimental data are still far from satisfactory, such that the question of the isovector EOS has to be considered still open. Generally, the investigations into the low density iso-EOS favor an asy-stiff EOS from the isospin diffusion and the neck fragmentation data, and rather an asy-soft behavior from the pre-equilibrium studies. Further intensive work, perhaps also with data from more asymmetric radioactive beams, is highly desirable.

This work has been supported by the BMBF, Germany, grant 06LM189, by the DFG Cluster of Excellence *Origin and Structure of the Universe*, and by the Romanian Min. of Educ. and Research, contract CEX-05-D10-02.

References

1. C. Fuchs and H. H. Wolter, *Eur. Phys. Jour. A* **30**, 5 (2006), and refs. therein.
2. V. Baran, M. Colonna, V. Greco and M. Di Toro, *Phys. Rep.* **410**, 335 (2005).
3. *Isospin Physics in Heavy-ion Collisions at Intermediate Energies*, Eds. B. A. Li and W. Udo Schröder, Nova Science Publishers, New York, 2001.
4. V. Baran, M. Colonna, M. Di Toro *et al.*, *Phys. Rev. C* **72**, 064620 (2005).
5. M. B. Tsang *et al.*, *Phys. Rev. Lett.* **92**, 062701 (2004).
6. L. Shi and P. Danielewicz, *Phys. Rev. C* **68**, 064604 (2003).
7. B. A. Li and L. W. Chen, *Phys. Rev. C* **72**, 064611 (2005); L. W. Chen, C. M. Ko and B. A. Li, *Phys. Rev. Lett* **94**, 032701 (2005).
8. M. Colonna, G. Fabbri *et al.*, *Nucl. Phys. A* **742**, 337 (2004).
9. P. Chomaz, M. Colonna and J. Randrup, *Phys. Rep.* **389**, 263 (2004).
10. J. Rizzo, M. Colonna and M. Di Toro, *Phys. Rev. C* **72**, 064609 (2005).
11. C. Gale, G. M. Welke, M. Prakash *et al.*, *Phys. Rev. C* **41**, 1545 (1990).
12. P. Danielewicz, R. Lacey and W. G. Lynch, *Science* **298**, 1592 (2002).
13. M. A. Famiano *et al.*, *Phys. Rev. Lett.* **97**, 052701 (2006).
14. Y. X. Zhang, P. Danielewicz, M. Famiano *et al.*, arXiv:0708.3684v1 [nucl-th].
15. M. Colonna, V. Baran, M. Di Toro and H. H. Wolter, arXiv:0707.3092v1 [nucl-th].
16. F. Rami *et al.*, *Phys. Rev. Lett.* **84**, 120 (2000).
17. G. A. Souliotis, M. Velselsky, D. W. Shetty and S. J. Yennello, *Phys. Lett. B* **588**, 35 (2004).
18. M. Di Toro, A. Olmi and R. Roy, *Eur. Phys. Jour. A* **30**, 65 (2006), and refs. therein.
19. M. Colonna and M. B. Tsang, *Eur. Phys. J. A* **30**, 165 (2006), and refs. therein.
20. E. De Filippo *et al.*, *Phys. Rev. C* **71**, 044602, 064604 (2005).

INSTABILITY AGAINST CLUSTER FORMATION IN NUCLEAR AND COMPACT-STAR MATTER

C. DUCOIN,[1,*] PH. CHOMAZ,[2] F. GULMINELLI[1] and J. MARGUERON[3]

[1] *LPC (IN2P3-CNRS/Ensicaen et Université), F-14076 Caen cédex, France*
[2] *GANIL (DSM-CEA/IN2P3-CNRS), F-14076 Caen cédex, France*
[3] *IPN-Orsay, Université Paris Sud, F-91406 Orsay cédex, France*
[*] *E-mail: ducoin@lpccaen.in2p3.fr*

We address nuclear liquid-gas instablitities in the mean-field framework, using a Skyrme-like density functional. These instabilities lead to the clusterization of nuclear and compact-star matter at sub-saturation density. In this contribution, we study the extension of the spinodal region, how it affects star matter at β-equilibrium and how it is affected by the choice of different Skyrme forces. The dynamics of cluster formation is also characterized, comparing a semi-classical approach to a quantal one.

Keywords: Nuclear-matter equation of state; Skyrme forces; Liquid-gas phase transition; Neutrino trapping; Spinodal decomposition; Thomas-Fermi; RPA.

The liquid-gas phase transition is a well-known feature of the nuclear-matter equation of state.[1] It induces instabilities against finite-size density fluctuations in nuclear systems at sub-saturation density. This could be the origin of the multifragmentation observed in heavy-ion collisions.[2] Matter clusterization is also an important aspect of compact-star physics: inhomogeneities (such as "pasta" phases) in neutron star crust and core-collapse supernovae should affect, in particular, neutrino transport. Crucial consequences are expected in supernova dynamics.[3–6] In the present contribution we study the onset of clusterization: in a spinodal-decomposition scenario,[2] the initial instability dominates the dynamics of cluster formation, and characteristic time and size can be derived analytically.

The thermodynamic liquid-gas instability in nuclear matter corresponds to a curvature anomaly in the entropy surface of the homogeneous system. At a given temperature, this is equivalent to a concavity of the free energy represented in the plane of neutron and proton densities: the spinodal region corresponds to a negative eigen-value of the free-energy curvature matrix. The region of instability of homogeneous matter against clusterization can be defined firstly following a static criterion: one system will be considered unstable if the introduction of a density fluctuation reduces the total free energy. We determine this in a mean-field calculation.

The mean field is derived from a Skyrme energy-density functional $\mathcal{H}(\rho_n, \rho_p, \tau_n, \tau_p)$ where ρ_i and τ_i are the local particle and kinetic densities of

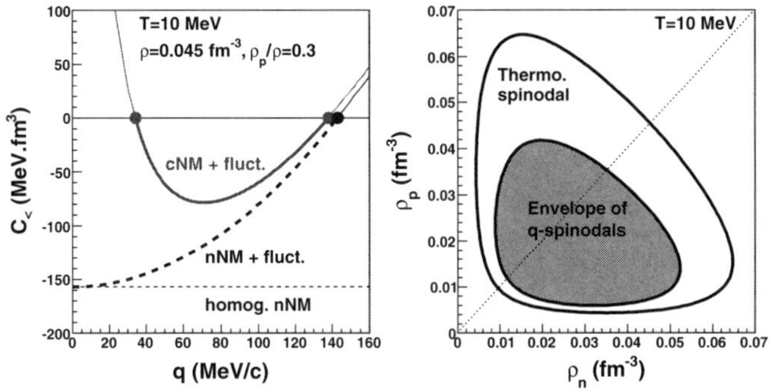

Fig. 1. Effect of Coulomb and surface contribution on the liquid-gas instability (SLy230a). Left (for fixed average densities): the dotted horizontal line "homog. nNM" (neutral nuclear matter) gives the minimal free-energy curvature in the thermodynamic case. The thick dotted line "nNM+fluct." shows the reduction of the instability due to the surface terme, and for the thick full line "cNM+fluct" (charged nuclear matter) Coulomb contribution is also included. Right: thermodynamic spinodal, compared to the region of finite-size instabilities given by the envelope of all q-spinodals.

neutrons n and protons p (we will also use the isoscalar and isovector notations $\rho = \rho_n + \rho_p$ and $\rho_3 = \rho_n - \rho_p$). The expression of this fonctional is given for exemple in the reference by Chabanat et al.[7] (a reduced form is used in the present work, where we consider spin-saturated matter, and calculate the Coulomb interaction separately). Having performed the thermodynamic study of nuclear matter in mean field, we obtain for each temperature the free-energy density $f(T, \rho_n, \rho_p)$. In a semi-classical Thomas-Fermi treatment[8,9] the introduction of a plane-wave density fluctuation such that $\rho_i = \rho_i^0 + (A_i e^{i\mathbf{q}\cdot\mathbf{r}} + c.c.)$ leads to the free-energy variation $\delta f = \tilde{A}^* C^f \tilde{A}$; in this expression, C^f is the free-energy curvature matrix in the space of fluctuation amplitudes $\tilde{A} = (A_n, A_p)$ and is given by:

$$C_{ij}^f = \partial\mu_j/\partial\rho_i + 2q^2 C_{ij}^\nabla + 4\pi e_i e_j/q^2 \,, \tag{1}$$

where μ_i is the chemical potential of particle i, $e_p = q_e/\sqrt{4\pi\epsilon_0}$, $e_n = 0$. For a given wave number q, the region of instability (called in a condensed way q-spinodal) corresponds to the occurrence of a negative eigen-value of this matrix, $C_<^f 0$. Considering star matter, the electron degree of freedom can be taken into account, adding a third dimension which is only perturbative due to the high incompressibility of the electron-gas.[9,10] The first term is identical to the free-energy curvature matrix in density space for the homogeneous system, the second term is a surface contribution due to the density-gradient dependence of the nuclear force ($\mathcal{H}_{fin} = \sum_{ij} C_{ij}^\nabla \nabla\rho_i \nabla\rho_j$), and the third term is the Coulomb contribution (neglecting the exchange part). The thermodynamic instability is recovered by switching off the Coulomb contribution and taking the infinite-wavelength limit $q \to 0$. The effects of surface and Coulomb terms on the thermodynamic instability are shown on Fig. 1, where we use the Skyrme parametrization SLy230a[7] (The choice

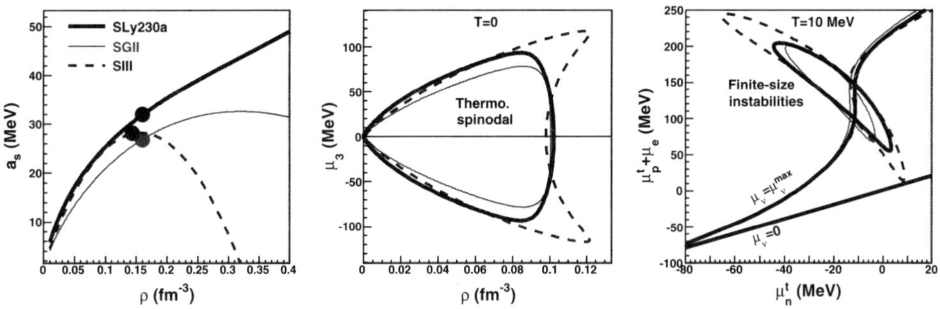

Fig. 2. Role of the symmetry energy. Left: density dependence of a_s for different Skyrme forces. Middle: extension in isovector chemical potential of the thermodynamic spinodal. Right: region of finite-size instabilities for star-matter, in chemical potential representation, compared to the region of possible β equilibrium (see text).

of Skyrme-Lyon forces is also made in other studies of star matter[11]).

Fig. 2 illustrates the dependence of the instability region on the different Skyrme forces, comparing the parametrizations SLy230a, SIII[12] and SGII.[13] Differences are essentially related to the symmetry energy and its density dependance, which is the object of many current investigations.[14] The symmetry energy characterizes the parabolic evolution of the (free-)energy per particle with the asymmetry $y = \rho_3/\rho$: $a_s = 1/2\partial^2(F/A)/\partial y^2$. In the parabolic approximation for the isovector part of the energy per particle, the behavior of the isovector chemical potential $\mu_3 = \mu_n - \mu_p$ is fully determined by $a_s(\rho)$:

$$f(\rho, \rho_3) = \rho \tfrac{F}{A} \simeq f_{sym}(\rho) + a_s(\rho)\tfrac{\rho_3^2}{\rho} \quad \rightarrow \quad \mu_3(\rho, \rho_3) = 2\tfrac{\partial f}{\partial \rho_3} \simeq 4y a_s(\rho)\,. \tag{2}$$

The differences affecting $a_s(\rho)$ between the different Skyrme forces are then reflected on the μ_3 extension of the thermodynamic spinodal, as shown on the middle part of Fig. 2.

An important consequence of the dependence on $a_s(\rho)$ concerns neutrino trapping in hot star matter, which could affect supernova dynamics. This is illustrated on the right part of Fig. 2. The region of star-matter instability against clusterization is represented in the space of chemical potentials, and compared to the region of possible β-equilibrated matter which depends on neutrino (ν) presence according to $\mu^t_n + \mu_\nu = \mu^t_p + \mu_e$ (for non-relativistic nucleons we use the notation $\mu^t_i = \mu_i + m_i$). The lower limit ($\mu_\nu = 0$) corresponds to neutrino transparency, the upper one ($\mu_\nu = \mu_\nu^{max}$) is calculated in the case all neutrinos produced by electron capture are trapped in the star.[9] Different rates of neutrino trapping are needed to reach the instability region, depending on its extension in the μ_3 direction.

Fig. 3 addresses the comparison between the semi-classical approach presented so-far (coupled to a hydrodynamic description of the early dynamics of cluster formation) and a quantum RPA approach.[15] The early dynamics of the unstable modes is characterized by a dispersion relation, which associates to an unstable fluctuation of wave number q a time constant τ such that $\delta\rho_i \propto e^{t/\tau}$. For given

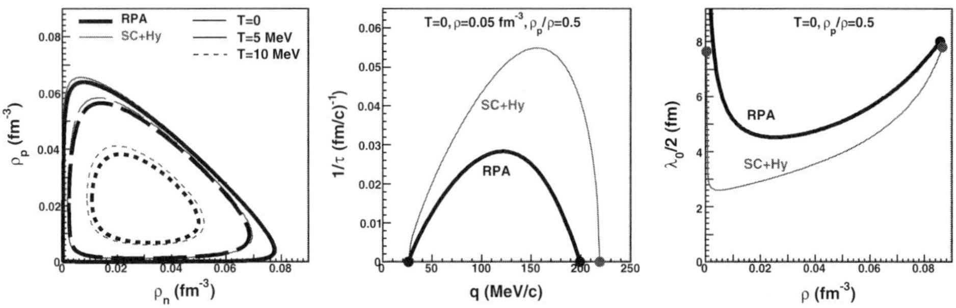

Fig. 3. Dynamics of cluster formation: comparison between a semi-classical hydrodynamical approach (SC+Hy) and quantum RPA approach. Left: region of finite-size instabilities. Middle: dispersion relation for unstable modes, linking the time constant τ to the transfered momentum q. Right: half-wavelength associated with the most unstable mode $\lambda_0/2 = \pi/q_0$.

densities, the most unstable mode (q_0, τ_0) is defined by the fastest amplification (maximum of $1/\tau$ on the middle part of the figure): it dominates the formation of clusters, whose expected size is then about $\lambda_0/2 = \pi/q_0$. The quantum quenching of the instability due to the kinetic cost of the fluctuation (which disfavors large values of q[16]) results in longer typical times and larger typical sizes for cluster formation, as shown on the middle and right parts of the figure. However, The left part shows that the global extension of the instability region is hardly affected by this quenching, and our earlier comments based on the semi-classical results are confirmed.

References

1. G. Bertsch and P. J. Siemens, *Phys. Lett. B*, 1269 (1983).
2. Ph. Chomaz, M. Colonna and J. Randrup, *Phys. Rep.* **389**, 263 (2004).
3. M. Prakash *et al.*, *Lect. Notes Phys.* **578**, 364 (2001).
4. R. Buras *et al.*, *Phys. Rev. Lett.* **90**, 241101 (2003).
5. C. J. Horowitz, M. A. Perez-Garcia and J. Piekarewicz, *Phys. Rev. C* **69**, 045804 (2004).
6. J. Margueron, J. Navarro and P. Blottiau, *Phys. Rev. C* **70**, 28801 (2004).
7. E. Chabanat *et al.*, *Nuclear Physics A* **627**, 710 (1997).
8. C. J. Pethick, D. G. Ravenhall and C. P. Lorentz, *Nucl. Phys. A* **584**, 675 (1995).
9. C. Ducoin, Ph. Chomaz and F. Gulminelli, *Nucl. Phys. A* **789**, 403 (2007).
10. C. Providencia *et al.*, *Phys. Rev. C* **73**, 025805 (2006).
11. F. Douchin and P. Haensel, *Phys. Lett. B* **485**, 107 (2000).
12. M. Beiner *et al.*, *Nucl. Phys. A* **238**, 29 (1975).
13. N. Van Giai and H. Sagawa, *Nucl. Phys. A* **371**, 1 (1981).
14. V. Baran *et al.*, *Phys. Rep.* **410**, 335 (2005).
15. C. Ducoin, J. Margueron and Ph. Chomaz, in preparation.
16. S. Ayik, M. Colonna and Ph. Chomaz, *Phys. Lett. B* **353**, 417 (1995).

MICROSCOPIC OPTICAL POTENTIALS OF NUCLEON-NUCLEUS AND NUCLEUS-NUCLEUS SCATTERING

ZHONG-YU MA* and JIAN RONG

China Institute of Atomic Energy,
Beijing 102413, P.R. of China
**E-mail: Mazy12@ciae.ac.cn*

YIN-QUN MA

Department of Physics, Taiyuan Normal University,
Taiyuan 030001, P.R. of China

The relativistic microscopic optical potential (RMOP) in the nucleon-nucleus scattering is studied in the framework of the Dirac Brueckner Hartree-Fock (DBHF) approach. The real part of the nucleon self-energy in asymmetric nuclear matter is calculated with the G-matrix, while the imaginary part is obtained from the polarization diagram. Nuclear optical potentials in finite nuclei are derived from the nucleon self-energies in asymmetric nuclear matter through a local density approximation. The RMOP is applied to study the nucleon scattering off stable nuclei and nucleon effective mass. A satisfactory agreement with the experimental data is found. The complex nucleon-nucleus optical potential is extended to microscopic optical potentials of nucleus-nucleus interaction by a folding method. The elastic scattering data of ^{6}He at 229.8 MeV on ^{12}C target are analyzed.

Keywords: Relativistic microscopic optical potential; Dirac Brueckner Hartree-Fock (DBHF) approach; Nucleon effective mass; Nucleus-nucleus scattering.

1. Introduction

The optical model potential is an essential tool in the study of nuclear reactions. Recently, the relativistic mean field (RMF) approach has been of great success in describing not only the ground state properties of stable nuclei, but also those of exotic nuclei. There have been also various applications to investigate the nucleon-nucleus scattering. Studies of the relativistic microscopic optical potential (RMOP) in the past years were based on several models, such as the RMF,[1,2] relativistic Hartree-Fock (RHF),[3] and Dirac-Brueckner Hartree-Fock (DBHF)[4,5] approaches. The nucleon self-energies were calculated in symmetric nuclear matter and the isospin dependence was usually neglected. A reasonable description of the differential cross sections, analyzing powers and spin rotation functions of proton scattering off stable nuclei has been achieved in an energy range of 50-500 MeV without readjusting parameters.[3,5]

With the development of experiments on newly constructed radioactive beam facilities, the investigation of the isospin dependence of the optical potential becomes more and more important. Therefore we shall study the RMOP microscopically in the framework of the DBHF approach. Recently, a new decomposition of the Dirac structure of nuclear self-energy in the DBHF has been extended to the calculations in asymmetric nuclear matter.[6] In the DBHF we investigate the proton and neutron self-energies in asymmetric nuclear matter, therefore the RMOP.[7] The real part of the nucleon self-energy in asymmetric nuclear matter is calculated with the G matrix in the Hartree-Fock approach and the imaginary part of the the nucleon self-energy is obtained by the G-matrix polarization diagram. The nucleon-nucleus optical potentials in finite nuclei are obtained by the local density approximation (LDA). The nucleon scattering off nucleus is analyzed in a Schrödinger-type equation, which is obtained by eliminating the lower component of the Dirac spinor in the Dirac equation. A satisfactory agreement with experimental data has been found.

The study of the nucleus-nucleus optical potentials is an object of importance not only for the description of elastic cross sections but also as an ingredient in the description of all the phenomena which occur when two nuclei collide.[12] The interest has been renewed recently due to the development of radioactive beam facilities in the world. In particular, it is very important to understand the complex optical potential for composite projectiles from a microscopic point of view not only to understand the relevant reaction dynamics involved but also to develop a practical tool for predicting optical potentials of colliding systems for which the elastic scattering measurement is absent or difficult, such as in the case of neutron-rich or proton-rich β-unstable nuclei. In this work we shall extend our method to investigate the microscopic optical potentials (MOP) of nucleus-nucleus scattering by a folding method. The elastic scattering data of ^{6}He at 229.8 MeV on ^{12}C target are analyzed within the standard optical model.

2. Theoretical Framework and Nucleon-Nucleus Scattering

Starting from a bare NN interaction V the NN effective interaction in the nuclear medium is calculated by summing up all ladder diagrams in the DBHF approach. The DBHF G can be decomposed into the bare NN interaction and a correlation term: $G = V + \Delta G$.[6] The correlation term is parameterized by four vertices: scalar, vector and isoscalar, isovector. Due to the characteristic of the short range correlation, they can be described by infinite masses and finite ratios of strengths to the corresponding masses. The nucleon self-energies in nuclear matter are calculated with V and ΔG in the relativistic Hartree-Fock approach. The exchange term produces weak energy dependence of the nucleon self-energy, which takes account of the antisymmetry of the nucleon with all nucleons in the nuclear medium. The DBHF nucleon self-energy in asymmetric nuclear matter has the general form:

$$\Sigma^i(k, k_F, \beta) = \Sigma^i_s(k, k_F, \beta) - \gamma_0 \Sigma^i_0(k, k_F, \beta) + \boldsymbol{\gamma} \cdot \boldsymbol{k} \Sigma^i_v(k, k_F, \beta) \ , \tag{1}$$

where i stands for proton or neutron. $\beta = (\rho_n - \rho_p)/(\rho_n + \rho_p)$, ρ_n, ρ_p are the asymmetry parameter, neutron and proton densities, respectively. Due to the isovector meson exchanges, the proton and neutron self-energies are distinguished. The imaginary part of the nucleon self-energy can be obtained by the G-matrix polarization diagram. An effective nucleon interaction was introduced in order to avoid difficulties caused by π-meson and simplify the calculation.[7] Four scalar and vector mesons with density dependent coupling constants were introduced to reproduce the saturation curves and nucleon self-energies at various densities and asymmetry parameters calculated with the DBHF G-matrix.

It is well known that the optical potential of a nucleon in the nuclear medium is equivalent to its self-energy. For finite nuclei the nucleon potential is obtained by means of the LDA, in which the space dependence of the relativistic microscopic optical potential is directly connected with the density of the target nucleus and asymmetry parameter in asymmetric nuclear matter. The Dirac equation of the projectile nucleon in the mean field of the target nucleus can be written as:

$$[\boldsymbol{\alpha} \cdot \boldsymbol{k} + \gamma_0(M + U_s^i) + U_0^i + V_c]\psi^i = E\psi^i, \tag{2}$$

where

$$U_s^i = \frac{\Sigma_s^i - M\Sigma_v^i}{1 + \Sigma_v^i}, \qquad U_0^i = \frac{-\Sigma_0^i + E\Sigma_v^i}{1 + \Sigma_v^i} . \tag{3}$$

In these expressions, M is the nucleons mass, U_s^i, U_0^i are Lorentz scalar potential and time like component of a four-vector potential, respectively. $E = \varepsilon + M$, and ε is the energy of the projectile in the center-of-mass system. V_c is the Coulomb potential.

By eliminating the lower component of the Dirac spinor in Eq. (2), a Schroedinger-type equation[7,13] is obtained for the upper component:

$$\left[\frac{p^2}{2E} + V_{eff}^i(r) + V_C(r) + V_{so}^i(r)\boldsymbol{\sigma} \cdot \boldsymbol{L}\right] \Phi^i(\boldsymbol{r}) = \frac{E^2 - M^2}{2E}\Phi^i(\boldsymbol{r}), \tag{4}$$

where V_{eff}^i and V_{so}^i are the central and spin-orbit part of the Schrödinger-equivalent potential, respectively. The explicit expressions for V_{eff}^i and V_{so}^i are

$$\begin{aligned} V_{eff}^i &= \frac{M}{E}U_s^i + U_0^i + \frac{1}{2E}\left[(U_s^i)^2 - (U_0^i + V_C)^2\right] + V_D^i \,, \\ V_{so}^i &= -\frac{1}{2ErD^i(r)}\frac{dD^i(r)}{dr} \,, \end{aligned} \tag{5}$$

where $D^i(r) = M + E + U_s^i - U_0^i - V_c$ and a small Darwin term V_D^i is usually neglected. The Schrödinger equivalent potential yields exactly the same scattering phase shifts as the original relativistic potential in the four-component Dirac equation. Therefore we have obtained the microscopic proton- and neutron-nucleus interaction in the framework of the DBHF approach, which are complex with both real and imaginary parts. Including the isovector meson exchanges the microscopic optical potential also depends on the neutron and proton density distributions of the target nucleus.[7]

We calculate the differential cross sections, analyzing powers and spin rotation functions of proton scattering off spherical nuclei, such as ^{40}Ca and ^{208}Pb at various energies of $E_p < 200$ MeV and compare with the experimental data. It is found that the calculated cross sections as well as spin observables are in rather good agreement with the experimental data at low energies.

3. Nucleon Effective Mass

The nucleon effective mass characterizes the propagation of a nucleon in the nuclear medium, which is adopted to describe an independent quasi-particle model in the nuclear many-body system.[8] The isospin dependence of the nucleon effective interaction as well as the nucleon effective mass is critically important for understanding properties of neutron stars and the dynamics of nuclear collisions induced by radioactive beams.[9,10] Unfortunately, up to now the knowledge about the isospin dependence of those quantities from experiments is very little. Study of the ROMP with isospin dependence may provide the information of the nucleon effective mass.

In the non-relativistic approach the nucleon effective mass can be derived by following two equivalent expressions,

$$\begin{aligned}\frac{M^*}{M} &= 1 - \frac{d}{d\varepsilon}V(k(\varepsilon),\varepsilon) \\ &= \left(1 + \frac{M}{k}\frac{d}{dk}V(k,\varepsilon(k))\right)^{-1}_{k=k(\varepsilon)} ,\end{aligned} \tag{6}$$

where $\varepsilon(k)$ is a function of the momentum k defined by the energy momentum relation, $\varepsilon = k^2/2M + V(k,\varepsilon)$. In the relativistic approach the nucleon effective mass is usually defined by the nucleon attractive scalar potential in the nuclear medium.

$$M_s^* = M + U_s , \tag{7}$$

which is called nucleon scalar mass. However, this definition of the effective mass is not directly related to that in Eq. (6). Actually the definitions in Eq. (6) and Eq. (7) denote different physical quantities. However, the Schrödinger equivalent central potential plays the same role as the potential $V(\varepsilon)$ in the non-relativistic approach. Therefore, the non-relativistic-type effective mass as defined in Eq. (6) can be derived in a similar way,

$$M_v^*/M = 1 - U_0/M - (1 + U_s/M)dU_s/d\varepsilon - (1 + \varepsilon/M - U_0/M)dU_0/d\varepsilon , \tag{8}$$

which may be called nucleon vector effective mass.[11] The neutron and proton effective mass are calculated with the Schrödinger-equivalent potential Eq. (5) as a function of the asymmetry parameter β at $E = 50$ MeV. The results are plotted in Fig. 1, where upper and lower curves correspond to nucleon scalar and vector effective masses, respectively. It is found that due to the stronger energy dependence of the neutron vector potential than the scalar potential the neutron vector effective mass becomes larger than that of proton in the neutron-rich nuclear matter in the neutron-rich nuclear matter.

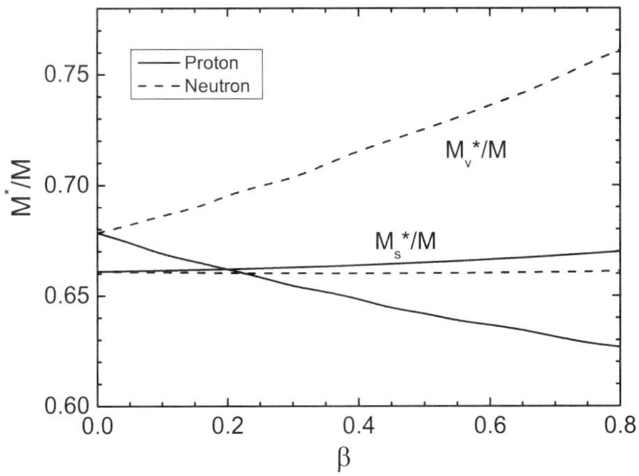

Fig. 1. Nucleon scalar and vector effective masses at E=50 MeV in the asymmetric nuclear matter with k_F=1.36 fm.

4. ^{6}He Scattering off ^{12}C

In a simple practical approach to deal with a composite particle scattering, one considers the target just a scatterer, and the nucleus-nucleus optical potential can be obtained by a folding method. The proton- and neutron-nucleus optical potentials are folded with the corresponding proton and neutron density distributions in the projectile.

$$V_{FM}(\boldsymbol{R}) = \sum_{i=p,n} \int \rho_i(\boldsymbol{r}_i) V^i_{eff}(\boldsymbol{s}_i) d\boldsymbol{r}_i \ , \tag{9}$$

where $\boldsymbol{R}$ is the separation distance between two centers of the colliding nuclei. $\boldsymbol{r}_i$ is the coordinate of the proton (neutron) at the center of mass frame of the projectile, while $\mathbf{s}_i$ the vector between the proton (neutron) in the projectile and the center of mass of the target, $\boldsymbol{s}_i = \boldsymbol{R} - \boldsymbol{r}_i$. ρ_i is the density distribution of proton (neutron) at the projectile. In this method the overlap of two colliding nuclei has not taken into account, which would in general overestimate the real potentials and underestimate the imaginary ones, respectively. We shall discuss this effect late.

With the folding method we could obtain the nucleus-nucleus microscopic optical potential with the real and imaginary part, simultaneously, which appears in a one-body standard Schrödinger equation,

$$\left[-\frac{\hbar^2}{2\mu}\nabla^2 + U_{opt}(\boldsymbol{R}) + U_C(\boldsymbol{R})\right] \chi(\boldsymbol{R}) = E_{c.m.}\chi(\boldsymbol{R}) \ . \tag{10}$$

Here, μ is the reduced mass of the pair, $E_{c.m.}$ is the center of mass energy of the relative motion. $U_C(\boldsymbol{R})$ is the Coulomb potential, which can be given as usual by

$$U_C(R) = \begin{cases} \frac{Z_P Z_T e^2}{R} & R > R_C \\ \frac{Z_P Z_T e^2}{2R_c}(3 - (\frac{R}{R_C})^2) & R \leq R_C \end{cases} , \tag{11}$$

where Z_P and Z_T are the charge numbers of the projectile and target, respectively. $R_C = r_0(A_P^{1/3} + A_T^{1/3})$ is the Coulomb radius of two charged spheres, $r_0 = 1.2$ fm, A_P, A_T are the mass numbers of the projectile and target, respectively.

Applying this method we analyze the recent experimental data on ^{6}He scattering off ^{12}C at the bombarding energy $E_{lab} = 229.8$ MeV at GANIL.[14] ^{6}He is a halo nucleus with a dispersed neutron distribution. We adopt a halo-type density, which was obtained by calculations in a three-body model.[15]

$$\rho_p(r) = 2e^{-(r/a)^2}/\pi^{3/2}a^3 \ , \tag{12}$$
$$\rho_n(r) = 2e^{-(r/a)^2}/\pi^{3/2}a^3 + 4(e^{-(r/b)^2}/3\pi^{3/2}b^5)r^2 \ ,$$

with $a = 1.55$ fm, $b = 2.24$ fm. A two parameter Fermi function is used for the density distribution of ^{12}C,[14]

$$\rho(r) = \frac{\rho_0}{1 + exp((r - r_0)/a)} \ , \tag{13}$$

with r_0=2.1545 fm and a=0.425 fm. The density is normalized to the atomic number of ^{12}C, then ρ_0 is determined as 0.207 fm^{-3}. The real and imaginary parts of the optical potential $U_{opt}(R)$ for ^{6}He+^{12}C reaction at 229.8 MeV are calculated by folding the isospin dependent microscopic optical potentials in $p(n)$+^{12}C with the density of ^{6}He.

The elastic differential cross section for ^{6}He scattering off ^{12}C at 229.8 MeV are calculated with the MOP. In order to reproduce the experimental data one introduces phenomenologically normalization factors on real and imaginary parts of the MOP. It is found that the imaginary part of the MOP obtained is too weak and an enhancing factor of $N_I = 3.0$ on the imaginary part of the optical potential is required when one keeps the real potential unchanged ($N_R = 1$). This can be understood that ^{6}He is a loosely bound nucleus, a breakup process is one of the main processes in the reaction, which is not considered in the microscopic optical potential of $p(n)$+^{12}C. The imaginary part of the $p(n)$+^{12}C optical potential was calculated from the polarization diagram of G matrix, which takes account only the processes of particle-hole type excitations.[7] Thus the calculated imaginary part of the optical potential in the reaction of ^{6}He+^{12}C is much too weak. In addition for a weakly bound or halo nucleus its particle threshold is close to the ground state, which implies a strong coupling to the continuum during the interaction of the nucleus with a target. A special treatment of the interaction potential to consider explicitly transitions to the low-lying excited states, to the resonances and breakup states may also required.

The differential cross section calculated directly by the folded potential with $N_I = 3.0$ is plotted in Figbb. 2 with a solid curve. The agreement between our theoretical calculation and the experimental data is impressive, which is rather encouraging without adjusting more free parameters. As mentioned above that the overlap effect between two colliding nuclei has not taken into account in our model. The results without normalization factor on the real part may illustrate that the

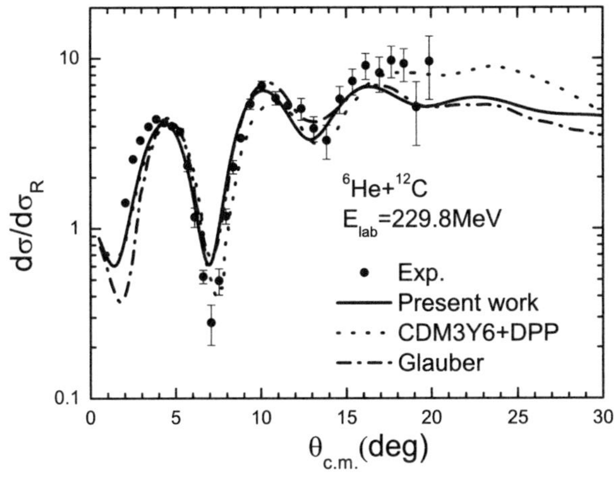

Fig. 2. Angular distribution of the elastic differential cross section for ^{6}He+^{12}C scattering at 229.8 MeV. The solid curve shows the results of the optical model calculations with an enhancing factor 3 on the imaginary potential. The dotted, and dot-dashed curves show the results of calculations by the double folding model with CDM3Y + DPP and Glauber-model, respectively.

overlap of two colliding nuclei is nonsignificant in the collision with a weakly bound nucleus.

In Fig. 2 we also plot the comparison with the results based on various theoretical models. The dotted curve was calculated with a double folding model (DFM).[14] In this model the real part of the optical potential was obtained by double folding of the CDM3Y6 effective nucleon-nucleon interaction, and the imaginary part was based on a phenomenological Woods-Saxon type potential with including a dynamic polarization potential (DPP). A frozen density approximation was adopted, where the total density of the two overlapping nuclei was the sum of the two densities. In their calculation a best fit to the experimental data was obtained by adding a complex surface potential with a repulsive real part designed to simulate the polarization effects caused by the projectile breakup. The dash-dotted curve corresponds to the calculation in a few-body Glauber model (FBGM),[16] where the ^{6}He wave function is factorized as a three-body wave function consisting of the α core and two neutrons. The target is considered just a scatterer and the effective phases in the Glauber model are calculated from a nucleon-target or a core-target optical potentials. It is found that all models could well reproduce the experimental data.

Actually, a fully microscopic investigation of the nucleus-nucleus scattering is much more complicated and difficult.[12] Inelastic and multi-reaction channels, absorption and breakup processes are involved in the nucleus-nucleus reactions. A proper treatment of the density overlap of two colliding nuclei is not clear, which depends on various factors, such as the structure of two colliding nuclei, bombarding energy, etc. In our present investigations we have not attempted to deal microscopically with these dynamic reaction process and the density overlap between two colliding nuclei. In comparison with the experimental data, we investigate phenomenologically the effect of the dynamic processes and overlap density in the nucleus-nucleus scattering, especially for loosely bound nuclei.

Acknowledgments

This work is supported by the National Natural Science Foundation of China under Grant Nos 10475116, 10535010, Major State Basic Research Development Program in China Under Contract Number 2007CB81500 and the European Community Project Asia-Europe Link in Nuclear Physics and Astrophysics, CN/ASIA-LINK/008(94791). ZYM gratefully acknowledges Nguyen Van Giai and Hermann Wolter for many stimulating discussions.

References

1. C. J. Horowitz, and B. D. Serot, *Nucl. Phys. A* **399**, 529 (1983).
2. C. J. Horowitz, *Nucl. Phys. A* **412**, 228 (1984).
3. Ma Zhong-yu, Zhu Ping, Gu Ying-qi *et al.*, *Nucl. Phys. A* **490**, 619 (1988).
4. Ma Zhong-yu and Chen Bao-qiu, *J. Phys. G* **18**, 1543 (1992).
5. B. Q. Chen and A. D. Mackellar, *Phys. Rev. C* **52**, 878 (1995).
6. E. Schiller and H. Müther, *Eur. Phys. J. A* **11**, 15 (2001).
7. J. Rong, Z. Y. Ma and N. V. Giai, *Phys. Rev. C* **73**, 014614 (2006).
8. C. Mahaux, P. F. Bortignon, R. A. Broglia and C. H. Dasso, *Phys. Rep.* **120**, 1 (1985).
9. J. Rizzo, M. Colonna, M. DiToro and V. Greco, *Nucl. Phys. A* **732**, 202 (2004).
10. B. A. Li, C. B. Das, D. Gupta and C. Gale, *Nucl. Phys. A* **735**, 563 (2004).
11. Zhong-Yu Ma, Jian Rong, Bao-Qiu Chen, Zhi-Yuan Zhu and Hong-Qiu Song, *Phys. Lett. B* **604**, 170 (2004).
12. G. R. Satchler and W. G. Love, *Phys. Rep.* **55**, 183 (1979).
13. M. Jaminon, C. Mahaux and P. Rochus, *Phys. Rev. C* **22**, 2027 (1980).
14. V. Lapoux, N. Alamanos, F. Auger *et al.*, *Phys. Rev. C* **66**, 034608 (2002).
15. J. S. Al-Khalili, J. A. Tostevin and I. J. Thompson, *Phys. Rev. C* **54**, 1843 (1996).
16. B. Abu-Ibrahim and Y. Suzuki, *Nucl. Phys. A* **728**, 118 (2003).

PART B

The Neutron Star Crust: Structure, Formation and Dynamics

NEUTRON STAR CRUST BEYOND THE WIGNER-SEITZ APPROXIMATION

N. CHAMEL

Institute of Astronomy and Astrophysics, Université Libre de Bruxelles, Brussels, Belgium
E-mail: nchamel@ulb.ac.be
www.astro.ulb.ac.be

For more than three decades, the inner crust of neutron stars, formed of a solid lattice of nuclear clusters coexisting with a gas of electrons and neutrons, has been traditionally studied in the Wigner-Seitz approximation. The validity of this approximation is discussed in the general framework of the band theory of solids, which has been recently applied to the nuclear context. Using this novel approach, it is shown that the unbound neutrons move in the crust as if their mass was increased.

Keywords: Neutron star crust; Band theory; Hartree-Fock; Skyrme; Wigner-Seitz approximation; Effective mass; Entrainment.

1. Introduction

The interpretation of many observational neutron star phenomena is intimately related to the properties of neutron star crusts:[1] pulsar glitches, giant flares and toroidal oscillations in Soft Gamma Repeaters, X-ray bursts and superbursts, initial cooling in quasi persistent Soft X-ray Transients, precession, gravitational wave emission. Besides the ejection of neutron star crust matter into the interstellar medium has been invoked as a promising site for r-process nucleosynthesis, which still remains one of the major mysteries of nuclear astrophysics.[2]

The outer layers of the crust, at densities above $\sim 10^6$ g.cm^{-3}, are formed of a solid Coulomb lattice of fully ionized atomic nuclei. Below $\sim 6 \times 10^{10}$ g.cm^{-3}, the structure of the crust is well established and is completely determined by the experimental atomic masses.[3] It is found that nuclei become increasingly neutron rich with increasing depth. At densities above $\sim 6 \times 10^{10}$ g.cm^{-3}, nuclei are so exotic that experimental data are lacking. Nevertheless, the masses of these nuclei could be determined in the near future by improving experimental techniques. At density $\sim 4.10^{11}$ g.cm^{-3}, calculations predict that neutrons "drip" out of nuclei forming a neutron liquid. Such an environment is unique and cannot be studied in the laboratory. We will refer to the nuclei in the inner crust as nuclear "clusters" since they do not exist in vacuum. The inner crust extends from $\sim 4.10^{11}$ g.cm^{-3} up to $\sim 10^{14}$ g.cm^{-3}, at which point the clusters dissolve into a uniform mixture

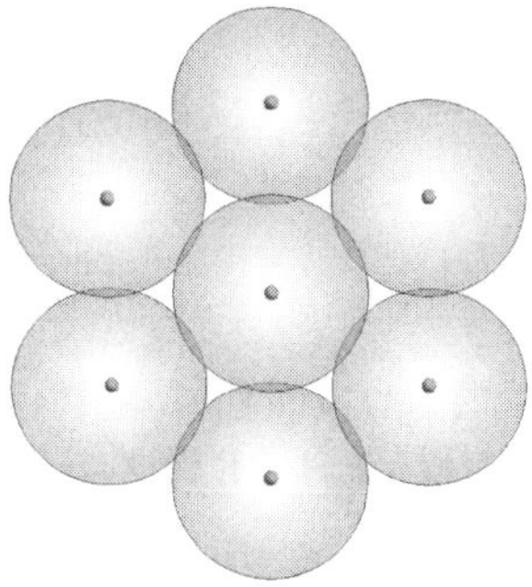

Fig. 1. In the Wigner-Seitz approximation, the crust (represented here as a two dimensional hexagonal lattice) is divided into independent identical spheres, centered around each site of the lattice. The radius of the sphere is chosen so that the volume of the sphere is equal to $1/n_{\rm N}$, where $n_{\rm N}$ is the density of lattice sites (clusters).

of electrons and nucleons. Near the crust-core interface, the clusters might adopt unusual shapes such as slabs, rods or more complicated structures.[1,4]

The structure of the inner crust has been studied using different approximations and nuclear models. The current state-of-the-art is represented by self-consistent quantum calculations which were pioneered by Negele&Vautherin in 1973.[5] In these calculations, the crust is decomposed into an arrangement of identical spheres centered around each cluster as illustrated in Fig. 1. Each sphere can thus be seen as one big "nucleus" so that the usual techniques of nuclear physics can be directly applied to study neutron star crust. However, it has been recently found that this approximation, proposed a long time ago by Wigner-Seitz[6] in solid state physics, leads to unreliable predictions of the structure and composition of the crust especially in the bottom layers.[7] Besides this approach does not allow the study of transport properties which are essential for interpreting observations. A more elaborate treatment of neutron star crust beyond the Wigner-Seitz approximation is therefore required.[8]

2. Band Theory of Solids

It is usually assumed that neutron stars are composed of cold catalyzed matter (for a discussion of deviations from this idealized situation, see for instance Ref. 1). Besides, it is rather well-established that the ground state of dense matter (at zero temperature) below saturation density possesses the symmetry of a perfect crystal.

A crystal lattice can be partitioned into identical primitive cells, each of which contains exactly one lattice site. The specification of the primitive cell is not unique. A particularly useful choice is the Wigner-Seitz or Voronoi cell defined by the set of points that are closer to a given lattice site than to any other. This cell reflects the local symmetry of the crystal. It follows from the translational symmetry that a primitive cell contains all the information about the system. The single particle states of any species q in a periodic system are characterized by a wave vector $\boldsymbol{k}$.

If the wavefunction $\varphi_{\boldsymbol{k}}^{(q)}(\boldsymbol{r})$ is known inside one cell, the wavefunction in any other cell can be deduced from the Floquet-Bloch theorem[9]

$$\varphi_{\boldsymbol{k}}^{(q)}(\boldsymbol{r}+\boldsymbol{T}) = e^{\mathrm{i}\boldsymbol{k}\cdot\boldsymbol{T}}\varphi_{\boldsymbol{k}}^{(q)}(\boldsymbol{r}), \tag{1}$$

where $\boldsymbol{T}$ is the corresponding lattice translation vector (which translates the initial cell to the other). Apart from the wave vector $\boldsymbol{k}$, the single particle states are labelled by a discrete index α (principal quantum number) so that the single particle energy spectrum consists in a series of "bands" or sheets in $\boldsymbol{k}$-space. This band index accounts for the local rotational symmetry of the lattice.

It can be shown that the single particle states (therefore the single particle energies) are periodic in $\boldsymbol{k}$-space

$$\varphi_{\boldsymbol{k}+\boldsymbol{K}}^{(q)}(\boldsymbol{r}) = \varphi_{\boldsymbol{k}}^{(q)}(\boldsymbol{r}), \tag{2}$$

where the reciprocal vectors $\boldsymbol{K}$ are defined by

$$\boldsymbol{K}\cdot\boldsymbol{T} = 2\pi N, \tag{3}$$

N being any positive or negative integer. The set of all possible reciprocal vectors define a reciprocal lattice in $\boldsymbol{k}$-space. Equation (2) entails that only the wave vectors $\boldsymbol{k}$ lying inside the so-called first Brillouin zone (i.e. Wigner-Seitz cell of the reciprocal lattice) are relevant. The first Brillouin zone can be further divided into irreducible domains by considering the rotational symmetry of the lattice.

The inner crust of neutron stars is constituted of neutrons, protons and electrons. Unlike the situation in ordinary solids (ordinary meaning under terrestrial conditions), the electronic properties in neutron star crust are very simple. The matter density is so high that the Coulomb energy of the electrons is negligible compared to their kinetic energy. The electrons can thus be treated as a degenerate relativistic Fermi gas.[4] Since all the protons are bound inside nuclear clusters (except possibly in the bottom layers of the crust), the proton single particle states are almost independent of $\boldsymbol{k}$ and to a good approximation can thus be described only by the discrete principal quantum number α like in isolated nuclei. In contrast, the effects of the nuclear lattice on the neutrons (i.e. dependence of the states on both $\boldsymbol{k}$ *and* α) have to be taken into account because some neutrons are unbound. In the following, we will consider the interface between the outer and the inner parts of neutron star crust where unbound neutrons appear. At densities below $\sim 10^{12}$ g.cm^{-3}, these unbound neutrons are not expected to be superfluid.[10] We will therefore treat the nucleons in the Hartree-Fock approximation with the effective Skyrme nucleon-nucleon interaction.[11] Let us however point out that the above considerations are very general and apply to any approximation of the many-body problem. Indeed the Bloch states form a complete set of single particle states into which the many-body wave function can be expanded (for instance, the application of band theory including pairing correlations has been discussed in Ref. 12).

The single particle wave functions are obtained by solving the self-consistent equations ($q = n, p$ for neutrons and protons respectively) inside the Wigner-Seitz cell

$$h_0^{(q)}\varphi_{\alpha\boldsymbol{k}}^{(q)}(\boldsymbol{r}) = \varepsilon_{\alpha\boldsymbol{k}}^{(q)}\varphi_{\alpha\boldsymbol{k}}^{(q)}(\boldsymbol{r})\,, \tag{4}$$

where the single particle Hamiltonian is defined by

$$h_0^{(q)} \equiv -\boldsymbol{\nabla}\cdot\frac{\hbar^2}{2m_q^{\oplus}(\boldsymbol{r})}\boldsymbol{\nabla} + U_q(\boldsymbol{r}) - \mathrm{i}\boldsymbol{W_q}(\boldsymbol{r})\cdot\boldsymbol{\nabla}\times\boldsymbol{\sigma}\,. \tag{5}$$

The effective masses $m_q^{\oplus}(\boldsymbol{r})$, mean fields $U_q(\boldsymbol{r})$ and spin-orbit terms $\boldsymbol{W_q}(\boldsymbol{r})$ depend on the occupied single particle wave functions.

It follows from the Floquet-Bloch theorem (1), that the wavefunctions can be expressed as

$$\varphi_{\alpha\boldsymbol{k}}^{(q)}(\boldsymbol{r}) = e^{\mathrm{i}\boldsymbol{k}\cdot\boldsymbol{r}}u_{\alpha\boldsymbol{k}}^{(q)}(\boldsymbol{r})\,, \tag{6}$$

where $u_{\alpha\boldsymbol{k}}^{(q)}(\boldsymbol{r})$ has the full periodicity of the lattice, $u_{\alpha\boldsymbol{k}}^{(q)}(\boldsymbol{r}+\boldsymbol{T}) = u_{\alpha\boldsymbol{k}}^{(q)}(\boldsymbol{r})$. Consequently, Eqs. (4) can be equivalently written as

$$(h_0^{(q)} + h_{\boldsymbol{k}}^{(q)})u_{\alpha\boldsymbol{k}}^{(q)}(\boldsymbol{r}) = \varepsilon_{\alpha\boldsymbol{k}}^{(q)}u_{\alpha\boldsymbol{k}}^{(q)}(\boldsymbol{r})\,, \tag{7}$$

where the $\boldsymbol{k}$-dependent Hamiltonian $h_{\boldsymbol{k}}^{(q)}$ is defined by

$$h_{\boldsymbol{k}}^{(q)} \equiv \frac{\hbar^2 k^2}{2m_q^{\oplus}(\boldsymbol{r})} + \boldsymbol{v_q}\cdot\hbar\boldsymbol{k}\,, \tag{8}$$

and the velocity operator $\boldsymbol{v_q}$ is defined by the commutator

$$\boldsymbol{v_q} \equiv \frac{1}{\mathrm{i}\hbar}[\boldsymbol{r}, h_0^{(q)}]\,. \tag{9}$$

The boundary conditions are completely determined by the assumed crystal symmetry. According to the Floquet-Bloch theorem (1), the wavefunctions $\varphi_{\alpha\boldsymbol{k}}^{(q)}(\boldsymbol{r})$ and $\varphi_{\alpha\boldsymbol{k}}^{(q)}(\boldsymbol{r}+\boldsymbol{T})$ between two opposite faces of the Wigner-Seitz cell ($\boldsymbol{T}$ being the corresponding lattice vector) are shifted by a factor $e^{\mathrm{i}\boldsymbol{k}\cdot\boldsymbol{T}}$ while $u_{\alpha\boldsymbol{k}}^{(q)}(\boldsymbol{r}) = u_{\alpha\boldsymbol{k}}^{(q)}(\boldsymbol{r}+\boldsymbol{T})$. Each cell is electrically neutral which means that it contains as many protons as electrons. Equations are solved for a given baryon density, assuming that the matter is in β-equilibrium.

3. Validity of the Wigner-Seitz Approximation

The three dimensional partial differential Eqs. (4) or (7) have to be solved for *each* wave vector $\boldsymbol{k}$ inside one irreducible domain of the first Brillouin Zone[a]. Such calculations are computationally very expensive (numerical methods that are applicable to neutron star crust have been discussed by Chamel[13,14]). Since the work of Negele&Vautherin,[5] the usual approach has been to apply the Wigner-Seitz approximation.[6] The complicated cell is replaced by a sphere of equal volume as shown in

[a]The Hartree-Fock equations have to be solved together with Poisson's equation for determining the Coulomb part of the proton mean field potential.

Fig. 1. In this approximation, the dependence on the crystal structure is therefore completely lost. Besides only the states with $\boldsymbol{k} = 0$ are considered or in other words the Hamiltonian $h_{\boldsymbol{k}}^{(q)}$ is neglected. Eqs. (4) and (7) are then similar and reduce to a set or ordinary differential equations. The price to be paid for such a simplification is that this approximation does not allow the study of transport properties (which depend on the $\boldsymbol{k}$-dependence of the states). As pointed out by Bonche&Vautherin,[15] two types of Dirichlet-Neumann mixed boundary conditions are physically plausible yielding a more or less constant neutron density outside the cluster: either the wave function or its radial derivative vanishes at the cell edge, depending on its parity.

In recent calculations,[16–18] the Wigner-Seitz cell has been replaced by a cube with periodic boundary conditions. While such calculations allow for possible deformations of the nuclear clusters, the lattice periodicity is still not properly taken into account since the $\boldsymbol{k}$-dependence of the states is ignored. Besides the Wigner-Seitz cell is only cubic for a simple cubic lattice and it is very unlikely that the equilibrium structure of the crust is of this type (the structure of the crust is expected to be a body centered cubic lattice[1]). It is therefore not clear whether these calculations, which require much more computational time than those carried out in the spherical approximation, are more realistic. This point should be clarified in future work by a detailed comparison with the full band theory.

It should be remarked that the states obtained in the Wigner-Seitz approximation do not coincide with those obtained in the full band theory at $\boldsymbol{k} = 0$.[8] The reason is that in the Wigner-Seitz approximation the shape of the exact cell is approximated by a sphere. By neglecting the $\boldsymbol{k}$-dependence of the states, the Wigner-Seitz approximation overestimates the neutron shell effects and leads to unphysical fluctuations of the neutron density, as discussed in details by Chamel *et al.*[8] Negele&Vautherin[5] proposed to average the neutron density in the vicinity of the cell edge in order to remove these fluctuations. However it is not *a priori* guaranteed that such *ad hoc* procedure did remove all the spurious contributions to the total energy. This issue is particularly important since the total energy difference between different configurations may be very small. It has thus been found that the dependence of the equilibrium crust structure on the choice of boundary conditions, increases with density.[7,15] In particular, the Wigner-Seitz approximation becomes very unreliable in the bottom layers of the crust where the clusters nearly touch.

While the Wigner-Seitz approximation is reasonable at not too high densities for determining the equilibrium crust structure, the full band theory is indispensable for studying transport properties.

4. Band Theory and Transport Properties

At low temperature the transport properties of the neutron liquid in neutron star crust are determined by the shape of the Fermi surface, defined by the locus of points in $\boldsymbol{k}$-space such that $\varepsilon_{\alpha\boldsymbol{k}} = \varepsilon_{\rm F}$ (we drop the superscript n since we only consider neutrons in this section). In general, the Fermi surface is composed of several sheets

Fig. 2. Neutron Fermi surface in the inner crust of neutron star at average mass density $\rho = 7 \times 10^{11}$ g.cm^{-3}. The crust is composed of a body centered cubic lattice of zirconium-like clusters ($Z = 40$) with $N = 160$ neutrons per lattice site (70 neutrons are unbound). The single particle states have been calculated by solving the Hartree-Fock equations with the Skyrme force SLy4 and by applying Bloch boundary conditions.[8] The Fermi surface is shown inside the first Brillouin zone. The different colors correspond to different bands.

corresponding to the different bands that intersect the Fermi level. An example of neutron Fermi surface is shown on Fig. 2.

The motion of the unbound neutrons is affected by the presence of the nuclear clusters. In particular, they can be Bragg-reflected by the crystal lattice. As a result of the momentum transfer with the lattice, a neutron in a state characterized by a wave vector $\boldsymbol{k}$ and a band α, move in the crystal as if its mass was replaced by an effective mass given by (considering cubic crystals)

$$m_n^\star|_{\alpha\boldsymbol{k}} = 3\hbar^2 \left(\Delta_{\boldsymbol{k}}\varepsilon_{\alpha\boldsymbol{k}}\right)^{-1}, \tag{10}$$

where $\Delta_{\boldsymbol{k}}$ denotes the Laplacian operator in $\boldsymbol{k}$-space. This concept of effective mass has been very useful in interpreting neutron diffraction experiments.[19] In the context of neutron star crust, the number of unbound neutrons can be very large and it is therefore more appropriate to introduce a "macroscopic" effective mass $m_n^\star$ averaged over all occupied single particle states

$$m_n^\star = n_{\rm f}/\mathcal{K}, \qquad \mathcal{K} = \sum_\alpha \int_{\rm F} \frac{{\rm d}^3\boldsymbol{k}}{(2\pi)^3} \left(m_n^\star|_{\alpha\boldsymbol{k}}\right)^{-1}, \tag{11}$$

where $n_{\rm f}$ is the number density of free neutrons and the integral is performed over all occupied states "inside" the Fermi surface[b]. Let us remark that the mobility coefficient $\mathcal{K}$ can be equivalently written as

$$\mathcal{K} = \frac{1}{3(2\pi)^3\hbar^2} \sum_\alpha \int_{\rm F} \boldsymbol{\nabla}_{\boldsymbol{k}}\varepsilon_{\alpha\boldsymbol{k}} \cdot \mathbf{d}\boldsymbol{S}, \tag{12}$$

[b]The occupied states are those for which the energy $\varepsilon_{\alpha\boldsymbol{k}}$ is lower than the Fermi energy $\varepsilon_{\rm F}$.

showing that the effective mass depends on the shape of the Fermi surface. This effective mass $m_n^\star$ has been found to be very large compared to the bare mass.[13,14] For instance, the effective mass corresponding to the Fermi surface shown in Fig. 2 is about $m_n^\star \simeq 4.3\, m_n$. On the contrary, this effective mass is slightly lower than the bare mass in the liquid core owing to the absence of clusters[20] (the effectice mass $m_n^\star$ still differ from m_n due to neutron-neutron and neutron-proton interactions). With the definition (11), it can be shown that in the crust frame the momentum $\boldsymbol{p_\mathrm{f}}$ of the neutron liquid is simply given by $\boldsymbol{p_\mathrm{f}} = m_n^\star \boldsymbol{v_\mathrm{f}}$, where $\boldsymbol{v_\mathrm{f}}$ is the neutron velocity.[21,22] This result is quite remarkable. It implies that in an *arbitrary* frame, the momentum and the velocity of the neutron liquid are not aligned! Indeed, if $\boldsymbol{v_\mathrm{c}}$ is the velocity of the crust, applying the Galilean transformation leads to

$$\boldsymbol{p_\mathrm{f}} = m_n^\star \boldsymbol{v_\mathrm{f}} + (m_n - m_n^\star)\boldsymbol{v_\mathrm{c}}\,. \tag{13}$$

These so-called entrainment effects are very important for the dynamics of the neutron liquid in the crust. For instance, it has been suggested that for sufficiently large effective mass $m_n^\star \gg m_n$, a Kelvin-Helmholtz instability could occur and might explain the origin of pulsar glitches.[23]

5. Conclusions

Since the work of Negele&Vautherin,[5] the Wigner-Seitz approximation has been widely applied to study neutron star crust. Nevertheless, the necessity to go beyond has become clear in the last few years. It has thus been shown that the results of the Wigner-Seitz approximation become less and less reliable with increasing density.[7] Besides this approach, which decomposes the crust into a set of independent spherical cells, does not allow the study of transport properties. A more realistic description of the crust requires the application of the band theory of solids.[8] By taking consistently into account both the nuclear clusters which form a solid lattice and the neutron liquid, this theory provides a unified scheme for studying the properties of neutron star crust. We have shown in this framework that the unbound neutrons move in the crust as if they had an effective mass much larger than the bare mass. The analogy between neutron star crust and periodic systems in condensed matter suggests that some layers of the crust might have properties similar to those of band gap materials like semiconductors or photonic crystals for instance. The existence of such "neutronic" crystals in neutron star crusts opens a new interdisciplinary field of research, at the crossroad between nuclear and solid state physics.

Acknowledgments

N.C. gratefully acknowledges financial support from a Marie Curie Intra European fellowship (contract number MEIF-CT-2005-024660).

References

1. P. Hansel, A. Y. Potekhin and D. G. Yakovlev, *Neutron Stars 1: Equation of State and Structure* (Springer, 2006).
2. M. Arnould, S. Goriely and K. Takahashi, *Phys. Rep.* **450**, 97 (2007).
3. S. B. Rüster, M. Hempel and J. Schaffner-Bielich, *Phys. Rev. C* **73**, 035804 (2006).
4. C. J. Pethick and D. G. Ravenhall, *Ann. Rev. Nucl. Part. Sci.* **45**, 429 (1995).
5. J. W. Negele and D. Vautherin, *Nucl. Phys. A* **207**, 298 (1973).
6. E. Wigner and F. Seitz, *Phys. Rev.* **43**, 804 (1933).
7. M. Baldo, E. E. Saperstein and S. V. Tolokonnikov, *Nucl. Phys. A* **775**, 235 (2006).
8. N. Chamel, S. Naimi, E. Khan and J. Margueron, *Phys. Rev. C* **75**, 055806 (2007).
9. G. Grosso and G. P. Parravicini, *Solid State Physics* (Elsevier, 2000).
10. C. Monrozeau, J. Margueron and N. Sandulescu, *Phys. Rev. C* **75**, 065807 (2007).
11. M. Bender, P. Heenen and P. Reinhard, *Rev. Mod. Phys.* **75**, 121 (2003).
12. B. Carter, N. Chamel and P. Haensel, *Nucl. Phys. A* **759**, 441 (2005).
13. N. Chamel, *Nucl. Phys. A* **747**, 109 (2005).
14. N. Chamel, *Nucl. Phys. A* **773**, 263 (2006).
15. P. Bonche and D. Vautherin, *Nucl. Phys. A* **372**, 496 (1981).
16. P. Magierski and P.-H. Heenen, *Phys. Rev. C* **65**, 045804 (2002).
17. W. G. Newton, J. R. Stone and A. Mezzacappa, *J. Phys. Conf. Series* **46**, 408 (2006).
18. P. Gögelein and H. Müther, *Phys. Rev. C* **76**, 024312 (2007).
19. A. Zeilinger, C. G. Shull, M. A. Horne and K. D. Finkelstein, *Phys. Rev. Lett.* **57**, 3089 (1986).
20. N. Chamel and P. Haensel, *Phys. Rev. C* **73**, 045802 (2006).
21. B. Carter, N. Chamel and P. Haensel, *Int. Mod. Phys. D* **15**, 777 (2006).
22. B. Carter, N. Chamel and P. Haensel, *Nucl. Phys. A* **748**, 675 (2005).
23. N. Andersson, G. L. Comer and R. Prix, *MNRAS* **354**, 101 (2004).

THE INNER CRUST OF A NEUTRON STAR WITHIN THE WIGNER–SEITZ METHOD WITH PAIRING: FROM DRIP POINT TO THE BOTTOM

E. E. SAPERSTEIN

Kurchatov Institute, Moscow 123182, Russia
E-mail: saper@mbslab.kiae.ru

M. BALDO

INFN, Sezione di Catania,
Via S.-Sofia 64, I-95123 Catania, Italy
E-mail:baldo@ct.infn.it

S. V. TOLOKONNIKOV

Kurchatov Institute, Moscow 123182, Russia
E-mail: tolk@mbslab.kiae.ru

Results of the semi-microscopic self-consistent approach to describe the ground state properties of the inner crust of a neutron star developed recently within the Wigner-Seitz method with pairing effects taken into account are briefly reviewed.

Keywords: Neutron stars; Inner crust; Pairing.

1. Introduction

The first self-consistent completely quantum approach to describe the cold inner crust of a mature neutron star was developed by Negele and Vautherin (N&V) in a classical paper.[1] It was based on the energy density functional method in combination with the spherical Wigner-Seitz (WS) approximation. Within this approach, the crystal matter of the crust is approximated with a set of independent spherical cells, the neutron single-particle wave functions obeying some boundary condition. The ground state properties of the inner crust were studied in Ref. 1 in a wide density region, from the drip point to the very bottom. At a fixed value of the average density ρ, the equilibrium configuration (Z, R_c) was found, where Z and R_c are the proton number and the radius of the WS cell respectively. It is determined minimizing the total binding energy $E_{\rm B}$ under the beta-stability condition. The latter requires the equality of the neutron chemical potential μ_n to the sum of the proton and electron chemical potentials $\mu_p + \mu_e$. N&V did not take into account the pairing correlation in view of the smallness of its contribution to $E_{\rm B}$. Only 30 years later the pairing correlations in the inner crust were taken into account in a self-consistent way.[2,3] In the investigations of our group[3–5] it was found that the pairing correlations could change the (Z, R_c) values significantly. Systematic calculations for the intermediate and lower parts of the crust[6,7] showed that the equilibrium Z values could sometimes differ from those by N&V by about a factor two. This effect originates mainly from the beta-stability condition. In fact the influence of pairing to the chemical potentials μ_n and μ_p is much stronger than to the total energy.

Only recently a consistent band theory was developed for the inner star crust.[8–10] In Ref. 11 this more fundamental method is compared with the WS approximation for the upper region of the inner crust. It is concluded that the WS approximation is well suited for describing the ground state properties of the system under consideration, whereas the description of some dynamical aspects needs the consistent band theory. Calculations within the band theory are quite cumbersome, and a systematic study of the inner crust within this consistent approach hardly will be carried out soon. Therefore examination of accuracy of the WS method and the development of approximate methods to improve some of its deficiencies seem to be quite actual. One of these drawbacks comes from the fact that the neutron single particle spectrum is necessarily discrete in the WS method. With neutron pairing correlation included, this effect can be smoothed and become irrelevant, provided the characteristic inter-level distance $\delta\varepsilon$ is much less than the neutron gap Δ_n. At intermediate densities, $k_{\rm F} \simeq 0.6 \div 0.8$ fm^{-1}, $\delta\varepsilon$ doesn't exceed $100 \div 200$keV, whereas the gap in neutron matter reaches the maximum values, $\Delta_n \gtrsim 1$ MeV, therefore errors due to the discrete form of the spectrum should be negligible. As far as $\delta\varepsilon$ depends on R_c as $\delta\varepsilon \propto 1/R_c^2$, a dangerous situation could occur at $k_{\rm F} \gtrsim 1$ fm^{-1}, where $\delta\varepsilon$ could exceed 1 MeV, and Δ_n becomes smaller. In some cases, the artificial "Shell effect" appears[12] when $\delta\varepsilon$ at the Fermi surface becomes of the order of $2\Delta_n$ and the neutron gap $\Delta_n(r)$ found for the WS cell could become anomalously small and even vanish.

In Ref. 12 the role of the boundary condition in the WS method was examined. N&V used the one with vanishing of the neutron radial wave functions at the cell radius, $R_{nlj}(r{=}R_c){=}0$, for odd orbital angular momentum l, and of the derivative, $R'_{nlj}(r{=}R_c){=}0$, for even l. In Ref. 12 it was named as BC1. The alternative boundary condition (BC2) was also used with $R_{nlj}(r{=}R_c){=}0$ for even l and $R'_{nlj}(r{=}R_c){=}0$ for odd l. These two kinds of the boundary condition *a priory* seem equivalent. Direct comparison of BC1 *versus* BC2 has shown that the difference between predictions of the two calculations is negligible only in the upper part of the inner crust. In the intermediate density region, $k_{\rm F}{=}0.6 \div 0.8$ fm^{-1}, it is rather small, but not negligible. E.g., the difference of the Z values δZ doesn't exceed 2 units (only even Z were considered), with $\delta R_c \leq 1$ fm, and the difference in the gap values is of the order of 10%. These uncertainties are inherent to the WS method itself, thus putting internal limits to accuracy of the approach. For lower layers of the crust, $k_{\rm F} \gtrsim 0.9$ fm^{-1}, these uncertainties grow, $\delta Z \leq 6$ with $\delta R_c \leq 2$ fm. Especially big uncertainties appear in the neutron gap $\Delta_n(r)$ due to the pseudo-Shell effect discussed above. Fortunately, if it occurs for the BC1 version, it doesn't take place for the BC2 one, and *vice versa*. In such a situation, an approximate recipe to improve the WS method was suggested in Ref. 12. It consists just in choosing the boundary condition for which the spurious effect under discussion is absent.

A more consistent method to overcome this deficiency of the WS method was developed in Ref. 13.

2. The Drip Point

The method used is described in detail in Refs. 5 and 7. Here we mention only that the Generalized Energy Density Functional (GEDF) method[14] is used in which the GEDF $\mathcal{E}$ depends on the equal footing on the normal densities ρ_n, ρ_p and the anomalous ones ν_n, ν_p as well. The semi-microscopic GEDF is used which is constructed with matching the phenomenological functional $\mathcal{E}^{\rm ph}$ DF3[14] to describe the nuclear-like cluster in the center of the WS cell and the microscopic one, $\mathcal{E}^{\rm mi}$, to describe the neutron surroundings. The matching function is chosen as a two-parameter Fermi function with the radius R_m and the diffuseness parameter d_m. The latter was taken once forever to be equal to d_m=0.3 fm, whereas R_m is chosen anew in any new case in such a way that the equality $\rho_p(R_m)$=$0.1\rho_p(0)$ holds.

The normal component of $\mathcal{E}^{\rm mi}$ was taken from Ref. 15 where it was found for neutron matter within the Brueckner theory with the Argonne NN-force $\rm v_{18}$. The anomalous component of $\mathcal{E}^{\rm mi}$ was calculated in Ref. 5 for the same NN-force within the BCS method. In Ref. 7 the many-body corrections to the BCS theory for neutron matter were taken into account in an approximate way by introducing a momentum and density independent suppression factor $f_{\rm m-b}$ which, as it is commonly excepted,[4] should be between 1/2 and 1/3. Here we present results obtained within the so-called P2-model of Ref. 7, with $f_{\rm m-b}$=1/2.

In Ref. 16 the method described above was used to study the so-called neutron drip point which separates the outer crust from the inner one, as well as the upper layers of the inner crust, $k_{\rm F}$=$0.2 \div 0.5$ fm^{-1}. To find the drip point, we used the WS method for the outer crust where the phenomenological component of GEDF plays the main role. Indeed, as far as all neutron are now bound, the term $\mathcal{E}^{\rm mi}$ is switched on only in the region where the neutron density falls. However, we can not *a priory* neglect this contribution. In the physical situation of the outer crust the use of the complete GEDF, which includes both $\mathcal{E}^{\rm ph}$ and $\mathcal{E}^{\rm mi}$, is practically equivalent to a modification of the surface components of the effective interaction in DF3, but it is well known that predictions of the self-consistent calculations for atomic nuclei are highly sensitive to details of the effective interaction at the surface. Therefore, for finding the drip point in Ref. 16 two sets of calculations were performed. In the first one the phenomenological nuclear DF3 functional was only used, in the second one, the complete semi-microscopic DEDF was used.

The drip point was originally calculated by Baym, Pethick, and Sutherland (BPS)[17] who developed a thermodynamic approach to describe the outer crust using a phenomenological nuclear equation of states based on an extrapolation of

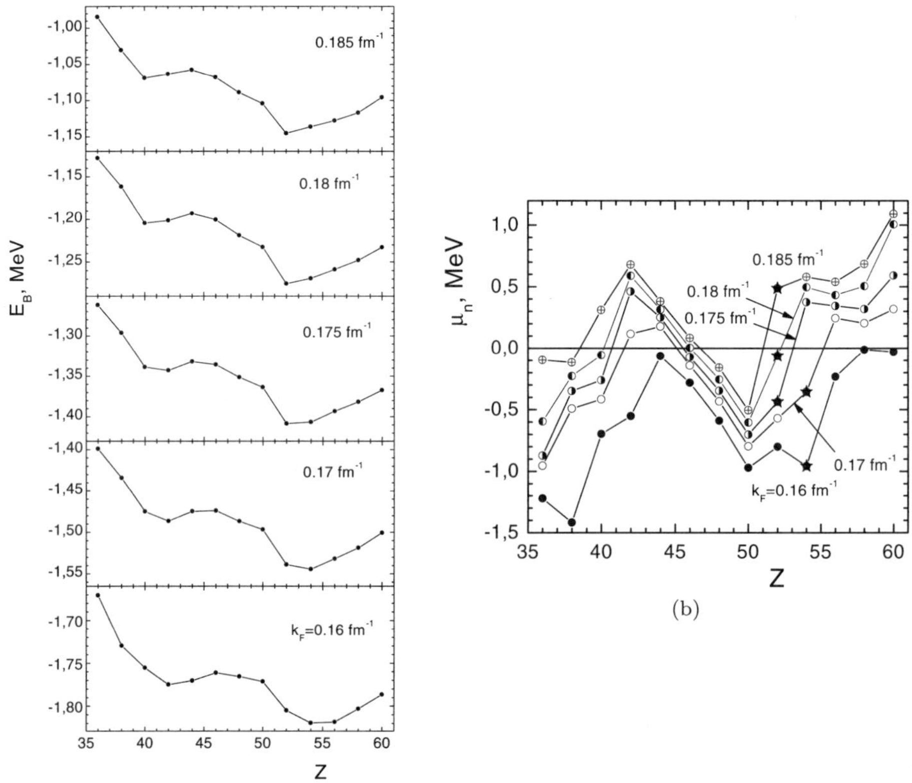

(a)

Fig. 1. For various k_{F} values in vicinity of the drip point, (a) binding energy per a nucleon and (b) the neutron chemical potential as a function of Z.

the nuclear mass data. Their prediction for the drip point was $\rho_{\mathrm{d}}{=}4.3 \cdot 10^{11}\mathrm{g/cm}^3$ which corresponds to the average Fermi momentum $k_{\mathrm{F}}^{\mathrm{d}}{=}0.1977\ \mathrm{fm}^{-1}$. N&V did not examine this point in detail, but their calculations agree qualitatively with this value, the neutron instability occurring for the nucleus $^{118}_{36}$Kr. Recently S. B. Ruster et al.[18] carried out a systematic analysis of the outer crust structure in the vicinity of the drip point within the thermodynamic BPS method with a set of different models for the nuclear equation of state (various kinds of Skyrme forces, modern versions of the droplet model, relativistic nuclear mean field theory and so on). All the predictions for the neutron drip point turned out to be rather close to the BPS one, $\rho_{\mathrm{d}} \simeq 4 \cdot 10^{11}\mathrm{g/cm}^3$. As to the drip Z values, $Z{=}34 \div 38$, they are quite close to the one by N&V.

Let us first consider in more detail the version with the pure phenomenological DF3 functional. As it was discussed in the Introduction, the equilibrium configuration (Z, R_c) at a fixed value of k_{F} is obtained by finding the minimum of the energy per nucleon E_{B} under the β-stability condition. The drip point is determined by the

Table 1. Main characteristics of the WS cell for the ground state configurations of the inner crust.

k_F ,fm^{-1}	Z	Z (N&V)	A	R_c, fm	x	x (N&V)
0.2	52	40	212	57.19	0.245	0.222
0.3	54	40	562	52.79	0.096	0.125
0.4	50	40	830	45.09	0.060	0.080
0.5	46	40	1020	38.64	0.045	0.042
0.6	56	50	1529	36.85	0.037	0.037
0.7	46	50	1351	30.31	0.034	0.037
0.8	40	50	1269	25.97	0.032	0.028
0.9*	26	50	946	20.93	0.027	0.028
1.0*	24	40	798	17.80	0.030	0.027
1.1*	24	40	794	16.16	0.030	0.027
1.2	20	40	633	13.73	0.032	0.027

k_F value for which the neutron chemical potential μ_n vanishes. The functions $E_B(Z)$ and $\mu_n(Z)$ are displayed in Figs. 1a and 1b, correspondingly, for different values of k_F in vicinity of the critical point by BPS. As one can see, the minimum of the function $E_B(Z)$ is shifted from the values by N&V or those in Ref. 18, Z=34 ÷ 38, to $Z = 52 \div 54$. In Fig. 1b, at each value of k_F, the asterisk indicates the point corresponding to the minimum of $E_B(Z)$, i.e. to the equilibrium configuration. As one can see, the drip point corresponds to the critical value of the Fermi momentum which is a little above k_F=0.18 fm^{-1}. A simple interpolation results in the value of k_F^d=0.181 fm^{-1}. The latter corresponds to the density ρ_d=3.30 · 10^{11}g/cm^3. Thus, the difference from the BPS prediction (and that of Ref. 18) for the drip point position in this case is not negligible. But the most essential deviation occurs for the ground state configuration at the drip point. They are (Z=52, $R_c \simeq 60$ fm). Note that the corresponding neutron number coincides with the magic one, N=126. Fig. 1b shows that there is another "drip region" with $Z \simeq 40$ (corresponding neutron number values are in vicinity of the magic number N=82). However, as it is seen from Fig. 1a, this region corresponds to higher values of the binding energy.

The repetition of these calculations with the complete semi-microscopic GEDF led to a small shift of the drip point to $k_F^d = 0.194$ fm^{-1} which corresponds to the density ρ_d=4.06 · 10^{11}g/cm^3. It is quite close to the value by BPS and is within the interval given in Ref. 18. As to the Z and N values, they coincide both with those for the phenomenological GEDF. The cell radius becomes a little less, $R_c \simeq 57.5$ fm, as a result of a small grow of the value of ρ_d.

3. Ground State Properties of the Inner Crust

In this section we present results of calculations of the ground state characteristics of the inner crust in the wide density region, from the drip point to almost the bottom. They are taken from Ref. 16 for k_F=0.2 ÷ 0.5 fm^{-1} and from Ref. 7 for k_F=0.6 ÷ 1.2 fm^{-1} Results are given in Table 1.

The range of density was limited up to k_F=1.2 fm^{-1}, since according to the commonly accepted point of view in the deep layers of the crust the "lasagna" or "spaguetti" structures of the crust matter are energetically favorable and the spherical WS method cannot be applied. In fact, these or other exotic configurations evidently occur already at k_F=1.2 fm^{-1} (Ref. 19) or even at k_F=1.1 fm^{-1} (Ref. 20). Therefore consideration of these two values of k_F is mainly of methodological interest. In any case, as it was discussed above, the accuracy of the WS method itself is doubtful for such high densities. All the results presented here are obtained within the P2-model of the Ref. 7 mentioned above. In this model, the microscopic component of the anomalous term of the GEDF is renormalized in order to produce a gap in infinite neutron matter equal to $\Delta_n=\Delta_n^{BCS}/2$. The value of k_F is marked with a star for the cases in which the BC2 boundary condition was used, in accordance with the recipe described in the Introduction. The average proton concentration $x=Z/A$ is presented in the last two columns. For comparison, the N&V values of Z and x are given. As we see, the difference for the Z values is, as a rule, significant (the corresponding relative difference in R_c is of the same order). As the analysis has shown,[6,7] they originate mainly from the pairing effects, which are taken into account in a self-consistent manner. The peculiarities of the DF3 functional play also an important role (see discussion in Ref. 16). It is interesting that the difference in the values of x is, as a rule, much less, i.e. this quantity is, evidently, more or less model independent.

Let us go now to the analysis of the neutron gap function $\Delta_n(r)$, which, as it is well known, is of primary importance for neutron star observable properties. The gap functions for different values of k_F are displayed in Figs. 2a and 2b. For a comparison, we draw also (with dashes) predictions of the simplest LDA version, where the $\Delta_n(r)$ function at the point r is taken from the neutron matter values at the density $\rho_n(r)$. As far as we deal with the P2 model, the BCS value of Δ_n is divided by 2. One can see that in the asymptotic region the "exact" gap and the LDA one are rather close to each other, but in the central region deviations are very strong. For all the values of k_F under consideration the LDA gap has a sharp maximum at the central cluster surface and is very small at $r = 0$, being almost independent of k_F. Such a pattern contradicts drastically to the results of the direct solution of the gap equation for the WS cell. Indeed, in the latter case, although at each value of k_F there is a non-regular behavior of $\Delta_n(r)$ at the surface, the absolute value of the gap inside the cluster is governed by the one in the asymptotic region, being essentially different for different k_F. Such the gap behavior demonstrates a strong inside-outside interplay in the gap equation, the so-called proximity effect. For brevity, we do not display here the neutron gap for higher values of k_F where the accuracy of the WS method is worse, but qualitatively the pattern in this case is similar, the difference between exact solution and the LDA one being even more pronounced (see Ref. 7).

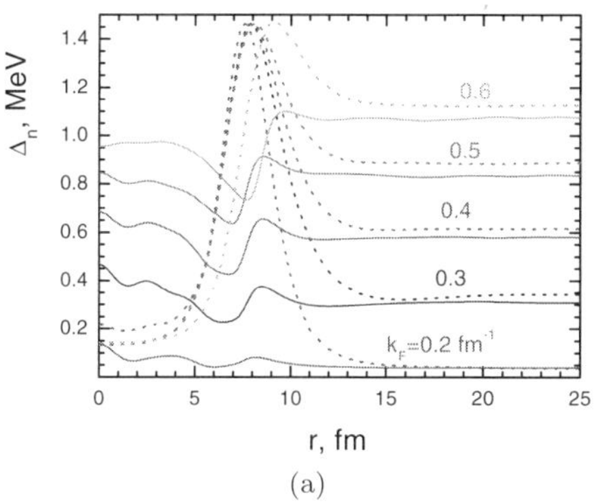

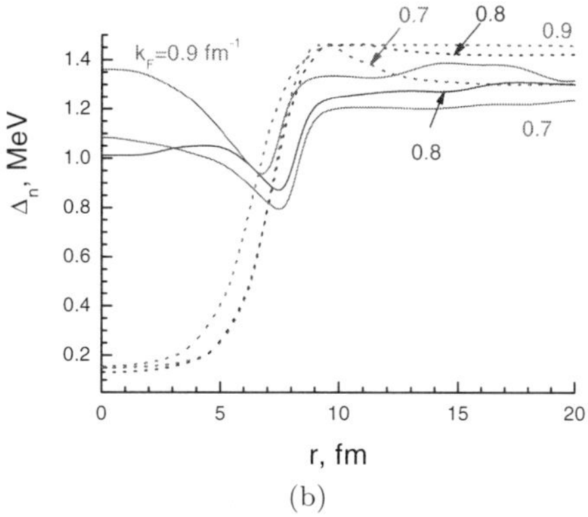

Fig. 2. Neutron gap function for (a) k_F=0.2 ÷ 0.6 fm^{-1} and (b) k_F=0.7 ÷ 0.9 fm^{-1}.

4. Conclusion

Results of the semi-microscopic self-consistent approach to describe the ground state properties of the inner crust of a neutron star developed recently (see Refs. 3–7,12, 16) are briefly reviewed. The theory is developed within the WS scheme and adopts a specific GEDF which is constructed by matching at the nuclear cluster surface the phenomenological nuclear DF3 functional[14] to the microscopic one calculated for neutron matter within the Brueckner approach with the Argonne force v_{18}. For the superfluid component of the latter, the model P2 of Ref. 7 is used which takes into account approximately the many-body corrections to the BCS approximation.

The drip point separating the inner crust from the outer one found within this approach turned out rather close to the commonly adopted one.[18] At the same time, the equilibrium Z=52 value exceeds significantly that of Ref. 18 (Z=34 ÷ 38). The corresponding neutron number is N=126 *versus* $N \simeq 82$ of Ref. 18. The equilibrium configurations of the inner crust for the wide density region corresponding to average Fermi momenta k_F=0.2÷1.2 fm^{-1} are presented and compared with those by N&V. The difference in the equilibrium Z values turned out to be significant for the upper layers of the inner crust ($Z \simeq 50$ *versus* Z=40 by N&V) and for the bottom ones (Z=20 ÷ 26 *versus* Z=40 by N&V). We explain this effect mainly with the pairing effects taken into account self-consistently. Peculiarities of the DF3 functional[14] play also evidently an important role. Our confidence to this form of the nuclear GDEF is based on its accuracy in reproducing the binding energies and the radii of long isotopic chains, including the odd-even effects. The high accuracy of the effective force by S.A. Fayans et al., in comparison with the popular version SLy4 of the Skyrme force, was recently confirmed also by Y.Yu and A.Bulgac.[21] In order to separate effects of the pairing from the influence of the particular form of the energy functional, it would be quite instructive to carry out systematic calculations along the same lines with different choices of the normal component of the DF3 functional, in particular with modern versions of the Skyrme force, provided they are of comparable accuracy.

Acknowledgements

This research was partially supported by the Grant NSh-8756.2006.2 of the Russian Ministry for Science and Education and by the RFBR grants 06-02-17171-a and 07-02-00553-a.

References

1. J. Negele and D. Vautherin, *Nucl. Phys. A* **207**, 298 (1973).
2. F. Montani, C. May and M. Müther, *Phys. Rev. C* **69**, 065801 (2004).
3. M. Baldo, U. Lombardo, E. E. Saperstein and S. V. Tolokonnikov, *JETP Lett.* **80**, 523 (2004).
4. M. Baldo, E. E. Saperstein and S. V. Tolokonnikov, *Nucl. Phys. A* **749**, 42 (2005).
5. M. Baldo, U. Lombardo, E. E. Saperstein and S. V. Tolokonnikov, *Nucl. Phys. A* **750**, 409 (2005).
6. M. Baldo, U. Lombardo, E. E. Saperstein and S. V. Tolokonnikov, *Phys. At. Nucl.* **68**, 1812 (2005).
7. M. Baldo, E. E. Saperstein and S. V. Tolokonnikov, *Eur. Phys. J. A* **32**, 97 (2007).
8. B. Carter, N. Chamel and P. Haensel, *Nucl. Phys. A* **748**, 675 (2005).
9. N. Chamel, *Nucl. Phys. A* **747**, 109 (2005).
10. N. Chamel, *Nucl. Phys. A* **773**, 263 (2006).
11. N. Chamel, S. Naimi, E. Khan and J. Margueron, arxiv preprint astro-ph/0701851.
12. M. Baldo, E. E. Saperstein and S. V. Tolokonnikov, *Nucl. Phys. A* **775**, 235 (2006).
13. J. Margueron, *Equation of state in the inner crust: discussion of the shell effects*, these proceedings.
14. S. A. Fayans, S. V. Tolokonnikov, E. L. Trykov and D. Zawischa, *Nucl. Phys. A* **676**, 49 (2000).
15. M. Baldo, C. Maieron, P. Schuck and X. Vinas, *Nucl. Phys. A* **736**, 241 (2004).
16. M. Baldo, E. E. Saperstein and S. V. Tolokonnikov, arxiv preprint nucl-th/0703099.
17. G. Baym, C. Pethick and P. Sutherland, *Ap. J* **170**, 299 (1971).
18. S. B. Ruster, M. Hempel and J. Schaffner-Bielich, *Phys. Rev. C* **73**, 035804 (2006).
19. P. Magierski and P.-H. Heenen, *Phys. Rev. C* **65**, 045804 (2002).
20. K. Oyamatsu, *Nucl. Phys. A* **561**, 431 (1993).
21. Y. Yu and A. Bulgac, *Phys. Rev. Lett.* **90**, 222501 (2003).

NUCLEAR SUPERFLUIDITY AND THERMAL PROPERTIES OF NEUTRON STARS

N. SANDULESCU

Institute of Physics and Nuclear Engineering, 76900 Bucharest, Romania

We discuss how the present uncertainty in the pairing properties of neutron matter affects the thermal response of the inner crust of neutron star in the case of a rapid cooling process. It is shown that the thermalisation time of the inner crust is changing by a factor of two if in the calculations one shifts between two posible pairing scenarios for neutron superfluidity, i.e., one corresponding to the BCS approximation and the other to many-body techniques which include polarisation effects.

Keywords: Neutron stars; Nuclear superfluidity; Cooling.

1. Introduction

Nuclear superfluidity plays a major role in the neutron stars cooling.[1] Thus, as a consequence of the gap in the excitation spectrum the superfluidity in the core matter suppresses the neutrino cooling while the superfluidity in the inner crust matter is shortening the core-crust thermalisation time.

The core-crust thermalisation time, which from now on will be refered to as the cooling time, is an important quantity in rapid cooling models. In these models the core cools so fast that a temperature inversion develops between the core and the crust. The crust keeps the surface warm until the cooling wave reaches the surface. When this happens, due to the big temperature gradient at the core-crust interface the surface temperature drops quickly to the temperature of the core. In this cooling scenario the cooling time is associated to the time needed by the cooling wave to arrive from the cold core to the surface of the star.

One of the first estimation of the cooling time was given by Brown et al.,[2] who considered the possibility of a rapid cooling induced by the strangeness condensation. Later on, using a direct Urca process as cooling mechanism, more realistic calculations of thermal evolution of neutron stars were performed.[3] The numerical simulations showed that the cooling time does not depend on the details of the rapid cooling mechanism but rather on the structure of the neutron star. Since then, a few quantum calculations of the inner crust matter superfluidity, including the effects of the nuclear clusters, have been done.[4,6,7] How the nuclear clusters could affect the cooling time of the inner crust was investigated by Pizzochero et al.[5] This study was based on the assumption that the pairing correlations in dilute neutron matter are given by the BCS approximation, neglecting thus the contribution of in-medium effects. The impact which in-medium effects could have on the specific heat of the inner cust matter was analysed later on in the framework of HFB approach at finite temperature.[7] It was shown that if the pairing correlations are reduced so that to take into account self-energy and screening effects, the specific heat of the inner crust matter can change by several orders of magnitude. The consequences of these changes in the specific heat upon the cooling time of inner crust matter were analysed recently in Ref. 8 Below we shall discuss what are the most important results of these studies.

2. Superfluid Properties of Inner Crust Matter

The inner crust consists of a lattice of neutron-rich nuclei immersed in a sea of unbound neutrons and relativistic electrons. In microscopic calculations the inner crust matter is divided in independent cells treated in the Wigner-Seitz approximation. Up to baryonic densities of the order of half the nuclear saturation density, considered up to now in cooling studies, each cell is supposed to contain in its center a spherical neutron-rich nucleus surrounded by unbound neutrons and immersed in a relativistic electron gas uniformly distributed inside the cell. The cell structure used in the studies presented here is taken from Ref. 9.

The pairing properties of the inner crust matter are analysed by using the HFB approach at finite temperature (FT-HFB).[7] In this approach the superfluid properties are described by the thermal averaged pairing density:

$$\kappa_T(r) = \frac{1}{4\pi}\sum_i (2j_i+1)U_i^*(r)V_i(r)(1-2f_i)\ ,$$

where $f_i = [1+exp(E_i/k_BT)]^{-1}$ is the Fermi distribution, k_B is the Boltzmann constant and T is the temperature.

For describing the pairing correlations we have employed a density dependent contact force of the following form:[10]

$$V(\mathbf{r}-\mathbf{r}') = V_0[1-\eta(\frac{\rho(r)}{\rho_0})^\alpha]\delta(\mathbf{r}-\mathbf{r}') \equiv \tilde{V}(\rho(r))\delta(\mathbf{r}-\mathbf{r}'), \tag{1}$$

where $\rho(r)$ is the baryonic density and $\rho_0 = 0.16$ fm^{-3}. With this force the thermal averaged pairing field, which enters in the FT-HFB equations, is given by $\Delta_T(r) = 1/2\ \tilde{V}\ \kappa_T(r)$, where $\kappa_T(r)$ is the thermal averaged pairing density.

In order to analyse how the uncertainty on the pairing gap in neutron matter could reflect upon the thermal response of the inner crust, we did calculations with two zero range pairing interactions which simulate the pairing gap in nuclear matter obtained either with the Gogny force, which gives a maximum gap value of about 3 MeV, or with models which take into account the in-medium effects. For the latter we consider a maximum gap of 1 MeV, as indicated by recent calculations.[14] The pairing force corresponding to a maximum gap of 3(1) Mc will be called below the strong (weak) pairing force.

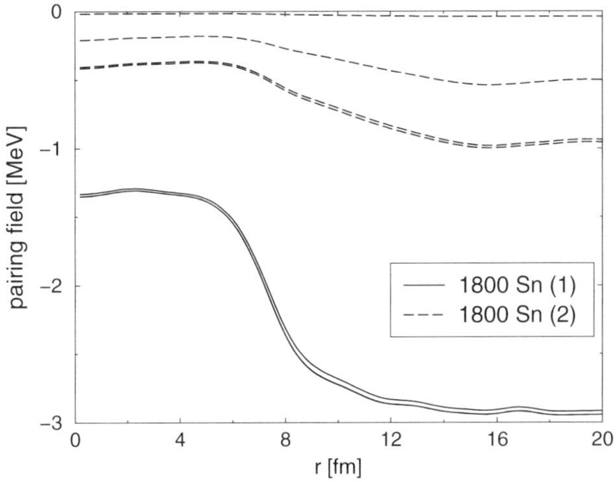

Fig. 1. Neutron pairing fields for the cell ^{1800}Sn. The full (dashed) lines indicates the results for strong (weak) pairing force. The curves corresponds, from bottom upwards, to the set of temperatures T={0.0, 0.1, 0.3, 0.5}MeV)

How the pairing field looks like for the weak and the strong pairing forces is shown in Figure 1 for the case of a Wigner-Seitz cell containing Z=50 protons and N=1750 neutrons.[7] We notice that for all temperatures the nuclear clusters modify significantly the profile of the pairing field. One can also see that for the weak pairing force the temperature dependence of the pairing field is significant. In the calculations presented here the mean field properties are described with the Skyrme force SLy4,[11] which was fixed to describe properly the mean field properties of neutron-rich nuclei and infinite neutron matter.

3. Thermal Properties of Inner Crust Matter

The thermal response of the inner crust matter depends on thermal diffusivity, defined as the ratio of the thermal conductivity to the heat capacity. The heat capacity of the inner crust has contributions from the electrons, the neutrons and the lattice. The heat capacity of the electrons, considered as a uniform and relativistic gas, has the standard form[18] while the contribution of the lattice to the specific heat is usually neglected.

In the normal phase, the specific heat of the neutrons exceeds the specific heat of the electrons by about two orders of magnitude. However, the onset of the neutron superfluidity reduces drastically the neutron specific heat, which could become smaller than the electron specific heat in some regions of the inner crust. How the neutron specific heat is affected by the superfluidity as well as by the temperature is discussed in the following subsection. The discussion is done for all relevant densities of the inner crust, starting from the neutron drip density up to about half the

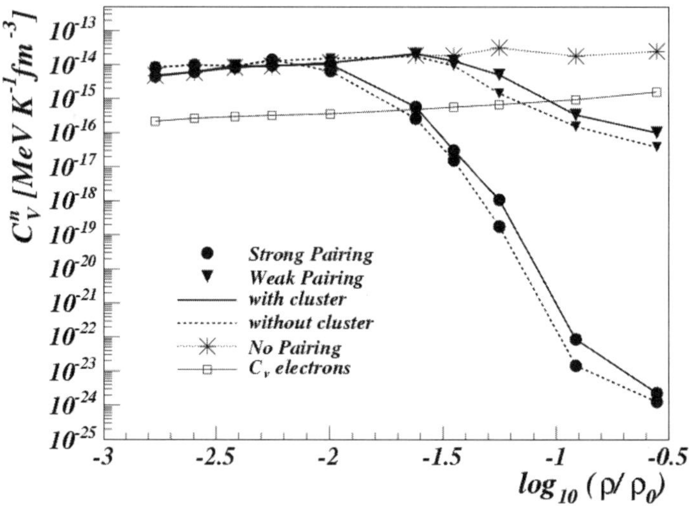

Fig. 2. Specific heat of the neutrons for the Wigner-Seitz cells listed in Ref. 9. The results correspond to the strong and the weak pairing forces (see the text) and for the cells with (without) the nuclear clusters. The specific heat of the non-uniform cells obtained when the pairing correlations are switched off are indicated by star symbols. The square symbols show the specific heat of the electrons.

nuclear saturation density. This region of the inner crust is supposed to give the largest contribution to the cooling time of the crust.[5]

3.1. *Specific heat*

The quasiparticle spectrum determined by solving the FT-HFB equations is used to calculate the specific heat of the neutrons inside the Wigner-Seitz cell.[7,8] The results are shown in Figure 2. In the same figure is also shown the specific heat of the electrons. The specific heats are calculated for a temperature of T=0.1 MeV, which is a typical temperature for the inner crust matter at the cooling stage analysed here (see the discussion below). From Figure 2 we can see that if the neutrons are in the normal phase, their specific heat is greater than the specific heat of the electrons in all the Wigner-Seitz cells. When the neutron superfluidity is turned on, the specific heat of the neutrons is suppressed due to the pairing gap in the excitation spectrum. Since the suppression depends exponentially on the pairing gap, the results obtained with the strong and the weak pairing forces are very different, as seen for the WS cells 1-5. For the second WS cell, in which the pairing gap in the neutron gas region has the maximum value, the specific heat obtained with the two pairing forces differs by about 7 orders of magnitude. In the WS cells 7-10 the neutron gas is in the normal phase at the temperature T=100 keV. Therefore both pairing forces give the same results for the specific heat.

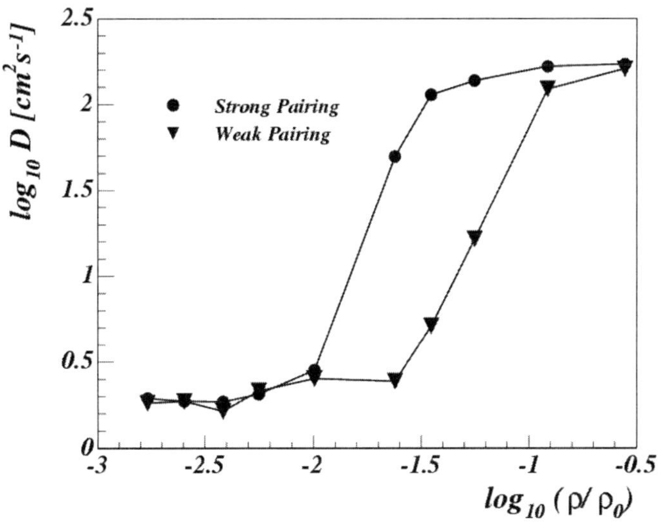

Fig. 3. Thermal diffusivity (neutrons plus electrons) corresponding to the Wigner-Seitz cells listed in Table I. The notations are the same as in Figure 2.

In the calculation presented above the specific heat was evaluated by considering only the non-interacting FT-HFB quasiparticles states. However, in the inner crust the residual interaction among the quasiparticles can also generates low-lying collective modes[15] which could have an important contribution to the specific heat of the crust matter.[16]

3.2. *Thermal diffusivity*

The specific heat enters in the heat transport through the thermal diffusivity, defined by $D = \frac{\kappa}{C_V}$, where κ is the thermal conductivity. In the inner crust, the latter is primarly determined by the electrons. In the calculations we have used the thermal conductivity parametrized by Lattimer et al.[3] (based on the results of Itoh et al.[17]). With this conductivity and the specific heat calculated in the FT-HFB approach one gets the thermal diffusivity shown in Figure 4. As expected from the behaviour of the specific heat, the diffusivity is much smaller for the weak pairing force, except for the last four WS cells. For both pairing forces one can see that the diffusivity is much smaller in the outermost layers of the inner crust. As seen below, these layers have an important contribution to the cooling time of the inner crust.

4. Cooling Time of Inner Crust Matter

In order to calculate the cooling time, i.e., the time needed for the cooling wave to propagate from the cold core to the surface, here have used the cooling model employed in Refs. 2,5. The model is based on the following assumptions: a) the spherical geometry for the heat transport is approximated by a planar geometry, i.e, one considers the heat diffusion through a one-dimensional piece of matter. This approximation is supported by the small thickness of the inner crust compared to the size of the core; b) the inner crust is divided in layers of constant thermal diffusivity. The diffusion time through a layer of thickness x_i and diffusivity D_i is calculated by the relation $t_i = \gamma \frac{x_i^2}{D_i}$,[18] where the factor γ, which depends on the boundary conditions of the problem, is taken equal to $4/\pi^2$;[5] c) the total diffusion time across the crust is obtained by summing up the contributions of the layers, i.e.,

$$t_{diff} = \gamma \sum_i \frac{x_i^2}{D_i}. \tag{2}$$

In the equation above the thermal diffusivity depends on density and temperature, $D_i = D(\rho(R_i), T(R_i))$, where R_i is the position of the layer i. The position corresponding to each cell can be found by solving the Tolman-Oppenheimer-Volkov (TOV) equations, which provides the density profile of the star. In the present calculations we use the solution of TOV equations corresponding to the following equations of state:[19] Baym-Pethick-Sutherland[20] for the outer crust, Negele-Vautherin[9] for the inner crust and Glendenning-Moszkowski[21] for the core. From the solution of the TOV equations one extracts the radii R_i corresponding to the densities of the cells. Then, doing a linear interpolation, we determine the size x_i of the layers considered for each cell.

The diffusivity depends also on the temperature profile. Numerical simulations indicate that before the core-crust thermalisation the temperature is increasing from about T=0.1 MeV to about T=0.2-0.3 MeV when one goes from the outer part to the inner part of the crust. Since the inner part zones of the inner crust have large diffusivities, they contribute less to the cooling time compared to the outermost zones. Therefore, following Ref. 5, we shall consider for all layers a flat temperature profile equal to T=0.1 MeV. The diffusion time across the inner crust obtained for this value of the temperature is shown in Figure 5. The most striking thing we can notice is the critical dependence of the cooling time on the pairing force. Thus, for a strong pairing force the cooling time is about 12 years. The largest contributions come from the outermost zones, as noticed also in Ref.[5] Concerning the effect of the clusters, one can see that is rather small for this temperature. In the case of the weak pairing force, the cooling time is increasing by about a factor two compared to the strong force. Moreover, if the neutron superfluidity is ignored completely, the cooling time is increasing to about 90 years.

5. Summary and Conclusions

We have discussed the effects of neutron superfluidity on thermal properties of inner crust of neutron stars. To analyse if the present uncertainty in the intensity of pairing correlations in infinite neutron matter has observational consequences on

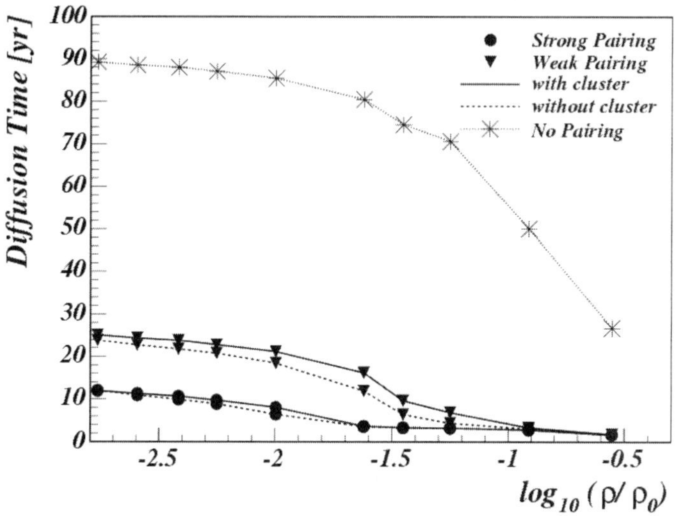

Fig. 4. The diffusion time across the inner crust. The notations are the same as in Figure 2.

neutron stars cooling, we have considered two pairing forces. They have been fixed to reproduce the pairing properties of infinite neutron matter given either by a Gogny force or by microscopic calculations which take into acocunt polarisation effects. For the latter we considered a maximum pairing gap in neutron matter equal to 1 MeV. With these two pairing forces we have estimated the heat diffusion time through the inner crust in the case of a rapid cooling of the core. These calculations show that the integrated diffusion time, which give the thermalisation time of the crust, depends drastically on the scenario used for the intensity of pairing correlations in infinite neutron matter. Thus, the thermalisation time increases by a factor of two if the maximum gap in infinite neutron matter is reduced by the in-medium effects by a factor of three compared to BCS approximation. This result show that the thermalisation time of the inner crust matter could put firm constraints on the neutron matter superfluidity.

The cooling times calculated in this section are based on the assumption of a flat temperature across the inner crust. This is a rather drastic approximation, especially for the scenario of a weak pairing force when. More realistic calculations of the cooling time should be based on dynamical solutions of the heat equations.

References

1. D. G. Yakovlev, *Ann. Rev. Astron. Astrophys.* **42**, 169 (2004).
2. G. E. Brown, K. Kubodera, D. Page and P. M. Pizzochero, *Phys. Rev. D* **37**, 2042 (1988).
3. J. M. Lattimer, K. A. Van Riper, Madappa Prakash and Manju Prakash, *Ap. J.* **425**, 802 (1994).

4. F. Barranco, R. Broglia, H. Esbensen and E. Vigezzi, *Phys. Rev. C* **59**, 1257 (1998).
5. P. M. Pizzochero, F. Barranco, E. Vigezzi and R. A. Broglia, *Ap. J.* **569**, 381 (2002).
6. N. Sandulescu, N. Van Giai and R. J. Liotta, *Phys. Rev. C* **69**, 045802 (2004).
7. N. Sandulescu, *Phys. Rev. C* **70**, 025801 (2004).
8. C. Monrozeau, J. Margueron and N. Sandulescu, *Phys. Rev. C* **75**, 065807 (2007).
9. J. W. Negele and D. Vautherin, *Nucl. Phys. A* **207**, 298 (1973).
10. G. F. Bertsch and H. Esbensen, *Ann. Phys. (N.Y.)* **209**, 327 (1991).
11. E. Chabanat, P. Bonche, P. Haensel, J. Meyer and R. Schaeffer, *Nucl. Phys. A* **627**, 710 (1997).
12. U. Lombardo and H.-J. Schulze, *Lectures Notes in Physics* **578**, 30 (2001).
13. J. Decharge and D. Gogny, *Phys. Rev. C* **21**, 1568 (1980).
14. C. Shen, U. Lombardo, P. Schuck, W. Zuo and N. Sandulescu, *Phys. Rev. C* **67**, 061302 (2003).
15. E. Khan, N. Sandulescu and N. Van Giai, *Phys. Rev. C* **71**, 042801 (2005).
16. N. Sandulescu, in "Collective Motion and Phase Transitions in Nuclear Systems," ed. A. A. Raduta (World Scientific, 2007); nucl-th/0612047.
17. N. Itoh, Y. Kohyama, N. Matsumoto and M. Seki, *Ap. J.* **285**, 758 (1984).
18. L. Landau and E. Lifshitz, *Physique Statistique* (1984).
19. I. Vidana, private comunication
20. G. Baym, C. Pethick and P. Sutherland, *Ap. J.* **170**, 299 (1971).
21. N. K. Glendenning and S. A. Moszkowski, *Phys. Rev. Lett.* **67**, 2414 (1991).

COLLECTIVE EXCITATIONS: FROM EXOTIC NUCLEI TO THE CRUST OF NEUTRON STARS

E. KHAN,* M. GRASSO and J. MARGUERON

Institut de Physique Nucléaire, Université Paris-Sud, IN2P3-CNRS, Orsay, F-91406, France

**E-mail: khan@ipno.in2p3.fr*

Low-energy quadrupole excitations are analyzed in nuclear Fermi systems. The crust of neutron stars provides a unique framework for studying the evolution of low-lying modes from neutron-rich nuclei to a pure neutron gas. We modelize the crust with Wigner-Seitz cells and focus our attention on the Zr and Sn cells with baryon densities equal to 0.02 fm^{-3} and 0.04 fm^{-3}, respectively. It is shown that the excitations with low multipolarities are concentrated almost entirely in one strongly collective mode which exhausts a very large fraction of the energy-weighted sum rule. Since these collective modes are located at very low energies compared to the giant resonances in standard nuclei, they may affect significantly the specific heat of baryonic inner crust matter of neutron stars.

Keywords: Collective excitations; Exotic nuclei; Neutron stars.

1. Introduction

Low-lying excitation modes can be studied in several many-body Fermi systems like ultra-cold trapped atomic gases,[1–3] metallic clusters[4] or atomic nuclei.[5] It has been recently pointed out that a general picture can be adopted to treat resonant or strongly interacting dilute Fermi systems due to the universality of their behavior, for instance atomic gases near a Feshbach resonance or neutron matter at densities lower than the neutron drip density (typical densities in the outer crust of a neutron star).[6] Systems in weakly interacting regime (atomic nuclei and nuclear systems in the inner crust of a neutron star where the baryon density is higher than the neutron drip density) can be described by mean field approaches. Fully microscopic approaches are the most accurate methods to properly treat many-body systems in a quantum framework.

Neutron stars provide a unique and fascinating scenario where the evolution from a microscopic to a hydrodynamic description can be analyzed. In the outer crust of these astrophysical objects, exotic nuclei form a Coulomb lattice in the presence of an electron gas.[9] When one goes deeply in the interior of the crust, density increases and nuclei are more and more neutron-rich up to the drip density $\rho \sim 4 \cdot 10^{11}$ g/cm^3.[10] Beyond this limit neutrons drip out and very exotic structures of clusters surrounded by relativistic electrons and superfluid neutrons form a Coulomb lattice (inner crust). These structures can be regarded as a bridge between a finite nucleus and an infinite nuclear gas. Low-lying collective excitations in the inner crust of neutron stars have been microscopically analyzed in Ref. 11. By modelizing the Coulomb lattice with spherical Wigner-Seitz (WS) cells,[12] each cell has been studied as independent of all the others within the Hartree-Fock-Bogoliubov[5,13,14] + quasiparticle random-phase approximation[2,3,15] (HFB + QRPA). The role of the low-lying collective modes, the supergiant resonances, has been for the first time underlined in Ref. 11 in connection with the evaluation of the specific heat and the cooling time[16] of a star. The WS cells which mostly contribute to the cooling time have a low-density neutron gas and are spherical.[17] It should be noted that these collective excitations can also have an impact on the propagation of neutrino in neutron stars.[18]

Apart from the evolution from finite to infinite systems, inner crust WS cells provide also the opportunity to analyze strong asymmetries, i.e. very high excesses of neutrons with respect to protons. The challenge of designing the nucleon-nucleon interaction for very neutron-rich systems is still an open question.[19] It is therefore necessary to develop a method involving experimental data on neutron-rich nuclei. A procedure to constrain the nucleon-nucleon interaction, aiming to accurately predict the properties of WS cells, is proposed. In this spirit these systems can be viewed as extreme extrapolations of very neutron-rich nuclei.

In the last years, several projects for a new generation of radioactive nuclear beam facilities have been started and are presently in progress.[20] These facilities will allow to explore more systematically regions closer to the drip lines. In the next decade more and more exotic nuclei will be produced and their properties will be experimentally accessible; the impact of strong neutron excess on low-lying excitations will be investigated. The aim of this work is to study the evolution of low-lying excitations in nuclear Fermi systems with respect to neutron excess, passing from exotic nuclei to pure neutron gas via a WS cell. We will treat the low-lying modes of neutron-rich isotopes as well as the supergiant resonances of a WS cell with a fully microscopic and self-consistent approach.

In all the calculations mentioned above the specific heat was evaluated by considering only non-interacting quasiparticles states. However, the specific heat can be also strongly affected by the collective modes created by the residual interaction between the quasiparticles, especially if these modes appear at low-excitation energies.

2. HFB and QRPA Approaches for WS Cells

Free neutrons are not distinguished from those which are bound. One can show that the most favorable configuration in the cell is obtained when protons are clustered at its center. We can in this sense picture it as composed of a central Zr cluster surrounded by an interacting gas of superfluid neutrons.

For both the nucleus and the WS cell we perform microscopic HFB + QRPA calculations at zero temperature and in spherical symmetry. We use the Skyrme interaction SLy4[21] for the mean field and a zero-range density dependent interaction for pairing. As in Ref. 11 the parameters of the pairing interaction are chosen in order to have the same gap as obtained with a Gogny force.[22]

We consider first ground state properties of the two systems. Within the HFB approach we have calculated the neutron and proton density profiles,

$$\rho_q(r) = \sum_{\nu l j} \frac{2j+1}{4\pi} v^2_{\nu l j q}(r), \tag{1}$$

where q stands for n (neutrons) or p (protons) and ν, l, j are the quantum numbers associated to the wave function v; v indicates the lower component of the HFB quasiparticle wave function, solution of the equations,

$$\begin{aligned} H(r)u_{\nu l j q}(r) + \Delta(r)v_{\nu l j q}(r) &= E_{\nu l j q}u_{\nu l j q}(r), \\ \Delta(r)u_{\nu l j q}(r) - H(r)v_{\nu l j q}(r) &= E_{\nu l j q}v_{\nu l j q}(r). \end{aligned} \tag{2}$$

In Eqs. (2), $H(r)$ contains the kinetic term and the Hartree-Fock mean field while $\Delta(r)$ represents the pairing field; u and v are the upper and lower components of the wave function and E the corresponding quasiparticle energy. We have imposed Dirichlet (Von Neuman) boundary conditions for the even-(odd-)l wave functions. We refer the reader to the contribution of N. Sandulescu for the HFB results providing the neutron and proton densities profiles of WS cells.

The QRPA equations, which correspond to the small-amplitude limit of the time-dependent HFB equations, are solved in coordinate representation. We integrate the Bethe-Salpeter equation for the QRPA Green function G,

$$G = (1 - G_0 V)^{-1} G_0, \tag{3}$$

where G_0 is the unperturbed Green function and V the residual interaction, and we evaluate the response function S,

$$S(\omega) = -\frac{1}{\pi} Im \int d\vec{r} d\vec{r'} F^*(\vec{r}) G(\vec{r}, \vec{r'}; \omega) F(\vec{r'}), \tag{4}$$

where F is the excitation operator. Only particle-hole components of G and F appear in Eq. (4).[15]

3. Supergiant Resonances in the WS Cells

Within the HFB+QRPA formalism presented above we first calculate the monopole neutron response in the cell ^{1800}Sn. The unperturbed HFB response, built by non-interacting quasiparticle states, and the QRPA response are shown in Figure 1. As can be clearly seen, when the residual interaction is introduced among the quasiparticles the unperturbed spectrum, distributed over a large energy region, is gathered almost entirely in a peak located at about 3 MeV. This mode is extremely collective[11] and is denoted as supergiant resonance. This underlines the fact that this WS

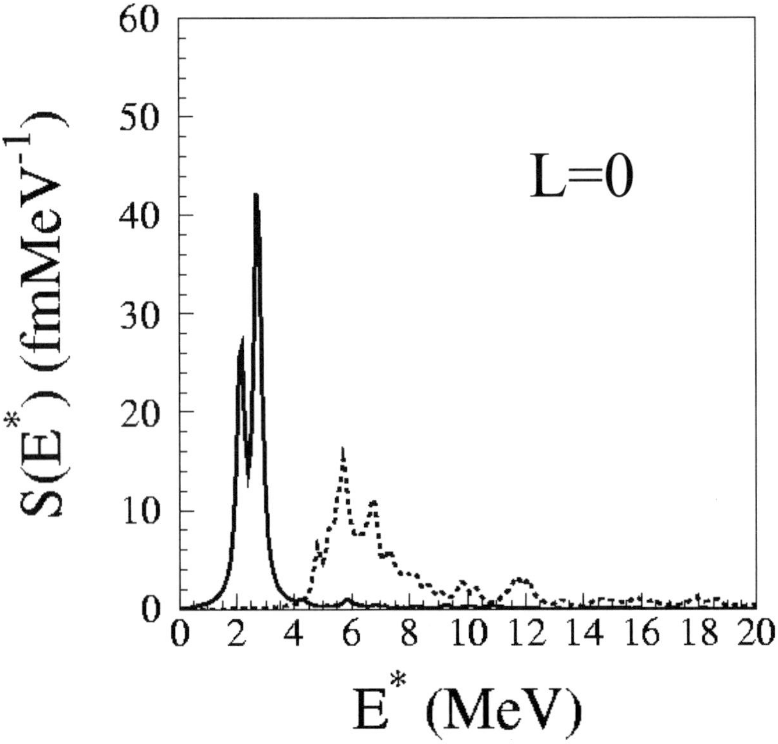

Fig. 1. Neutron monopole strength distributions for ^{1800}Sn. The solid and dashed lines are the QRPA and unperturbed strengths, respectively.

cell cannot be simply considered as a giant nucleus. As discussed below, the reason is that in this cell the collective dynamics of the neutron gas dominates over the cluster contribution. Apart from the monopole mode discussed above, we have also investigated the response of the WS cell to the dipole and quadrupole excitations: they display similar features, leading to the same qualitative conclusions.

This response can be compared with the one obtained in very neutron-rich nuclei. Let us take the example of ^{122}Zr and the ^{1500}Zr WS cell. In the neutron quadrupole spectrum of ^{122}Zr we identify a low-lying 2^+ state and a giant resonance region at higher energies. Due to the residual interaction, the low-lying 2^+ state is split into two peaks at about 2 MeV and 3.2 MeV. In comparison the strength of the cell is concentrated around a very collective low-lying state of about 3.5 MeV (supergiant mode). It should be noted that the strength of the giant resonance is negligible with respect to the strength associated to the low-lying mode. This is due to the fact that the giant resonance is built with the contribution of particles bound to the central cluster while a huge number of particles belonging to both the cluster and the free gas participate to the low-energy supergiant mode. To analyze the respective contributions to the mode, we have evaluated the transition density associated to the supergiant resonance in the cell, $\delta\rho(r) = \rho(r) - \rho_0(r)$, where ρ_0 represents the

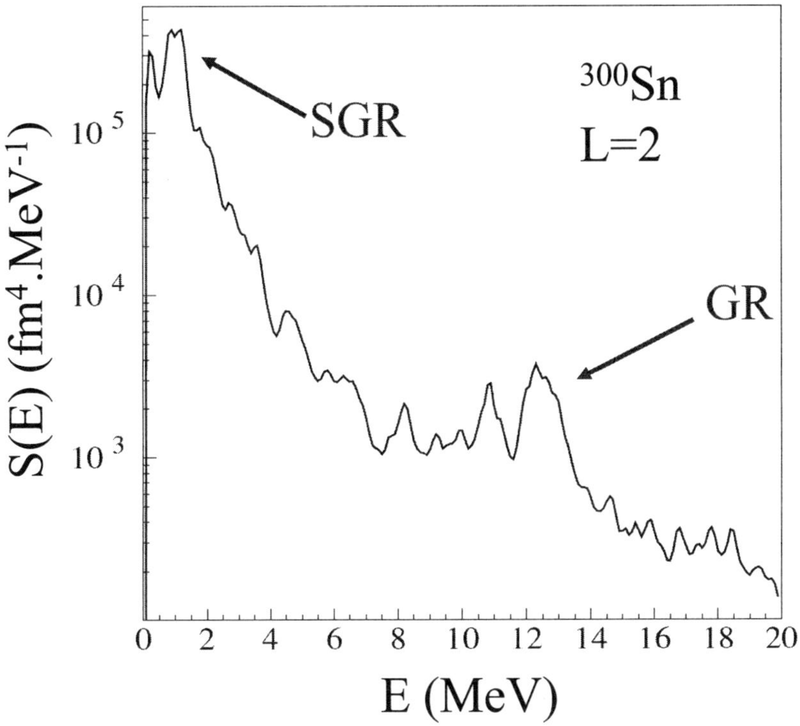

Fig. 2. Neutron quadrupole QRPA strength distributions for ^{300}Sn.

equilibrium density and ρ the density of the excited system. The analysis of the radial profile of the transition density gives indications on the location of the particles which mainly participate to the excitation. Apart from a strong contribution in the external region where the free neutron gas is located, there is an important contribution at the cluster surface where the transition density is peaked. This means that the supergiant excitation is partly composed of two-quasiparticle configurations localized at the cluster surface. The internal part of neutron and proton transition densities are very similar to typical nuclei transition densities.[15]

In order to trace the evolution of the low-lying excitations from neutron-rich nuclei to WS cells, QRPA calculations have been performed for systems located in between, that is beyond the drip-line. Fig. 2 shows the quadrupole response in ^{300}Sn. One observe the giant resonance, and already a strong low lying state. The magnitude of this state is due, as mentioned above, to the neutron of the continuum, which can have large spatial delocalisation. Hence supergiant resonances cannot be considered as the limit of low energy components of giant resonances but are rather generated by the neutron of the continuum. However their energy position depends on the single quasiparticle spectrum. As a conclusion, we have identified two contributions to the supergiant mode: the gas and the cluster surface.

On this basis, the following method should be observed in order to predict low-energy modes in the WS cells. The nucleon-nucleon interaction should be designed in order to reproduce the measured transition densities of the analogous neutron-rich nuclei (^{122}Zr in this case or lighter Zr isotopes). The combination between Coulomb excitation and proton scattering has proven to be an accurate tool to probe transition densities in neutron-rich nuclei, using the present HFB+QRPA approach ([23] and references therein). Once validated, this nucleon-nucleon interaction can be used to perform microscopic calculations in the WS cells. Note that this method is similar to that proposed to deduce the nuclear matter incompressibility from the measurements of the giant monopole resonances in nuclei.[24]

The contribution due to the neutron gas in the external region suggests that a neutron gas filling the same volume of the cell should display a very similar quadrupole response. We have actually verified that decreasing the proton number by one order of magnitude the resulting neutron response has a peak shifted of only $\sim$ 100 keV with respect to the low-lying peak in the cell.

One can wonder whether the contribution to the collective motion due to the free neutron gas may be eventually described with a semiclassical picture by using the hydrodynamic model. This description can be performed within the limit of validity of the semiclassical description which is based on the local-density approximation (LDA). One should thus verify whether the free gas can be treated locally as if it was homogeneous, i.e. if finite-size effects are negligible. Hence, we should check whether the coherence length ξ_0[13] ($\xi_0 = \hbar v_F/\pi\Delta$, where v_F is the Fermi velocity and Δ the pairing gap) is much smaller than the size of the system. The Fermi velocity can be evaluated in the case of a pure neutron gas, $v_F = \pi\hbar/m(3\rho/\pi)^{1/3}$, where the spin degeneracy has been taken into account, m is the neutron mass and ρ the density of the free gas. The pairing gap Δ obtained within the HFB approach is 3 MeV, which gives $\xi_0 \sim 4$ fm. Since, in the case of ^{1500}Zr, the radius R of the WS cell is 19.6 fm, the condition $\xi_0 << R$ is not strongly satisfied. Indeed, the energy evaluated with the hydrodynamic linear dispersion law for a homogeneous Fermi superfluid,[25] $E = v_F p/\sqrt{3}$, is equal to 7.5 MeV. This is in bad agreement with the microscopic QRPA result. One should be aware of this limit of the hydrodynamic model when evaluating the energy of the low-energy mode in a WS cell. The model is less and less adequate when one goes deeper in the interior of the crust of a neutron star, i.e. when the radius of the WS cells becomes smaller and smaller. On the contrary, in the ^{1800}Sn cell ($\rho = 0.018$ fm^{-3}, $R = 27.6$ fm) the coherence length is 3.7 fm and the energy of the low-lying mode is 4 MeV. The QRPA microscopic calculation gives a peak located at about 3 MeV.[11] The hydrodynamic result is in better agreement with the microscopic energy in this latter case as the condition $\xi_0 << R$ is now better satisfied.

4. Conclusions

We have discussed low-energy quadrupole modes in two Fermi systems, exotic nuclei and WS cells which modelize the nuclear crystal in the neutron star crust. The supergiant mode of the WS cells is built on two contributions, one due to neutrons in the cluster surface and the other due to neutrons in the external free gas. To predict the supergiant mode of the WS cell, we propose a method based on the experimental analysis of the low-lying 2^+ mode in neutron-rich Zr and Sn isotopes and on designing the nucleon-nucleon interaction for neutron-rich systems. Measurements of 2^+ modes for nuclei with extended neutron skins are therefore relevant for the prediction of neutron stars cooling properties. The measurement of such exotic nuclei properties will be achieved by the new generation of radioactive beam facilities available in the next decades.

References

1. J. Kinast *et al.*, *Phys. Rev. Lett.* **92**, 150402 (2004); M. Bartenstein *et al.*, *Phys. Rev. Lett.* **92**, 203201 (2004); S. Stringari, *Europhys. Lett.* **65**, 749 (2004).
2. G. M. Bruun and B. R. Mottelson, *Phys. Rev. Lett.* **87**, 270403 (2001).
3. M. Grasso, E. Khan and M. Urban, *Phys. Rev. A* **72**, 043617 (2005).
4. A. Pohl, P.-G. Reinhard and E. Suraud, *Phys. Rev. Lett.* **84**, 5090 (2000); V. O. Nesterenko *et al.*, *Phys. Rev. Lett.* **85**, 3141 (2000).
5. P. Ring and P. Schuck, *The Nuclear Many-Body Problem* (Springer-Verlag, Berlin, 1980).
6. H. Heiselberg, *Phys. Rev. A* **63**, 043606 (2001); A. Schwenk and C. J. Pethick, *Phys. Rev. Lett.* **95**, 160401 (2005).
7. A. L. Fetter and J. D. Walecka, *Theoretical Mechanics of Particles and Continua* (McGraw-Hill, New York, 1980).
8. M. Grasso and M. Urban, *Phys. Rev. A* **68**, 033610 (2003).
9. E. E. Salpeter, *Astrophys. J.* **134**, 669 (1961).
10. G. Baym, C. Pethick and P. Sutherland, *Astrophys. J.* **170**, 299 (1971).
11. E. Khan, N. Sandulescu and Nguyen Van Giai, *Phys. Rev. C* **71**, 042801(R) (2005).
12. J. W. Negele and D. Vautherin, *Nucl. Phys. A***207**, 298 (1973).
13. P. G. de Gennes, *Superconductivity of Metals and Alloys* (Addison-Wesley, Reading, MA, 1989).
14. J. Dobaczewski, H. Flocard and J. Treiner, *Nucl. Phys. A* **422**, 103 (1984); M. Grasso, N. Sandulescu, N. Van Giai and R. J. Liotta, *Phys. Rev C* **64**, 064321 (2001).
15. E. Khan, N. Sandulescu, M. Grasso and Nguyen Van Giai, *Phys. Rev. C* **66**, 024309 (2002).
16. J. M. Lattimer, K. A. Van Riper, M. Prakash and M. Prakash, *Astro. J.* **425**, 802 (1994).
17. P. M. Pizzochero, F. Barranco, E. Vigezzi and R. A. Broglia, *Astrophys. J.* **569**, 381 (2002).
18. J. Margueron, J. Navarro and P. Blottiau, *Phys. Rev. C* **70**, 028801 (2004).
19. T. Otsuka *et al.*, *Phys. Rev. Lett.* **95**, 232502 (2005).
20. Proceedings of the 17th International Conference on Cyclotrons and their Applications (October 2004), Tokyo.
21. E. Chabanat *et al.*, *Nucl. Phys. A* **635**, 231 (1998).
22. G. F. Bertsch and H. Esbensen, *Ann. Phys. (NY)* **209**, 327 (1991); E. Garrido, P. Sarriguren, E. Moya de Guerra and P. Schuck, *Phys. Rev. C* **60**, 064312 (1999).
23. E. Becheva *et al.*, *Phys. Rev. Lett.* **96**, 012501 (2006).
24. J. P. Blaizot, *Phys. Rep.* **64**, 171 (1980).
25. A. Bulgac, *Phys. Rev. Lett.* **95**, 140403 (2005).

MONTE CARLO SIMULATION OF THE NUCLEAR MEDIUM: FERMI GASES, NUCLEI AND THE ROLE OF PAULI POTENTIALS

M. ÁNGELES PÉREZ-GARCÍA

Department of Fundamental Physics,
University of Salamanca, Spain
and
Instituto Universitario de Física Fundamental y Matemáticas,
Facultad de Ciencias, Plaza de la Merced s/n,
Salamanca, E-37008, Spain
**E-mail: mperezga@usal.es*
www.usal.es

The role of Pauli potentials in the semiclassical simulation of Fermi gases at low temperatures is investigated. An alternative Pauli potential to the usual bivariate Gaussian form by Dorso *et al*[5] is proposed. This new Pauli potential allows for a simultaneous good reproduction of not only the kinetic energy per particle but also the momentum distribution and the two-body correlation function. The reproduction of the binding energies in finite nuclei in the low and medium mass range is also analyzed. What is found is that given a reasonable short-range atractive nuclear interaction one can include correlation effects in a suitable chosen density dependent Pauli potential.

Keywords: Fermi gas; Pauli potential; Many-body simulations; Nuclear pasta.

1. Formalism

Nuclear many-body simulations are a useful tool to study the relevant properties of the nuclear medium in the thermodynamic conditions arising in matter in the aftermath of a Supernova event or in Neutron Stars. Examples of this are, for instance, nuclear *pastas*[1,2] at densities in the range $0.01\rho_0 \le \rho \le 0.5\rho_0$ ($\rho_0 = 0.148\, fm^{-3}$) and temperatures of decens of MeV or in heavy ion collisions.[3] This type of simulations based on Monte Carlo or Molecular Dynamics techniques allow for a dynamical description of the nuclear medium usually by using an effective interaction hamiltonian in a semiclassical treatment. In fermionic systems the genuine antisymmetrization of the wave function is considered trough the inclusion of a Pauli potential. Pioneering works on this line include those of Wilets *et al.*[4] In this work the hamiltonian used to study the low temperature nucleon systems consists of a kinetic energy term and a Pauli effective potential (V_{Pauli}).

$$H = \sum_{i=1}^{N} \frac{\mathbf{p}_i^2}{2m_N} + \sum_{i=1,j>i}^{A} V_{Pauli}(r_{ij}, p_{ij})\delta_{\tau_i\tau_j}\delta_{\sigma_i\sigma_j}, \tag{1}$$

where $\delta_{\tau_i\tau_j}$ ($\delta_{\sigma_i\sigma_j}$) is the Kronecker's delta for the nucleon isospin (spin) third-component. $\mathbf{p}_i$ is the 3-momentum of i-th nucleon and $r_{ij} = |\mathbf{r}_i - \mathbf{r}_j|$ ($p_{ij} = |\mathbf{p}_i - \mathbf{p}_j|$) the relative distance (momentum) of the i-th and j-th nucleons.

We will consider, for the sake of comparison, two ways to implement this potential.

i) A Gaussian form introduced by Dorso *et al.*,[5]

$$V_{Pauli}(r_{ij}, p_{ij}) = V_S \ \exp\left(-\frac{r_{ij}^2}{2q_0^2} - \frac{p_{ij}^2}{2p_0^2}\right), \tag{2}$$

Here p_0 and q_0 are momentum and length scales related to the excluded phase-space volume that is used to mimic fermionic correlations and V_S is the Pauli potential strength. All three parameters have been adjusted to reproduce only the kinetic energy of a low temperature Fermi gas.

ii) A new form proposed, based on spatial and momentum-dependent, two-body terms of the following form[6]

$$V_{Pauli}^{new}(r_{ij}, p_{ij}) = V_q \exp(-r_{ij}/q_0) + V_p \exp(-p_{ij}/p_0) + V_\Theta \, \Theta_\eta(q_i) \,, \tag{3}$$

where $q_i = |\mathbf{p}_i|/p_{\mathrm{F}}$, and Θ_η is a smeared Heaviside-step function, $\Theta_\eta(q) \equiv \frac{1}{1+\exp[-\eta(q^2-1)]}$ and $\Theta_\eta(q) \longrightarrow \Theta(q)$ when η is sufficiently big. The parameters of the new Pauli potential V_q, V_p, V_Θ and q_0, p_0, η will be adjusted to reproduce the kinetic energy per particle and both the momentum distribution and two-body correlation function of a low-temperature Fermi gas. The first and second terms in the potential penalize two particles with the same quantum numbers coming together either in space or momentum. This retains the essence of the fermionic wave function given by the Slater determinant. The third term forbids any particle from having a momentum significantly larger than the Fermi momentum.

In both cases the potential parameters will depend on the density of the system and this will be crucial in reproducing experimental binding energies when simulating low and medium mass nuclei as will be shown later. The values for the parameters at saturation density ρ_0 and reduced temperature $\tau = T/T_F = 0.05$ are given in Table 1.

Table 1. Pauli potential parameters.

Pauli potential	Potential strength (MeV)	q_0 (fm)	p_0 (MeV/c)
Dorso *et al*[5]	$V_S = 207$	1.644	120
This work	$V_q = 13.517$, $Vp = 1.260$, $V_\Theta = 3.560$ ($\eta = 30$)	0.66	49.03

2. Results

The simulations are performed in a NVT system with $\tau = T/T_F$, and N fermions in a cubic box of volume $V = L^3 = N/\rho$. Then, using the Metropolis algorithm the system is thermalized until the stage where configurations are sampled in order to calculate the statistical averages for the magnitudes discussed below.

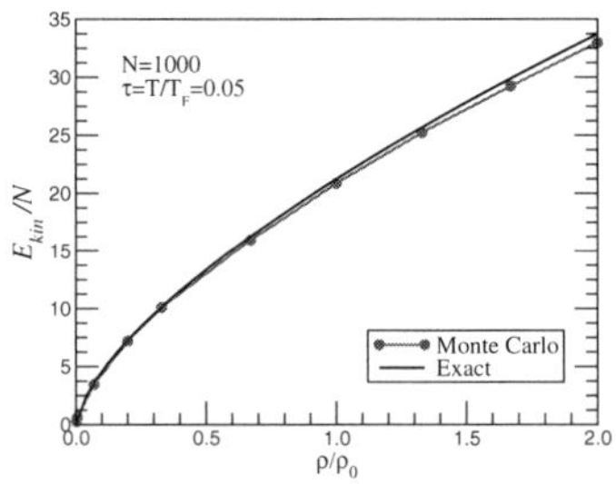

Fig. 1. Energy per particle of a Fermi gas simulated with $N = 1000$ particles at $T = 0.05\ T_F$ as a function of density.

In Fig. 1 the red line shows the kinetic energy per particle for a Fermi gas system with $N = 1000$ particles at $\tau = 0.05$ as a function of density calculated using the new form of the Pauli potential Eq. (3). Also plot with a black line is the exact result. We can see that there is a good reproduction of the kinetic energy.

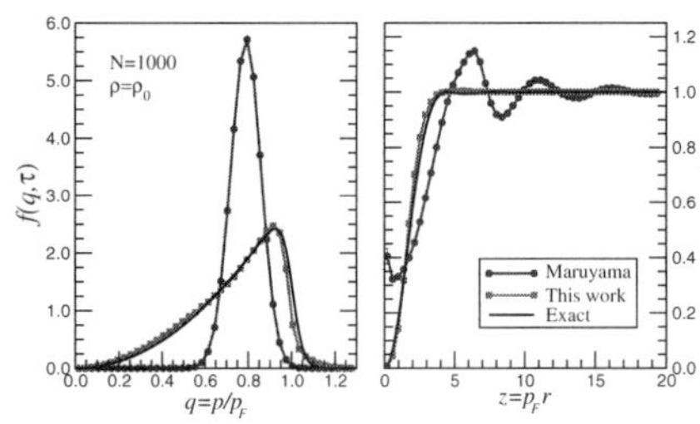

Fig. 2. Momentum distribution function and two-body correlation function of a Fermi gas. See text for details.

In Fig. 2 we can see on the left side the momentum distribution function $f(q, \tau)$ and, on the right side, the two-body correlation function $g(z, \tau)$ with $z = p_F r$. The blue curves correspond to the Dorso potential Eq. (2) and the red curves to the new Pauli potential proposed Eq. (3). Again the black line shows the exact result. We can see that a simultaneous good reproduction of both magnitudes is achieved with the alternative new potential but not with the Dorso version. Particularly the "Fermi hole" fails to be reproduced at small distances with the Dorso potential. This should be emphasized since these models are used in nuclear many-body simulations as in nuclear *pastas* as, for instance, in the work by Maruyama *et al.*[2] The velocity

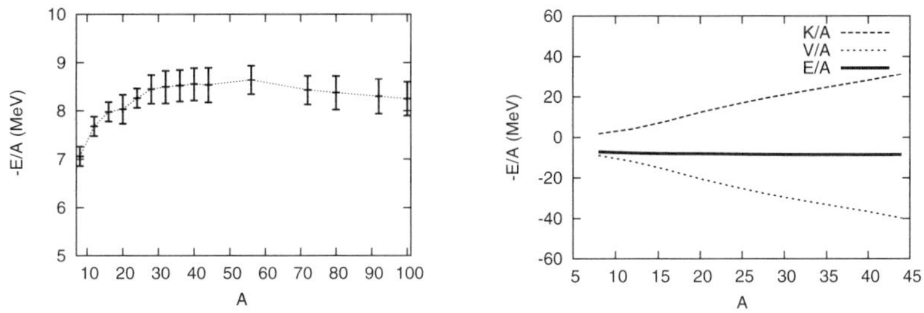

Fig. 3. Application to nuclei. (a) Binding energy. (b) Kinetic and potential contributions to the binding energy.

distribution,[6] not shown here, however peaks at lower values than the momentum distribution due to the fact that canonical and kinematical momentum are not the same quantities.[7] This is a genuine feature in this treatment with momentum dependent Pauli potentials in a hamiltonian formalism.

We now show finite nuclei simulation[8] results calculated with a simplified square-well nuclear potential with $V_{well} = -3$ MeV of width 2 fm and a core with $V_{core} = 10$ MeV and width 1 fm. Coulomb interaction is also included. In Fig. 3(a) binding energy per particle for a low to medium mass set of spin saturated symmetric nuclei of A nucleons. As can be seen in Fig. 3(b) kinetic(dashed line) and potential (dotted line) energy balance to obtain the total binding energy (solid line) per particle. The density dependence of the parameters of the Pauli potential is crucial to provide enough positive contribution to the linearly A-growing negative potential energy[8] and reproduce the experimental binding energy curve.

Acknowledgments

We acknowledge J. Piekarewicz, J. Taruna, K. Tsushima and A. Valcarce who are collaborators in this work. Partial funding has been provided by project DGI-FIS2006-05319.

References

1. G. Watanabe *et al.*, *Phys. Rev. Lett.* **94** (2005); C. J. Horowitz, M. A. Pérez-García and J. Piekarewicz, *Phys. Rev. C* **69** (2004).
2. Toshiki Maruyama *et al.*, *Phys. Rev. C* **72** (2005).
3. G. Peilert, J. Randrup, H. Stöcker and W. Greiner, *Phys. Lett. B* **260** (1991).
4. L. Wilets, E. M. Henley, M. Kraft and A. D. MacKellar, *Nucl. Phys. A* **282** (1977).
5. C. Dorso, S. Duarte and J. Randrup, *Phys. Lett. B* **188** (1987).
6. J. Taruna, J. Piekarewicz and M. A. Pérez-García, arXiv:nucl-th/0702086.
7. J. J. Neumann and G. I. Fai, *Phys. Lett. B* **329** (1994).
8. M. A. Pérez-García, K. Tsushima and A. Valcarce, arXiv:nucl-th/0706.0958, arXiv:nucl-th/0707.1951

LOW-DENSITY INSTABILITIES IN RELATIVISTIC HADRONIC MODELS

C. PROVIDÊNCIA,* L. BRITO and A. M. SANTOS

Centro de Física Teórica, Department of Physics, University of Coimbra, P-3004-516, Portugal
**E-mail: cp@teor.fis.uc.pt*

S. S. AVANCINI and D. P. MENEZES

CFM, Departamento de Física, Universidade Federal de Santa Catarina, Florianópolis - SC - CP. 476 - CEP 88.040 - 900 - Brazil

Dynamical instability modes and thermodynamical instabilities of low-density asymmetric nuclear matter (ANM) neutralised by electrons as found in supernova core and neutron star crust are studied in the framework of relativistic mean-field hadron models with the inclusion of electron and photon fields. Both models with constant and density dependent coupling constants are considered. It is shown that the Coulomb field quenches large structure formation but has little effect on medium and small-size fluctuations. Implications on the crust of compact stars are discussed.

Keywords: Asymmetric nuclear matter; Stellar matter; Distillation effect.

1. Introduction

The instabilities and phase transitions appearing in asymmetric nuclear matter (ANM) are important quantities in the understanding of the physics underlying isospin distillation, multifragmentation and fractionation effects. Experimental results of multifragmentation in relativistic heavy ion collisions are consistent with equilibration of the excited systems prior to decay and, therefore, justify a statistical and thermodynamical treatment of the reaction.

In the present paper we study the region of instability at subsaturation densities both from a thermodynamical and a dynamical point of view. We will determine the model and temperature dependence of the spinodal curve and discuss the isospin fractionation effect, the distillation/antidistillation effect and will calculate physical quantities of astrophysical interest such as the density of the inner edge of the crust. Density dependent relativistic models (DDRM) and non linear Walecka models (NLWM) are compared and astrophysical implications are discussed.

2. Relativistic Mean-Field Lagrangian for *npe* Matter

In order to describe stellar matter at low density it is important to include electrons and take into account explicitly the electromagnetic interaction. The lagrangian density is given by $\mathcal{L} = \mathcal{L}_N + \mathcal{L}_{mesons} + \mathcal{L}_\gamma + \mathcal{L}_e$ including the electron contribution. The nucleon and electron contributions are given by

$$\mathcal{L}_N = \bar{\psi}\left[\gamma_\mu D^\mu - (M - g_s\phi)\right]\psi,$$

$$\mathcal{L}_e = \bar{\psi}_e\left[\gamma_\mu\left(i\partial^\mu + eA^\mu\right) - m_e\right]\psi_e.$$

where $D^\mu = i\partial^\mu - g_v V^\mu - \frac{g_\rho}{2}\boldsymbol{\tau}\cdot\mathbf{b}^\mu - eA^\mu\frac{1+\tau_3}{2}$. The electromagnetic contribution enters through a pure electromagnetic term, $\mathcal{L}_\gamma = -\frac{1}{4}F_{\mu\nu}F^{\mu\nu}$, and the interaction term involving protons and electrons and the electromagnetic field. The meson lagrangian density includes the σ, ω and ρ mesons represented respectively by the fields ϕ, V^μ and $\mathbf{b}^\mu$.[1] We will consider the NL3 parametrisation of the NLWM[2] as well as the TW[3] parametrisation of the DDRM.[4]

3. Thermodynamical Stability Conditions

We define the free energy density of the system in terms of the densities of protons and neutrons

$$\mathcal{F}(T,\rho_i) = -p(T,\mu_i) + \sum_{i=p,n}\mu_i\rho_i, \quad \text{with } \mu_i = \left(\frac{\partial\mathcal{F}(T,\rho_j)}{\partial\rho_i}\right)_{T,\rho_j\neq\rho_i}.$$

The system is stable against separation into two phases if the free energy density $\mathcal{F}$, at constant volume and temperature, is a convex function of ρ_p and ρ_n, i.e. the symmetric free energy curvature matrix $\mathcal{F}$

$$\mathcal{F}_{ij} = \left(\frac{\partial^2\mathcal{F}}{\partial\rho_i\partial\rho_j}\right)_T,$$

is positive.[5] The stability conditions for ANM take the form $\mathrm{Tr}(\mathcal{F}) > 0$ and $\mathrm{Det}(\mathcal{F}) > 0$, which are equivalent to

$$\left(\frac{\partial P}{\partial\rho}\right)_{T,y_p} > 0, \qquad \left(\frac{\partial P}{\partial\rho}\right)_{T,y_p}\left(\frac{\partial\mu_p}{\partial y_p}\right)_{T,P} > 0.$$

The instability region is defined by a mixture of baryon density and concentration fluctuations and cannot be separated into mechanical and chemical instabilities.[6] The eigenvalues of the stability matrix are given by

$$\lambda_\pm = \frac{1}{2}\left(\mathrm{Tr}(\mathcal{F}) \pm \sqrt{\mathrm{Tr}(\mathcal{F})^2 - 4\mathrm{Det}(\mathcal{F})}\right),$$

with the eigenvectors $\boldsymbol{\delta\rho}_\pm$

$$\frac{\delta\rho_i^\pm}{\delta\rho_j^\pm} = \frac{\lambda_\pm - \mathcal{F}_{jj}}{\mathcal{F}_{ji}}, \qquad i,j = p,n.$$

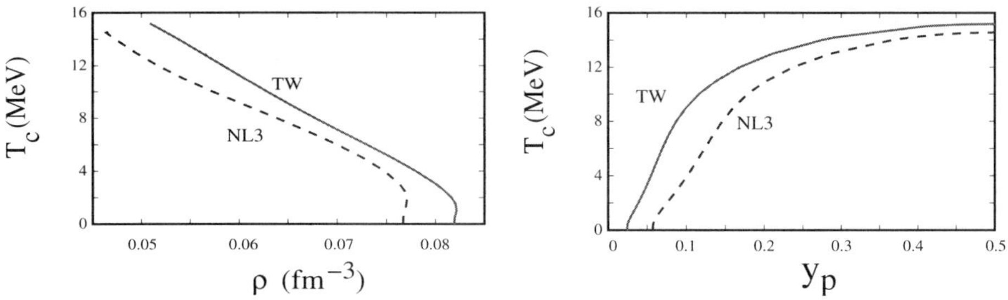

Fig. 1. Critical temperatures for NL3 and TW: a) versus density, b) versus proton fraction.

For the nuclear force the largest eigenvalue λ_+ is always positive.[6] The smallest λ_- becomes negative at subsaturation densities. The spinodal is the surface on which the eigenvalue λ_- becomes zero.

In Ref. 7 the spinodal sections were compared for several models, both NLWM and DDRM parametrisations, at different temperatures. At $T = 0$ MeV all models give similar results. This is not anymore true for finite T and large isospin asymmetry when the differences are more pronounced. In Fig. 1 we compare the critical temperatures obtained with NL3 and with TW. The line of critical points for these two models satisfies the conditions[5]

$$L_1 = \begin{vmatrix} \mathcal{F}_{pp} & \mathcal{F}_{pn} \\ \mathcal{F}_{np} & \mathcal{F}_{nn} \end{vmatrix} = 0, \qquad M_1 = \begin{vmatrix} \mathcal{F}_{pp} & \mathcal{F}_{pn} \\ \frac{\partial}{\partial \rho_p} L_1 & \frac{\partial}{\partial \rho_n} L_1 \end{vmatrix} = 0\,.$$

For the same temperature the critical density and critical proton fraction occur, respectively, at larger densities and smaller values of y_p in TW.[5] In Fig. 2 we plot the ratio $\delta\rho_p^-/\delta\rho_n^-$ as a function of the density for $T = 0$ MeV and $y_p = 0.2$ for two NLWM (NL3 and TM1), one density dependent model (TW) and a non-relativistic model (SLy230a). For $y_p = 0.2$ the ratio $\rho_p/\rho_n = 0.25$ is well below the values the ratio of the density fluctuations $\delta\rho_p^-/\delta\rho_n^-$ takes. This is the so called distillation effect: the system tries to recover isospin symmetry increasing the proton fraction of the denser partitions and decreasing in the less dense region. The density dependent TW model shows a $\delta\rho_p^-/\delta\rho_n^-$ smaller than NL3 and TM1 and therefore the reposition of isospin symmetry in liquid nuclear matter is not so efficient. It was shown in Ref. 5 that all NLWM models behave in a similar way, predicting a high distillation effect. It is seen that for Skyrme interaction SLy230a the distillation effect is not so strong and the behaviour of this model is closer to the behaviour of TW.[8]

Stellar matter is neutral and at low densities it is constituted by neutrons, protons and electrons. The electrical charge neutrality implies $\rho_e = \rho_p$. The free energy including electrons is $\mathcal{F}(\rho_n, \rho') = \mathcal{F}_N(\rho_n, \rho_p = \rho') + \mathcal{F}_e(\rho_e = \rho')$. The inclusion of a uniform background of electrons stabilises nuclear matter completely for TW and almost completely for models like NL3 and TM1.[5] However, small density fluctuations will drive the system into a liquid-gas phase transition.

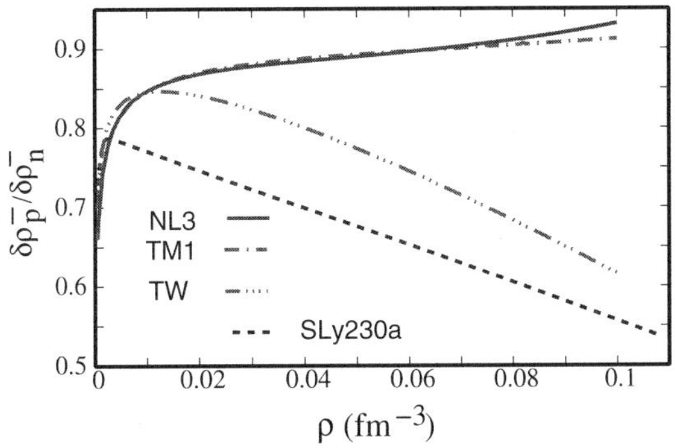

Fig. 2. Distillation effect in relativistic and non-relativistic models.

4. Dynamical Instabilities

In the present section we investigate the dynamical instability region of ANM in the presence of electrons. The dynamical response of the system will be studied in the framework of the relativistic Vlasov formalism.[9] In particular, we want to understand the role of isospin and the modification of the distillation phenomenon due to the presence of the Coulomb field and electrons.

At zero temperature, the equilibrium state of neutron-proton-electron (npe) matter is characterised by the Fermi momenta of neutrons, protons and electrons, P_{Fn}, P_{Fp}, P_{Fe} and given by the diagonal matrix $f_0(\mathbf{r},\mathbf{p}) = \mathrm{diag}\left(\Theta(P_{Fp}^2-p^2),\, \Theta(P_{Fn}^2-p^2),\, \Theta(P_{Fe}^2-p^2)\right)$. Charge neutrality relates the Fermi momenta of protons and electrons $P_{Fe} = P_{Fp}$.

In order to determine the normal modes of the system we consider a small perturbation of the system. The perturbed fields are written in terms of the equilibrium fields plus a small perturbation, $\phi = \phi_0 + \delta\phi$, $V_0 = V_0^{(0)} + \delta V_0$, $V_i = \delta V_i$, $b_0 = b_0^{(0)} + \delta b_0$, $b_i = \delta b_i$, $A_0 = \delta A_0$, $A_i = \delta A_i$. We express the perturbed distribution function, $f_i = f_{0i} + \delta f_i$, in terms of a generating function $S(\mathbf{r},\mathbf{p},t) = \mathrm{diag}\,(S_p,\, S_n,\, S_e)$, so that $\delta f_i = \{S_i, f_{0i}\}$ The time evolution of f_i is described by the Vlasov equation

$$\frac{\partial f_i}{\partial t} + \{f_i, h_i\} = 0, \qquad i = p,\, n,\, e,$$

which corresponds to the semiclassical approximation of the time dependent Hartree-Fock equation. Since we are only interested in small perturbations we linearise the Vlasov equation for protons, neutrons and electrons, and look for the solutions of these equations which determine the collective modes of the system. In terms of the generating functions they are given by

$$\frac{\partial S_e}{\partial t} + \{S_e, h_{0e}\} = \delta h_e = -e\left[\delta A_0 - \frac{\mathbf{p}\cdot\delta\mathbf{A}}{\epsilon_{0e}}\right], \tag{1}$$

$$\frac{\partial S_i}{\partial t} + \{S_i, h_{0i}\} = \delta h_i = -g_s\delta\phi\frac{M^*}{\epsilon_0} + \delta\mathcal{V}_{0i} - \frac{\mathbf{p}\cdot\delta\boldsymbol{\mathcal{V}}_i}{\epsilon_0}, \quad i = p, n, \tag{2}$$

where $\delta\mathcal{V}_{0i} = g_v \delta V_0 + \tau_i \frac{g_\rho}{2}\, \delta b_0 + e\, \frac{1+\tau_i}{2}\, \delta A_0$, $\delta\boldsymbol{\mathcal{V}}_i = g_v \delta \mathbf{V} + \tau_i \frac{g_\rho}{2}\, \delta \mathbf{b} + e\, \frac{1+\tau_i}{2}\, \delta \mathbf{A}$, and the single particle hamiltonian is $h_{0i} = \epsilon_{0i} + \mathcal{V}_{0i}^{(0)}$, for protons and neutrons and $h_{0e} = \epsilon_{0e}$ for electrons, with $\epsilon_{0i} = \sqrt{p_{Fi}^2 + m^{*2}_i}$, and $m_e^* = m_e$, $m_{n,p}^* = M^*$.

The longitudinal normal modes, with momentum $\mathbf{k}$ and frequency ω, are described by the ansatz

$$\left(S_j(\mathbf{r},\mathbf{p},t),\, \delta\phi,\, \delta B_0,\, \delta B_i \right) = \left(\mathcal{S}^j_\omega(\cos\theta),\, \delta\phi_\omega,\, \delta B^0_\omega,\, \delta B^i_\omega \right) \mathrm{e}^{i(\omega t - \mathbf{k}\cdot\mathbf{r})}\,,$$

with $j = e,\, p,\, n$ and $B = V,\, b,\, A$. θ is the angle between the particle momentum $\mathbf{p}$ and the direction of propagation of the perturbation defined by $\mathbf{k}$. This ansatz is replaced in Eqs. (1) and (2) and in the linearised equations of the fields.

Eliminating the fields from Eqs. (1) and (2) we obtain

$$\begin{pmatrix} 1 + (F^{pp} - C_A^{pp})\, L(s_p) & F^{pn}\, L(s_p) & C_A^{pe}\, L(s_p) \\ F^{np}\, L(s_n) & 1 - F^{nn}\, L(s_n) & 0 \\ C_A^{ep}\, L(s_e) & 0 & 1 - C_A^{ee}\, L(s_e) \end{pmatrix} \begin{pmatrix} A_{\omega p} \\ A_{\omega n} \\ A_{\omega e} \end{pmatrix} = 0\,, \qquad (3)$$

where the amplitudes $A_{\omega i}$ are directly related with the transition densities $\delta\rho_i = \frac{3}{2}\frac{k}{P_{Fi}} \rho_{0i} A_{\omega i}$, and $L(s_i)/2$ is the Lindhard function where $L(s_i) = 2 - s_i \ln\left[(s_i+1)/(s_i-1)\right]$. At low densities the system has unstable modes characterised by an imaginary frequency. The dynamical spinodal surface is defined by the points for which the energy of the normal mode goes to zero. In Fig. 3 we compare the dynamical spinodals obtained with NL3 and TW for $k = 11$ MeV and $k = 80$ MeV. The value $k = 80$ MeV defines, except for small corrections, the envelope of the spinodals for k values. Due to the $p - e$ infinite range interaction, proton and electron fluctuations have to be in phase at $k = 0$. The effect of electrons decreases as k increases up to $70 - 80$ MeV/c due to the $1/k^2$ behaviour of the Coulomb interaction. For values of $k > 100$ MeV/c, the *npe* results coincide with the *np* matter result, i.e. the instability region decreases with the increase in k due to the finite range of the nuclear interaction.[9] The spinodal region is strongly asymmetric at small k because of the electron stabilising effect and it is almost symmetric at large k due to the reduction of the coupling with electrons. For $k = 11$ MeV we consider the spinodals obtained in three different situations: only neutron-proton (*np*) matter excluding the Coulomb field, together with *np* matter and *npe* matter including the Coulomb interactions. While the first situation is not realistic, but allows a comparison with the thermodynamical limit, the second describes neutron-proton matter and the third stellar matter. The results for *np* matter with no Coulomb field for $k = 11$ MeV practically coincide to the thermodynamical spinodal (which corresponds to $k = 0$ MeV). For $k = 11$ MeV the spinodal for *np* matter with Coulomb is much smaller than the corresponding spinodals for *npe* matter due to the attractive force between protons and electrons; for *npe* symmetric matter ($\rho_p = \rho_n$) the two models considered, NL3 and TW, have quite similar results but differences occur for asymmetric matter, TW showing the largest asymmetries (except for small k values).

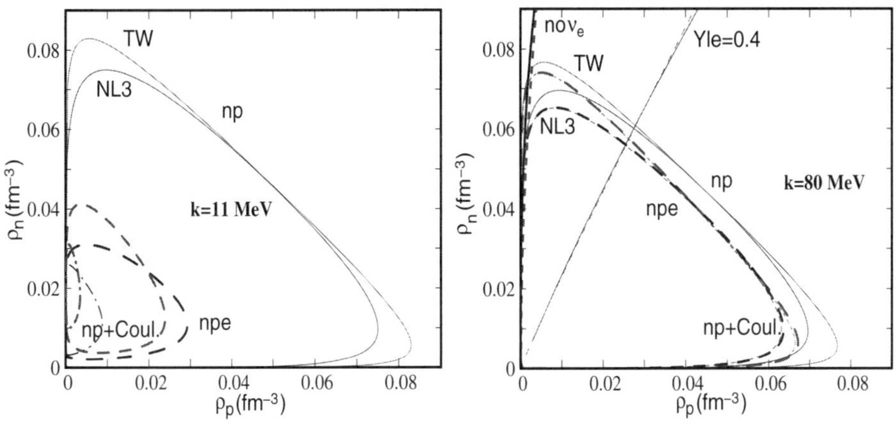

Fig. 3. *npe* matter: dynamical spinodal for $k = 11$ MeV (left) and $k = 80$ MeV (right).

In Fig. 3 for $k = 80$ MeV we include the β-equilibrium EoS for *npe* neutrino-free matter and for *npe*ν matter as in supernovae with a constant lepton fraction $Y_L = Y_e + Y_\nu = 0.4$. In fact, neutrino trapping occurs in warm stellar matter but the EoS is not very much affected by temperature. The crossing of these EoS with the spinodal tells us that there is a non-homogeneous region in the star, at low densities. The density at the inner edge of the crust for neutrino-free stellar matter is 0.050 fm^{-3} for NL3 and 0.075 fm^{-3} for TW. For the neutrino trapped matter the corresponding values 0.082 fm^{-3} for NL3 and 0.084 fm^{-3} for TW, are only an upper limit because for warm matter the spinodal is smaller. For neutrino trapped matter both models give similar results because the matter considered has a very high proton fraction, $y_p \sim 0.3$. However for neutrino free matter, the density value at the inner edge of the crust is very sensitive to the model because we are probing highly asymmetric matter where the larger differences between models arise.

The system is driven to the non-homogeneous phase by the mode with the largest growth-rate. In Fig. 4 (left panel) half of the wavelength associated to the most unstable mode is plotted as a function of density for the proton fractions 0.1 and 0.3. The size of the instabilities that drive the system is $4 - 10$ fm.

The largest differences between NL3, TW occur at densities and proton fractions of interest for β-equilibrium stellar matter. The size of the clusters calculated agree with the results of a density functional calculation with relativistic mean-fields coupled with the electric field.[10] An extension of the present formalism to finite temperature will allow the determination of the cluster sizes in warm stellar matter. The formalism discussed in this section may be easily generalised to finite temperatures.[11] The spinodal region decreases very fast with an increase of temperature. For smaller values of k the critical temperatures are lower in *npe* matter, see Fig. 4 (right panel). For large k values the spinodal is symmetric while for small k values there is a larger contribution for $\rho_p < \rho_n$. The electron effect on critical

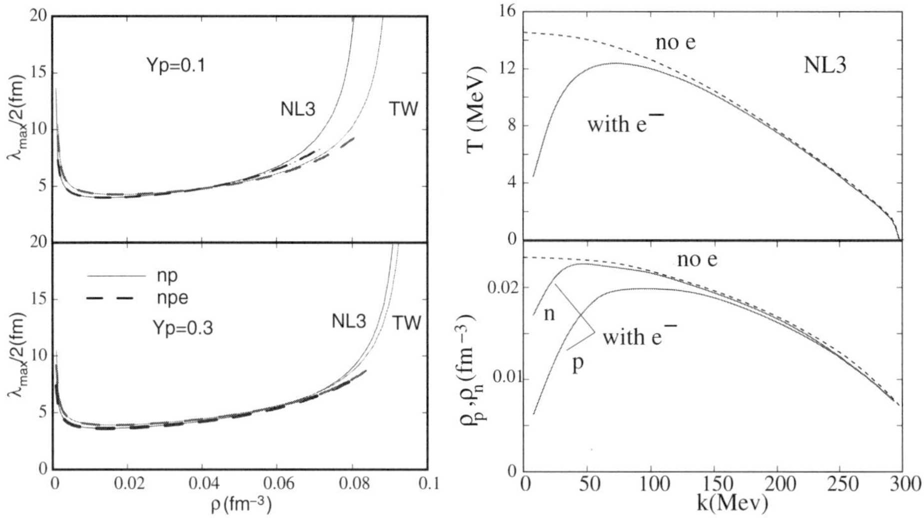

Fig. 4. Cluster size versus density for cold matter (left), and critical temperatures and densities for a finite temperature dynamical response (right).

values occurs only for $k < 150$ MeV. Critical temperature and density values are drastically reduced for small k. At small k, the critical proton densities have a larger reduction and critical temperatures occur for asymmetric matter.

In Ref. 11 we have discussed the effect of temperatures in the instability region. In particular, we have seen that the β-equilibrium EoS crosses the instability region at $T = 0$ MeV, for $6 < k < 150$ MeV, therefore the star crust will present a nucleation phase with inhomogeneities 8 - 200 fm. At $T = 10$ MeV there are no instabilities in neutrino-free EoS but for neutrino trapped EoS instabilities for $30 < k < 137$ MeV are present, corresponding to clusters with 5-20 fm. The neutrino diffusion out of the star will be affected by these inhomogeneities.[11]

The ratio $\delta\rho_n/\delta\rho_p$ gives us information on how the system behaves in the instability region. For np matter this ratio is almost independent of k and satisfies $\delta\rho_n/\delta\rho_p < \rho_n/\rho_p$. This corresponds to the distillation effect.

For npe matter at low k the ratio $\delta\rho_n/\delta\rho_p > \rho_n/\rho_p$, i.e. an anti-distillation effect occurs.[11] However this exotic behaviour is present only for wave numbers smaller than the ones corresponding to the most unstable modes and, therefore, dominating partitions will not show an abnormal isospin distillation. Only large clusters are expected to present this exoticity enhancement.

At finite temperature the distillation effect is not so efficient at large k, and for $k < 50$ MeV the anti-distillation effect settles at larger values of k.[11] The anti-distillation effect will be important in some situations. In fact, in a supernovae explosion 99% of the energy is radiated away in neutrinos. Neutrinos interact strongly with neutrons (large weak vector charge of the neutron) and therefore the way neutrons clusterize is important to determine the neutrino mean free path. Neutrinos may couple strongly to the neutron-rich matter low-energy modes present in this explosive environment and revive the stalled supernovae shock.

5. Conclusions

We summarise the main conclusions. For large k transfers protons and electrons behave independently. There is no screening effect and it is enough to consider a homogeneous distribution of electrons. For small k transfers, electrons and protons stick together corresponding to a large screening effect. A more realistic description of electrons should be considered in this limit. The most unstable modes indicate generally weak electron-screening effects.

We point out that the distillation effect within DDRM is smaller than the values obtained with NLWM and gives results closer to the predictions of Skyrme interactions. Also the anti-distillation effect at low k sets on at higher k values for DDRM. This has astrophysical consequences. The behaviour of the instability region defines the extent of the crust and its constitution. Densities of the inner edge of the crust are model dependent, larger in TW than in NL3 model. The crust formation and neutrino diffusion during the cooling process are affected by instabilities: neutron richer clusters obtained within DDRM couple more strongly to neutrinos.

Acknowledgements

This work was partially supported by FEDER and FCT (Portugal) under the grant SFRH/BPD/29057/2006, and projects POCI/FP/63918/2005, PDCT/FP/64707/2006, and by CNPq (Brazil).

References

1. B. D. Serot and J. D. Walecka, *Ad. Nucl. Phys.* **16**, 1 (1995); J. Boguta and A. R. Bodmer, *Nucl. Phys. A* **292**, 413 (1977).
2. G. A. Lalazissis, J. König and P. Ring, *Phys. Rev. C* **55**, 540 (1997).
3. S. Typel and H. H. Wolter, *Nucl. Phys.* **A656**, 331 (1999).
4. C. Fuchs, H. Lenske and H. Wolter, *Phys. Rev. C* **52**, 3043 (1995).
5. S. S. Avancini, L. Brito, P. Chomaz, D. P. Menezes and C. Providência, *Phys. Rev. C* **74**, 024317 (2006).
6. J. Margueron and P. Chomaz, *Phys. Rev. C* **67**, 041602(R) (2003); P. Chomaz, M. Colonna and J. Randrup, *Phys. Rep.* **389**, 263 (2004).
7. S. S. Avancini, L. Brito, D. P. Menezes and C. Providência, *Phys. Rev. C* **70**, 015203 (2004).
8. A. M. Santos, L. Brito and C. Providência, unpublished.
9. S. S. Avancini, L. Brito, D. P. Menezes and C. Providência, *Phys. Rev. C* **71**, 044323 (2005).
10. T. Maruyama, T. Tatsumi, D. N. Voskresensky, T. Tanigawa and S. Chiba, *Phys. Rev. C* **72**, 015802 (2005).
11. L. Brito, C. Providência, A. M. Santos, S. S. Avancini, D. P. Menezes and P. Chomaz, *Phys. Rev. C* **74**, 045801 (2006); C. Providência, L. Brito, S. S. Avancini, D. P. Menezes and P. Chomaz, *Phys. Rev. C* **73**, 025805 (2006).

QUARTETTING IN NUCLEAR MATTER AND α PARTICLE CONDENSATION IN NUCLEAR SYSTEMS

G. RÖPKE

Institut für Physik, Universität Rostock, D-18051 Rostock, Germany

P. SCHUCK

Institut de Physique Nucléaire, CNRS, UMR8608, Orsay, F-91406, France
Université Paris-Sud, Orsay, F-91505, France

H. HORIUCHI and A.TOHSAKI

Research Center for Nuclear Physics (RCNP), Osaka University,
Ibaraki, Osaka 567-0047, Japan

Y. FUNAKI

The Institute of Physical and Chemical Research (RIKEN), Wako, Saitama 351-0198, Japan

T. YAMADA

Laboratory of Physics, Kanto Gakuin University, Yokohama 236-8501, Japan

Alternatively to pairing, four-particle correlations may become of importance for the formation of quantum condensates in nuclear matter. With increasing density, four-particle correlations are suppressed because of Pauli blocking. Signatures of α-like clusters are expected to occur in low-density nuclear systems. The famous Hoyle state ($0_2{}^+$ at 7.654 MeV in ^{12}C) is identified as being an almost ideal condensate of three α-particles, hold together only by the Coulomb barrier. It, therefore, has a ^8Be-α structure of low density. Transition probability and inelastic form factor together with position and other physical quantities are correctly reproduced without any adjustable parameter from our two parameter wave function of α-particle condensate type. The possibility of the existence of α-particle condensed states in heavier $n\alpha$ nuclei is also discussed.

Keywords: Quartetting; α-particle condensation; Hoyle state; THSR wave function.

1. Introduction

Quantum condensation of particles is one of the most amazing phenomena of many body systems. Striking well known examples are superconducting metals and superfluid ^4He. Also nuclear matter may become superfluid. Pairing is well established not only in infinite nuclear systems like neutron matter, but also in finite nuclei. However, in nuclear systems, the most tightly bound light cluster is not a pair but a quartet. Therefore, α-like clusters will dominate the structure of low-temperature,

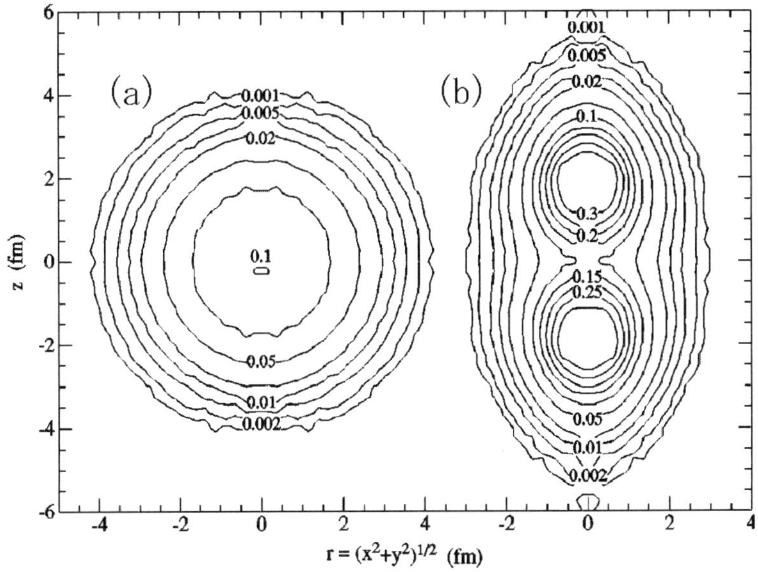

Fig. 1. Contours of constant density (taken from Ref. 4), plotted in cylindrical coordinates, for ^{8}Be(0^+). (a) is in the "laboratory" frame, (b) is in the intrinsic frame.

low-density nuclear matter.[1] As a consequence of Pauli blocking, with increasing density the four-particle correlations will disappear.[2,3]

Similar to pairing, are there also signatures of α-particle condensation in nuclei? The only nucleus which in its ground state has a pronounced α-cluster structure is ^{8}Be. In Fig. 1(a) we show the result of an exact calculation with a realistic N-N interaction for the density distribution in the laboratory frame, whereas in Fig. 1(b) we see the same in the intrinsic, deformed frame where in addition the question has been asked where to find the second α particle when the first is placed at a given position. So we see that the two α's are $\sim$ 4 fm apart giving raise to a very low average density $\rho \sim \rho_0/3$ as seen on Fig. 1(a) where ρ_0 is the nuclear saturation density. ^{8}Be is also a very large object with an rms radius of $\sim$ 3.7 fm to be compared with the nuclear systematics of $R = r_0 A^{1/3} \sim 2.44$ fm.

2. Correlations in Dense Fermion Systems

The quantum statistical approach to the many-nucleon system starts from a Hamiltonian

$$H = \sum_1 E(1) a_1^\dagger a_1 + \frac{1}{2} \sum_{121'2'} V(12, 1'2') a_{1'}^\dagger a_{2'}^\dagger a_2 a_1 \,, \tag{1}$$

where $\{1\} = \{\vec{p}_1, \sigma_1, \tau_1\}$, $E(1) = p_1^2/(2m_1)$. The interaction is described by an effective potential, fitted to empirical properties of nucleons.

The evaluation of properties of dense systems at finite temperature can be given using methods of quantum statistics, see Ref. 5. The nucleon density $\rho(T, \mu)$ as a function of temperature $T = 1/k_B\beta$ and chemical potential μ is related to the spectral function and the self-energy. Different approximations have been performed for the self-energy or the corresponding T matrix which allow to take into account the formation of clusters in a self-consistent way.

We discuss first the inclusion of two-particle states. Evaluating the self-energy containing the two-particle T matrix in the low-density limit, from the single-particle spectral function we obtain the Beth-Uhlenbeck formula[5]

$$\rho(T,\mu) = \frac{1}{\Omega}\sum_1 f(E(1)) + \frac{1}{\Omega}\sum_{\alpha P} f_2(E_{\alpha P}) + \frac{1}{\Omega}\sum_{\alpha P}\int \frac{dE}{\pi} f_2(E+\frac{P^2}{4m})\frac{d}{dE}\delta_{\alpha P}(E)\,, \tag{2}$$

where $f(1) = (\exp[\beta(E(1)-\mu_1)]+1)^{-1}$, $f_2(E) = (\exp[\beta(E-\mu_1-\mu_2)]-1)^{-1}$ denote the Fermi or Bose distribution functions, respectively.

In mean-field approximation, the influence of the medium is represented by the Hartree-Fock self-energy shift $E^{\rm mf}(1) = p_1^2/2m_1 + \sum_2 V(12,12)_{\rm ex} f(2)$. To be consistent, in the interaction kernel of the two-particle Bethe-Salpeter equation also the Pauli blocking has to be included so that

$$\begin{aligned}&\left[E^{\rm mf}(1) + E^{\rm mf}(2) - E^{\rm mf}_{nP}\right]\psi^{\rm mf}_{\alpha P}(12)\\ &+[1-f(1)-f(2)]\sum_{1'2'} V(12,1'2')\psi^{\rm mf}_{\alpha P}(1'2') = 0\,.\end{aligned} \tag{3}$$

Using the solution of the in-medium two-particle Schrödinger equation (3) to calculate the self-energy, we derive a generalized Beth-Uhlenbeck equation.[5] Besides medium-dependent quasiparticles for the single-particle contribution, also the two-particle states become medium dependent. In particular, the bound states are dissolved with increasing density due to Pauli blocking. Furthermore, the Bose pole in the correlated density signals the onset of a quantum condensate. As is well known, for the bound-state (deuteron channel) contribution, the T-matrix approach breaks down when the pole corresponding to the bound-state energy coincides with twice the chemical potential. This is the Thouless condition, embodied in

$$\mathrm{T}(1,2,1'',2'';2\mu) = \frac{(1-f(1)-f(2))}{2\mu - E(1) - E(2)}\sum_{1'2'} V(12,1'2')\;\mathrm{T}(1',2',1'',2'';2\mu)\,. \tag{4}$$

The same condition also holds for the contribution of scattering states. Consequently, the transition temperature for the onset of a quantum condensate appears as a smooth function of density[6] with a crossover from BEC to BCS. The corresponding wave equation (3), where $E^{\rm mf}_{nP}$ coincides with 2μ is the well-known Gor'kov equation.

In our approximation correlations in the medium are neglected, and the Fermi function occurring in the self-energy and the Pauli blocking is adapted to the full nucleon density. A more detailed description of the medium effects is given by the cluster mean field approach,[7] where in addition to the mean field produced by the single-particle states, the mean field produced by clusters (bound states) is also taken into account.

3. Four-Particle Condensates and Quartetting in Nuclear Matter

The Green function approach allows the inclusion of higher clusters in the self-energy and the resulting equation of state, extending the Beth-Uhlenbeck equation to higher clusters.[7] We are interested in the contribution of four-particle correlations (α-particles) in nuclear matter. In correspondence to Eq. (3), the effective wave equation contains in mean field approximation the Hartree-Fock self-energy shift of the single-particle energies as well as the Pauli blocking of the interaction,

$$\begin{aligned}&\left[E^{\rm mf}(1)+E^{\rm mf}(2)+E^{\rm mf}(3)+E^{\rm mf}(4)-E^{\rm mf}_{4,n,P}\right]\psi_{4,n,P}(12)\\&+\sum_{i<j}\sum_{1'2'3'4'}[1-f(i)-f(j)]V(ij,i'j')\prod_{k\neq i,j}\delta_{k,k'}\psi_{4,n,P}(1'2'3'4')=0.\end{aligned}\quad(5)$$

An extended Thouless condition based on the relation

$$\begin{aligned}&\mathrm{T}_4(1234,1''2''3''4'',4\mu)\\&=\sum_{1'2'3'4'}\left\{\frac{V(12,1'2')[1-f(1)-f(2)]}{4\mu-E^{\rm mf}(1)-E^{\rm mf}(2)-E^{\rm mf}(3)-E^{\rm mf}(4)}\delta_{3,3'}\delta_{4,4'}+\text{perm.}\right\}\\&\qquad\times\mathrm{T}_4(1'2'3'4',1''2''3''4'',4\mu)\end{aligned}\quad(6)$$

serves to determine the onset of Bose condensation of α-like clusters, noting that the existence of a solution of this relation signals a divergence of the four-particle correlation function. An approximate solution has been obtained by a variational approach, in which the wave function is taken as Gaussian incorporating the correct solution for the two-particle problem.[1] In the low-density region, the critical density tracks that for Bose-Einstein condensation of ideal α particles; hence the Bose condensation of deuterons considered previously becomes irrelevant. As expected, with increasing density the transition temperature is depressed from that of the ideal Bose gas of α's due to medium corrections. The critical density at which the α condensate disappears is estimated to be about $\rho_0/3$. However, the variational approach of Ref. 1 on which this estimate is based represents only a first attempt to describe the transition from quartetting to pairing. The detailed nature of this fascinating transition remains to be clarified.

Below the critical temperature for the quartetting transition, in particular at zero temperature, the condensate fraction is suppressed compared with the ideal Bose gas because of correlation effects. A simple estimate based on the concept of excluded volume leads to a vanishing condensate fraction at saturation density. A more detailed evaluation of the off-diagonal long-range order of an α condensate at zero temperature is given in Ref. 8.

4. α-Condensate States in Self-Conjugate $4n$-Nuclei

We now will show that α-particle condensation most likely exists in $n\alpha$ nuclei around energies of α-particle break-up thresholds. We want to give the demonstration here that, due to the existence of ample experimental data, we have identified at least one nucleus where such an α-particle condensed state exists and then we discuss the indications and the likelyhood that such states very naturally also are present in other nuclei and that it may be a quite general phenomenon in nuclear systems.

The nucleus we want to draw our attention to is ^{12}C. We, indeed, will give strong arguments that the 0_2^+ state at 7.654 MeV in ^{12}C is a state of α-particle condensate nature. As ^{8}Be, this state is unstable and situated about 300 keV above the three α-break up threshold, only stabilised by the Coulomb barrier. It has a width of 8.7 eV and a corresponding lifetime of 7.6×10^{-17} s. It is well known that this Hoyle state, as it is called now, is a notoriously difficult state for any nuclear theory and for example the most modern no-core shell model calculations predict the 0_2^+ state in ^{12}C to occur at around 17 MeV, that is more than two times its actual value.[9] This fact alone may tell us that the Hoyle state must have a very unusual structure and it is easy to understand that, should it indeed have a loosely bound three α-particle structure, a shell model type of calculation would have great difficulties to reproduce its properties. It also was recognised that the three α's in the Hoyle state form sort of gas like state, a feature which had already been pointed out by H. Horiuchi[10] and was confirmed by the works of Ref. 11 and 12 .

Though these authors all had already stressed the somewhat α-gas like nature of the Hoyle state, eventually two important aspects were missed at that time. First comes the fact that, because all three α's move in identical S-wave orbits, this forms an α-condensate state, albeit not in the macroscopic sense. Second is that the complicated three body wave function can be replaced by a structurally and conceptually very simple microscopic three α wave function of the condensate type which has practically 100 percent overlap with the previous ones.[13]

We now shortly want to describe this condensate wave function. We start with an analogy to the BCS wave function of ordinary pairing

$$\langle \vec{r}_1 ... \vec{r}_N | \mathrm{BCS} \rangle = \mathcal{A}[\phi(\vec{r}_1, \vec{r}_2)\phi(\vec{r}_3, \vec{r}_4)...\phi(\vec{r}_{N-1}, \vec{r}_N)] \,, \tag{7}$$

where $\phi(\vec{r}_1, \vec{r}_2)$ is the Cooper pair wave function, including spin and isospin, which is being determined variationally by the well known BCS equations, $\mathcal{A}$ is the antisymmetriser. The condensate character of the BCS ansatz is born out by the fact that we have a product of $N/2$ times the same pair wave function ϕ. Formally it now is a simple matter to generalise (7) to α-particle condensation. We write

$$\langle \vec{r}_1 ... \vec{r}_N | \Phi_{n\alpha} \rangle = \mathcal{A}[\phi_\alpha(\vec{r}_1, \vec{r}_2, \vec{r}_3, \vec{r}_4)\phi_\alpha(\vec{r}_5, .., \vec{r}_8)...\phi_\alpha(\vec{r}_{N-3}, .., \vec{r}_N)] \,, \tag{8}$$

where ϕ_α is the wave function common to all condensed α-particles. Of course, in general, the variational solution for $\phi_\alpha(\vec{r}_1, .., \vec{r}_4)$ from (8) is extraordinarily more complicated than to find the Cooper pair wave function of (7). However, in the case of the α-particle and for relatively light nuclei, the complexity of the problem can be reduced dramatically. This stems from the fact that, as was already recognised by the authors in Ref. 11,12, an intrinsic wave function of the α-particle of Gaussian form with only the size parameter b to be determined, is an excellent variational ansatz (see above). In addition, and here resides the essential and crucial novelty of our wave function, even the center of mass motion of the various α-particles can very well be described by a Gaussian wave function with, this time, a size parameter $B \gg b$ to account for the motion over the whole nucleus. We therefore write

$$\phi_\alpha(\vec{r}_1, \vec{r}_2, \vec{r}_3, \vec{r}_4) = e^{-\vec{R}^2/B^2} \varphi_0(\vec{r}_1 - \vec{R})\, \varphi_0(\vec{r}_2 - \vec{R})\, \varphi_0(\vec{r}_3 - \vec{R})\, \varphi_0(\vec{r}_4 - \vec{R}), \quad (9)$$

where $\vec{R} = (\vec{r}_1 + \vec{r}_2 + \vec{r}_3 + \vec{r}_4)/4$ is the c.m. coordinate of one α-particle. Of course, in (8) the center of mass $\vec{X}_{\rm cm}$ of the three α's, i.e. of the whole nucleus, should also be eliminated what is easily achieved by replacing $\vec{R}$ by $\vec{R} - \vec{X}_{\rm cm}$ in each of the α wave functions in (8). The α-particle condensate wave function (8) with (9), proposed in[14] and henceforth called THSR-wavefunction, now depends only on two parameters, B and b. The expectation value of the microscopic Hamiltonian

$$\mathcal{H}(B, b) = \langle \Phi_{n\alpha}(B, b)|H|\Phi_{n\alpha}(B, b)\rangle / \langle \Phi_{n\alpha}|\Phi_{n\alpha}\rangle \quad (10)$$

can be evaluated and the corresponding two dimensional energy surface quantised in using the two parameters B and b as Hill-Wheeler coordinates. The energy land scapes $\mathcal{H}(B, b)$ for various $n\alpha$ nuclei are interesting by themselves[16] but for brevity not shown here.

Before coming to the results, let us discuss this THSR- wave function (8), (9) a little more. One should realise that it contains two limits exactly: if $B = b$, then (8) boils down to a standard Slater determinant with harmonic oscillator wave functions with oscillator length b as the single variational parameter. This holds because (9), with $B = b$, becomes a product of four identical Gaussians and the antisymmetrisation creates all the necessary P, D, etc. harmonic oscillator wave functions automatically.[14] On the contrary, when $B \gg b$, the density of α-particles is very low and in the limit $B \to \infty$, the average distance between α-particles is so large that the antisymmetrisation between α's can be neglected, i.e. the operator '$\mathcal{A}$' in front of (8) can be taken off. Our wave function then becomes an ideal gas of independent condensed α-particles, i.e. a pure product state of α's! On the other hand, in realistic cases, the antisymmetriser $\mathcal{A}$ can not be neglected and the evaluation of the expectation values in (10) becomes an analytical (but non-trivial) task.

5. Results for Finite Nuclei

The THSR wave function constructed from the Hill-Wheeler equation based on Eqs. (8)-(10), has practically 100 percent overlap with the ones in Refs. 11,12, once the same Volkov force is used.[13] It is, therefore, not astonishing that we also get very similar results to theirs. For ^{12}C we obtain two eigenvalues: the ground state and the Hoyle state. Theoretical values for positions, rms values, transition probabilities, compared to the data, are given in Table 1. From the comparison of the rms radii we see that the volume of the Hoyle state is a factor 3 to 4 larger than the one of the ground state of ^{12}C. This is the aspect of dilute gas state we were talking about above. Constructing an α-particle density matrix $\rho(\vec{R}, \vec{R}')$, in integrating out of the

Table 1. Comparison of the binding energies, rms radii ($R_{\rm r.m.s.}$), and monopole matrix elements ($M(0_2^+ \to 0_1^+)$) for ^{12}C given by solving Hill-Wheeler equation based on Eq. (8) and by RGM.[11] Volkov No. 2 force as the effective two-nucleon force is adopted in the two cases, for which the 3α threshold energy is calculated to be −82.04 MeV.

		THSR	RGM[11]	Exp.
E (MeV)	0_1^+	−89.52	−89.4	−92.2
	0_2^+	−81.79	−81.7	−84.6
$R_{\rm r.m.s.}$ (fm)	0_1^+	2.40	2.40	2.44
	0_2^+	3.83	3.47	(Fig.2)
$M(0_2^+ \to 0_1^+)$ (fm^2)		6.45	6.7	5.4

total density matrix all intrinsic α-particle coordinates, we find in diagonalising this density matrix that the corresponding $0S$ α-particle orbit is occupied to 70 percent by the three α-particles.[17,18] This is a huge percentage, underlining the almost ideal α-particle condensate aspect of the Hoyle state. In this regard one should remember that superfluid ^{4}He has only 10 percent of the particles in the condensate! Let us also mention that for the ground state of ^{12}C the α-particle occupations are equally shared between $0S$, $0D$, and $0G$ orbits, thus invalidating a condensate picture for the ground state. Please notice that also the ground state energy of ^{12}C is reasonably reproduced by our theory.

Let us now discuss the, to our mind, most convincing feature that our description of the Hoyle state is the correct one. As the authors of Ref. 11, we reproduce very accurately the inelastic form factor $0_1^+ \to 0_2^+$ of ^{12}C. This is shown in Fig. 2. The agreement with experiment is as such already quite impressive. Additionally, however, we made the following study shown in Fig. 3. We artificially varied the extension of the Hoyle state and studied the influence on the form factor. We found that the overall shape of the form factor only varies little, for instance in what concerns the position of the minimum. On the contrary, we found a strong dependence on the absolute magnitude of the form factor and in Fig. 3 we also plot the variation of the height of the first maximum of the inelastic form factor as a function of the percentage change of the rms radius of the Hoyle state.[20] It can be seen that a 20 percent increase of the rms radius decreases the maximum by a factor of two! This strong dependence of the magnitude of the form factor makes us firmly believe that the agreement with the actual measurement is practically a proof that our calculated wide extension of the Hoyle state corresponds to reality. In a recent calculation[21] the low value of the nucleon density in the Hoyle state has been confirmed.

The Hoyle state can be considered as the ground state of the α-particle condensate. Exciting one α-particle out of the condensate and putting it into the $0D$ orbit reproduces the experimentally measured position of the 2_2^+ state in ^{12}C. Without going into details, we also state that the width of this state is correctly reproduced.[22] It is tempting to imagine that the 0_3^+ state which experimentally is almost

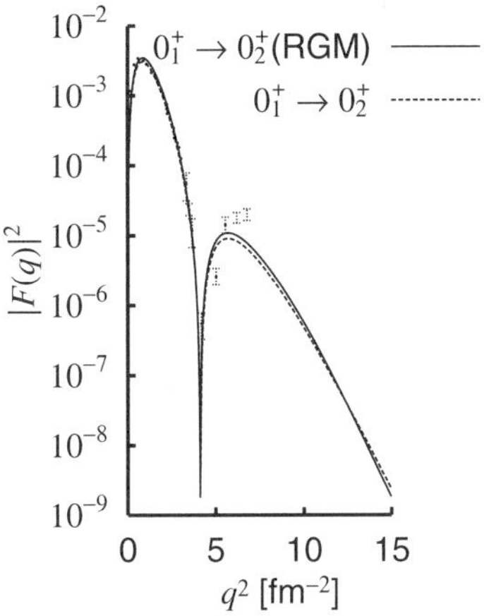

Fig. 2. Experimental values of inelastic form factor in ^{12}C to the Hoyle state[19] are compared with the THSR values (dotted) and those given by Kamimura et al.[11] (RGM). The THSR wave function is obtained by solving the Hill-Wheeler equation based on (8).

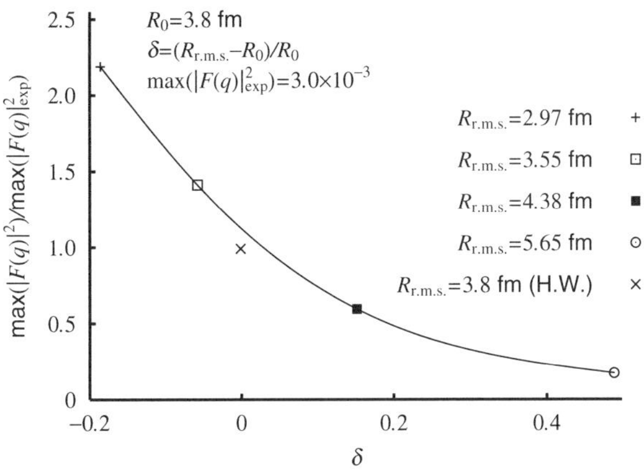

Fig. 3. THSR result of the inelastic form factor compared with experiment. The ratio of the value of maximum height, theory versus experiment, for the inelastic form factor, i.e. $\max|F(q)|^2$ / $\max|F(q)|^2_{\rm exp}$, is plotted as a function of δ, which is defined as $\delta=(R_{\rm r.m.s.}-R_0)/R_0$. $R_{\rm r.m.s.}$ and R_0 are the rms radii corresponding to the wave function (8) and the one obtained by solving the Hill-Wheeler equation based on (8), respectively.

degenerate with the 2_2^+ state is obtained by lifting one α-particle into the $1S$ orbit. First theoretical studies[23] indicate that this view might indeed be true. However, its width ($\sim$ 3MeV) is very broad what makes a theoretical treatment rather delicate. Further investigations are necessary to validate this picture. At any rate, it would be very satisfying, if the the triplet of states, i.e. $0_2^+, 2_2^+, 0_3^+$, could all be explained from the α-particle point of view, since those three states are precisely the ones which can not be explained within a (no core) shell model approach.[9]

In conclusion, in what concerns ^{12}C, we think we have accumulated enough facts to become convinced that the Hoyle state is, indeed, what one can call an α-particle condensate state, being aware of the fact that 'condensate' for only three particles constitutes a certain abuse of the word. We, however, should remember in this context that also in the case of nuclear Cooper pairing, only a few Cooper pairs are sufficient to obtain clear signatures of superfluidity in nuclei!

What about α-particle condensation in heavier nuclei? Of course, once one accepts the idea that the Hoyle state is essentially a state of three free α-particles hold together only by the Coulomb barrier, it is hard to believe that analogous states should not also exist in heavier $n\alpha$ nuclei like ^{16}O, ^{20}Ne, ^{24}Mg, ... At least our calculations systematically always show a 0^+-state close to the α-particle disintegration threshold. For example in ^{16}O we obtain three 0^+-states: the ground state at $E_0 = -124.8$ MeV (experimental value: -127.62 MeV), a second state at excitation energy $E_{0_2^+} = 8.8$ MeV and a third one at $0_3^+ = 14.1$ MeV. The threshold in

^{16}O is at 14.4 MeV. Unfortunately the experimental situation in ^{16}O is by far not so complete as the one in ^{12}C. For example no transition probability measurement of 0^+-states around threshold in ^{16}O nor inelastic form factors do exist. Recently Wakasa et al.[24] identified a new 0^+ state at 13.5 MeV in ^{16}O which is the 5-th 0^+-state. There are indications that it might be the α-condensate state.[25]

An intersting question is how many α's can maximally be in a self bound α-gas state. For answering this question, a schematic investigation using an effective α-α interaction in an α-gas mean field calculation of the Gross-Pitaevsky type was performed. Our estimate yields[26] a maximum of ten α-particles which can be held together in a condensate. However, a couple of extra neutrons may stabilise larger condensates.

Another interesting idea concerning α-particle condensates was put forward by von Oertzen and collaborators.[27,28] Adding more and more α-particles to e.g. the ^{40}Ca core, one sooner or later will arrive at the α-particle drip. Therefore it may need little further excitation energy to shake loose further α-particles, so that an $n\alpha$-condensate could be created on top of an inert ^{40}Ca core. Similar ideas also have been advanced by Ogloblin[29] who imagines a three α-particle condensate on top of ^{100}Sn and earlier by Brenner and Gridnev who think having detected gaseous α-particles in ^{28}Si and ^{32}S on top of an inert ^{16}O core.[30]

References

1. G. Röpke, A. Schnell, P. Schuck and P. Nozières, *Phys. Rev. Lett.* **80**, 3177 (1998).
2. G. Röpke, L. Münchow and H. Schulz, *Nucl. Phys. A* **379**, 536 (1982).
3. M. Beyer, S. A. Sofianos, C. Kuhrts, G. Röpke and P. Schuck, *Phys. Lett. B* **478**, 86 (2000).
4. R. B. Wiringa, S. C. Pieper, J. Carlson and V. R. Pandharipande, *Phys. Rev. C* **62**, 014001 (2000).
5. M. Schmidt, G. Röpke and H. Schulz, *Ann. Phys. (NY)* **202**, 57 (1990).
6. H. Stein, A. Schnell, T. Alm and G. Röpke *Z. Phys. A* **351**, 295 (1995).
7. G. Röpke, T. Seifert, H. Stolz and R. Zimmermann, *Phys. Stat. Sol. (b)* **100**, 215 (1980); G. Röpke, M. Schmidt, L. Münchow and H. Schulz, *Nucl. Phys. A* **399**, 587 (1983).
8. M. T. Johnson and J. W. Clark, *Kinam* **2**, 3 (1980) (PDF available at Faculty web page of J. W. Clark at http://wuphys.wustl.edu); see also J. W. Clark and T. P. Wang, *Ann. Phys. (N.Y.)* **40**, 127 (1966) and G. P. Mueller and J. W. Clark, *Nucl. Phys. A* **155**, 561 (1970).
9. P. Navrátil, J. P. Vary and B. R. Barrett, *Phys. Rev. Lett.* **84**, 5728 (2000); *Phys. Rev. C* **62**, 054311 (2000); B. R. Barrett, B. Mihaila, S. C. Pieper and R. B. Wiringa, *Nucl. Phys. News* **13**, 17 (2003).
10. H. Horiuchi, *Prog. Theor. Phys.* **51**, 1266 (1974); **53**, 447 (1975).
11. Y. Fukushima and M. Kamimura, *Proc. Int. Conf. on Nuclear Structure*, Tokyo, 1977, ed. T. Marumori *Suppl. of J. Phys. Soc. Japan* **44**, 225 (1978)); M. Kamimura, *Nucl. Phys. A* **351**, 456 (1981).
12. E. Uegaki, S. Okabe, Y. Abe and H. Tanaka, *Prog. Theor. Phys.* **57**, 1262 (1977); E. Uegaki, Y. Abe, S. Okabe and H. Tanaka, *Prog. Theor. Phys.* **59**, 1031 (1978); **62**, 1621 (1979).

13. Y. Funaki, A. Tohsaki, H. Horiuchi, P. Schuck and G. Röpke, *Phys. Rev. C* **67**, 051306(R) (2003).
14. A. Tohsaki, H. Horiuchi, P. Schuck and G. Röpke, *Phys. Rev. Lett.* **87**, 192501 (2001).
15. A. Tohsaki, *Phys. Rev. C* **49**, 1814 (1994).
16. A. Tosaki, H. Horiuchi, P. Schuck and G. Röpke, *Proc. of the 8th Int. Conf. on Clustering Aspects of Nuclear Structure and Dynamics*, Nara, Japan, 2003, ed. K. Ikeda, I. Tanihata and H. Horiuchi, *Nucl. Phys. A* **738**, 259 (2004)).
17. T. Yamada and P. Schuck, *Eur. Phys. J. A* **26**, 185 (2005).
18. H. Matsumura and Y. Suzuki, *Nucl. Phys. A* **739**, 238 (2004).
19. I. Sick and J. S. McCarthy, *Nucl. Phys. A* **150**, 631 (1970); A. Nakada, Y. Torizuka and Y. Horikawa, *Phys. Rev. Lett.* **27**, 745 (1971); and 1102 (Erratum); P. Strehl and Th. H. Schucan, *Phys. Lett. B* **27**, 641 (1968).
20. Y. Funaki, A. Tohsaki, H. Horiuchi, P. Schuck and G. Röpke, *Eur. Phys. J. A* **28**, 259 (2006).
21. M. Chernykh, H. Feldmeier, T. Neff, P. von Neumann-Cosel and A. Richter, *Phys. Rev. Lett.* **98**, 032501 (2007).
22. Y. Funaki, H. Horiuchi, A. Tohsaki, P. Schuck and G. Röpke, *Eur. Phys. J. A* **24**, 321 (2005).
23. C. Kurakowa and K. Kato, *Phys. Rev. C* **71**, 021301 (2005).
24. T. Wakasa *et al.*, *Phys. Lett. B* **653**, 173 (2007).
25. Y. Funaki *et al.*, RCNP workshop, April, 2006, Osaka, to appear in *Mod. Phys. Lett. A*, World Scientific.
26. T. Yamada and P. Schuck, *Phys. Rev. C* **69**, 024309 (2004).
27. Tz. Kokalova, N. Itagaki, W. von Oertzen and C. Wheldon, *Phys. Rev. Lett.* **96**, 192502 (2006).
28. W. von Oertzen *et al.*, *Eur. Phys. J. A* **29**, 133 (2006).
29. A. A. Ogloblin *et al.*, Proceedings of the International Nuclear Physics Conference, Peterhof, Russia, June 28-July 2, 2005.
30. M. W. Brenner *et al.*, Proceedings of the International Conference "Clustering Phenomena in Nuclear Physics," St. Petersburg, published in *Physics of Atomic Nuclei (Yadernaya Fizika)*, 2000.

PART C

Neutron Star Structure and Dynamics

SHEAR VISCOSITY OF NEUTRON MATTER FROM REALISTIC NUCLEAR INTERACTIONS

O. BENHAR and M. VALLI

INFN, Sezione di Roma
Department of Physics, Università "La Sapienza"
I-00185 Roma, Italy

The calculation of transport properties of Fermi liquids, based on the formalism developed by Abrikosov and Khalatnikov, requires the knowledge of the probability of collisions between quasiparticles in the vicinity of the Fermi surface. We have carried out a study of the shear viscosity of pure neutron matter, whose value plays a pivotal role in determining the stability of rotating neutron stars, in which these processes are described using a state-of-the-art nucleon-nucleon potential. Medium modifications of the scattering cross section have been consistently taken into account, through an effective interaction obtained from the matrix elements of the bare interaction between correlated states. Medium effects produce a large increase of the viscosity at densities $\rho \gtrsim 0.1$ fm^{-3}.

Keywords: Neutron matter; Transport coefficients; Neutron-star oscillations.

1. Introduction

Viscosity plays a pivotal role in determining the stability of rotating neutron stars. As Chandrasekhar first pointed out,[1] emission of gravitational radiation (GR) following the excitation of non radial oscillation modes may lead to the instability of rotating stars. While this effect would make all perfect fluid rotating stars unstable, in presence of viscosity dissipative effects damp the oscillations, and may prevent the onset of the instability. As a consequence, a quantitative understanding of the viscosity of neutron star matter is required to determine whether a mode is stable or unstable (for a recent review see, e.g., Ref. 2 and references therein).

Early estimates of the shear viscosity coefficients of neutron star matter were obtained in the 70s by Flowers and Itoh, who used the measured scattering phase shifts to estimate the neutron-neutron scattering probability.[3,4] Based on the these results, Cutler and Lindblom carried out a systematic study of the effect of the viscosity on neutron-star oscillations, using a variety of different equations of state (EOS) of neutron star matter.[5]

The procedure followed by the authors of Ref. 5, while allowing for a quantitative analysis of the damping of neutron-star oscillations, cannot be regarded as fully consistent. Ideally, the calculation of transport properties of neutron star matter and the determination of its EOS should be carried out using the *same* dynamical model. The work discussed in this paper is aimed at making a first step towards this goal.

We have computed the shear viscosity of pure neutron matter using a realistic nucleon-nucleon (NN) potential, the Argonne v_{18} model,[6] previously employed to obtain the state-of-the-art EOS of Akmal, Pandharipande and Ravenhall.[7] Within our approach, based on the formalism of Correlated-Basis-Function (CBF) perturbation theory,[8,9] medium modifications of the NN scattering cross section are also consistently taken into account, through an effective interaction derived from the same NN potential.[10]

2. Formalism

The theoretical description of transport properties of normal Fermi liquids is based on Landau theory.[11] Working within this framework and including the leading term in the low-temperature expansion, Abrikosov and Khalatnikov[12] obtained the approximate expression of the shear viscosity coefficient

$$\eta_{AK} = \frac{1}{5}\rho m^\star v_F^2 \tau \, \frac{2}{\pi^2(1-\lambda_\eta)} \ , \tag{1}$$

where ρ is the density, v_F is the Fermi velocity and $m^\star$ and τ denote the quasiparticle effective mass and lifetime, respectively. The latter can be written in terms of the angle-averaged scattering probability $\langle W \rangle$ according to

$$\tau T^2 = \frac{8\pi^4}{m^{*3}} \, \frac{1}{\langle W \rangle} \ , \tag{2}$$

where T is the temperature and

$$\langle W \rangle = \int \frac{d\Omega}{2\pi} \, \frac{W(\theta,\phi)}{\cos(\theta/2)} \ . \tag{3}$$

Note that the scattering process involves quasiparticles on the Fermi surface. As a consequence, for any given density ρ, W depends only on the angular variables θ and ϕ, the magnitude of all quasiparticle momenta being equal to the Fermi momentum $p_F = (3\pi^2\rho)^{1/3}$. Finally, the quantity λ_η appearing in Eq. (1) is defined as

$$\lambda_\eta = \frac{\langle W[1 - 3\sin^4(\theta/2)\sin^2\phi]\rangle}{\langle W \rangle} \ . \tag{4}$$

The exact solution of the equation derived in Ref. 12, obtained by Brooker and Sykes,[13] reads

$$\eta = \eta_{AK} \, \frac{1-\lambda_\eta}{4} \sum_{k=0}^{\infty} \frac{4k+3}{(k+1)(2k+1)[(k+1)(2k+1)-\lambda_\eta]} \ , \tag{5}$$

the size of the correction with respect to the result of Eq. (1) being $0.750 < (\eta/\eta_{AK}) < 0.925$.

Eqs. (1)-(5) show that the key element in the determination of the viscosity is the in-medium NN scattering cross section. In Ref. 14, the relation between NN scattering in vacuum and in nuclear matter has been analyzed under the assumption that the nuclear medium mainly affects the flux of incoming particles and the phase space available to the final state particles, while leaving the transition probability unchanged. Within this picture $W(\theta, \phi)$ can be extracted from the NN scattering cross section measured in free space, $(d\sigma/d\Omega)_{\text{vac}}$, according to

$$W(\theta, \phi) = \frac{16\pi^2}{m^{\star 2}} \left(\frac{d\sigma}{d\Omega} \right)_{\text{vac}} , \tag{6}$$

where $m^\star$ is the nucleon effective mass and θ and ϕ are related to the kinematical variables in the center of mass frame through $E_{cm} = p_F^2(1 - \cos\theta)/(2m)$, $\theta_{cm} = \phi$.[4]

The above procedure has been followed in Ref. 15, whose authors have used the available tables of vacuum cross sections obtained from partial wave analysis.[16] In order to compare with the results of Ref. 15, we have first carried out a calculation of the viscosity using Eqs. (1)-(6) and the free space neutron-neutron cross section obtained from the Argonne v_{18} potential, which is written in the form

$$v_{ij} = \sum_{n=1}^{18} v_n(r_{ij}) O_{ij}^n . \tag{7}$$

In the above equation

$$O_{ij}^{n \leq 6} = [1, (\boldsymbol{\sigma}_i \cdot \boldsymbol{\sigma}_j), S_{ij}] \otimes [1, (\boldsymbol{\tau}_i \cdot \boldsymbol{\tau}_j)] , \tag{8}$$

where $\boldsymbol{\sigma}_i$ and $\boldsymbol{\tau}_i$ are Pauli matrices acting in spin and isospin space, respectively, and

$$S_{ij} = \frac{3}{r_{ij}^2} (\boldsymbol{\sigma}_i \cdot \mathbf{r}_{ij})(\boldsymbol{\sigma}_j \cdot \mathbf{r}_{ij}) - (\boldsymbol{\sigma}_i \cdot \boldsymbol{\sigma}_j) . \tag{9}$$

The operators corresponding to $p = 7, \ldots, 14$ are associated with the non static components of the NN interaction, while those corresponding to $p = 15, \ldots, 18$ account for charge symmetry violations. Being fit to the full Nijmegen phase shifts data base, as well as to low energy scattering parameters and deuteron properties, the Argonne v_{18} potential provides an accurate description of the scattering data by construction.

3. Results

In the left panel of Fig. 1, we show the quantity ηT^2 as a function of density. Our results are represented by the solid line, while the dot-dash line corresponds to the results obtained from Eqs. (43) and (46) of Ref. 15 using the same effective masses, computed from the effective interaction discussed below. The differences between the two curves are likely to be ascribed to the correction factor of Eq. (5), not taken into acount by the authors of Ref. 15, and to the extrapolation needed to determine the cross sections at small angles within their approach.

To gauge the model dependence of our results, we have replaced the full Argonne v_{18} potential with its simplified form, referred to as v_8',[17] which only includes the

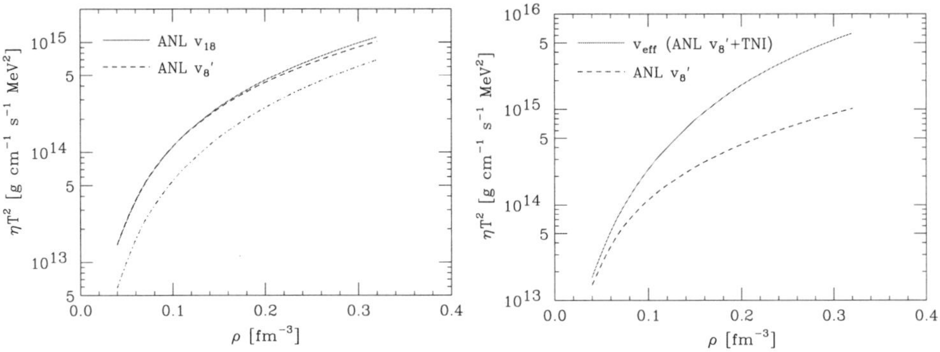

Fig. 1. Neutron matter ηT^2 as a function of density. Left panel. Solid line: results obtained from Eqs. (1)-(6) using the Argonne v_{18} potential and $m^\star$ computed from the effective interaction described in the text. Dot-dash line: results obtained from Eqs. (43) and (46) of Ref. 15 using the same $m^\star$. Dashed line: same as the solid line, but with the Argonne v_{18} replaced by its reduced form v_8'. Right panel. Solid line: results obtained using the effective interaction described in the text. Dashed line: ηT^2 obtained from the free space cross section corresponding to the v_8' potential.

six static operators of Eq. (8) and the two spin-orbit operators $\mathbf{L}\cdot\mathbf{S}\otimes[1,(\boldsymbol{\tau}_i\cdot\boldsymbol{\tau}_j)]$. These eight operators are the minimal set required to describe NN scattering in S and P states. The corresponding results, represented by the dashed line, show that using the v_8' potential leads to a few percent change of ηT^2 over the density range corresponding to $1/4 < (\rho/\rho_0) < 2$, $\rho_0 = 0.16$ fm^{-3} being the equilibrium density of symmetric nuclear matter.

To improve upon the approximation of Eq. (6) and include the effects of medium-modifications of the NN scattering amplitude, we have replaced the bare NN potential with an *effective interaction*, derived within the CBF approach as discussed in Ref. 18.

The *correlated* states of neutron matter are obtained from the Fermi gas (FG) states through the transformation

$$|n\rangle = F|n_{FG}\rangle \ , \tag{10}$$

where the operator F, embodying the correlation structure induced by the NN interaction, is written in the form

$$F = \mathcal{S}\prod_{ij} f_{ij} \ , \tag{11}$$

$\mathcal{S}$ being the symmetrization operator. The two-body correlation functions f_{ij}, whose operatorial structure reflects the complexity of the NN potential, read

$$f_{ij} = \sum_{n=1}^{6} f^n(r_{ij})O^n_{ij} \ , \tag{12}$$

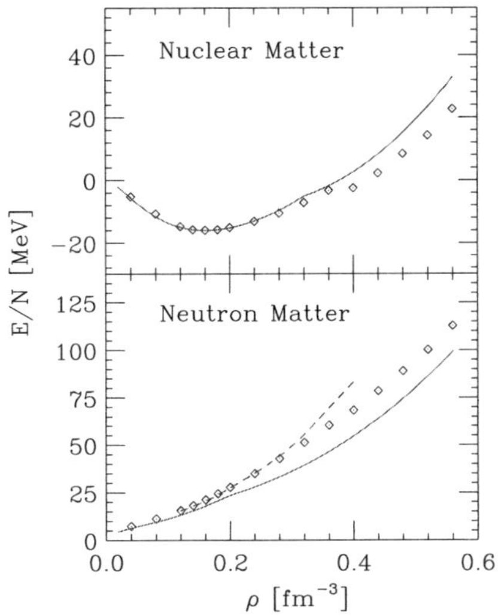

Fig. 2. Energy per particle of symmetric nuclear matter (upper panel) and pure neutron matter (lower panel). The diamonds represent the results obtained using the effective interaction discussed in the text in first order perturbation theory with the FG basis, whereas the solid lines correspond to the results of Akmal, Pandharipande and Ravenhall.[7] The dashed line of the lower panel represents the results of the AFDMC approach or Ref. 21.

with the O^n_{ij} given by Eq. (8). The effective interaction $v_{\rm eff}$ is defined by the relation

$$\frac{\langle n|H|n\rangle}{\langle n|n\rangle} = \langle n_{FG}|K + v_{\rm eff}|n_{FG}\rangle \ , \tag{13}$$

where H is the full nuclear hamiltonian and K is the kinetic energy operator. Realistic models of H include, in addition to the NN potential v_{ij}, a three-nucleon potential V_{ijk} needed to account for the measured binding energies of the few-nucleon systems.[19] In this work, we follow a somewhat different approach, originally proposed in Ref. 20, in which the main effect of three- and many-body forces is taken into account through a density dependent modification of the NN potential v_{ij} at intermediate range. Moreover, in view of the weak model dependence of ηT^2 (see Fig. 1), the full v_{18} potential is replaced by its reduced form v_8', and the contribution of the non static components is disregarded.[18]

In order to obtain $v_{\rm eff}$ from Eq. (13) the expectation value of H in the correlated ground state is evaluated at the two-body level of the cluster expansion.[18] The resulting effective interaction reads

$$v_{\rm eff} = \sum_{i<j} f^\dagger_{ij} \left[-\frac{1}{m}(\nabla^2 f_{ij}) - \frac{2}{m}(\boldsymbol{\nabla} f_{ij}) \cdot \boldsymbol{\nabla} + v_{ij} f_{ij} \right] \ . \tag{14}$$

For any given density, the radial functions $f^n(r_{ij})$ of Eq. (12) are solutions of a set of Euler-Lagrange equations satisfying the boundary conditions $f^1(r_{ij} \geq d) = 1$,

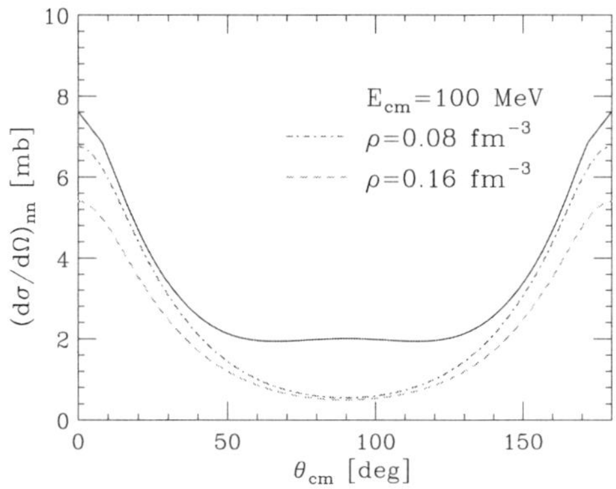

Fig. 3. Differential neutron-neutron scattering cross section at $E_{cm} = 100$ MeV, as a function of the scattering angle in the center of mass frame. Solid line: cross section in vacuum, calculated with the v_8' potential. Dot-dash line: medium modified cross section obtained from the effective interaction described in the text at $\rho = 0.08$ fm^{-3}. Dashed line: same as the dot-dash line, but for $\rho = 0.16$ fm^{-3}.

$f^n(r_{ij} \geq d) = 0$, for $n = 2, 3$ and 4, and $f^n(r_{ij} \geq d_t) = 0$, for $i = 5, 6$ (see, e.g., Ref. 20).

The effective interaction of Eq. (14) was tested by computing the energy per particle of symmetric nuclear matter and pure neutron matter in first order perturbation theory using the FG basis. In Fig. 2 our results are compared to those of Refs. 7 and 21. The calculations of Ref. 7 (solid lines) have been carried out using a variational approach based on the FHNC-SOC formalism, with a hamiltonian including the Argonne v_{18} NN potential and the Urbana IX three-body potential.[19] The results of Ref. 21 (dashed line of the lower panel) have been obtained using the v_8' and the same three-body potential within the framework of the Auxiliary Field Diffusion Monte Carlo (AFDMC) approach.

The results of Fig. 2 show that the effective interaction provides a fairly reasonable description of the EOS over a broad density range. Note that empirical equilibrium properties of symmetric nuclear matter are accounted for without including the somewhat *ad hoc* density dependent correction of Ref. 7. This is probably to be ascribed to the fact that, unlike the Urbana IX potential, the three-nucleon interaction (TNI) model of Ref. 20 also takes into account the contribution of many-body forces. It should also be emphasized that, using $v_{\rm eff}$ of Eq. (14) and the TNI model, one effectively includes the contribution of clusters involving more than two nucleons.

Note that our approach does not involve adjustable parameters. The correlation ranges d and d_t have been taken from Ref. 22, while the parameters entering the definition of the TNI have been determined by the authors of Ref. 20 through a fit of nuclear matter equilibrium properties. Knowing the effective interaction, the in-medium scattering probability can be readily obtained from Fermi's golden rule.

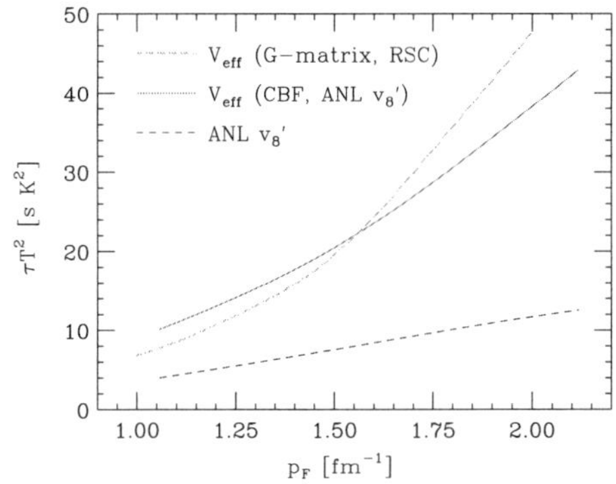

Fig. 4. Fermi momentum dependence of τT^2, τ being the quasiparticle lifetime, computed from Eq. (2) using the free-space (dashed line) and in-medium (solid line) scattering probabilities. The dot-dash line corresponds to the results of Ref. 24, obtained using G-matrix perturbation theory and the Reid soft-core potential.

The corresponding cross section at momentum transfer $\mathbf{q}$ reads

$$\frac{d\sigma}{d\Omega} = \frac{m^{\star 2}}{16\pi^2} \, |\hat{v}_{eff}(\mathbf{q})|^2 \ , \tag{15}$$

$\hat{v}_{eff}$ being the Fourier transform of the effective potential. The effective mass can also be extracted from the quasiparticle energies computed in Hartree-Fock approximation. For symmetric nuclear matter at equilibrium, we find $m^\star(p_F)/m = 0.65$, in close agreement with the lowest order CBF result of Ref. 23.

In Fig. 3 the in-medium neutron-neutron cross section at $E_{cm} = 100$ MeV obtained from the effective potential, with $\rho = \rho_0$ and $\rho_0/2$, is compared to the corresponding free space result. As expected, screening of the bare interaction leads to an appreciable suppression of the scattering cross section.

Rcplacing thc cross section in vacuum with the one defined in Eq. (15), the medium modified scattering probability can be obtained from Eq. (6). The resulting $\langle W \rangle$ can then be used to calculate the quasiparticle lifetime τ, from Eq. (2) and ηT^2, from Eq. (5).

Figure 4 shows the Fermi momentum dependence of the product τT^2, computed from Eq. (2) using both the free space and medium modified scattering probabilities. It clearly appears that the suppression of the cross section (see Fig. 3) results in a significant increase of the quasiparticle lifetime. For comparison, we also report the results of Ref. 24, whose authors derived an effective interaction using G-matrix perturbation theory and the Reid soft-core NN potential. The predictions of the two approaches based on effective interactions are close to one another for $p_F \lesssim 1.7$ fm^{-1}, corresponding to $\rho \lesssim \rho_0$. The differences observed at larger density are likely to be ascribed to the fact that the calculations of Ref. 24 does not include three-nucleon interactions, whose effects become more and more important as the density increases.

The effect of using the medium modified cross section is illustrated in the right panel of Fig. 1. Comparison between the solid and dashed lines shows that inclusion of medium modifications leads to a large increase of the viscosity, ranging between $\sim 75\%$ at half nuclear matter density to a factor of ~ 6 at $\rho = 2\rho_0$. Such an increase is likely to produce appreciable effects on the damping of neutron-star oscillations.

In conclusion, we have computed the shear viscosity of pure neutron matter using an effective interaction derived from a dynamical model that can also be used to obtain the EOS. While our results are interesting in their own right, as they can be employed in a quantitative analysis of the effect of viscosity on neutron-star oscillations, we emphasize that the work described in this paper should be seen as a first step towards the development of a general approach, allowing for a consistent calculation of the properties of neutron star matter.

The authors are grateful to V. Ferrari, for drawing their attention to the subject of this paper, and to R. Schiavilla, for providing a code for the calculation of the NN scattering cross section. Useful discussions with I. Bombaci are also gratefully acknowledged.

References

1. S. Chandrasekhar, *Phys. Rev. Lett.* **24**, 611 (1970).
2. L. Lindblom, in *Gravitational Waves: a Challenge to Theoretical Astrophysics*, Eds. V. Ferrari, J. C. Miller and L. Rezzolla, (ICTP, Trieste, 2001).
3. E. Flowers and N. Itoh, *Astrophys. J.* **206**, 218 (1976).
4. E. Flowers and N. Itoh, *Astrophys. J.* **230**, 847 (1979).
5. C. Cutler and L. Lindblom, *Astrophys. J.* **314**, 234 (1987).
6. R. B. Wiringa, V. G. J. Stoks and R. Schiavilla, *Phys. Rev. C* **51**, 38 (1995).
7. A. Akmal, V. R. Pandharipande and D. G. Ravenhall, *Phys. Rev. C* **58**, 1804 (1998).
8. J. W. Clark, *Prog. Part. Nucl. Phys.* **2**, 89 (1979).
9. S. Fantoni and V. R. Pandharipande, *Phys. Rev. C* **37**, 1697 (1988).
10. O. Benhar and M. Valli, arXiv:0707.2681, *Phys. Rev. Lett.*, in press.
11. G. Baym and C. Pethick, *Landau Fermi-Liquid Theory* (John Wiley & Sons, New York, 1991).
12. A. A. Abrikosov and I. M. Khalatnikov, *Soviet Phys. JETP* **5**, 887 (1957); *Rep. Prog. Phys.* **22**, 329 (1959).
13. G. A. Brooker and J. Sykes, *Phys. Rev. Lett.* **21**, 279 (1968).
14. V. R. Pandharipande and S. C. Pieper, *Phys. Rev. C* **45**, 791 (1992).
15. D. A. Baiko and P. Haensel, *Acta Phys. Pol.* **30**, 1097 (1999).
16. R. A. Arndt, L. D. Roper, R. A. Bryan, R. B. Clark, B. J. VerWest and P. Signell, *Phys. Rev. D* **28**, 97 (1983).
17. B. S. Pudliner, V. R. Pandharipande, J. Carlson, S. C. Pieper and R. B. Wiringa, *Phys. Rev. C* **56**, 1720 (1997).
18. S. Cowell and V. R. Pandharipande, *Phys. Rev. C* **73**, 025801 (2006).
19. B. S. Pudliner, V. R. Pandharipande, J. Carlson and R. B. Wiringa, *Phys. Rev. Lett.* **74**, 4396 (1995).
20. I. Lagaris and V. R. Pandharipande, *Nucl. Phys. A* **359**, 349 (1981).
21. A. Sarsa, S. Fantoni, K. E. Schmidt and F. Pederiva, *Phys. Rev. C* **68**, 024308 (2003).
22. A. Akmal and V. R. Pandharipande, *Phys. Rev. C* **56**, 2261 (1997).
23. S. Fantoni, B. L. Friman and V. R. Pandharipande, *Nucl. Phys. A* **399**, 51 (1983).
24. J. Wambach, T. L. Ainsworth and D. Pines, *Nucl. Phys. A* **555**, 128 (1993).

PROTONEUTRON STAR DYNAMO: THEORY AND OBSERVATIONS

ALFIO BONANNO

INAF, Osservatorio Astrofisico di Catania, Via S. Sofia 78, 95123 Catania, Italy
INFN, Sezione di Catania, Via S. Sofia 64, 95123 Catania, Italy

VADIM URPIN

A.F. Ioffe Institute of Physics and Technology
Isaac Newton Institute of Chile, Branch in St. Petersburg, 194021 St. Petersburg, Russia

We briefly review the turbulent mean-field dynamo action in protoneutron stars that are subject to convective and neutron finger instabilities during the early evolutionary phase. By solving the mean-field induction equation with the simplest model of α-quenching we estimate the strength of the generated magnetic field. If the initial period of the protoneutron star is short, then the generated large-scale field is very strong ($> 3\times10^{13}$G) and exceeds the small-scale field at the neutron star surface, while if the rotation is moderate, then the pulsars are formed with more or less standard dipole fields ($< 3 \times 10^{13}$G) but with surface small-scale magnetic fields stronger than the dipole field. If rotation is very slow, then the mean-field dynamo does not operate, and the neutron star has no global field.

1. Introduction

A protoneutron star (PNS) is a very hot ($T \sim 10^{11}$ K), rapidly rotating, lepton rich object that has been formed from the collapse of a massive stellar progenitor. It is believed that lepton and negative entropy gradients generates hydrodynamical instabilities which can play a significant dynamical role in the early stage of the PNS evolution.[1,2] While convective instability is presumably connected to the entropy gradient, the so-called neutron-finger instability is instead generated by a negative lepton gradient. The latter is due to dissipative processes which are rather fast in PNSs and it grows on a timescale $\sim 30 - 100$ ms, that is one or two orders of magnitude longer than the growth time of convection.[2] Turbulent motions caused by hydrodynamic instabilities in combination with rotation make turbulent dynamo one of the most plausible mechanism of the pulsar magnetism.

The character of turbulent dynamo depends on the Rossby number, $Ro = P/\tau$, where P is the PNS spin period and τ the turnover time of turbulence. If $Ro \gg 1$, the effect of rotation on turbulent motions is weak and the mean-field dynamo is inefficient. The small-scale dynamo can be operative, however, even at very large Rossby numbers. If $Ro \leq 1$ and the turbulence is strongly influenced by the rotation, then the PNS can be subject to a large-scale mean-field dynamo action. Note that the small-scale dynamo can still operate in this case. The Rossby number is typically large in the convective region, $Ro \sim 10 - 100$, and the mean-field dynamo is not likely to work in this region. On the contrary, except very slowly rotating PNSs, the Rossby number is of the order of unity, $Ro \sim 1$, in the neutron-finger unstable region,[3,4] where turbulent motions are slower, and the turbulence is strongly modified by rotation. This favors the efficiency of mean-field dynamos in the neutron-finger unstable region. This dynamo mechanism is then very different from the one proposed in Ref. 5, who argued that only small-scale dynamos can operate in most PNSs.

2. The Model

The problem for dynamo modelling in PNSs has been described in Refs. 3,4 We model the PNS as a sphere of radius R with two different turbulent zones separated at $R_c < R$. The inner part ($r < R_c$) corresponds to the convective region, while the outer one ($R_c < r < R$) to the neutron-finger unstable region. The mean-field equation reads

$$\frac{\partial \vec{B}}{\partial t} = \nabla \times (\vec{v} \times \vec{B} + \alpha \vec{B}) - \nabla \times (\eta \nabla \times \vec{B}) \ , \tag{1}$$

where η and α are the turbulent magnetic diffusivity and α-parameter of the dynamo theory, respectively (For details see Ref. 6). We assume that the rotation is the only large-scale motion and $\vec{v} = \vec{\Omega}(\vec{r}) \times \vec{r}$. We can basically distinguish two different dynamo regime: in the $\alpha - \Omega$ regime the dynamo action is mainly due to the

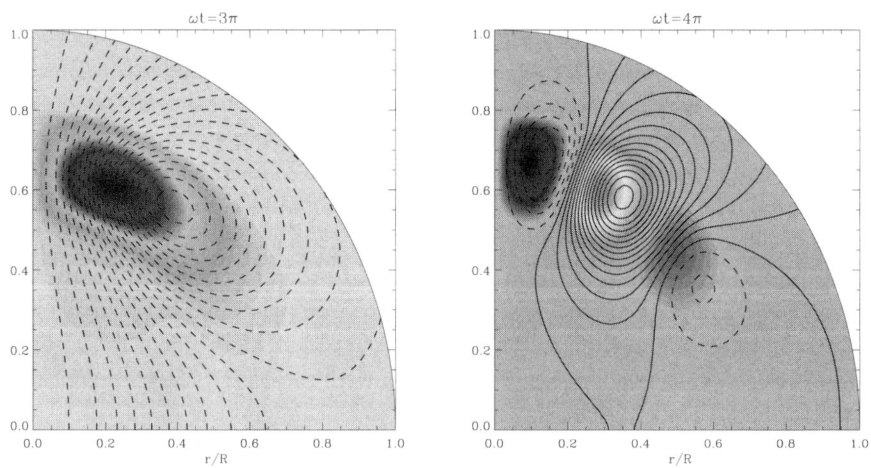

Fig. 1. On the left panel is depicted a typical field configuration for α^2-dynamo while in the right panel is depictied a field configuration in the presence of differential rotation. Color levels are for the toroidal field, while dashed lines represents poloidal field lines.

differential rotation, while in the α^2 regime the dynamo action is mainly due to the turbulent helicity. We found that most of the PNS do generate a large scale mean field dynamo provided the initial period is shorter than a critical period P_c which is of the order of $0.1 - 1$s (See Ref. 3). The typical field configuration in presence of differential rotation is depicted in Fig. 1 where we have supposed that the boundary of the NF unstable region is located at 0.6 stellar radii. As it is possible to note the region of the maximum strength of the generated field is inside the region of instability.

3. Results

The critical period that determines the onset of the mean-field dynamo action is rather long, and dynamos should be effective in most PNSs. The unstable stage lasts ~ 40 s, and this is sufficient for the dynamo to reach saturation. We can estimate a saturation field assuming the simplest α-quenching with non-linear α given by $\alpha(\tilde{B}) = \alpha_{nf}(1 + \tilde{B}^2/B_{eq}^2)^{-1}$, where $\tilde{B}$ is the characteristic value of the generated field, and B_{eq} is the equipartition small-scale magnetic field. The generated field reduces the α-parameter and slows down the generation. The saturation is reached when non-linear α becomes equal to the critical value α_0 corresponding to the marginal dynamo stability. Then, the saturation field is $B_s \approx B_{eq}\sqrt{P_c/P - 1}$. A subsequent activity of the PNS as a radiopulsar is determined by the poloidal component of the saturation field, B_{ps} and by the ratio $\xi = |B_t/B_p|$. Using the estimate $B_s \sim B_{ps}(1 + \xi)$, for B_{ps} we obtain $B_{ps} \approx B_{eq}(1 + \xi)^{-1} \sqrt{P_c/P - 1}$, The equipartition field, $B_{eq} \approx 4\pi\rho v_T^2$, varies during the unstable stage, rising rapidly soon after collapse, reaching a quasi-steady regime, and then going down when the temperature and lepton gradients are smoothed. We can estimate B_{eq} at a peak of instabilities as $\sim 10^{16}$ G in the convective zone, and $\sim (1 - 3) \times 10^{14}$ G in the neutron finger unstable zone.[7] However, the temperature and lepton gradients are progressively reduced as the PNS cools down, hence, v_T and B_{eq} decrease whereas the turnover time τ increases. The meand field dynamo description can be applied as the quasi-steady condition $\tau_{cool} \gg \tau$ is fulfilled, τ_{cool} being the cooling timescale. The final equipartition field is the same for the both the unstable zones. We thus have $B_{eq} \sim (1 - 3) \times 10^{13}$ G for the largest turbulent scale ($\ell_T = L \sim 1 - 3$ km) if $\tau_{cool} \sim$ few seconds.

We can distinguish few types of neutron stars which possess different magnetic characteristics: (i) *Strongly magnetized neutron stars.* If the initial period satisfies the condition $P < P_m \equiv P_c[1 + (1 + \xi)^2]^{-1}$ then the dynamo action leads to the formation of a strongly magnetized PNS with $B_{ps} > B_{eq} \sim 3 \times 10^{13}$G. Since $Ro \leq 1$ for fast rotators, we can expect that the rotation of such stars is almost rigid. For a rigid rotation, the α^2-dynamo is operative, hence, $\xi \sim 1 - 2$ in the neutron-finger unstable region. Therefore, strongly magnetized stars can be formed if the initial period is shorter than $\sim 0.1P_c$. Then for the surface field of such PNSs we have $B_{ps} \sim 0.3B_{eq}\sqrt{P_c/P - 1} \sim 10^{13}\sqrt{P_c/P - 1}$ G. The shortest possible period is likely $\sim$ 1 ms, hence, the strongest mean magnetic field that can be generated by the dynamo action is $\sim 3 \times 10^{14}$ G. Since the small-scale magnetic field is weaker than the large-scale field at the surface of such stars, one can expect that radiopulsations from them may have a more regular structure than those from low-field pulsars. (ii) *Neutron stars with moderate magnetic fields.* If $P_c > P > P_m$ then the generated mean poloidal field (for example, dipolar) is weaker than the small-scale field, $B_{ps} < B_{eq} \sim 3 \times 10^{13}$G. The Rossby number is ~ 1 or slightly larger for such PNSs, and their rotation can be differential. Departures from a rigid rotation are weak if P is close to P_m but can be noticeable for P close to P_c. As a result, the parameter ξ can vary within a wider range for this group of PNSs, $\xi \sim 5 - 60$, and the toroidal field should be essentially stronger than the poloidal one in the neutron-finger unstable region. The toroidal field is stronger than the small-scale field if $P_c/2 > P > P_m$, and weaker if $P_c > P > P_c/2$. Note that both differential rotation and strong toroidal field can influence the thermal evolution of such neutron stars. Heating caused by the dissipation of the differential rotation is important during the early evolutionary stage because viscosity operates on a relatively short timescale $\sim 10^2 - 10^3$ yrs. On the contrary, ohmic dissipation of the toroidal field is a slow process and can maintain the surface temperature $\sim (1-5) \times 10^5$K during $\sim 10^8$ yrs.[2] (iii) *Neutron stars with no large-scale field.* If the initial period is longer than P_c then a large scale mean-field dynamo does not operate in the PNS but the small-scale dynamo can still be efficient. We expect that such neutron stars have only small-scale fields with the strength $B_{eq} \sim 3 \times 10^{13}$ G and no dipole field. Likely, such slow rotation is rather difficult to achieve if the angular momentum is conserved during the collapse, and the number of such exotic PNSs is small.

Likely, the most remarkable property of these neutron stars is a discrepancy between the magnetic field that can be inferred from spin-down measurements and the field strength obtained from spectral observations. Features in X-ray spectra may indicate the presence of rather a strong magnetic field $\sim 3 \times 10^{13}$ G (or a bit weaker because of ohmic decay during the early evolution) associated to sunspot-like structures at the surface of these objects. The field inferred from spin-down data should be essentially lower.

In particular the recent discovery of young slowly spinning radio pulsar 1E 1207.4-5209[8] with a very weak magnetic field seems to support our theory.

References

1. S. Bruenn and T. Dineva, *Ap. J.* **458**, L71 (1996).
2. J. Miralles, J. Pons and V. Urpin, *Ap. J.* **543**, 1001 (2000).
3. A. Bonanno, L. Rezzolla and V. Urpin, *A&A* **410**, L33 (2003).
4. A. Bonanno, V. Urpin and G. Belvedere, *A&A* **440**, 149 (2005).
5. C. Thompson and R. Duncan, *Ap. J.* **408**, 194 (1993).
6. A. Bonanno, V. Urpin and G. Belvedere, *A&A* **451**, 1049 (2006).
7. V. Urpin and J. Gil, *A&A* **415**, 305 (2004).
8. E. V. Gotthelf and J. P. Halpern, *Ap. J.* **664**, L35 (2007).

MAGNETIC FIELD DISSIPATION IN NEUTRON STARS: FROM MAGNETARS TO ISOLATED NEUTRON STARS

JOSE A. PONS

Departament de Física Aplicada, Universitat d'Alacant,
Ap. Correus 99, 03080 Alacant, Spain
E-mail: jose.pons@ua.es

We present the first long term simulations[1] of the nonlinear magnetic field evolution in realistic neutron star crusts with a stratified electron number density and temperature dependent conductivity. We show that Hall drift influenced Ohmic dissipation takes place in neutron star crusts on a timescale of 10^6 years. When the initial magnetic field has magnetar strength, the fast Hall drift results in an initial rapid dissipation stage that lasts $\sim 10^4$ years. The interplay of the Hall drift with the temporal variation and spatial gradient of conductivity tends to favor the displacement of toroidal fields toward the inner crust, where stable configurations can last for $\sim 10^6$ years. The decay of magnetic fields in young neutron stars is also supported by the existence of a strong trend between neutron star surface temperature and the magnetic field strength.[2] This trend can be explained by the decay of currents in the crust over a time scale of $\sim 10^6$ yr. The implication of these results on the interpretation of cooling curves and previous constraints on the existence of exotic phases in neutron star interiors will be outlined.

References

1. J. A. Pons and U. Geppert, *A&A* (2007), in press; (astro-ph/0703267).
2. J. A. Pons *et al.*, *Phys. Rev. Lett.* **98**, 071101 (2007).

GRAVITATIONAL RADIATION AND EQUATIONS OF STATE IN SUPER-DENSE CORES OF CORE-COLLAPSE SUPERNOVAE

KEI KOTAKE

Division of Theoretical Astronomy, National Astronomical Observatory Japan,
2-21-1, Osawa, Mitaka, Tokyo, 181-8588, Japan
Max-Planck-Institut für Astrophysik, Karl-Schwarzschild Str. 1, D-85741 Garching, Germany,
E-mail: kkotake@th.nao.ac.jp

Core-collapse supernovae have been supposed to be one of the most plausible sources of gravitational waves. Based on a series of our magnetohydrodynamic core-collapse simulations, we found that the gravitational amplitudes at core bounce can be within the detection limits for the currently running laser-interferometers for a galactic supernova if the central core rotates sufficiently rapidly. This is regardless of the difference of the realistic equations of state and the possible occurence of the QCD phase transition near core bounce. Even if the core rotates slowly, we point out that the gravitational waves generated from anisotropic neutrino radiation in the postbounce phase due to the standing accretion shock instability (SASI) could be within the detection limits of the detectors in the next generation such as LCGT and the advanced LIGO for the galactic source. Since the waveforms significanly depend on the exploding scenarios, our results suggest that we can obtain the information about the long-veiled explosion mechanism from the gravitational wave signals when the supernova occurs near to us.

Keywords: Core collapse supernovae; Gravitational waves; Neutrinos.

1. Introduction

The gravitational astronomy is now gaining practicality. In fact significant progress has been made on the gravitational wave (GW) detectors, such as TAMA300, LIGO, VIGRO, GEO600, and AIGO with their international network of the observatories. For the detectors, core-collapse supernovae especially in our Galaxy, have been supposed to be the most plausible sources of GWs (see, for example, Ref. 1 for a review). Since the GWs (plus neutrinos) are the only tool that gives us the information in the innermost part of an evolved massive stars, the detection is important not only for the direct confirmation of general relativity but also for unveiling the physics of supernovae itself such as the equation of state (EOS) in the central cores.

Since massive stars are generally rotating, the stellar rotation has been long supposed to play an important role in the GWs from core collapse supernovae. The large-scale asphericities at core bounce induced by rotation can convert the part of the gravitational energy into the form of the GWs, which we start to discuss at first below.

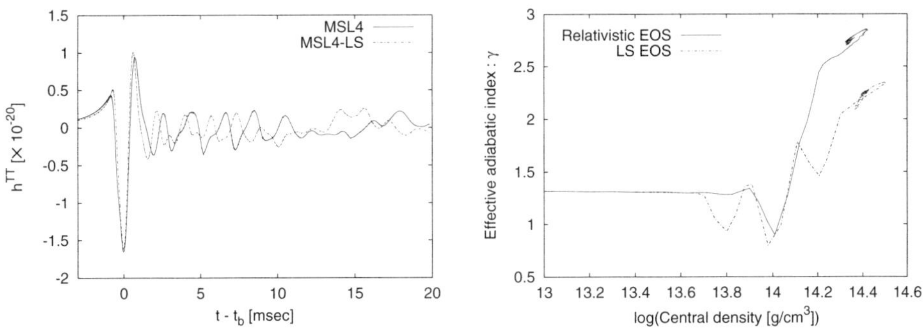

Fig. 1. Waveforms (left panel) for the models with the relativistic EOS (the solid line labeled as MSL4) and with LS EOS (the dashed line labeled as MSL4-LS) and the relation between the central density and the effective adiabatic index γ near core bounce (right panel).

2. Effect of Realistic Equations of State

Needless to say, the equation of state (EOS) is an important microphysical ingredient for determining the dynamics of core collapse and, eventually, the GW amplitude. As a realistic EOS, Lattimer-Swesty (LS) EOS has been used in recent papers discussing gravitational radiations from the rotational core collapse. It has been difficult to investigate the effect of EOS's on the gravitational signals because available EOS's based on different nuclear models are limited. Recently, a new complete EOS for supernova simulations has become available[5] bbased on the relativistic mean field theory. By implementing these two realistic EOS's, we looked into the difference of the GW signals.[6] The left panel of Fig. 1 shows the waveforms for the models with the relativistic EOS (model MSL4) or the LS EOS (model MSL4-LS). The maximum amplitudes for the two models do not differ significantly (left panel). The important difference of the two EOS's is the stiffness. As seen from the right panel of Fig. 1, LS EOS is softer than the relativistic EOS, which makes the central density larger at core bounce and thus results in the shorter time interval between the subsequent bounces. On the other hands, softer EOS results in the smaller lepton fraction in the inner core, which reduces the mass quadrupole moments at core bounce. By the competition of these factors, the maximum amplitude remains almost the same between the two realistic EOS's, while the typical frequencies of the GW become slightly higher for the softer equation of state. Furthermore it was found that the type III waveform observed in a very soft EOS polytropic EOS does not appear when the realistic EOSs are employed.

3. Possible QCD Phase Transition and its Impact on the Waveforms

As has been presented, the quark matter might appear during supernova explosions (see Refs. 2). Current supernova studies demonstrate that the stellar collapse of stars below $\sim 25 M_\odot$ in the main sequence stage leads to the formation of neutron stars, while in the case of more massive stars, to the formation of the black hole.

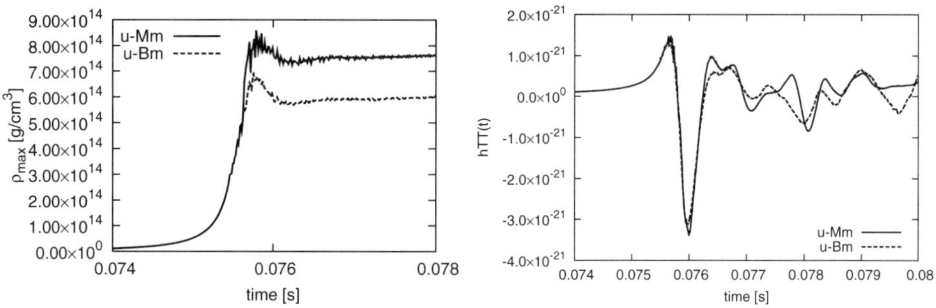

Fig. 2. Time evolution of the central density (left) and the gravitational waveforms for the model with (u-Mm) and without (u-Bm) the QCD phase trantion. The bounce occurs around 0.075-0.076 s. The initial angular velocity of 2.3 rad s^{-1} with the uniform rotation is imposed on the $40M_{\odot}$ model by Nomoto and Hashimoto.

In the latter case, quark matters might appear because the central density could exceed the density of the QCD phase transition. Performing two-dimensional, magnetohydrodynamical core-collapse simulations of massive stars accompanying the QCD phase transition, we studied how the phase transition affects the gravitational waveforms near the epoch of core-bounce. Here the first order phase transition is assumed, where the conversion of bulk nuclear matter to a chirally symmetric quark-gluon phase is described by the MIT bag model. From the left panel of Fig. 2, one can see the slight increase of the central density at the moment of core bounce, but the resultant enhancements in the amplitudes are small (see the right panel), or can be up to $\sim$ 10 percents than the ones without the phase transition, even when we assumed the QCD phase at the relatively low density and explored the very soft EOS before the transition. Furthermore we found that the maximum amplitudes become smaller up to $\sim$ 10 percents owing to the phase transition, when the degree of the differential rotation becomes larger.[2]

4. Gravitational Wave Signals at Postbounce Phase

In addition to the bounce signals, two other sources of the gravitational-wave emissions have been considered to be important in the later phases after core bounce, namely convective motions and anisotropic neutrino radiation. Both of them can contribute to the non-spherical part of the energy momentum tensor of the Einstein equations.

We performed long-term two dimensional axisymmetric simulations in the post-bounce phase of core-collapse supernovae to study how the asphericities induced by the growth of the standing accretion shock instability (SASI)[3] produce the gravitational waveforms.[4] It is found that the waveforms due to the anisotropic neutrino emissions show the monotonic increase with time (see the inset of the left panel of Fig. 3), whose amplitudes are up to two order-of-magnitudes larger than the ones from the convective matter motions outside the protoneutron stars . From the

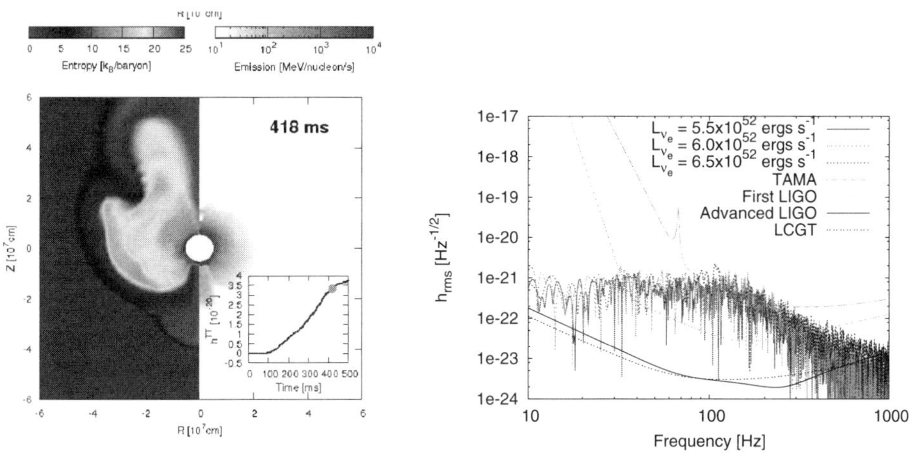

Fig. 3. Left panel is the entropy (left half of the left panel) and net neutrino emissivity (right half of the left panel) distributions in the meridian section, showing the growth of the shock instability. The inset shows the wave amplitudes. Right panel shows the GW spectrum contributed for different luminosities models with the detection limits of TAMA, first LIGO, advance LIGO, and LCGT.

spectrum analysis of the waveforms (the right panel of Fig. 3), we find that the amplitudes can be within the detection limits of the detectors in the next generation such as LCGT and the advanced LIGO for a supernova at 10 kpc. It is noted that the GWs from neutrinos are dominant over the ones from the matter motions at the frequency below ~ 100 Hz. It is noted that the waveforms from the explosions excited by the g-mode oscillations of the protoneutron star[7] are significantly different from the ones of the neutrino driven explosions presented here. Thus our results suggest that we could obtain the information about the long-unsettled explosion mechanism if the supernova occurs near to us.

References

1. K. Kotake, K. Sato and K. Takahashi, *Rep. Prog. Phys.* **69**, 971 (2006).
2. N. Yasutake, K. Kotake, M. A. Hashimoto and S. Yamada, *Phys. Rev. D* **75**, 084012 (2007).
3. N. Ohnishi, K. Kotake and S. Yamada, *Astrophys. J.* **641**, 1018 (2006).
4. K. Kotake, N. Ohnishi and S. Yamada, *Astrophys. J.* **655**, 406 (2007).
5. H. Shen, H. Toki, K. Oyamatsu and K. Sumiyoshi, *Nucl. Phys. A* **637**, 435 (1998).
6. K. Kotake *et al.*, *Phys. Rev. D* **69**, 124004 (2004).
7. C. D. Ott, A. Burrows, L. Dessart and E. Livne, *Phys. Rev. Lett.* **96**, 201102 (2006).

JOULE HEATING IN THE COOLING OF MAGNETIZED NEUTRON STARS

DEBORAH N. AGUILERA,* JOSÉ A. PONS and JUAN A. MIRALLES

Departamento de Física Aplicada, Universidad de Alicante,
Apartado Correos 99, 03080 Alicante, Spain
** E-mail: deborah.aguilera@ua.es*

We present 2D simulations of the cooling of neutron stars with strong magnetic fields ($B \geq 10^{13}$ G). We solve the diffusion equation in axial symmetry including the state of the art microphysics that controls the cooling such as slow/fast neutrino processes, superfluidity, as well as possible heating mechanisms. We study how the cooling curves depend on the the magnetic field strength and geometry. Special attention is given to discuss the influence of magnetic field decay. We show that Joule heating effects are very large and in some cases control the thermal evolution. We characterize the temperature anisotropy induced by the magnetic field for the early and late stages of the evolution of isolated neutron stars.

1. Introduction

The observed thermal emission of neutron stars (NSs) can provide information about the matter in their interior. Comparing the theoretical cooling curves with observational data[1,2] one can infer not only the physical conditions of the outer region (atmosphere) where the spectrum is formed but also of the poorly known interior (crust, core) where high densities are expected.

There is increasing evidence that most of nearby NSs whose thermal emission is visible in the X-ray band have a non uniform temperature distribution.[3,4] There is a mismatch between the extrapolation to low energy of the fits to X-ray spectra, and the observed Rayleigh Jeans tail in the optical band (*optical excess flux*), that cannot be addressed with a unique temperature (e.g. RX J1856.5−3754,[5] RBS 1223,[6] and RX J0720.4−3125[7]).

A non uniform temperature distribution may be produced not only in the low density regions,[8] but also in intermediate density regions, such as the solid crust. Recently, it has been proposed that crustal confined magnetic fields with strengths larger than 10^{13} G could be responsible for the surface thermal anisotropy.[9,10] In the crust, the magnetic field limits the movement of electrons (main responsible for the heat transport) in the direction perpendicular to the field and the thermal conductivity in this direction is highly suppressed, while remains almost unaffected along the field lines.

Moreover, the observational fact that most thermally emitting isolated NSs have

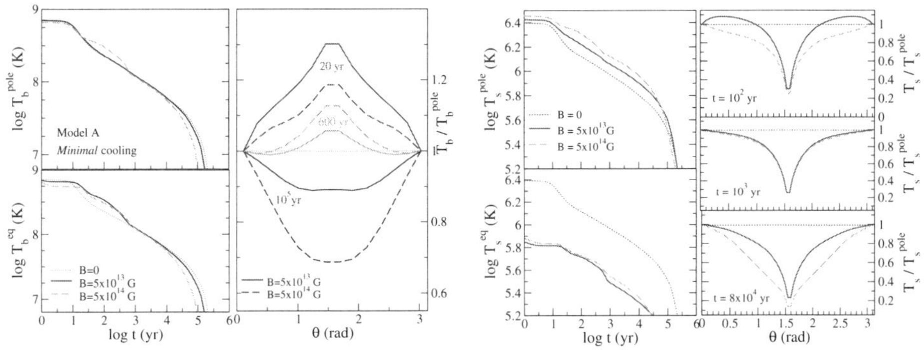

Fig. 1. Cooling of strongly magnetized NSs for the Model A. Temperatures T_b (left figure) and T_s (right figure) at the pole and at the equator as a function of the age t (left panels). Corresponding evolution of T/T^{pole} vs. θ in the right panels.

magnetic fields larger than 10^{13} G implies that a realistic cooling model must include magnetic field effects. In a recent work,[11] first 2D simulations of the cooling of magnetized NSs have been presented. In particular, it has been stated that magnetic field decay, as a heat source, could strongly affect the thermal evolution and the observations should be reinterpreted in the light of these new results. We present the main conclusions of this work next.

2. Non-Uniform Temperature Distribution Induced by Magnetic Field

We consider two baseline models:[11] Model A, a low mass NS with $M = 1.35\ M_\odot$ and Model B, a high mass NS with $M = 1.63\ M_\odot$. These two models correspond to the *minimal cooling* scenario (controlled by modified Urca neutrino emission) and the *fast cooling* scenario (where direct Urca operates), respectively. We consider crustal confined magnetic fields, keeping the geometry fixed and varying the magnetic field strength at the pole (B).

For a given magnetic field, we solve the diffusion equation in axial symmetry, considering an anisotropic thermal conductivity tensor $\hat{\kappa}$. The ratio of its components along and perpendicular to the field can be defined in terms of the magnetization parameter ($\omega_B\tau$) as $\kappa_e^{\parallel}/\kappa_e^{\perp} = 1 + (\omega_B\tau)^2$, where τ is the electron relaxation time and ω_B is the electron cyclotron frequency. When $\omega_B\tau \gg 1$ the magnetic field effects on the transport properties are crucial. We show the results in Fig. 1 (on the left), where the temperature at the base of the envelope T_b (at 10^9 g/cm^3) is shown as a function of the age t. We see an *inverted temperature distribution* with cooler polar caps and a warmer equatorial belt at $t \lesssim 500$ yr for the Model A. Similar qualitative results are found for Model B.

In Fig. 1 (on the right) we show for Model A the corresponding cooling curves but for the surface temperature T_s. We see that the anisotropy found at the level of T_b does not automatically result in a similar T_s distribution: the blanketing effect of the envelope overrides the inverted temperature distribution found at intermediate

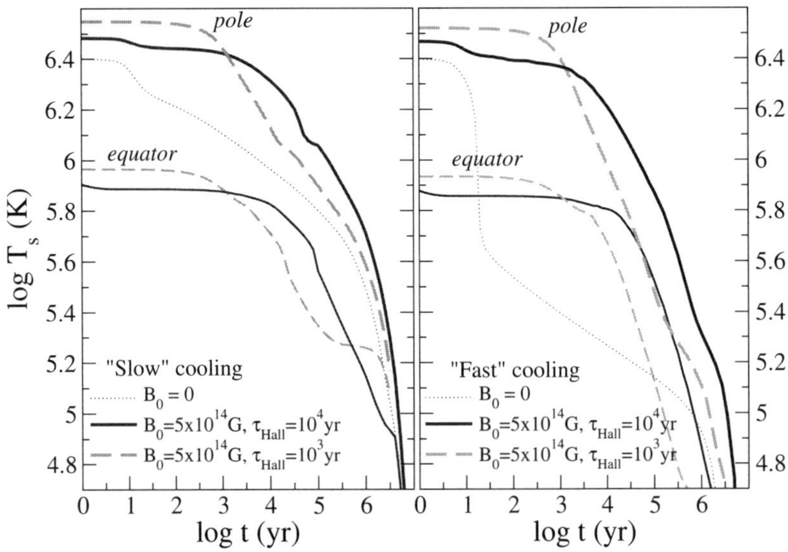

Fig. 2. Cooling of strongly magnetized NSs with Joule heating with $B_0 = 5 \times 10^{14}$ G for Model A (left panel) and Model B (right panel). We have set $\tau_{\rm Ohm} = 10^6$ yr.

ages. Thus, the equator remains always cooler than the pole, and only at early times and for strong fields we find larger surface temperatures in middle latitude regions.

3. Magnetic Field Decay and Joule Heating

For the magnetic field decay, we assume the approximate solution of the diffusion equation

$$B = B_0 \frac{\exp\left(-t/\tau_{\rm Ohm}\right)}{1 + \frac{\tau_{\rm Ohm}}{\tau_{\rm Hall}}\left(1 - \exp\left(-t/\tau_{\rm Ohm}\right)\right)}, \tag{1}$$

where $\tau_{\rm Ohm}$ is the Ohmic characteristic time, and the typical timescale of the fast, initial stage is defined by $\tau_{\rm Hall}$.

In the cooling curves, there is a huge effect due to the decay of such a large field: as a consequence of the heat released, T_s remains much higher than in the case of constant field (Fig. 2). The strong effect of the field decay is evident for all of the pairs of parameters chosen. Notice that T_s of the initial plateau is higher for shorter $\tau_{\rm Hall}$, but the duration of this stage with nearly constant temperature is also shorter. After $t \simeq \tau_{\rm Hall}$, there is a drop in T_s due to the transition from the fast Hall stage to the slower Ohmic decay.

4. Conclusion: Towards a Coupled Magneto-Thermal Evolution

The main result of this work is that, in magnetized NSs with $B > 10^{13}$ G, the decay of the magnetic field affects strongly their cooling. In particular, there is a huge

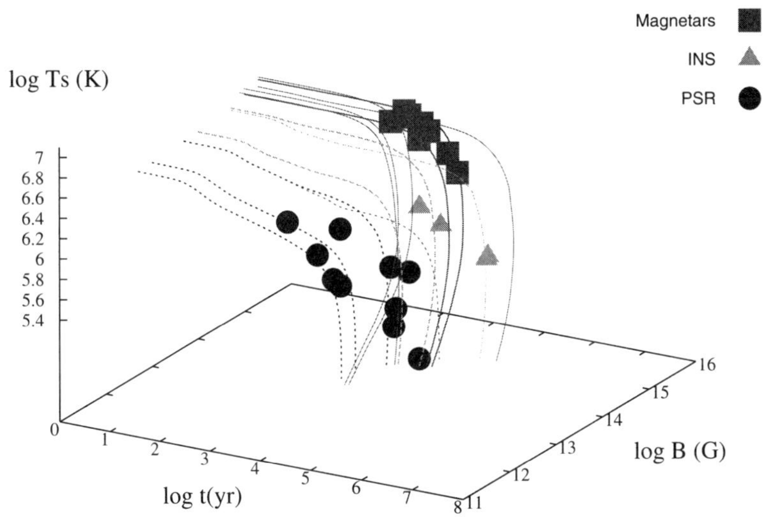

Fig. 3. Coupled magneto-thermal evolution of isolated neutron stars:[12] T_s for the hot component as a function of B and t. Observations: squares for magnetars ($B > 10^{14}$ G), triangles for intermediate-field isolated NSs (10^{13} G$< B < 10^{14}$ G) and circles for radio pulsars ($B < 10^{12}$ G). Corresponding cooling curves in solid, dashed and dotted lines, respectively.

effect of Joule heating on the thermal evolution. In NSs born as magnetars, this effect plays a key role in maintaining them warm for a long time. Moreover, it can also be important in high magnetic field radio pulsars and in radio–quiet isolated NSs. As a conclusion, the thermal and magnetic field evolution of a NS is at least a two parameter space (Fig. 3), and a first step towards a coupled magneto-thermal evolution has been given in this work.

References

1. D. G. Yakovlev and C. J. Pethick, *ARA&A* **42**, 169 (2004).
2. D. Page, U. Geppert and F. Weber, *Nucl. Phys. A* **777**, 497 (2006).
3. V. E. Zavlin, ArXiv, astro-ph/0702426 (2007).
4. F. Haberl, *Ap&SS* **308**, 181 (2007).
5. J. A. Pons *et al.*, *ApJ* **564**, 981 (2002).
6. A. D. Schwope, V. Hambaryan, F. Haberl and C. Motch, *Ap&SS* **308**, 619 (2007).
7. J. F. Pérez-Azorín, J. A. Pons, J. A. Miralles and G. Miniutti, *A&A* **459**, 175 (2006).
8. G. Greenstein and G. J. Hartke, *ApJ* **271**, 283 (1983).
9. U. Geppert, M. Küker and D. Page, *A&A* **426**, 267 (2004).
10. J. F. Pérez-Azorín, J. A. Miralles and J. A. Pons, *A&A* **451**, 1009 (2006).
11. D. N. Aguilera, J. A. Pons and J. A. Miralles, arXiv:0710.0854 [astro-ph] (2007).
12. D. N. Aguilera, J. A. Pons and J. A. Miralles, in preparation, (2007).

EXOTIC FERMI SURFACE OF DENSE NEUTRON MATTER

M. V. ZVEREV* and V. A. KHODEL+

Russian Research Centre Kurchatov Institute,
Moscow, 123182, Russia
**E-mail: zverev@mbslab.kiae.ru*
+E-mail: vak@wuphys.wustl.edu

J. W. CLARK

McDonnell Center for the Space Sciences
and Department of Physics, Washington University,
St. Louis, MO 63130, USA
E-mail: jwc@wuphys.wustl.edu

We demonstrate that the Fermi surface of dense neutron matter experiences a rearrangement near the onset of pion condensation, due to strong momentum dependence of the effective interaction induced by spin-isospin fluctuations. We show that new filling $n(p)$ has a hole adjacent to the center of the Fermi sphere, and its appearance activates the normally forbidden Urca cooling mechanism.

Keywords: Spin-isospin fluctuations; Topological transition; Multi-connected Fermi surface.

1. Introduction

In the Landau-Migdal theory of Fermi liquids,[1,2] the ground state of a homogeneous Fermi system is described in terms of a quasiparticle momentum distribution $n_F(p,T)$ that coincides with the momentum distribution of the ideal Fermi gas. This theory has been remarkably successful in advancing our qualitative and quantitative understanding of a broad spectrum of Fermi systems, including bulk liquid ^{3}He, conventional superconductors, and nucleonic subsystems in neutron stars. However, the theory is known to fail in the strongly correlated electron systems of solids. Certain experimental results obtained recently[3–5] may prove decisive to understanding of this failure. The systems involved are a dilute two-dimensional (2D) electron gas and 2D liquid ^{3}He. The experiments show how, under variation of the density, these systems progress from conditions of moderate correlations to the regime of very strong correlations. A striking feature is that both systems appear to experience a divergence of the effective mass M^* as the density approaches a certain critical value ρ_*. Beyond the point of divergence of the effective mass, the conventional Fermi liquid theory fails. In this paper, we will analyze a similar failure of this theory in dense neutron matter due to exchange of spin-isospin fluctuations, strongly

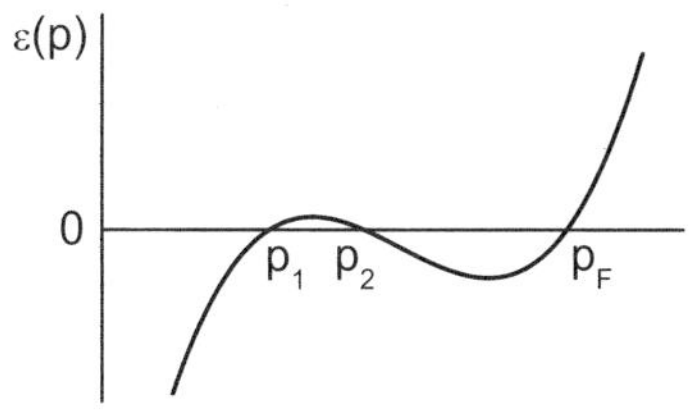

Fig. 1. Illustration of the emergence of additional roots p_1, p_2 of equation (3).

enhanced near the onset of pion condensation.

2. Topological Instability of the Landau State

We base our analysis on a necessary condition for the stability of the Landau state, which reads that for any admissible variation from the FL quasiparticle distribution $n_F(p) = \theta(p_F - p)$, preserving the particle number, the change δE_0 of the Landau ground state energy E_0 must be *positive*. Explicitly, this condition reads

$$\delta E_0 = \int \epsilon(p; n_F)\delta n_F(\mathbf{p})\, d\tau > 0 \ , \tag{1}$$

for any variation $\delta n_F(\mathbf{p})$ satisfying

$$\int \delta n_F(\mathbf{p})\, d\tau = 0 \ . \tag{2}$$

In these equations, $d\tau$ is the volume element in momentum space, while $\epsilon(p)$ is the single-particle spectrum, measured from the chemical potential μ and evaluated with the distribution $n_F(p)$.

The condition (1) holds provided the equation

$$\epsilon(p) = 0 \tag{3}$$

has the single root $p = p_F$. Otherwise, it is violated, the Landau state loses its stability, and the ground state alters due to a rearrangement of single-particle degrees of freedom. In weakly correlated Fermi systems, $\epsilon(p)$ is a monotonic function of p, so that equation (3) has no extra roots. However, as correlations build up, in many cases, (see e.g. Refs. 6–9), the character of the curve $\epsilon(p)$ changes, it becomes non-monotonic. In particular, the spectrum becomes non-monotonic in the vicinity of an impending second-order phase transition, specified by critical fluctuations of the wave number $q_c > 0$. Let the second-order phase transition occur at a critical density ρ_c. As shown in Ref. 10, there is another critical density ρ_b, at which a bifurcation arises in equation (3), resulting in the emergence of a two additional roots p_1, p_2 (see Fig. 1), the distance between which increases linearly from zero in proportion to $|\rho - \rho_b|$. This means that the stability condition (1) is violated when applied to variations of the quasiparticle distribution $n(p)$ for momenta lying within the interval $p_1, p_2]$. The instability of the ground state results in its alteration due to a *topological* phase transition,[11] in which the Fermi surface becomes

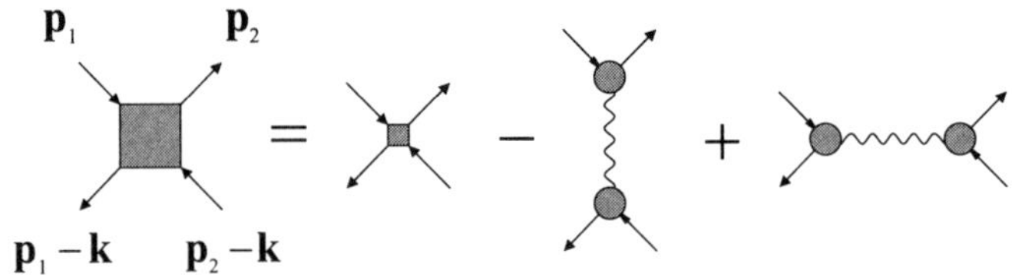

Fig. 2. Regular and singular contributions to the scattering amplitude in the vicinity of the second-order phase transition.

multi-connected but no violation of any symmetry occurs. In fact, this and related phenomena were studied in model problems more than 20 years ago.[12,13] It should be emphasized that in the case at issue, a rearrangement that entails a major alteration of the ground state in configuration space, is disfavored energetically and therefore irrelevant to the present study.

It is worth noting that any change of the momentum distribution $n_F(p)$ entails an increase of the kinetic energy of the quasiparticle system. Therefore the anticipated rearrangement of $n_F(p)$ becomes possible only if it is accompanied by a counterbalancing reduction of potential energy that requires a substantial momentum dependence of the effective interaction between quasiparticles. The emergence of such a strong momentum dependence is exactly what one expects to occur as the density ρ is increased toward the critical value ρ_c for a second-order phase transition in which a branch of the spectrum $\omega_s(q)$ of collective excitations of the Fermi system collapses at a nonzero value q_c of the wave vector q.

3. Landau Quasiparticle Interaction Function near the Point of Pion Condensation

To justify this expectation, we follow Dyugaev[14] and consider the behavior of the quasiparticle scattering amplitude $F(\mathbf{p}_1, \mathbf{p}_2, \mathbf{k}) \equiv z^2\Gamma(\mathbf{p}_1, \mathbf{p}_2; \mathbf{k}, \omega = 0)M^*/M$ near of the phase-transition point. Here $\Gamma(\mathbf{p}_1, \mathbf{p}_2; \mathbf{k}, \omega)$ is the ordinary (in-medium) scattering amplitude, M^* is the effective mass, and z is the renormalization factor specifying the weight of the quasiparticle pole. The amplitude F can be written as the sum $F^r + F^s$ of a regular part F^r and a singular part F^s, with the latter taking the universal form

$$F^s_{\alpha\delta;\beta\gamma}(\mathbf{p}_1, \mathbf{p}_2, \mathbf{k}; \rho \to \rho_c) = -O_{\alpha\delta}O_{\beta\gamma}D(k) + O_{\alpha\gamma}O_{\beta\delta}D(|\mathbf{p}_1 - \mathbf{p}_2 + \mathbf{k}|) \qquad (4)$$

in terms of the propagator $D(k)$ of the collective excitation. This form has been derived with due attention to the antisymmetry of the two-particle wave function under exchange of the particle coordinates (spatial, spin, isospin).

The vertex O appearing in Eq. (3) determines the structure of the collective-mode operator and is normalized by $\mathrm{Tr}(OO^\dagger) = 1$. Specifically, the choice $O = 1$ is made in treating the rearrangement of the quasiparticle distribution due to collapse of density oscillations, while $O = \vec{\sigma}$ is appropriate when studying the rearrangement of $n_F(p)$ triggered by the softening of the spin collective mode. In the present investigation we will be concerned with dense, homogeneous neutron matter in which abnormal occupation is induced by spin-isospin fluctuations; thus the pertinent operator is $O = (\vec{\sigma}\cdot\mathbf{k})\vec{\tau}$.

The propagator D is given by the Migdal formula

$$-D^{-1}(q,\omega) = q^2 + m_\pi^2 + \Pi_{NI}(q,\omega=0) + \Pi_{NN}(q,\omega) \ , \tag{5}$$

made up of the ordinary part

$$\Pi_{NN}(q,\omega) = -Bq^2 - iq|\omega|M^2/(2\pi m_\pi^2) \tag{6}$$

of the pion polarization operator Π, along with the term

$$\Pi_{NI}(q,\omega=0) = -q^2\rho/\rho_I(1+q^2/q_I^2) \tag{7}$$

arising from pion conversion into a Δ-isobar and neutron hole. In the domain of critical fluctuations, we have[15,16]

$$-D^{-1}(q\to q_c;\rho\to\rho_c;\omega=0) = \gamma^2\frac{(q^2-q_c^2)^2}{q_I^2} + \eta\kappa^2 q_I^2 \ , \tag{8}$$

where $\eta = (\rho_c-\rho)/\rho_c$. The parameters γ and κ are determined from the obvious relations

$$(1-B)q_c^2 + m_\pi^2 - q_c^2 r_c\zeta_c = 0 \tag{9}$$

and

$$1 - B - r_c\zeta_c^2 = 0 \ , \tag{10}$$

with $r_c = \rho_c/\rho_I$ and $\zeta_c = (1+q_c^2/q_I^2)^{-1}$. Simple algebra leads to

$$q_c = (m_\pi q_I)^{1/2}(1-B)^{-1/4} \ , \tag{11}$$

$$r_c = \left(m_\pi/q_I + \sqrt{1-B}\right)^2 \ , \tag{12}$$

$$\gamma^2 = r_c\zeta_c^3 \ , \tag{13}$$

and

$$\kappa^2 = r_c\zeta_c q_c^2/q_I^2 \ . \tag{14}$$

Employing the parameter set

$$q_I^2 = 5\,m_\pi^2, \quad \rho_I \simeq 1.8\,\rho_0, \quad B \simeq 0.7 \tag{15}$$

from Refs. 15,16, one arrives at

$$q_c \simeq 1.9\,m_\pi, \quad \rho_c \simeq 2\,\rho_0, \quad \gamma \simeq 0.4, \quad \kappa \simeq 0.7 \ . \tag{16}$$

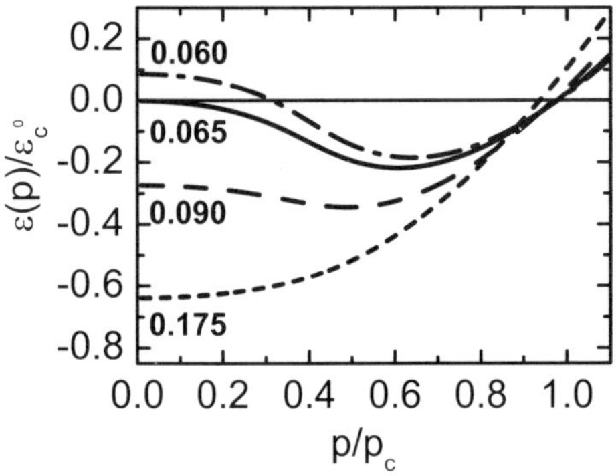

Fig. 3. Neutron quasiparticle spectrum $\varepsilon(p)$ in units of $\varepsilon_c^0 = p_c^2/2M = (3\pi^2\rho_c)^{1/3}/2M$, plotted for different densities. Numbers near the curves give the corresponding values of $\eta = (\rho_c - \rho)/\rho_c$.

4. Numerical Evaluation of the Neutron Single-Particle Spectrum

Eqs. (4) and (8) furnish a suitable basis for efficient evaluation of the single-particle spectrum $\epsilon(p)$ near the second-order phase transition. We exploit a straightforward connection between $\epsilon(p)$ and the scattering amplitude $F(\mathbf{p}_1, \mathbf{p}_2, \mathbf{k} = 0)$, thereby circumventing the frequency integration that would be encountered in an RPA approach. This connection is made through the Landau relation[17,18]

$$\frac{\partial\epsilon(p)}{\partial\mathbf{p}} = \frac{\mathbf{p}}{M} + \frac{1}{2}\int F_{\alpha\beta;\alpha\beta}(\mathbf{p}, \mathbf{p}_1)\frac{\partial n(p_1)}{\partial\mathbf{p}_1}\, d^3p_1/(2\pi)^3 \,. \tag{17}$$

As far as in the neighborhood of the soft-mode phase-transition point, the second term in Eq. (4) for the function $F^s(\mathbf{p}_1, \mathbf{p}_2, \mathbf{k} = 0)$, proportional to $D(\mathbf{p}_1 - \mathbf{p}_2)$, dominates, the contribution to Eq. (17) from the singular part of F can be easily integrated over the momentum $\mathbf{p}$ to obtain an equation

$$\epsilon(p) = \frac{p^2 - p_F^2}{2M_r^*} + \frac{1}{2}\int D(\mathbf{p} - \mathbf{p}_1)\, n(p_1)\, d^3p_1/(2\pi)^3 \,, \tag{18}$$

well suited to investigation of stability of the Landau state as the density climbs to the critical value ρ_c. In stating this result, we assume that the contribution to $\epsilon(p)$ from the *regular, nonsingular* part of F can be simulated by replacing of the bare mass M appearing in Eq. (18) by an effective mass M_r^*. The generally accepted values for this effective mass are in the range 0.7–0.8 for the pertinent densities in the neutron-star interior.

Results of numerical calculations of the neutron quasiparticle spectrum $\epsilon(p)$ based on Eq. (18), depicted in Fig. 4, indicate that a bifurcation in Eq. (3) emerges at $\eta = (\rho_c - \rho)/\rho_c \simeq 0.065$. At $\eta < 0.065$ the additional zero p_1 of the quasiparticle spectrum $\epsilon(p)$ moves away from $p = 0$ and the Landau state becomes unstable against spontaneous generation of a quasihole in the interval $[0, p_1]$ and a quasiparticle at $p > p_F$. Hence, the Landau state should be rearranged, and the search of a new ground state of the quasiparticle system should be done in a self consistent way.

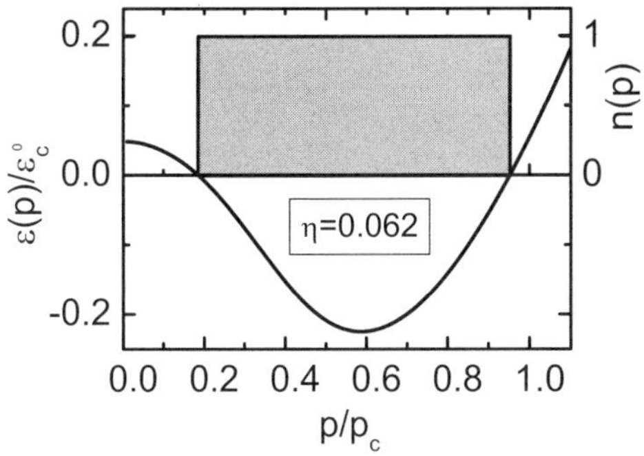

Fig. 4. Neutron quasiparticle spectrum $\varepsilon(p)$ in units of ε_c^0 evaluated in a self-consistent way for $\eta = 0.062$ together with the associated quasiparticle momentum distribution $n(p)$.

5. Topological Scenario of a Rearrangement of the Landau State

Eq. (18) may serve as an effective tool for a self-consistent evaluation of the quasiparticle spectrum $\epsilon(p)$ provided it is considered at temperature $T > 0$. Then upon substitution of the Fermi-Dirac distribution

$$n(p) = \frac{1}{1 + e^{\epsilon(p)/T}} \tag{19}$$

into the r.h.s. of Eq. (18) one arrives at the equation

$$\epsilon(p) + \mu = \frac{p^2}{2M_r^*} + \frac{1}{2}\int \frac{D(\mathbf{p} - \mathbf{p}_1)}{1 + e^{\epsilon(p_1)/T}}\, d^3p_1/(2\pi)^3 \ , \tag{20}$$

which, together with the normalization condition

$$\int n(p)\, d^3p/(2\pi)^3 = \rho \ , \tag{21}$$

provides a set for a self-consistent evaluation of the quasiparticle spectrum $\epsilon(p)$, the chemical potential μ and the quasiparticle momentum distribution $n(p)$.

Numerical results for $\eta = 0.062$ obtained at $T/\epsilon_c^0 = 10^{-5}$ are shown in Fig. 4. We see that the neutron Fermi surface changes its topology: an empty sphere appears at momenta $p < p_1 \simeq 0.2\, p_c$. This topological transition is similar to the one found in numerical evaluation of the quasiparticle spectrum of 2D liquid ^{3}He[8] and 2D electron gas.[9] Further topological metamorphoses of the neutron Fermi surface with approaching the pion-condensation point are shown in Fig. 5. As the parameter η decreases, the inner Fermi sphere gradually inflates as shown in upper right panel. However, at $\eta \simeq 0.018$ a new topological transition occurs, namely, a new sheet of the Fermi surface arises which surrounds a new small occupied sphere inside the empty one (see lower left panel). Two last panels show next topological transitions resulting in appearance of a multi-connected Fermi surface.

The topological phase transition giving rise to the alteration of the Fermi-Dirac formula for the neutron momentum distribution, described above, furnishes an opportunity to elucidate enhanced cooling of the Vela, Geminga and 3C58 pulsars.[21]

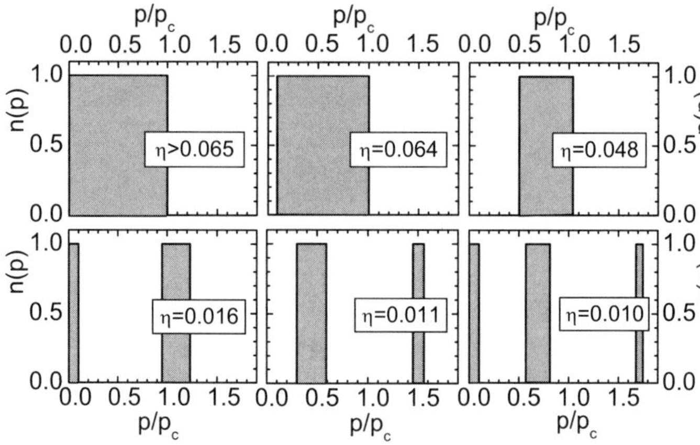

Fig. 5. Neutron momentum distributions $n(p)$ for different values of η.

The matter is that a standard scenario for neutron-star cooling, where the modified Urca process, neutrino bremsstrahlung from NN collision and neutrino emission due to Cooper pairing are of first importance, fails to explain cooling rates of these stars. The direct-Urca (DU) process, in which beta decay and inverse beta decay operate in tandem on thermally activated nucleons and electrons, could produce rapid cooling. However, in majority of neutron stars, this scenario is forbidden due to the violation of momentum/energy conservation because of the mismatch between neutron and proton Fermi momenta. The emergence in dense neutron matter of a new sheet of the Fermi surface, positioned at low momentum values, lifts the ban on the DU process due to recovering momentum/energy conservation in the direct-Urca reactions. The associated cooling scenario[19,20] provides the accelerated cooling tracks that do not conflict with observational data on the internal temperatures of the Vela, Geminga and 3C58 pulsars. Opening of the DU mechanism of a neutron-star cooling due to the topological rearrangement of the neutron Fermi surface may also explain the existence of very cold neutron stars, SAX J1808.4–3658 and SXT 1H 1905+000.

6. Conclusion

We have studied the stability of the neutron quasiparticles near the point of pion condensation, and we have found that the topological instability of the Landau state precedes the onset of pion condensation. This instability resolves in a topological quantum phase transition which expresses itself in the rearrangement of single-particle degrees of freedom with formation of a new sheet of a neutron Fermi surface at small momenta. No symmetry is broken in this transition. The topological instability in the neutron matter is similar to the one found in numerical evaluation[8,9] of the quasiparticle spectrum of 2D liquid ^{3}He and 2D electron gas. However, while in ^{3}He and electron gas the instability emerges just at the Fermi surface and effective mass diverges, in the neutron matter the rearrangement of the Landau state affects a region of momenta far from the Fermi surface and the effective mass does not dramatically change. As system further approaches the pion-condensation point, the set of new topological transitions occurs resulting in emergence of a multi-connected neutron Fermi surface. Appearance of one or more sheets of the Fermi surface at small momenta activates the normally forbidden Urca cooling mechanism.

We thank M. Alford, M. Baldo, and E. E. Saperstein for fruitful discussions. This research was supported by the McDonnell Center for the Space Sciences, by Grant No. NS-8756.2006.2 from the Russian Ministry of Education and Science, and by Grants Nos. 06-02-17171 and 07-02-00553 from the Russian Foundation for Basic Research.

References

1. L. D. Landau, *Zh. Eksp. Teor. Fiz.* **30**, 1058 (1956); **32**, 59 (1957); [*Sov. Phys. JETP* **3**, 920 (1957); **5**, 101 (1957)].
2. A. B. Migdal, *Theory of Finite Fermi Systems and Applications to Atomic Nuclei* (Wiley, New York, 1967).
3. A. Casey, H. Patel, J. Nyeki, B. P. Cowan and J. Saunders, *Phys. Rev. Lett.* **90**, 115301 (2003).
4. V. M. Pudalov, *J. Phys. IV France* **12**, Pr9-331 (2002).
5. S. V. Kravchenko and M. P. Sarachik, *Rep. Prog. Phys.* **67**, 1 (2004).
6. M. V. Zverev, V. A. Khodel and V. R. Shaginyan, *JETP* **82**, 567 (1996).
7. M. V. Zverev and M. Baldo, *JETP* **87**, 1129 (1998); *J. Phys.: Condens. Matter* **11**, 2059 (1999).
8. J. Boronat, J. Casulleras, V. Grau, E. Krotscheck and J. Springer, *Phys. Rev. Lett.* **91**, 085302 (2003).
9. V. V. Borisov and M. V. Zverev, *JETP Letters* **81**, 503 (2005).
10. M. Baldo, V. V. Borisov, J. W. Clark, V. A. Khodel and M. V. Zverev, *J. Phys.: Cond. Matter* **16**, 6431 (2004).
11. G. E. Volovik, *Springer Lecture Notes in Physics* **718**, 31 (2007) [cond-mat/0601372].
12. M. De Llano and J. P. Vary, *Phys. Rev. C* **19**, 1083 (1979); M. De Llano, A. Plastino and J. G. Zabolitsky, *Phys. Rev. C* **20**, 2418 (1979).
13. V. C. Aguilera-Navarro, M. De Llano, J. W. Clark and A. Plastino, *Phys. Rev. C* **25** 560 (1982).
14. A. M. Dyugaev, *Sov. Phys. JETP* **43**, 1247 (1976).
15. A. B. Migdal, *Rev. Mod. Phys.* **50**, 107 (1978).
16. A. B. Migdal, E. E. Saperstein, M. A. Troitsky and D. N. Voskresensky, *Phys. Rep.* **192**, 179 (1990).
17. L. D. Landau and E. M. Lifshitz, *Statistical Physics*, Vol. 2 (Pergamon Press, Oxford, 1980).
18. A. A. Abrikosov, L. P. Gor'kov and I. E. Dzialoshinskii, *Methods of Quantum Field Theory in Statistical Physics* (Prentice-Hall, Englewood Cliffs, NJ, 1965).
19. D. N. Voskresensky, V. A. Khodel, M. V. Zverev and J. W. Clark, *Ap. J. Lett.* **533**, 127 (2000).
20. V. A. Khodel, J. W. Clark, M. Takano and M. V. Zverev, *Phys. Rev. Lett.* **93**, 151101 (2004).
21. D. G. Yakovlev, O. Y. Gnedin, A. I. Kaminker and A. Y. Potekhin, arXiv: 0710.2047.

COUPLING OF NUCLEAR AND ELECTRON MODES IN RELATIVISTIC STELLAR MATTER

A. M. S. SANTOS,* C. PROVIDÊNCIA and L. BRITO

Department of Physics - University of Coimbra,
P-3004 - 516 - Coimbra - Portugal
**E-mail: magno@teor.fis.uc.pt*

D. P. MENEZES and S. S. AVANCINI

Depto de Física - CFM - Universidade Federal de Santa Catarina,
Florianópolis - SC - CP. 476 - CEP 88.040 - 900 - Brazil

The conditions under which nuclear collective modes couple to plasmon modes in asymmetric nuclear matter (ANM) neutralized by electrons, which is of interest for the study of neutron stars and supernovae, are investigated. We take a mean-field approach to nuclear matter, and the Coulomb field is included. We show that the coupling may be so strong that it affects the onset of the nuclear mode and it may also change its isovector/isoscalar character.

Keywords: Plasmon; Collective modes; Relativistic models; Neutron stars.

1. Introduction

The understanding of compact stars, supernova cores and neutron stars requires a multidisciplinary theoretical effort including astrophysics, nuclear and particle physics and thermodynamics. Not only the equation of state of stellar matter has to be understood, but also the neutrino mean free path in the medium has to be well described. It has been shown that the neutrino opacity is affected by nucleon-nucleon interactions due to coherent scattering off density fluctuations.[1]

In the present work we investigate the role of isospin and the presence of the Coulomb field and electrons on the collective nuclear modes. At low densities, an isovector-like mode is obtained. At higher densities, 2-3 times the saturation density, this mode changes to an isoscalar-like mode and the authors of Ref. 2 have even suggested that the experimental observation of the *neutron wave* would identify the transition density. In what follows we will discuss the behaviour of the modes.

We restrict ourselves to the zero temperature case, and we call *npe* matter, neutral matter composed of protons, neutrons and electrons and *np* matter, charged matter composed of protons and neutrons only.

2. The Formalism

We consider a system of nucleons interacting with and through an isoscalar-scalar field ϕ with mass m_s, an isoscalar-vector field V^μ with mass m_v and an isovector-vector field $\mathbf{b}^\mu$ with mass m_ρ. We will work within the formalism of the non-linear Walecka model (NLWM) with NL3.[3] Protons and electrons interact through the electromagnetic field A^μ. In order to include a system of electrons with mass m_e, we add to the NLWM Lagragrangian the electron contribution:

$$\mathcal{L}_e = \bar{\psi}_e \left[\gamma_\mu \left(i\partial^\mu + eA^\mu\right) - m_e\right] \psi_e .$$

In the present approach collective modes are described as small oscillations around the equilibrium state. The energies and the amplitudes of such modes are obtained from the linearized equations of motion. The time evolution of the perturbed system is described by the mesons equations of motion and the Vlasov equations for the nucleons and electrons:

$$\frac{\partial f_i}{\partial t} + \{f_i, h_i\} = 0, \qquad i = p,\, n,\, e\,, \tag{1}$$

where $\{,\}$ denotes the Poisson brackets, and f_i, h_i are the distribution function and the one-body hamiltonian for fermion type i, respectively. Eq. (1) expresses the conservation of the number of particles in phase space and is, therefore, covariant. Moreover, at zero temperature, the equilibrium distribution function is a step function, so the energy states are either unoccupied, or occupied by one particle, the latter having the Fermi energy.

Using the meson equations of motion to eliminate the meson fields from Eqs. (1), we obtain the dispersion relations. One might want to refer to Ref. 4 and references therein, for a more detailed description of the formalism.

2.1. *Plasmon modes*

We focus now on the collective modes of the electrons, so called plasmon modes. The longitudinal response of a relativistic degenerate gas of electrons was first studied by Jancovici.[5] In our formalism, if we discard the nucleon degrees of freedom the dispersion relation reduces to

$$1 - C_A^{ee}\, L(s_e) = 0, \tag{2}$$

where $C_A^{ee} = -\frac{e^2}{2\pi^2}\frac{1}{k^2} P_{Fe}\epsilon_{Fe}$, P_{Fe} and ϵ_{Fe} are the Fermi momentum and energy of the electron, respectively. $L(s_e)/2$ is the Lindhard function for argument $s_e = \omega/\omega_{0e}$, where $\omega_{0e} = kv_{Fe}$ and v_{Fe} is the sound velocity in units of the electron Fermi units.

The left-hand side term of equation (2) gives the dielectric constant of the electron gas in the limit $k \ll k_{Fe}$ and $\omega \ll \epsilon_{Fe}$,[5] which corresponds to the range of validity of the Vlasov equation.

In order to understand the coupling of the nuclear modes to the plasmon modes given by (2), we represent by thin lines the response of the free electron gas in Fig. 1(a). There are two modes: a sound-like mode and a plasmon mode with a

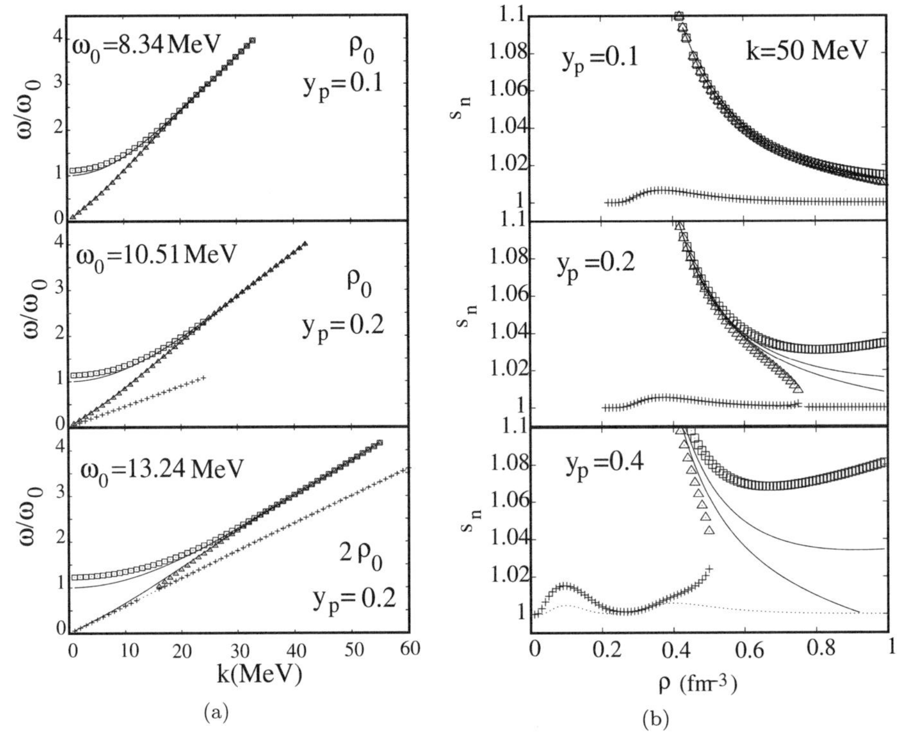

Fig. 1. (a) Collective modes as a function of the momentum transfer for two different densities ρ_0 and $2\rho_0$ and two proton fractions $y_p = 0.1$ and 0.2. The plasmon frequency at zero momentum ω_0 is defined in the text. (b) Sound velocities versus baryon density for $k = 50$ MeV, and different proton fractions: $y_p = 0.1$, 0.2 and 0.4.

frequency $\omega_0 = \sqrt{\frac{e^2 \rho_e}{\epsilon_{Fe}}}$, at zero momentum transfer, where $e = \sqrt{4\pi/137}$ is the electromagnetic coupling constant.

3. Results and Conclusion

Fig. 1(a) shows, for high isospin asymmetry, the dependence of the energy of the collective modes on the momentum transfer. The plasmon-like modes are represented by squares (higher) and triangles (lower energies), and the nuclear (low-lying) mode by crosses. We also include the results for np matter (thin dotted line) and for a relativistic gas of free electrons (full thin lines). The existence of low-lying nuclear modes depends on the density and on the proton fraction: for $\rho = \rho_0$ and $y_p = 0.1$ they do not exist for any value of k; for $\rho = \rho_0$ and $y_p = 0.2$ they propagate until a $k_{max} = 25$ MeV; for $\rho = 2\rho_0$ and $y_p = 0.2$, the nuclear and plasmon mode couple and they only exist as independent modes above $k = 14$ MeV. Then the plasmon-like mode propagates up to about $k = 55$ MeV and the nuclear mode propagates in npe matter just as in np matter.

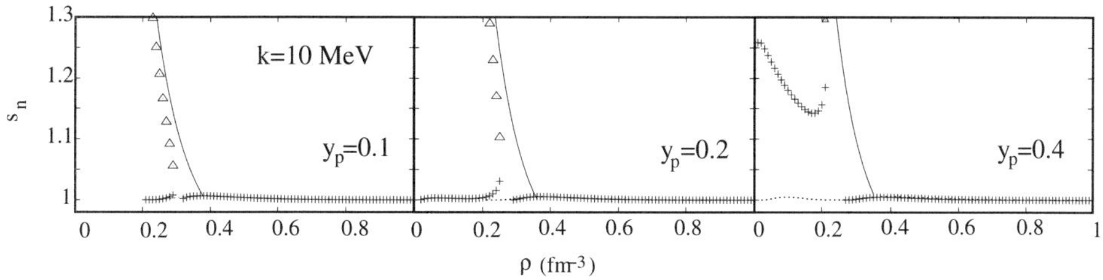

Fig. 2. Same as Fig. 1(b), now with k=10 MeV.

Fig. 1(b) shows the behaviour of the sound velocities of the nuclear collective modes for $k = 50$ MeV and different proton fractions: $y_p = 0.1$, 0.2 and 0.4, as a function of the density. For the lowest values of proton fraction the coupling of the nuclear to the plasmon mode only occurs at large densities (0.7 fm^{-3} for $y_p = 0.2$ and above 1 fm^{-3} for $y_p = 0.1$). For lower values of momentum transfer however, the picture changes, as seen from Fig. 2: the threshold density (where the coupling of nuclear-to-plasmon mode takes place) is not so much isospin-dependent. However, the sound velocity of the plasmon-like nuclear mode has a much more dramatic dependence on y_p, although it occurs for densities below the threshold value, only. For higher values of ρ (above ~ 0.3 fm^{-3}), this mode behaves like in np matter, even for lower values of k. We can see that the influence of the electrons on the collective modes depends on both the asymmetry and the momentum transfer: a small proton fraction implies a small electron fraction and therefore a weaker coupling; larger values of the momentum transfer also contribute for a smaller effect of the presence of the electrons.

The present results have implications mainly in astrophysical objects and the related transport properties, since one of the possible mechanisms of the neutrino emissions involves the decay of a plasmon into a neutrino-antineutrino pair.

Acknowledgements

CP would like to thank the stimulating discussions with J. da Providência. This work was partially supported by FEDER and FCT (Portugal) under the grant SFRH/BPD/29057/2006, and projects POCI/FP/63918/2005, and by CNPq (Brazil).

References

1. R. F. Sawyer, *Phys. Rev. D* **11**, 2740 (1975).
2. V. Greco, M. Colonna, M. Di Toro and F. Matera, *Phys. Rev. C* **67**, 015203 (2003).
3. G. A. Lalazissis, J. König and P. Ring, *Phys. Rev. C* **55**, 540 (1997).
4. L. Brito, C. Providência, A. M. Santos, S. S. Avancini, D. P. Menezes and Ph. Chomaz, *Phys. Rev. C* **74**, 045801 (2006).
5. B. Jancovici, *Il Nuovo Cimento* **25**, 428 (1962).
6. S. S. Avancini, L. Brito, D. P. Menezes and C. Providência, *Phys. Rev. C* **70**, 015203 (2004).

NEUTRON STARS IN THE RELATIVISTIC HARTREE-FOCK THEORY AND HADRON-QUARK PHASE TRANSITION

B. Y. SUN

School of Physics, Peking University,
Beijing, 100871, China
E-mail: sunbaoyuan@gmail.com

U. LOMBARDO and G. F. BURGIO

Istituto Nazionale di Fisica Nucleare, Laboratori Nazionali del Sud
Sezione di Catania, Via Santa Sofia 64, 95123, Italy

J. MENG

School of Physics, Peking University,
Beijing, 100871, China
Institute of Theoretical Physics, Chinese Academy of Sciences,
Beijing, 100080, China
Center of Theoretical Nuclear Physics, National Laboratory of Heavy Ion Accelerator,
Lanzhou, 730000, China
E-mail: mengj@pku.edu.cn

Based on the density-dependent relativistic Hartree-Fock theory (DDRHF) for hadronic matter, the properties of neutron stars have been studied and compared with the results from the density-dependent relativistic mean field theory (DDRMF). Though similar equations of state are obtained, DDRHF calculations give larger fractions of proton, electron and muon at high baryon density for neutron star matter than the ones from DDRMF. The maximum masses of neutron stars lie between 2.3 $M_\odot$ and 2.5 $M_\odot$, and the corresponding radii between 11.7 km and 12.5 km. In addition, the phase transition from hadronic matter to quark matter in neutron stars is studied by using the MIT bag model for the quark phase. The transition is studied in both Gibbs and Maxwell constructions.

Keywords: Many-body theory; Relativistic models; Nuclear matter; Neutron stars; Bag model; Phase transitions.

1. Introduction

Neutron stars provide a natural laboratory for exploring the baryonic matter at high densities, well exceeding in the center the nuclear saturation density of $\rho_{\rm sat} = 0.16$ fm^{-3}. Recently, new results from the observations of neutron star properties have been reported which provide stringent constraints on the equation of state (EoS) of strongly interacting matter at high densities, see Ref. 1,2 and references therein.

On the description of nuclear matter and finite nuclei, within the relativistic scheme, the mean field theory has achieved great success during the past years. One of the most outstanding models is the relativistic Hartree approach with the no-sea approximation, namely the relativistic mean field (RMF) theory.[3–5] In recent years, RMF models with density dependent nucleon-meson couplings (DDRMF) have been developed.[6–9] However, in the framework of the RMF approach, the Fock terms are dropped out, which may have remarkable effects on nuclear matter especially at high density.

During the past decades, there were several attempts to include the Fock term in the relativistic descriptions of nuclear systems.[10–13] Recently, a new RHF method, so called, density-dependent relativistic Hartree-Fock (DDRHF) theory[14–16] has brought us a new in-sight into this problem. With the effective Lagrangians of Refs. 14,15, the DDRHF theory can quantitatively describe the ground state properties of many nuclear systems on the same level as RMF.

The appearance of quark matter in the interior of massive neutron stars is one of the main issues of astrophysics. Many EoS have been used to describe the interior of neutron stars.[17–19] Due to the impact of recent experiments in heavy-ion collisions[20] and new observational limits for the mass and the mass-radius relationship of compact stars,[21–24] affected by large theoretical uncertainties for quark matter, the question whether a pure quark phase exists in the interior of neutron stars or not still have not yet received a clear answer.

The paper is organized as follows: In Sec. 2 we review the nuclear EoS within the relativistic Hartree-Fock theory and the quark matter EoS within the MIT bag model. In Sec. 3 we present the results for the neutron star structure in DDRHF and the hadron to quark phase transition. Section 4 contains our conclusions.

2. Theoretical Framework

2.1. *Hadronic phase: the relativistic Hartree-Fock theory*

In the present work, we study nuclear matter properties based on the new developed density dependent relativistic Hartree-Fock theory. The details of the DDRHF theory can be found in Ref. 14–16. The DDRHF theory starts from an effective Lagrangian density where nucleons are described as Dirac spinors interacting via exchange of several mesons (σ, ω, π and ρ) and the photons. Using the Legendre transformation and the equations of motion for the mesons and photon field operators, the Hamiltonian can be written in a form which includes only nucleon degree of freedom,

$$\begin{aligned} H = &\int d^3x \left[\bar{\psi}[-i\gamma \cdot \boldsymbol{\nabla} + M]\psi\right] \\ &+ \frac{1}{2}\int d^3x d^4y \sum_{i=\sigma,\omega,\rho,\pi,A} \bar{\psi}(x)\bar{\psi}(y)\Gamma_i(x,y)D_i(x,y)\psi(y)\psi(x)\,, \end{aligned} \tag{1}$$

where $\Gamma_i(x,y)$ is the interaction vertex of the respective mesons, and $D_i(x,y)$ is the corresponding meson propagator.

Generally, we can expand the nucleon field operator ψ into a complete set of Dirac spinors $u(p,s,\tau)$

$$\psi(x) = \sum_{p,s,\tau} u(p,s,\tau)e^{-ipx}c_{p,s,\tau}\,, \tag{2}$$

where $c_{p,s,\tau}$ denote annihilation operators for nucleons in the state (p,s,τ). Notice that the no-sea approximation is assumed here. On such a basis, we can construct the trial ground state $|\Phi_0\rangle = \prod_{i=1}^{A} c^\dagger_{p,s,\tau}|0\rangle$, where $|0\rangle$ is the physical vacuum. From the above trial state , we can build up the energy functional by taking the expectation value of Hamiltonian (1):

$$E \equiv \langle\Phi_0|H|\Phi_0\rangle = E_k + \left(E^D_\sigma + E^D_\omega + E^D_\rho\right) + \left(E^E_\sigma + E^E_\omega + E^E_\rho + E_\pi\right)\,, \tag{3}$$

wherein the exchange terms are given by

$$E^E_i = -\frac{1}{2}\sum_{p_1,s_1,\tau_1}\sum_{p_2,s_2,\tau_2}\bar{u}(p_1,s_1,\tau_1)\bar{u}(p_2,s_2,\tau_2)\frac{\Gamma_i(1,2)}{m_i^2+\boldsymbol{q}^2}u(p_1,s_1,\tau_1)u(p_2,s_2,\tau_2)\,.$$

Then, the self-energy can be determined by the self-consistent variation of the energy functional, namely

$$\Sigma(p)u(p,s,\tau) = \frac{\delta}{\delta\bar{u}(p,s,\tau)}\sum_{\sigma,\omega,\rho,\pi}\left[E^D_i + E^E_i\right]\,. \tag{4}$$

Generally, it can also be written as

$$\Sigma(p,p_F) = \Sigma_S(p,p_F) + \gamma_0\Sigma_0(p,p_F) + \boldsymbol{\gamma}\cdot\hat{\boldsymbol{p}}\Sigma_V(p,p_F)\,, \tag{5}$$

where $\hat{\boldsymbol{p}}$ is the unitary vector along $\boldsymbol{p}$, and p_F is the Fermi momentum. Here, the tensor term $\gamma_0\boldsymbol{\gamma}\cdot\hat{\boldsymbol{p}}\Sigma_T(p,p_F)$ is omitted because it does not appear in the Hartree-Fock approximation for the nuclear matter.

In this work, density-dependent meson-nucleon couplings will be used as introduced in Ref. 6. For the coupling constant g_π, the exponential density dependence is adopted as $g_\pi(\rho_v) = g_\pi(0)\,e^{-a_\pi x}$. Three new DDRHF parameter sets PKO1, PKO2, PKO3[14–16] have been used in recent calculations.

The chemical potential can be calculated from self-energies

$$\mu = E_F = \Sigma_0(p_F) + \sqrt{\left[p_F + \Sigma_V(p_F)\right]^2 + \left[M + \Sigma_S(p_F)\right]^2}\,. \tag{6}$$

In cold neutron star matter, the chemical potentials must fulfill the condition equilibrium under weak interaction, i.e., $\mu_p = \mu_n - \mu_e$ and $\mu_\mu = \mu_e$. Moreover, the baryon number conservation, $\rho_b = \rho_n + \rho_p$, as well as the condition of charge neutrality, $\rho_p = \rho_n + \rho_e$, must be satisfied. Then the pressure can be obtained from the thermodynamic relation

$$P(\rho_v) = \rho_v^2\frac{d}{d\rho_v}\frac{E}{\rho_v} = \sum_{i=n,p,e,\mu}\rho_i\mu_i - E \tag{7}$$

2.2. *Quark phase: the MIT bag model*

We now turn to briefly describe the bulk properties of uniform quark matter, deconfined from the β-stable hadronic matter as mentioned in the previous section. Here we use the MIT bag model.[25] The thermodynamic potential of $q = u, d, s$ quarks can be expressed as a sum of the kinetic term and the one-gluon-exchange term,[26,27]

$$\Omega_q(\mu_q) = -\frac{3m_q^4}{8\pi^2}\left[\frac{y_q x_q}{3}\left(2x_q^2 - 3\right) + \ln(x_q + y_q)\right] + \alpha_s \frac{3m_q^4}{2\pi^3}\left\{\left[y_q x_q - \ln(x_q + y_q)\right]^2 - \frac{2}{3}x_q^4 + \ln y_q + 2\ln\left(\frac{\sigma_{\rm ren}}{m_q y_q}\right)\left[y_q x_q - \ln(x_q + y_q)\right]\right\}, \tag{8}$$

where m_q and μ_q are the q current quark masses and the chemical potential, respectively, and $y_q = \mu_q/m_q$, $x_q = \sqrt{y_q^2 - 1}$. α_s denotes the QCD fine structure constant, whereas $\sigma_{\rm ren} = 313$ MeV is the renormalization point. In this work we will consider massless u and d quarks (together with $m_s = 150$ MeV), and choose $\alpha_s = 0$ since it has no remarkable influence on neutron star bulk properties.[28]

The number density ρ_q of q quarks is related to Ω_q via $\rho_q = -\partial\Omega_q/\partial\mu_q$. Then the total energy density and pressure for the quark system are given by

$$\epsilon_Q(\rho_u, \rho_d, \rho_s) = \sum_q (\Omega_q + \mu_q\rho_q) + B\,, \quad P_Q(\rho) = \sum_q \mu_q\rho_q - \epsilon_Q\,. \tag{9}$$

where B is the energy density difference between the perturbative vacuum and the true vacuum, i.e., the bag constant. In the following we present results based on the MIT model using constant values of the bag constant, $B = 90, 120, 150$ MeV/fm^3.

The composition of β-stable quark matter is determined by imposing the condition of equilibrium of the chemical potentials under weak interaction: $\mu_d = \mu_s = \mu_u + \mu_e$. As in baryonic matter, the equilibrium relations between the chemical potentials must be supplemented with the charge neutrality condition and the total baryon number conservation in order to determine the chemical composition $\rho_f(\rho)$.

3. Results and Discussion

3.1. *Neutron star properties in the DDRHF theory*

We now study the neutron star properties in the DDRHF theory. For comparison, we also perform the calculations with four DDRMF interactions: TW99,[6] DD-ME1,[7] DD-ME2,[8] PKDD.[9] In Fig. 1(a) are shown the symmetry energies as a function of the baryon density for different DDRMF and DDRHF EoSs. At high density, three DDRHF interactions PKO1, PKO2 and PKO3 give sizable enhancement for the symmetry energies than DDRMF ones, while PKDD give different results from other DDRMF EoSs. These remarkable distinctions would have a large influence on the cooling behavior of neutron stars. According to recent analyses,[1,29] an acceptable EoS shall not allow direct Urca processes to occur in neutron stars with masses below 1.5 $M_\odot$, otherwise it will be in disagreement with modern observational soft

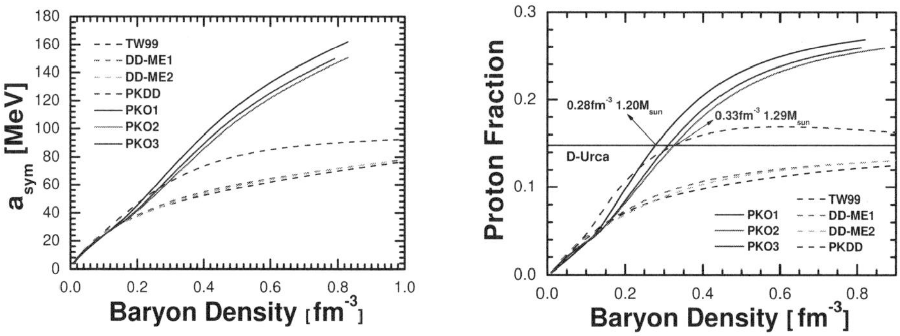

(a)

(b)

Fig. 1. (a) The symmetry energies for different EoSs. (b) Proton fractions in neutron star matter for different EoSs. The line labeled with D-Urca is threshold for happening direct Urca process.

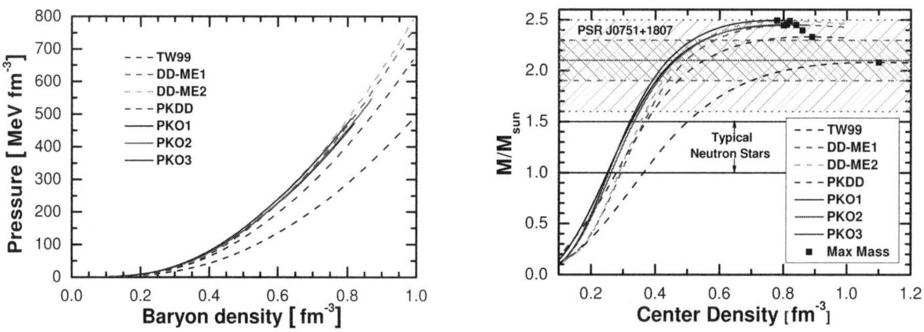

(a)

(b)

Fig. 2. (a) Pressure vs. baryon density of the neutron star matter for different nuclear EoSs. (b) Mass vs. central density of compact stars for different nuclear EoSs. Filled squares denote the maximum mass configurations. The observational constraints are taken from Fig. 2 in Ref. 1.

X-ray data in the temperature-age diagram. This constrains the density dependence of the nuclear asymmetry energy which should not be too strong. From Fig. 1(b), one can see that the DDRHF results cannot be in agreement with this constraint. In particular, for PKO1, D-Urca will occur at fairly low mass 1.2 $M_\odot$. However, there are still several uncertainties concerning the cooling mechanism of neutron stars,[30] and this constraint may not be considered as a stringent one and there could be other mechanisms, e.g. appearance of quark phase, to solve this problem.

The EoSs of β-stable neutron star matter are shown in Fig. 2(a). Three DDRHF interactions give the similar trend to DD-ME1 and DD-ME2 in DDRMF, while PKDD give a little softer one and TW99 have softest EoS. As a result, the maximum masses of neutron stars lie between 2.3 $M_\odot$ and 2.5 $M_\odot$ in DDRHF, as shown

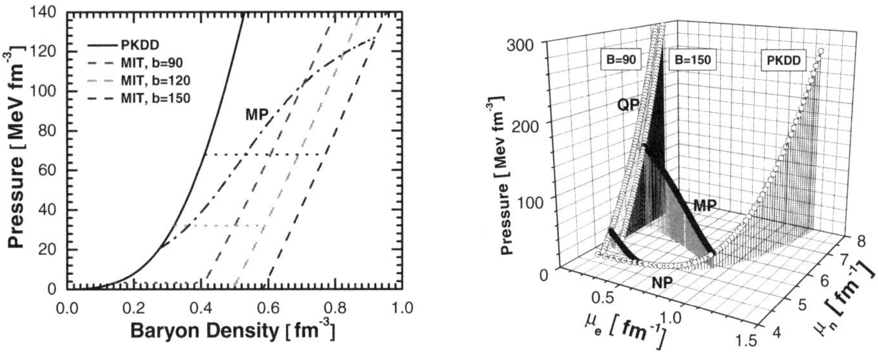

(a)

(b)

Fig. 3. (a) Pressure vs. baryon density of β-stable neutron star matter for different EoSs. The dotted lines show the hadron-quark phase transition in Maxwell construction for different bag constants, while the dot-dashed line show the result in Gibbs construction for $B = 150$ MeV/fm^3. (b) Gibbs phase construction of a two-component system. The red circle NP and QP show the pressure of the hadronic and the quark phase under the condition of charge neutrality. The solid black curves MP correspond to the mixed phase for different bag constants.

in Fig. 2(b), consisting with observational constraint from PSR J0751+1807.[21] Furthermore, mass-radius relations of pure neutron stars are shown in Fig. 4(a). One can see DDRHF have a better aggreement with three observational limits[22–24] than DDRMF, especially for EXO 0748-676.[24] In DDRMF, TW99 gives smaller maximum mass and radius than others, which is not consistent with constraints.

3.2. *Hadron-quark phase transition in neutron star matter*

We now consider the hadron-quark phase transition in neutron stars. Both the simple Maxwell construction and the more general Gibbs (Glendenning) construction[31] are adopted to treat the phase transition. In the Maxwell construction, the transition is determined by the intersection points between the hadronic and the quark phase in the plot of pressure versus baryonic (neutron) chemical potential. After projecting this crossing point onto the plot of density versus baryonic (neutron) chemical potential, one can get the corresponding transition densities from low-density baryonic matter, ρ_H, to high-density quark matter, ρ_Q. So between these two densities, there will be a plateau for pressure in the EoS curve, seen in Fig. 3(a). This sudden density increase would make stars to be unstable when the gravitational interaction is taken into account. In addition, from the figure we find that, when bag constant of the MIT bag model increases, the pressure of phase transition occurring will also increase, and the transition density will go up, too.

In the Gibbs construction, there are two conserved charges. Hence, a mixed phase will emerge where both hadron phase and quark phase coexist. From Fig. 3(a) we

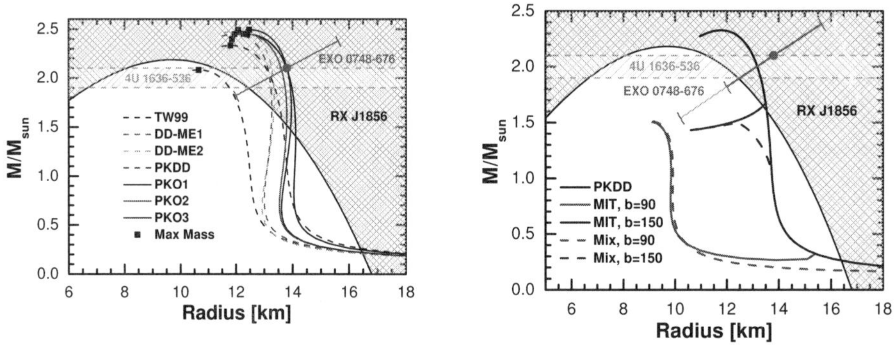

(a) (b)

Fig. 4. Mass-radius relations for compact stars corresponding to (a) different nuclear matter EoSs. (b) different hybrid star EoSs (DDRMF+MIT). The observational constraints are taken from Ref. 1,2 and therein.

can see the plateau for pressure has been destroyed when the Gibbs construction is considered. In the mixed phase, the pressure is the same in the two phases to ensure mechanical stability, and goes up continuously with increasing baryon density. In Fig. 3(b) are shown the pressure routes in two chemical potential component plane. While the neutron density goes up, the electron chemical potential increases merely in pure hadron phase. The emergence of quark matter makes it to decrease in mixed phase and approach to zero in pure quark phase. During the total process, the pressure always raises monotonously up. Moreover, from the figure one can see, given the lager bag constants, the transition to occur at higher baryon density.

Fig. 4(b) are shown the mass-radius relations using different hybrid star EoSs. Both two constructions for the phase transition are used. It has been found that different constructions influence very little the final mass-radius relation of massive neutron stars. Within the MIT bag model with constant B, the maximum mass of a neutron star never exceeds a value of about 1.6 $M_\odot$, and the star radius will increase when B goes up. These results are not in agreement with recent observational constraints. Therefore, more refined quark models are necessary in further studies.

4. Conclusions

In conclusion, we studied neutron star properties based on the DDRHF theory, and compared with recent observational data of neutron stars. For maximum masses of neutron stars, theoretical results 2.3 $M_\odot \leq M_{max} \leq 2.5\ M_\odot$ could reproduce recent observational constraint. However, DDRHF gives larger proton fractions at high baryon density for neutron star matter than ones from DDRMF, which are not consistent with D Urca constraint. Then the hadron quark phase transition is studied by using the MIT bag model. Different transition constructions give similar maximum masses. While bag constant decreases, the maximum mass will increase but can never be larger than 1.6 $M_\odot$, meanwhile the radius of stars will be reduced.

Acknowledgments

We thank H.-J. Schulze for the stimulating discussions at Catania University. This work is partly supported by Asia-Europe Link Program in Nuclear Physics and Astrophysics (CN/ASIA-LINK/008 094-791), Major State Basic Research Developing Program 2007CB815000 as well as the National Natural Science Foundation of China under Grant No. 10435010, 10775004 and 10221003.

References

1. T. Klähn, D. Blaschke, S. Typel *et al.*, *Phys. Rev. C* **74**, 035802 (2006).
2. T. Klähn, D. Blaschke, F. Sandin *et al.*, arXiv:nucl-th/0609067v2.
3. B. D. Serot and J. D. Walecka, *Adv. Nucl. Phys.* **16**, 1 (1986).
4. P. G. Reinhard, *Rep. Prog. Phys.* **52**, 439 (1989).
5. P. Ring, *Prog. Part. Nucl. Phys.* **37**, 193 (1996).
6. S. Typel and H. H. Wolter, *Nucl. Phys. A* **656**, 331 (1999).
7. T. Nikšić, D. Vretenar, P. Finelli *et al.*, *Phys. Rev. C* **66**, 024306 (2002).
8. G. A. Lalazissis, T. Nikšić, D. Vretenar *et al.*, *Phys. Rev. C* **71**, 024312 (2005).
9. W. H. Long, J. Meng, N. V. Giai *et al.*, *Phys. Rev. C* **69**, 034319 (2004).
10. A. Bouyssy, S. Marcos, J. F. Mathiot *et al.*, *Phys. Rev. Lett.* **55**, 1731 (1985).
11. A. Bouyssy, J. F. Mathiot, N. Van Giai and S. Marcos, *Phys. Rev. C* **36**, 380 (1987).
12. P. Bernardos, V. N. Fomenko, N. V. Giai *et al.*, *Phys. Rev. C* **48**, 2665 (1993).
13. S. Marcos, L. N. Savushkin, V. N. Fomenko *et al.*, *J. Phys. G* **30**, 703 (2004).
14. W. H. Long, N. V. Giai and J. Meng, *Phys. Lett. B* **640**, 150 (2006).
15. W. H. Long, N. Van Giai and J. Meng, in preparation; W. H. Long, Ph.D. thesis, Université Paris-Sud (2005, unpublished).
16. W. H. Long, N. Van Giai and J. Meng, arXiv:nucl-th/0608009.
17. J. M. Lattimer and M. Prakash, *Astrophys. J.* **550**, 426 (2001).
18. F. Weber, *Prog. Part. Nucl. Phys.* **54**, 193 (2005).
19. M. Buballa, *Phys. Rep.* **407**, 205 (2005).
20. P. Danielewicz, R. Lacey and W. G. Lynch, *Science* **298**, 1592 (2002).
21. D. J. Nice, E. M. Splaver, I. H. Stairs *et al.*, *Astrophys. J.* **634**, 1242 (2005).
22. J. E. Trümper, V. Burwitz, F. Haberl *et al.*, *Nucl. Phys. Proc. Suppl.* **132**, 560 (2004).
23. D. Barret, J. F. Olive and M. C. Miller, *Mon. Not. Roy. Astron. Soc.* **361**, 855 (2005).
24. F. Özel, *Nature* **441**, 1115 (2006).
25. A. Chodos, R. L. Jaffe, K. Johnson *et al.*, *Phys. Rev. D* **9**, 3471 (1974).
26. E. Witten, *Phys. Rev. D* **30**, 272 (1984); G. Baym, E. W. Kolb, L. McLerran, T. P. Walker and R. L. Jaffe, *Phys. Lett. B* **160**, 181 (1985); N. K. Glendenning, *Mod. Phys. Lett. A* **5**, 2197 (1990).
27. E. Fahri and R. L. Jaffe, *Phys. Rev. D* **30**, 2379 (1984).
28. G. F. Burgio, M. Baldo, P. K. Sahu *et al.*, *Phys. Lett. B* **526**, 19 (2002); G. F. Burgio, M. Baldo, P. K. Sahu and H.-J. Schulze, *Phys. Rev. C* **66**, 025802 (2002).
29. D. Blaschke, H. Grigorian and D. Voskresensky, *Astron. Astrophys.* **424**, 979 (2004).
30. D. Blaschke and H. Grigorian, *Prog. Part. Nucl. Phys.* **59**, 139 (2007).
31. N. K. Glendenning, *Phys. Rev. D* **46**, 1274 (1992).

PART D

Prospects of Present and Future Observations

MEASUREMENTS OF NEUTRON STAR MASSES

D. G. YAKOVLEV[1,2]

[1] *Ioffe Physico-Technical Institute,*
26 Politekhnicheskaya Street, Saint-Petersburg 194021, Russia
[2] *Joint Institute for Nuclear Astrophysics,*
210a Nieuwland Science Hall, Notre Dame, IN 46556-5670, USA

Current data (by September 2007) on masses of neutron stars in binary systems (X-ray binaries, double neutron star binaries, pulsar–white dwarf binaries, and pulsar–nondegenerate star binaries) are reviewed.

Keywords: Neutron stars; Mass measurements; Radio pulsars.

1. Introduction

Neutron star mass measurements are important in many ways, particularly, for constraining the equation of state of superdense matter in neutron star cores and for determining the critical mass that separates neutron stars and black holes. Here we summarize the measurements of masses of neutron stars which enter binary systems (we collect the data obtained by September 2007). More details can be found in Ref. 1 (although the data there are restricted by August 2006).

Binary systems containing neutron stars can be divided into X-ray binaries, double neutron star binaries, neutron star–white dwarf binaries, and neutron star–nondegenerate star binaries. We consider all of them; M_1 will denote a gravitational neutron star mass, and M_2 a gravitational mass of its companion. Any accurate neutron star mass measurement is important. For the problem of equation of state of superdense matter, most massive neutron stars are especially valuable.[1,2]

2. X-Ray Binaries

Masses of neutron stars in X-ray binaries are usually measured using the standard formalism of Keplerian motion of point-like masses M_1 and M_2 (with conserved total energy and angular momentum). At the first step it is necessary to measure a radial velocity curve of one binary companion (a neutron star or an optical star) which gives the orbital period P_b, eccentricity e, radial velocity semi-amplitude, and the mass function (f_1 or f_2). For instance, $f_1 = (M_2 \sin i)^3/M^2$, where $M = M_1 + M_2$ is the total mass and i is the orbit inclination angle. To reconstruct the orbit and determine M_1 and M_2 one still needs two equations to be obtained from additional

Table 1. Masses of stars in X-ray binaries.

System	P_b [d]	$M_1/M_\odot$	$M_2/M_\odot$	Signif.	Ref.
2A 1822−371	0.232	0.97±0.24	0.33±0.05	1σ	3
4U 1538−52	3.73	$1.06^{+0.41}_{-0.34}$	$16.4^{+5.2}_{-4.0}$	2σ	4
SMC X-1	3.89	$1.06^{+0.11}_{-0.10}$	$15.7^{+1.5}_{-1.4}$	1σ	5
Cen X-3	2.09	$1.34^{+0.16}_{-0.14}$	$20.2^{+1.8}_{-1.5}$	1σ	5
XTE J2123−058	0.248	$1.46^{+0.30}_{-0.39}$	$0.53^{+0.28}_{-0.39}$	1σ	6
LMC X-4	1.41	$1.25^{+0.11}_{-0.10}$	$14.5^{+1.1}_{-1.0}$	1σ	5
Her X-1	1.70	1.5±0.3	2.3±0.3	1σ	7
Cyg X-2	9.84	1.78±0.23	0.60±0.13	1σ	9
Vela X-1	8.96	1.88±0.13	23.1±0.2	1σ	8
4U 1700−37	3.41	2.44±0.27	58±11	1σ	10
V395 Car/2S 0921–630	9.02	1.44 ± 0.10	0.35 ± 0.03	1σ	11

measurements (e.g., the second mass function; observations of eclipses; rotational broadening of spectral lines).

X-ray binaries were discovered in the 1960s, in the beginning of X-ray astronomy. Some of them (low-mass and high-mass X-ray binaries, containing X-ray pulsars, X-ray bursters and other accreting neutron stars emitting in X-rays) were thought to be good for neutron star mass measurements.

The main results are summarized in Table 1. One can see a large scatter of masses M_1 (from $\sim M_\odot$ to more than twice values) measured (constrained) with large uncertainties. The mass of the X-ray pulsar Vela X-1 seems to be high but uncertain due to strong deviations from pure Keplerian velocity curve of the companion star, GP Vel, owing to GP Vel pulsations. In Table 1 we present conservative M_1 values[8] obtained assuming the maximum orbit inclination $i = 90°$. A compact star in 4U 1700–37 can be very massive ($M_1 \sim 2.4 M_\odot$) but it can be a black hole (the source is not an X-ray pulsar). Even if it is a neutron star, its mass can be lower ($\sim 1.7 M_\odot$[12]) if the data are interpreted with the Roche lobe model of the optical star. The application of that model to other sources[13] can also change inferred values of M_1.

3. Double Neutron Star Binaries

Mass measurements in these binaries can be more accurate; both stars can be treated as point masses. As a rule, one neutron star is observed as a radio pulsar and the second star is not observed at all. Observations consist in precise pulsar timing (detection of pulsar spin frequency distorted by orbital motion). At the first stage, the pulsar radial velocity curve is measured, and P_b, e, and f_1 are determined (but two independent equations are still missed).

At the next stage post-Keplerian (General Relativity – GR) parameters can be inferred. They are associated with the emission of gravitational waves (circu-

Table 2. Masses of stars in double neutron star binaries (at 2σ level).

System	P_b [d]	GR parameters	$M_1/M_\odot$	$M_2/M_\odot$	Ref.
J1518+4904	8.63	$\dot\omega$, $\dot P_b$	$1.56^{+0.20}_{-1.20}$	$1.05^{+1.21}_{-0.14}$	14
B1534+12	0.421	$\dot\omega$, γ, $\dot P_b$, s, r	1.3332±0.0020	1.3452±0.0020	15
B1913+16	0.323	$\dot\omega$, γ, $\dot P_b$	1.4408±0.0006	1.3873±0.0006	16
B2127+11C	0.335	$\dot\omega$, γ, $\dot P_b$	1.358±0.020	1.354±0.020	17
J0737–3039	0.102	$\dot\omega$, γ, s, r	1.337±0.010	1.250±0.010	18
J1756–2251	0.320	$\dot\omega$, γ, s	$1.40^{+0.04}_{-0.06}$	$1.18^{+0.06}_{-0.04}$	19

larization of orbital motion; a slow fall of neutron stars onto each other). The five main GR parameters are: the periastron advance $\dot\omega$; the parameter γ (that combines quadratic Doppler effect and the leading-order gravitational redshift of pulsar signals by a companion star); the orbital decay $\dot P_b$ due to gravitational wave emission; and the shape and range parameters (s and r) of Shapiro time dilatation of pulsar signals by a companion star.

The GR effects are most pronounced in compact binaries (say, $P_b \lesssim$ a few days). An eccentric binary is good for an accurate determination of $\dot\omega$ and γ. A binary observed edge-on (i is close to 90°) is favorable for detecting the Shapiro effect (s and r). A measurement of $\dot\omega$, s, or r gives, respectively, M, $\sin i$, or M_2. A determination of $\dot P_b$ or γ gives a combination of orbital parameters.

A measurement of any GR parameter adds a new equation to determine M_1 and M_2. It is sufficient to measure two parameters to find the masses. If other GR parameters are measured, they give extra equations to check the solution.

The results of mass measurements in double neutron star binaries are presented in Table 2. Column 3 lists measured GR parameters. The system J1518+4904 is too wide and the masses are still rather uncertain. All other systems are compact and eccentric; M_1 and M_2 are determined with excellent accuracy and lie in a narrow mass range (1.18–1.44) $M_\odot$ (probably because of similar evolution scenarios). Neutron stars in these binaries will merge during the time shorter than the Universe age. Such merging events are accompanied by strongest emission of gravitational waves (the best targets for modern gravitational observatories). The Hulse-Taylor pulsar B1913+16 was the first pulsar of such a type discovered in 1974.[20] Three GR parameters are accurately determined for this system; M_1 and M_2 are measured with very high accuracy. The PSR B1913+16 is still the most massive neutron star ($M_1 = 1.441\,M_\odot$) with an accurately measured mass. The system B1534+12 is the ideal double neutron star binary, where all five GR parameters have been measured. The system J0737–3039 is unique because both neutron stars are observed as radio pulsars. It is the most compact binary (among those listed in Table 2), with the strongest GR effects. For instance, it has a huge periastron advance, $\dot\omega \approx 17$ deg yr^{-1}, and its lifetime (before merging) is as short as 86 Myr.

Table 3. Masses of stars in pulsar–white dwarf binaries.

System	P_b [d]	GR par.	$M_1/M_\odot$	$M_2/M_\odot$	Ref.
J0437−4715	5.74	r, s	1.58±0.18 (1σ)	0.24	21
J0621+1002	8.32	$\dot{\omega}$	$1.70^{+0.09}_{-0.46}$ (2σ)	~ 1	22
J0751+1807	0.263	$\dot{P}_b$	$1.26^{+0.27}_{-0.28}$ (2σ)	~0.2	22
J1012+5307	0.605	r, s	1.7±1.0 (2σ)	0.165–0.215	14
J1045−4509	4.08	—	< 1.48	~ 0.13	14
J1141−6545	0.198	$\dot{\omega}, \gamma, \dot{P}_b$	1.30±0.02 (1σ)	0.99±0.02	23
J1713+0747	67.8	r, s	$1.53^{+0.08}_{-0.06}$ (1σ)	0.30–0.35	24
B1802−07	2.62	$\dot{\omega}$	$1.26^{+0.15}_{-0.67}$ (2σ)	0.36	14
J1804−2718	11.1	—	< 1.73	~ 0.2	14
B1855+09	12.33	r, s	$1.57^{+0.25}_{-0.20}$ (2σ)	0.25–0.28	25
J1909–3744	1.53	r, s	1.438±0.024 (1σ)	0.204 ± 0.002	26
J2019+2425	76.5	—	< 1.51	0.32–0.35	27
B2303+46	12.34	$\dot{\omega}$	1.24–1.44	1.2–1.4	28
J1911–5958A	0.837	—	$1.40^{+0.16}_{-0.10}$ (1σ)	0.18 ± 0.02	29,30
J1741+1351	16	r, s	1.8 ± 0.3 (1σ)	0.30 ± 0.07	31
J0514–4002A	18.78	—	< 1.5	> 0.96	32

In some cases, one observes other GR effects, for example, geodetic precession of the pulsar spin axis (with the period of $\approx$ 300 yr for the Hulse-Taylor pulsar). There are several other compact double neutron star binaries, where M_1 and M_2 are still not determined but will be determined soon.

All in all, double neutron star binaries are excellent objects to study GR effects and measure neutron star masses but the masses are restricted. It is difficult to expect that these objects contain really massive neutron stars.

4. Pulsar–White Dwarf Binaries

In these systems, the approximation of binary components by point-like masses is almost as good as in double neutron star binaries. The mass measurement is based on pulsar timing. At the first stage, the pulsar radial velocity curve is obtained (two equations are still missing). At the next stage, GR effects in pulsar motion can be detected giving additional equations. Other equations can be provided by observations of white dwarfs. For instance, one can estimate M_2 in rather wide binaries which contain millisecond pulsars in nearly circular orbits. The estimation is based on a specific evolutionary $M_2(P_b)$ relation; M_2 can also be measured directly from white dwarf observations.

The measured masses of neutron stars and white dwarfs in compact binaries are listed in Table 3. The binaries are nonuniform, and the neutron star mass spectrum is wide, from $M_1 \sim 1.25 M_\odot$ to essentially higher values.

The "ideal" binary of this type is J1141–6545, where three GR parameters are measured.[23] The pulsar mass is determined very accurately but appears to be not high, $M_1 \approx 1.30\,M_\odot$. Also, the mass of the PSR J1909–3744, $M_1 \approx 1.44\,M_\odot$, is measured with high accuracy; its orbit is almost circular ($\dot{\omega}$ and γ cannot be determined) but the system is observed edge-on, and one detected the Shapiro effect. The system J1911–5858A contains a white dwarf whose radial velocities and mass have been determined from optical spectroscopic observations. This gives two equations to find $M_1 \approx 1.4\,M_\odot$ without detecting GR effects.

Several binaries can contain massive neutron stars. First of all, they are the systems J0621+1002, J1012+5307, and J1741+1351, where the centrals values of $M_1 \gtrsim 1.7\,M_\odot$. However, the accuracy of these measurements is still insufficient to state that the neutron stars are definitely massive. Previous claims[33] that the PSR J0751+1807 is very massive have been disproved[22] after a reliable detection of the Shapiro effect. There are several other pulsar–white dwarf binaries, where the masses are still not measured but will be measured soon.

5. Pulsar–Nondegenerate Star Binaries

Masses of two pulsars in such binaries have been measured although with high uncertainties.

One system contains PSR J0045–7319 and a B1 V star (in the Small Magellanic Cloud). The pulsar timing gave $P_b = 51.2$ d and $e = 0.81$ ($f_1 = 2.17\,M_\odot$). It is possible to measure the radial velocity of the optical companion and estimate $M_2 = (10 \pm 1)\,M_\odot$ from its detailed observations. This gives $M_1(1\sigma) = (1.58 \pm 0.34)\,M_\odot$.[14]

The second system, containing the PSR J1740–5340, was discovered in the globular cluster NGC 6397. The timing yields $P_b = 1.354$ d, the circular orbit, and $f_1 = 0.00264\,M_\odot$. The optical companion is bright and variable. Its observations give[34] the orbit inclination angle, the radial velocities, and the masses $M_1(1\sigma) = (1.53 \pm 0.19)\,M_\odot$ and $M_2 \approx 0.3\,M_\odot$.

6. Conclusions

The results of neutron star mass measurements in compact binaries are very impressive (Fig. 1). It is especially true for binaries containing radio pulsars and demonstrating GR effects. Since the middle of the 1980s, one–two such systems are being discovered every year.

Masses of ten neutron stars in double neutron star binaries have been measured very accurately and lie in a narrow range from $1.18 M_\odot$ to $1.44 M_\odot$. The Hulse-Taylor pulsar is the most massive neutron star among them and the most massive neutron star with precisely measured mass.

Masses of several neutron stars in binaries with white dwarfs are also measured with very good accuracy (and fall in the same narrow mass range). There are at least three pulsars (J0621+1002, J1012+5307, J1741+1351) in binaries with white dwarfs whose masses seem to be $\gtrsim 1.7 M_\odot$. However, these masses are still rather

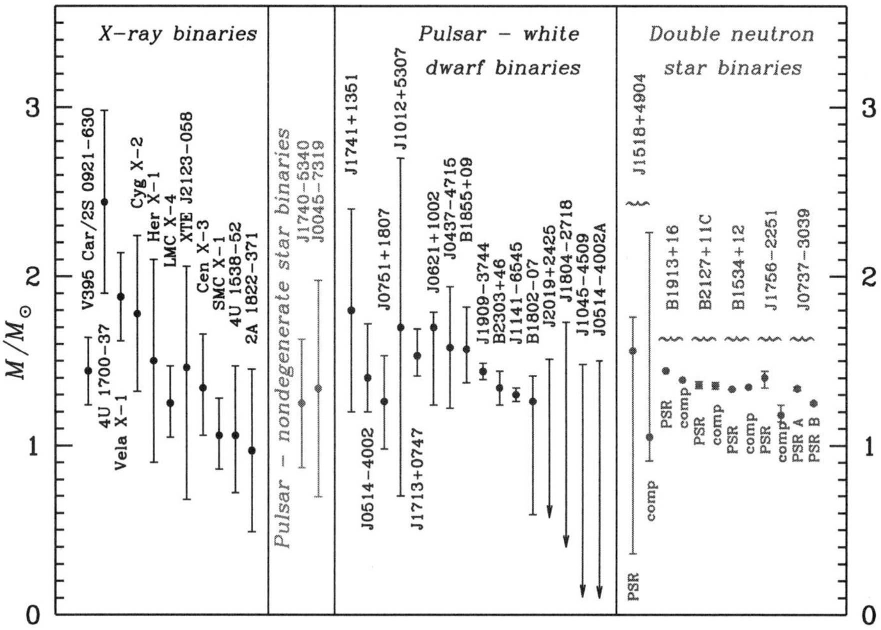

Fig. 1. Masses of neutron stars in binary systems (at the 2σ level).

uncertain although the accuracy will be definitely improved in the future.

The accuracy of neutron star mass measurements in X-ray binaries and in binaries with nondegenerate stars is still insufficiently high.

Even by observing the presently discovered neutron star binaries with the available technique, masses of many neutron stars will be measured accurately in forthcoming 5–10 years. In this way the accurately explored mass spectrum will be extended from $\approx 1.44\,M_\odot$ to at least $1.7\,M_\odot$. There is no doubt, that, in addition, new neutron star binaries will be discovered and new observational technique will be developed, accelerating the progress.

Acknowledgments

This work was partially supported by the Russian Foundation for Basic Research (grants 05-02-16245 and 05-02-22003); by the Russian Federal Agency for Science and Innovations (grant NSh 9879.2006.2); and by the Joint Institute for Nuclear Astrophysics (grant NSF PHY 0216783).

References

1. P. Haensel, A. Y. Potekhin and D. G. Yakovlev, *Neutron Stars. 1. Equation of State and Structure* (Springer, New York, 2007).
2. J. M. Lattimer, this volume.

3. P. G. Jonker, M. van der Klis and P. J. Groot, *Mon. Not. R. Astron. Soc.* **339**, 663 (2003).
4. M. H. van Kerkwijk, J. van Paradijs and E. J. Zuiderwijk, *Astron. Astrophys.* **303**, 497 (1995).
5. A. van der Meer, L. Kaper, M. H. van Kerkwijk, M. H. M. Heemskerk and E. P. J. van den Heuvel, *Astron. Astrophys.* **473**, 523 (2007).
6. J. A. Tomsick, W. A. Heindl, D. Chakrabarty and P. Kaaret, *Astrophys. J.* **581**, 570 (2002).
7. A. P. Reynolds, H. Quaintrell, M. D. Still, P. Roche, D. Chakrabarty and S. E. Levine, *Mon. Not. R. Astron. Soc.* **288**, 43 (1997).
8. H. Quaintrell, A. J. Norton, T. D. C. Ash, P. Roche, B. Willems, T. R. Bedding, I. K. Baldry and R. P. Fender, *Astron. Astrophys.* **401**, 313 (2003).
9. J. A. Orosz and E. Kuulkers, *Mon. Not. R. Astron. Soc.* **305**, 132 (1999).
10. J. S. Clark, S. P. Goodwin, P. A. Crowther, L. Kaper, M. Fairbairn, N. Langer and C. Brocksopp, *Astron. Astrophys.* **392**, 909 (2002).
11. D. Steeghs and P. G. Jonker, *Astrophys. J.* **669**, L85 (2007).
12. M. K. Abubekerov, *Astron. Rep.* **48**, 649 (2004).
13. M. K. Abubekerov, E. A. Antokhina and A. M. Cherepashchuk, *Astron. Rep.* **48**, 89 (2004).
14. S. E. Thorsett and D. Chakrabarty, *Astrophys. J.* **512**, 288 (1999).
15. I. H. Stairs, S. E. Thorsett, J. H. Taylor and A. Wolszczan, *Astrophys. J.* **581**, 501 (2002).
16. J. M. Weisberg and J. H. Taylor, The relativistic binary pulsar B1913+16, in *Radio Pulars* (Astron. Soc. Pacific, San Francisco, 2003).
17. B. A. Jacoby, P. B. Cameron, F. A. Jenet, S. B. Anderson, R. N. Murty and S. R. Kulkarni, *Astrophys. J.* **644**, L113 (2006).
18. A. G. Lyne, M. Burgay, M. Kramer, A. Possenti, R. N. Manchester, F. Camilo, M. A. McLaughlin, D. R. Lorimer, N. D'Amico, B. C. Joshi, J. Reynolds and P. C. C. Freire, *Science* **303**, 1153 (2004).
19. A. J. Faulkner, M. Kramer, A. G. Lyne, R. N. Manchester, M. A. McLaughlin, I. H. Stairs, G. Hobbs, A. Possenti, D. R. Lorimer, N. D'Amico, F. Camilo and M. Burgay, *Astrophys. J.* **618**, L119 (2005).
20. R. A. Hulse and J. H. Taylor, *Astrophys. J.* **195**, L51 (1975).
21. W. van Straten, M. Bailes, M. C. Britton, S. R. Kulkarni, S. B. Anderson, R. N. Manchester and J. Sarkissian, *Nature* **412**, 158 (2001).
22. D. J. Nice, Talk given at the Conference *40 Years of Pulsars. Millisecond Pulsars, Magnetars and More* (Montreal, August 12–17, 2007).
23. M. Bailes, S. M. Ord, H. S. Knight and A. W. Hotan, *Astrophys. J.* **595** L49 (2003).
24. E. M. Splaver, D. J. Nice, I. H. Stairs, A. N. Lommen and D. C. Backer, *Astrophys. J.* **620**, 405 (2005).
25. D. J. Nice, E. M. Splaver and I. H. Stairs, Neutron star masses from Arecibo timing observations of five pulsar–white dwarf binary systems, in *Radio Pulars* (Astron. Soc. Pacific, San Francisco, 2003).
26. B. A. Jacoby, A. Hotan, M. Bailes, S. Ord and S. R. Kulkarni, *Astrophys. J.* **629**, L113 (2005).
27. D. J. Nice, E. M. Splaver and I. H. Stairs, *Astrophys. J.* **549**, 516 (2001).
28. M. H. van Kerkwijk and S. R. Kulkarni, *Astrophys. J.* **516**, L25 (1999).
29. C. G. Bassa, M. H. van Kerkwijk, D. Koester and F. Verbunt, *Astron. Astrophys.* **456**, 295 (2006).
30. G. Cocozza, F. R. Ferraro, A. Possenti and N. D'Amico, *Astrophys. J.* **641**, L129

(2006).
31. P. Freire, B. Jacoby, M. Bailes, I. Stairs, A. Mott, R. Ferdman, D. Nice and D. C. Backer, Discovery and timing of the PSR J1741+1351 binary pulsar, *AAS Meeting* **208** No. 72.06 (2006).
32. P. C. C. Freire, S. M. Ransom and Y. Gupta, *Astrophys. J.* **662**, 1177 (2007).
33. D. J. Nice, E. M. Splaver, I. H. Stairs, O. Löhmer, A. Jessner, M. Kramer and J. M. Cordes, *Astrophys. J.* **634**, 1242 (2005).
34. J. Kaluzny, S. M. Rucinski and I. B. Thompson, *Astron. J.* **125**, 1546 (2003).

DENSE NUCLEAR MATTER: CONSTRAINTS FROM NEUTRON STARS

J. M. LATTIMER

Department of Physics & Astronomy, Stony Brook University,
Stony Brook, NY 11733, USA
E-mail: lattimer@astro.sunysb.edu

Neutron stars are a natural laboratory for the exploration of dense nuclear matter. Analytic constraints from general relativity, causality and stability provide a general framework into which specific constraints concerning dense matter properties can be made from measurements of masses, spin rates, thermal emissions, and quasi-periodic oscillations observed in accreting and bursting sources. Focus is placed upon recent mass and spin rate measurements and their implications for theory.

Keywords: Neutron stars; Equation of state.

1. Introduction

A series of several recent observations have suggested constraints for the structure of neutron stars and, in turn, the dense matter nuclear equation of state. Some of these observations could constrain the following properties of neutron stars:

- the maximum mass
- the minimum mass
- the minimum rotational period
- the radius (or, more precisely, the radiation radius)
- the moment of inertia
- the binding energy
- the fundamental vibration frequencies, related to the shear modulus and thickness of the crust
- the crustal cooling timescale
- the core cooling timescale (*i.e.*, whether or not the direct URCA process operates or not)

A detailed discussion of possible links between observables, neutron star structure and the equation of state is contained in a recent review.[1] Note that with the exception of the neutron star maximum mass, all of the above properties are highly sensitive to the nuclear symmetry energy, both its magnitude and density dependence. I will focus on the most recent developments for the above constraints.

2. Global Theoretical Constraints

Before touching upon the observations, it is worthwhile to summarize global constraints on neutron star structure from the fundamental theoretical perspectives of general relativity and causality. Assuming the equation of state is known up to the fiducial mass-energy density ϵ_f, that the associated pressure $p_f << \epsilon_f$, and the pressure is causally constrained above ϵ_f (*i.e.*, $p \le p_f + \epsilon - \epsilon_f$),[2,3]

$$M_{max} \simeq 4.2(\rho_s/\rho_f)^{1/2}\ \mathrm{M}_\odot. \tag{1}$$

For a given mass, causality also constrains the smallest possible radius[4–6]

$$R_{min} \simeq 2.9GM/c^2 = 4.3(M/\mathrm{M}_\odot)\ \mathrm{km}, \tag{2}$$

the largest possible central density[7]

$$\rho_c < 4.5 \times 10^{15}(\mathrm{M}_\odot/M)^2\ \mathrm{g\ cm}^{-3}, \tag{3}$$

and the smallest possible spin period[6]

$$P_{min} \simeq 0.74(\mathrm{M}_\odot/M_{sph})^{1/2}(R_{sph}/10\ \mathrm{km})^{3/2}\ \mathrm{ms}. \tag{4}$$

Note that Eq. (3) may be applied to the largest measured mass in order to set an absolute upper limit to the density possible in cold, static structures. For realistic baryonic equations of state (*i.e.*, not strange quark matter stars) not near M_{max}, even tighter constraints on the minimum spin period are realized[8]

$$P_{min} \simeq (0.96 \pm 0.03)(\mathrm{M}_\odot/M_{sph})^{1/2}(R_{sph}/10\ \mathrm{km})^{3/2}\ \mathrm{ms}, \tag{5}$$

which naturally leads to a minimum neutron star central density

$$\rho_c > 0.44 \times 10^{15}(1\ \mathrm{ms}/P_{min})^2\ \mathrm{g\ cm}^{-3}. \tag{6}$$

The combination of Eqs. (3) and (6) therefore constrain the overall range of neutron star central densities.

3. Neutron Star Spin Periods

The most rapidly rotating millisecond pulsar is the object PSR J1748-2446ad, with a frequency of 716 Hz, discovered by Hessels et al. in 2006.[9] This object spins rapidly enough to effectively contrain neutron star radii, especially if the star's mass (which is unknown) is in the typical pulsar range 1.2–1.4 $\mathrm{M}_\odot$, as shown in Fig. 1.

Recently, six type I X-ray bursts from the neutron star X-ray transient XTE J1739-285 showed oscillations with a frequency of 1122 Hz.[11] This detection was significat at the 99.96% level, and, if confirmed, might indicate that this object contains the fasted spinning neutron star yet. The limit on radius would be profound; for a 1.4 $\mathrm{M}_\odot$ star, $R < 11$ km is implied which would also suggest that the density derivative of the nuclear symmetry energy near nuclear density is either very small, *i.e.*, near zero, or that significant softening due to a Bose condensate or deconfined quarks occurs just above the nuclear saturation density. In the latter case, as Fig. 1 indicates, the neutron star maximum mass must also be relatively small.

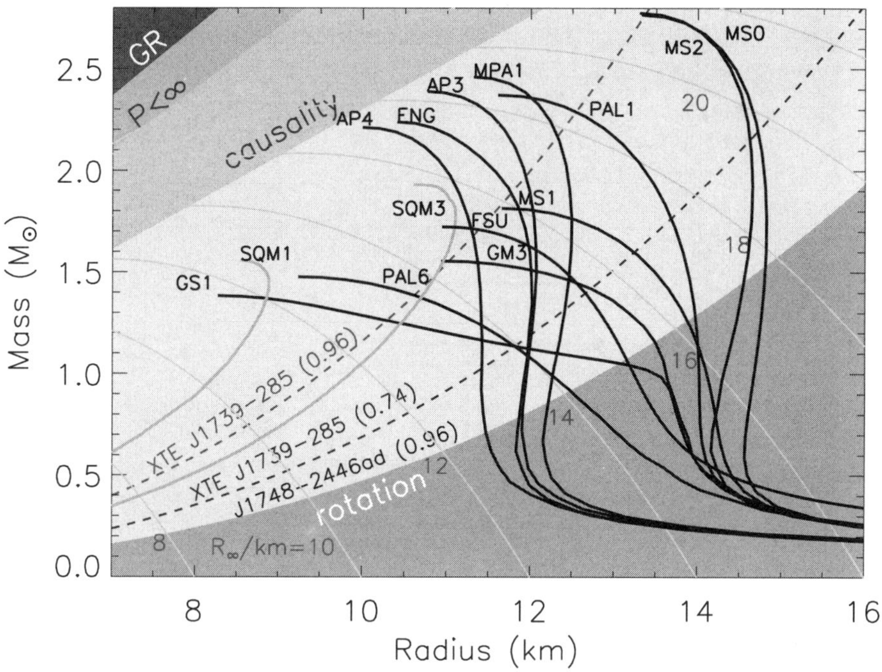

Fig. 1. Mass-radius diagram for neutron and strange quark matter stars. Regions excluded by general relativity and causality are in the upper left corner. Baryonic (solid black) and strange quark (solid green, extending to origin) $M-R$ curves, as labelled in Ref. 10, are displayed. Lighter curves (solid orange) are $R_\infty = R/\sqrt{1-2GM/Rc^2}$ contours. The 716 Hz pulsar[9] rotational limit, from Eq. (5) [Eq. (4)], is indicated by the green region [blue dashed curve]. Valid equations of state should enter the region to the left of the limit. The red dashed line illustrates a corresponding limit if claimed[11] 1122 Hz oscillations from XTE J1739-285 are due to its spin frequency.

4. Neutron Star Masses

Our knowledge of neutron star masses is summarized in Fig. 2. Since the June meeting in Catania, a few revisions in mass estimates have occurred: for these cases, the original estimates as shown in Catania are indicated in red (lighter shading) and the revised estimates in black.

The uppermost of the four regions contains mass estimates from X-ray binary sources, which are characterized by relatively large systematic errors. A recent analysis[12] of the source 2S 0921-630 resulted in a very large reduction from a mean estimated mass of 2.5 $M_\odot$ to 1.4 $M_\odot$. Other changes[13] to SMC X-1, LMC X-4 and Cen X-3, by comparison, are small. As Fig. 2 shows, the weighted mean of all X-ray binary masses is about 1.38 $M_\odot$, although a few sources, especially Vela X-1,[14,15] imply relatively large masses.

The lowest three regions are mass measurements from binary radio pulsars, which are characterized by less systematic uncertainties and hence higher accruacies. Nevertheless, even here, longer data streams can change mass values consid-

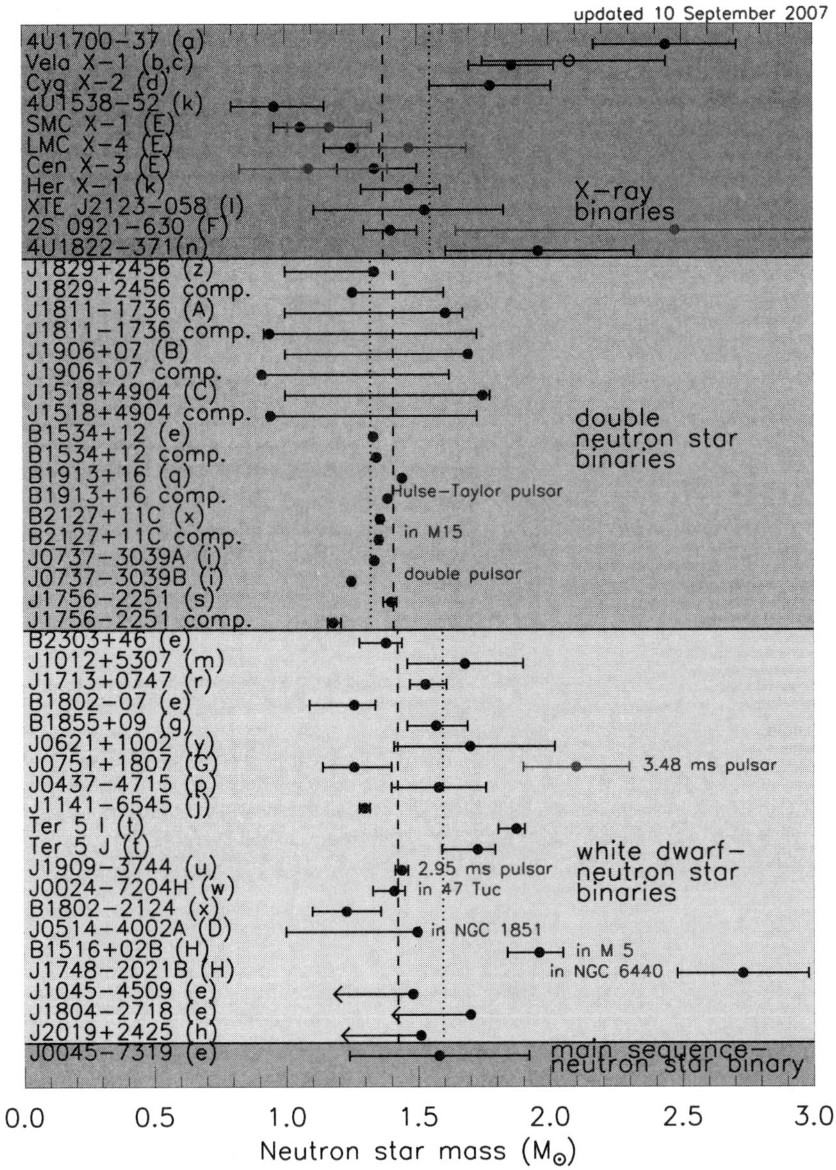

Fig. 2. Neutron star mass measurements with 1-σ uncertainties. Uppermost (green) region is for X-ray binaries, lowermost regions are for pulsar timing measurements. Dotted lines in each region indicate simple mass averages, dashed lines indicate weighted averages. For selected sources, red measurements were those reported in my Catania talk but have since been revised.

erably, although the changes remain within 3 or 4 sigma. During the Montreal "40 years of pulsars" anniversary meeting, two significant developments occurred: one of the largest mass measurements, that of J0751+1807,[16] was lowered from 2.1 ± 0.2 $M_\odot$ to $1.26^{+0.14}_{-0.12}$ $M_\odot$ (one-sigma uncertainties), and two more "high-mass" pulsars,

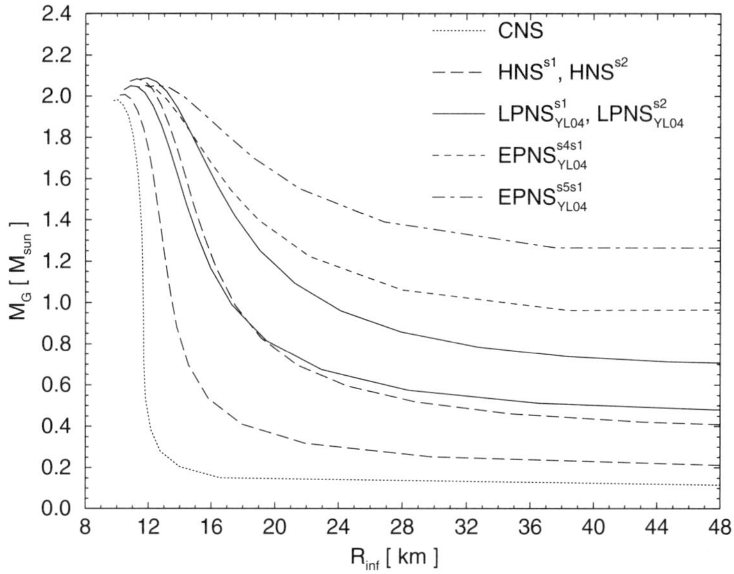

Fig. 3. Gravitational mass versus R_∞ for non-rotating proto-neutron stars in various evolutionary stages. In reverse order of evolution, CNS means a cold neutron star, HNS means hot deleptonized star, LPNS means late neutrino-trapped proto-neutron star, EPNS means early neutrino-trapped proto-neutron star with a hot-shocked mantle.

in the globular clusters M 5 (B1516+02B) and NGC 6440 (J1748-2021B) were reported.[17] Nevertheless, to the $3-\sigma$ level, all current measurements are consistent with $M_{max} < 1.5$ $M_\odot$.

It is also interesting to note the minimum observed neutron star mass, which is about 1.2 $M_\odot$ for the companion to J1756-2251.[18] This is of interest because it is near the minimum mass[19] of a proto-neutron star[20] early in its life. It is believed that a proto-neutron star is born with a core of approximately 0.7 $M_\odot$ that has a specific entropy $s \sim 1$ and a trapped lepton (electron + neutrino) fraction $Y_L \sim 0.4$ which is surrounded by a high-entropy shocked region ($s > 4-5$) (EPNS). Within a second, the outer region deleptonizes because of its relatively small neutrino mean free paths, and the entire star can be approximated with the core conditions (LPNS). After 10-20 s, the star deleptonizes due to neutrino diffusion, but the star becomes warmer ($s \sim 2$) because of neutrino heating (HNS). After 50-100 s, the star cools to essentially a cold neutron star (CNS). For a given equation of state, Fig. 3 shows mass-radius trajectories for each stage. Minimum masses of the EPNS configurations should therefore set the minimum possible mass for a cold neutron star which appears to be in excess of 1 $M_\odot$. Future measurements of neutron star masses might probe the minimum mass even more closely, and a sub-solar mass neutron star measurement will shed serious doubt on the current formation and/or proto-neutron star evolution paradigm.

5. Radiation Radius

Following their birth, neutron stars continue to cool via both neutrino and photon radiation. After an age of 50 or so s, the star becomes transparent to neutrinos and they escape without scattering within the star. The core, which is nearly isothermal, cools more quickly than the crust and reaches a lower temperature. The crust cools both by surface emission and heat conduction into the core, but becomes isothermal with the core only after tens of years.[22] Dimensionally, the crust cooling time scales as $\Delta^2 C_V(1-2GM/Rc^2)^{-3/2}/\dot{\epsilon}$, where Δ is the crust thickness, C_V is the crustal specific heat and $\dot{\epsilon}$ is the neutrino emissivity of core material. The cooling time is possibly observable either following neutron star birth from a supernova remnant, or following crustal heat during accretion episodes as X-ray transients. In the latter case, the entire crust might not be heated.

During the period 100 to 1 million years after birth, neutrino emission from the neutron star dominates surface photon fluxes but will be too small to observe. The star will be visible as an X-ray (and, if the star is near enough, as an optical) source. Several such cooling neutron stars are observed. To zeroth order, thermal emission from neutron stars is similar to a blackbody, so a measure of the integrated flux and the temperature yields an estimate of the radius/distance ratio. However, the flux is redshifted twice and the temperature once, so the measured radius is actually the radiation radius, $R_\infty = R/\sqrt{1-2GM/Rc^2}$. Principal uncertainties in extracting radii from observations include

- Distance uncertainties ($R_\infty \propto d$)
- Interstellar H absorption (most hard UV and an appreciable fraction of X-rays are absorbed between the star and the Earth)
- Atmospheric composition and magnetic field strength

The best chances of an accurate measurement are either from

- nearby isolated neutron stars, for which parallax distances are available, but which have unknown atmospheric compositions and field strengths, or
- quiescent X-ray binaries in globular clusters, which have reliable distances and, due to recent accretion episodes, low magnetic fields and, almost certainly, H-dominated atmospheres.

The best-studied isolated neutron star is RX J1856-3754,[23] with a parallax distance of 120–170 pc. Fitting the X-ray and optical spectra with a low-field, heavy-element atmosphere,[24] the estimated radius is $R_\infty > 19.5(d/140 \text{ pc})$ km, which is uncomfortably large. Assuming a magnetic field so high that the atmosphere condenses suggests a radius about 15% smaller,[26] which is still large. However, this model still requires a thin atmosphere of just the right density to explain the optical luminosity of this source. As is apparent from Fig. 1, this result together with a distance of 160 pc[25] can not be easily reconciled with an equation of state that also predicts maximum masses less than 2 $M_\odot$.

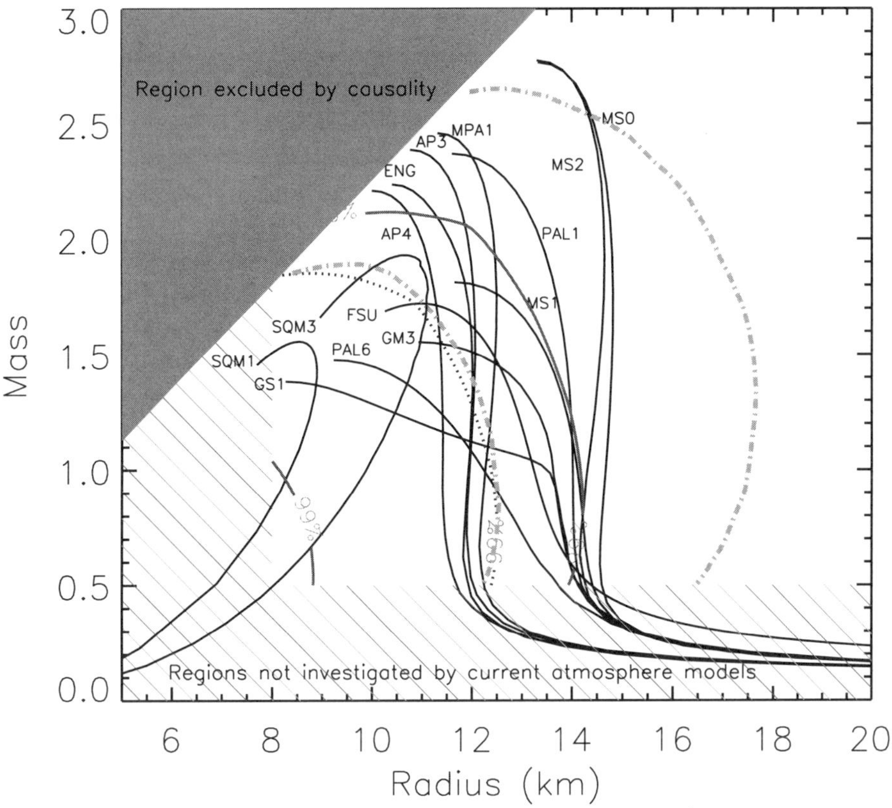

Fig. 4. Allowed regions of M and R for three globular cluster neutron stars from X-ray observations.[21] Solid (red), dashed (green) and dotted (blue) curves are 99% confidence contours for stars in ω Cen, 47 Tuc and M 13, respectively. Equation of state labels are given in Ref. 10.

Because of the uncertainties concerning neutron star atmospheres, globular cluster sources have advantages. Results[21] for sources in 3 globular clusters are summarized in Fig. 4. There is no overlap in all three 99% confidence contours, but the target neutron stars may have significantly different masses. At face value, the results suggest an equation of state that both has a relatively small radius for 1–1.5 $M_\odot$ and predicts a maximum mass greater than 1.8 $M_\odot$.

6. Conclusions

Space does not permit discussion of other exciting developments, including seismology from X-ray burst oscillations and thermal cooling times following X-ray bursts, which constrain neutron star crustal thicknesses, and moments of inertia from pulsar timing which tightly constrain radii. These are, however, described in Ref. 1 and references therein. Although definitive limits are not yet possible, it is impressive how the array of neutron star observations has multiplied in recent years.

Acknowledgements

I wish to profoundly thank the organizers of this meeting for the opportunity to present this work and to discover the (g)astronomic pleasures of Sicily. This work has been supported by the US Dept. of Energy under grant DE-AC02-87ER40317.

References

1. J. M. Lattimer and M. Prakash, *Phys. Rep.* **442**, 109 (2007).
2. C. E. Rhoades Jr. and R. Ruffini, *Phys. Rev. Lett.* **32**, 324 (1974).
3. J. B. Hartle and A. G. Sabbadini, *Ap. J.* **213**, 831 (1977).
4. L. Lindblom, *Ap. J.* **278**, 364 (1984).
5. N. K. Glendenning, *Phys. Rev.* **D46**, 4161 (1992).
6. S. Koranda, N. Stergioulas and J. L. Friedman, *Ap. J.* **488**, 799 (1997).
7. J. M. Lattimer and M. Prakash, *Phys. Rev. Lett.* **94**, 1101 (2005).
8. J. M. Lattimer and M. Prakash, *Science* **304**, 536 (2004).
9. J. W. T. Hessels, S. M. Ransom, I. H. Stairs, P. C. C. Stairs, V. M. Kaski and F. Camilo, *Science* **311**, 1901 (2006).
10. J. M. Lattimer and M. Prakash, *Ap. J.* **550**, 426 (2001).
11. P. Kaaret, Z. Prieskorn, J. J. M. In't Zand, S. Brandt, N. Lund, S. Mereghetti, D. Gotz, E. Kuulkers and J. A. Tomsick, *Ap. J.* **657**, L97 (2007).
12. D. Steeghs and P. G. Jonker, *Ap. J.* in press; astro-ph/0707.2067 (2007).
13. A. van der Meer, L. Kaper, M. H. van Kerkwijk, M. H. M. Heemskerk and E. P. J. van den Heuvel, *A&A* in press; astro-ph/0707.2802 (2007).
14. O. Barziv, L. Karper, M. H. van Kerkwijk, J. H. Telging and J. van Paradijs, *A&A* **377**, 925 (2001).
15. H. Quaintrell, A. J. Norton, T. D. C. Ash, P. Roche, B. Willems, T. R. Bedding, I. K. Baldry and R. P. Fender, *A&A* **401**, 303 (2003).
16. D. Nice, in *40 Year of Pulsars* (AIP, New York, 2007).
17. P. C. C. Freire, J. W. T. Hessels, S. M. Ransom, I. H. Stairs, S. Begin and A. Wolszczan, *ibid.*.
18. A. J. Faulkner *et al.*, *Ap. J.* **618**, L119 (2004).
19. K. Strobel, C. Schaab and M. Weigel, *A&A* **350**, 497 (1999).
20. A. Burrows and J. M. Lattimer, *Ap. J.* **307**, 178 (1986).
21. N. A. Webb and D. Barret, *Ap. J.* in press, astro-ph/0708.3816 (2007).
22. J. M. Lattimer, K. A. van Riper, M. Prakash and M. Prakash, *Ap. J.* **425**, 802 (1994).
23. F. M. Walter, S. J. Wolk and R. Neuhauser, *Nature* **379**, 233 (1996).
24. J. A. Pons, F. M. Walter, J. M. Lattimer, M. Prakash, R. Neuhauser and P. H. An, *Ap. J.* **564**, 981 (2002).
25. M. H. van Kerkwijk and K. L. Kaplan, in *Isolated Neutron Stars: From the Interior to the Surface*, eds. D. Page, R. Turolla and S. Zane; astro-ph/0607320 (2006).
26. W. C. G. Ho, D. L. Kaplan, P. Chang, M. van Adelsberg and A. Y. Potekhin, *MNRAS* **375**, 821 (2007).

NEUTRON STAR VERSUS HEAVY-ION DATA: IS THE NUCLEAR EQUATION OF STATE HARD OR SOFT?

JÜRGEN SCHAFFNER-BIELICH,* IRINA SAGERT and MIRJAM WIETOSKA

Institut für Theoretische Physik/Astrophysik, J. W. Goethe Universität, D-60438 Frankfurt am Main, Germany
**E-mail: schaffner@astro.uni-frankfurt.de*

CHRISTIAN STURM

Institut für Kernphysik, J. W. Goethe Universität, D-60438 Frankfurt am Main, Germany

Recent astrophysical observations of neutron stars and heavy-ion data are confronted with our present understanding of the equation of state of dense hadronic matter. Emphasis is put on the possible role of the presence of hyperons in the interior of compact stars. We argue that data from low-mass pulsars provide an important cross-check between high-density astrophysics and heavy-ion physics.

Keywords: Nuclear equation of state; Neutron stars; Pulsars; Kaon production in heavy-ion collisions.

1. Introduction

The research areas of high-density astrophysics, as the physics of compact stars, and relativistic heavy-ion collisions are probing matter at extreme densities. The properties of neutron stars are determined by the nuclear equation of state (EoS), as well as microphysical reactions in dense matter. The stiffness of the high-density matter controls the maximum mass of compact stars. New measurements of the global properties of pulsars, rotation-powered neutron stars, point towards large masses and correspondingly to a rather stiff equation of state (for a recent review on the equation of state for compact stars see Ref. 1). In a recent analysis of the x-ray burster EXO 0748–67 it was even claimed that soft nuclear equations of state are ruled out.[2] Note that this analysis, if confirmed, would not rule out the presence of quark matter in the core of compact stars.[3]

On the other side, strange particles (kaons) produced in relativistic heavy-ion collisions just below the threshold of the elementary reaction are sensitive to medium effects due to the created high-density matter (see e.g. Ref. 4). Recent investigations conclude that the systematics of kaon production can only be explained by an extremely soft nuclear equation of state above normal nuclear matter density.[5–8]

There seems to be conflict in determining the nuclear equation of state, which we will discuss in detail in the following. We investigate the impacts of the compression modulus and symmetry energy of nuclear matter on the maximum mass of neutron stars in view of the recent constraints from heavy-ion data on kaon production in dense matter. In particular, we delineate the different density regions probed in the mass-radius diagram of compact stars. We outline the importance of the Schrödinger equivalent potentials for subthreshold production of kaons. The possible effects from the presence of hyperons in dense neutron star matter are confronted with pulsar mass measurements.

2. The Nuclear EoS from Astrophysical and Heavy-Ion Data: a Soft or Hard EoS?

The properties of high-density nuclear matter is intimately related to the phase diagram of quantum chromodynamics (QCD), for a review see e.g. Ref. 9. The regime of high temperatures and nearly vanishing baryochemical potential is probed by present and ongoing heavy-ion experiments at BNL's Relativistic Heavy-Ion Collider and CERN's Large Hadron Collider and is related to the physics of the early universe. A rapid crossover transition due to chiral symmetry restoration and deconfinement is found in lattice gauge simulations, see e.g. Ref. 10. The QCD phase diagram at large baryochemical potential and moderate temperatures constitutes the region of the chiral phase transition and the high-density astrophysics of core-collapse supernovae and compact stars (see e.g. Ref. 11 for a recent treatise). Terrestrial heavy-ion experiments, as the Compressed Baryonic Matter (CBM) experiment at GSI's Facility for Antiproton and Ion Research (FAIR) will investigate this fascinating and largely unknown terrain of the QCD phase diagram.[12]

The nuclear equation of state serves as a crucial input for simulations of core-collapse supernovae,[13] neutron star mergers,[14,15] proto-neutron star evolution[16] and, of course for determining the properties of cold neutron stars.[17] Pulsar mass measurements provide constraints on the stiffness of the nuclear equation of state. Unfortunately, out of the more than 1600 known pulsars, only a few precise mass measurements from binary pulsars are currently available (see Ref. 18 and references therein). Still, the undoubtedly upper mass limit is given by the Hulse-Taylor pulsar of $M = (1.4414 \pm 0.0002)M_\odot$,[19] the lightest pulsar known is J1756-225 with a mass of $M = (1.18 \pm 0.02)M_\odot$.[20] New data on pulsar masses has been presented at the Montreal conference on pulsars. The mass of the pulsar J0751+1807, originally with a median above two solar masses with $M = 2.1 \pm 0.2 M_\odot$ (1σ),[21] is now corrected and below the Hulse-Taylor mass limit.[22] However, the mass of the pulsar J0621+1002 was determined to be between 1.53 to 1.80 solar masses (2σ).[23] Combined data from the pulsars Terzan 5I and J[24] with the pulsar B1516+02B.[25] results in a mass limit of 1.77 solar masses for at least one of these pulsars. Measurement of the pulsar J1748–2021B arrives at a lower mass limit of $M > 2M_\odot$[25] but that could be the mass of a two neutron star system. The analysis of x-ray burster is much more model dependent. For EXO 0748–676 a mass-radius constraint of $M \geq 2.10 \pm 0.28 M_\odot$ and

$R \geq 13.8 \pm 1.8$ km has been derived,[2] for a critical discussion on the analysis I refer to Ref. 26. That constraint would actually rule out soft nuclear equations of state but not the presence of quark matter, as quark matter is an entirely new phase which can be rather stiff.[3]

High-density nuclear matter is produced in the laboratory for a fleeting moment of time in the collisions of heavy nuclei at relativistic bombarding energies. The properties of kaons can change substantially in the high-density matter created. The in-medium energy of kaons will increase with density (basically due to the low-density theorem, see however Ref. 27 which arrives at a somewhat stronger repulsive potential). Kaons are produced by the associated production mechanism NN→ NΛK, NN→NNK$\overline{\mathrm{K}}$, and most importantly by the in-medium processes πN → ΛK, $\pi\Lambda$ → N$\overline{\mathrm{K}}$, which are rescattering processes of already produced particles. The effective energy of kaons in the medium will change the Q-values of the direct production and rescattering processes, therefore affecting the net production rate.[4] As kaons have long mean free paths, they can leave the high-density region and serve as an excellent tool to probe its properties. Indeed, detailed transport simulations find that nuclear matter is compressed up to $3n_0$ for a typical bombarding energy of 1 to 1.5 AGeV and that the produced kaons are dominantly produced around $2n_0$, where n_0 stands for the normal nuclear matter saturation density.[6,7] Kaons are produced below the elementary threshold energy due to multi-step processes which increase with the maximum density achieved in the collisions. The double ratio of the multiplicity per mass number for the C+C collisions and Au+Au collisions turns out to be rather insensitive to the input parameters (elementary cross sections, in-medium potential) which scale linearly with mass number or density. Only calculations with a compression modulus of $K \approx 200$ MeV can describe the trend of the kaon production data.[5–8] Hence, the analysis of heavy-ion experiments points towards a rather soft nuclear equation of state.

3. The Different Density Regimes of Neutron Stars

In the following we discuss the different densities encountered in neutron stars and the corresponding regions in the mass-radius diagram. While the standard lore is that the crust of a neutron star consists of nuclei, neutrons and electrons, the composition of the interior of a neutron star is basically unknown. At about $2n_0$ hyperons can appear as a new hadronic degree of freedom. Kaons can be formed as Bose-Einstein condensate. Finally, chirally restored quark matter can be present as an entirely new phase in the core of compact stars. After considering pure nucleonic matter, we focus on the role of hyperons and their importance for the properties of neutron stars (see also Ref. 28 and references therein).

First, let us consider just nucleonic matter. Its equation of state can be modelled by a Skyrme-type ansatz for the energy per nucleon. The parameters are fixed by the nuclear matter properties, as the saturation density, binding energy, compression modulus and asymmetry energy.[29] In addition, we explore effects from the asymmetry term having a density dependence which scales with a power α as extracted from heavy-ion collision measurements where α is between 0.7 and 1.1.[30–32] The pressure is determined by a thermodynamic relation, which fixes completely the EoS used in transport simulations of heavy-ion collisions. With that EoS at hand one can check now, whether the low compressibilities found in describing the kaon production data of $K \approx 200$ MeV[5–8] are ruled out by neutron star measurements.

Solving the Tolman-Oppenheimer-Volkoff equation gives the result that rather large maximum neutron star masses can be reached even for such low values of the compression modulus.[29] The maximum mass is greater than $M > 2M_\odot$ for a compression modulus of $K > 160$ MeV ($\alpha = 1.0$, 1.1) and for the case $\alpha = 0.7$ greater than $M > 1.6M_\odot$ for $K > 160$ MeV and greater than $M > 2M_\odot$ for $K > 220$ MeV. Changing the asymmetry energy within reasonable values ($S_0 = 28$ to 32 MeV) shifts the maximum mass by at most $\Delta M = \pm 0.1M_\odot$ for low values of K. The maximum central density is about $n_c = (7 \div 8)n_0$ for $\alpha = 1.0, 1.1$ and can hit even $10n_0$ for $\alpha = 0.7$. The EoS is causal up to a compression modulus of $K = 340$, corresponding to a maximum mass of $M = 2.6M_\odot$, for $\alpha = 1.0, 1.1$ and up to $K = 280$ MeV for $\alpha = 0.7$. Hence, we conclude that even a pulsar mass of $2M_\odot$ would be compatible with the 'soft' EoS as extracted from heavy-ion data. This statement is corroborated by more advanced many-body approaches to the nuclear EoS for kaon production in heavy-ion collisions and neutron star mass limits.[33]

For a field-theoretical investigation on the nuclear equation of state in heavy-ion collisions and for neutron stars one has to consider the Schrödinger equivalent potential, which is the actual input to the transport simulation codes, not the nuclear equation of state. There is a one-to-one correspondence between the energy per baryon and the nucleon potential for the non-relativistic Skyrme model as studied above. However, this direct relation is lost in a relativistic field theoretical approach as for example the relativistic nucleon potential exhibits now a scalar and vector part. We note also, that the direct relation between the compression modulus K and the stiffness of the nuclear equation of state at supra-nuclear densities is also lost. The stiffness of the EoS in the standard relativistic mean-field (RMF) model is controlled by the effective mass of the nucleon at saturation density not by the compression modulus, which is actually well known for quite some time, see e.g. Ref. 34. The Schrödinger equivalent potential for a sample of parameter sets of the relativistic mean-field model is depicted in Fig. 1. The line marked 'KaoS' stands for the nucleon potential as used in transport simulations for a compression modulus of $K = 200$ MeV. In order to be in accord with the KaoS data, the potential of the relativistic mean-field parameter set should be below the curve labelled 'KaoS' at a density region of around $2n_0$ where most of the kaons are produced at subthreshold collision energies. The parameter sets used for the standard nonlinear RMF model are 'bmw85' with an effective mass of $m^*/m = 0.85$ and $K = 300$ MeV,[34] 'gm1' with $m^*/m = 0.7$ and $K = 300$ MeV, and 'gl78' with $m^*/m = 0.78$ and $K = 240$ MeV.[35] Note, that the values chosen for the effective nucleon mass are quite high so that the nuclear EoS becomes soft. Typical fits to properties of nuclei arrive at values of

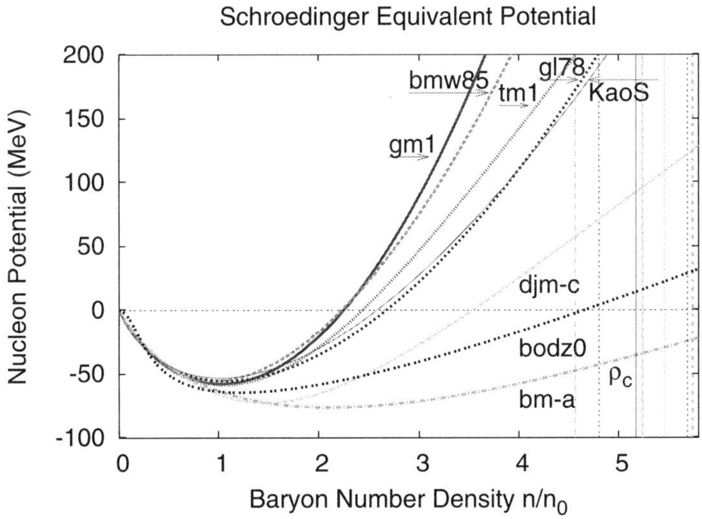

Fig. 1. The Schrödinger equivalent potential versus the baryon number density given in n_0 for various parameters of the relativistic mean-field model.

$m^*/m \approx 0.6$ as for the parameter set 'tm1' which is fitted to properties of spherical nuclei.[36] The set 'tm1' has an additional selfinteraction term for the vector fields which results in an overall similar behaviour of the nucleon potential in comparison to the other RMF parameter sets with a soft EoS. Those vector selfinteractions were introduced in Ref. 37 where the set 'bodz0' with $m^*/m = 0.6$ and $K = 300$ MeV is taken from. One motivation of introducing this vector selfinteraction term is to describe the nucleon vector potential as computed in more advanced many-body approaches which are based on nucleon-nucleon potentials. For the sets 'bm-a' and 'djm-c', the vector selfenergy of the nucleon in the RMF calculation was adjusted to the ones of Dirac-Brueckner-Hartree-Fock calculations.[38] The minima of the nucleon potential of those latter three parameter sets are located at larger densities than the saturation density, in particular for the set 'bm-a'. We stress that the nuclear equation of state for those sets, however, gives the right properties of saturated nuclear matter.[38] Figure 1 shows also the maximum density reached in the center of the maximum mass configuration of the neutron star sequence by vertical lines, which are surprisingly close lined up between 4.5 to $6n_0$ in view of the large differences in the nucleon potential at high densities.

The mass-radius diagram for the RMF parameter sets giving a small Schrödinger equivalent potential, i.e. one which is at or below the potential used in transport simulations ($K = 200$ MeV), is plotted in Fig. 2 for neutron star matter consisting of nucleons and leptons only. The sets 'gl78' and 'djm-c' reach maximum masses of $2.04M_\odot$ and $1.98M_\odot$, respectively, even though the nucleon potential for the set 'djm-c' is well below the limit given from the heavy-ion data analysis (see Fig. 1).

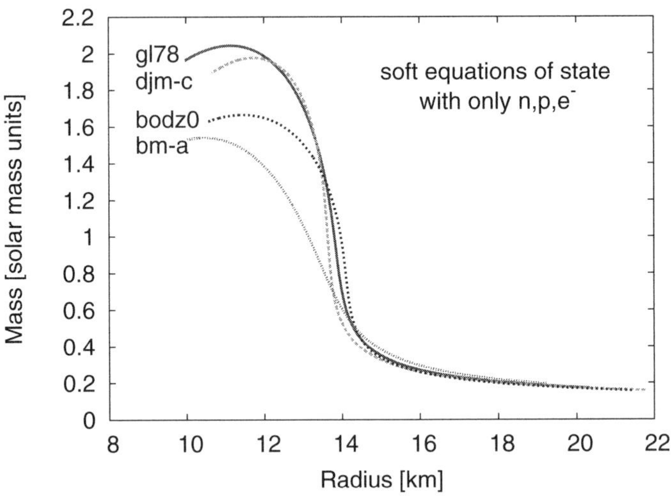

Fig. 2. The mass-radius plot for various parameter sets of the relativistic mean-field model with nucleons and leptons only.

The sets 'bodz0' and 'bm-a' just arrive at maximum masses of $1.66M_\odot$ and $1.54M_\odot$, respectively, which could be ruled out with a confirmed measurement of a heavy pulsar but so far can not be excluded. It seems, that the constraint from heavy-ion data on the nucleon potential alone is in agreement with pulsar mass measurements for relativistic mean-field approaches, even if masses of about $2M_\odot$ will be measured in the future. An important point to stress here is that the heavy-ion data and the determination of the maximum mass of neutron stars addresses completely different density regimes. While the heavy-ion data on kaon production probes at maximum 2 to $3n_0$, the central density of the most massive neutron stars tops $5n_0$. Hence, the maximum mass of neutron stars probes the high-density regime of the nuclear equation of state which is not constrained by the heavy-ion data presently available. In other words, if the pressure, or better the nucleon potential, rises slowly at densities up to $2n_0$, it could increase rapidly at larger densities so as to comply with astrophysical data on neutron star masses. Moreover, new particles and phases could certainly appear at such large densities which change the equation of state for massive neutron stars substantially, as hyperon matter, to which we turn now for making our argumentation more explicit.

The in-medium properties of hyperons are constrained by hypernuclear data. In particular, the Λ potential at n_0 is quite well determined to be -30 MeV. Other hyperon potential are much less well known, unfortunately. Hyperons, if present, have a strong impact on the properties of compact stars (see Ref. 28 for a recent outline). Λ hyperons constitute a new hadronic degree of freedom in neutron star matter at and above about $2n_0$. The population of other hyperons, Σ and Ξ hyperons, is highly sensitive to their in-medium potentials. For a slightly repulsive

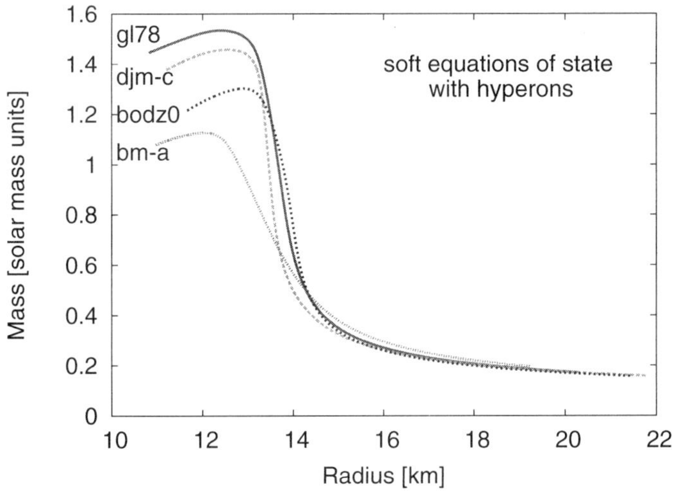

Fig. 3. The mass-radius plot for various parameter sets of the relativistic mean-field model including the effect of hyperons.

potential for the Σ these hyperons do not appear in compact star matter at all. The Ξ hyperons will be present, as their in-medium potential is likely to be attractive. The presence of hyperons changes drastically the properties of compact stars. The new degree of freedom lowers the pressure for a given energy density, so that the EoS is considerably softened at large densities. There is a substantial decrease of the maximum mass due to the appearance of hyperons in compact stars: the maximum mass for such "giant hypernuclei" can drop down by $\Delta M \approx 0.7 M_\odot$ compared to the case of neutron star matter consisting of nucleons and leptons only.[35] For the RMF parameter sets studied here, we add hyperons as outlined in Ref. 39 by fixing the hyperon vector coupling constants via SU(6) symmetry relations and the hyperon scalar coupling constant to the (relativistic) hyperon potentials as determined in Ref. 40 from hypernuclear data and hyperonic atoms. The resulting mass-radius plot when including hyperons is pictured in Fig. 3. Note the different mass scales of Figs. 2 and 3, the maximum mass with hyperons included is now substantially decreased to $1.53 M_\odot$ for the set 'gl78', to $1.46 M_\odot$ for the set 'djm-c', to $1.30 M_\odot$ for the set 'bodz0', and to $1.27 M_\odot$ for the set 'bm-a'. The latter two cases are now even below the Hulse-Taylor mass limit and can be ruled out. The former two cases are just above the Hulse-Taylor mass limit of $1.44 M_\odot$ and could be ruled out if measurements of heavy neutron stars masses of $1.6 M_\odot$ or more will be confirmed in future astrophysical observations. Clearly, the presence of hyperons in compact stars could be severely constrained by combining the heavy-ion data analysis with the measurement of a heavy neutron star. The limit on the nucleon potential from heavy-ion data seems to make it quite difficult to reach neutron star masses above say $1.6 M_\odot$ for the RMF model when hyperons are included via SU(6) symmetry and by adopting the presently (sometimes poorly) known hyperon potentials from hypernuclear data. Of course, a much more systematic analysis needs to be done, a firm statement can not be drawn from our sample of parameter sets. Certainly, the situation could be changed for other many-body approaches. Within the relativistic Hartree-Fock approach, for example, maximum masses of about $1.9M_\odot$ are possible even when effects from hyperons are added to the equation of state.[41] But one rather robust conclusion can be drawn from our analysis: the high-density EoS above $2n_0$, where hyperons appear and modify the EoS in the models used here, is crucial in determining the maximum mass of a neutron star. Hence, we are probing this density region when looking at the maximum mass configurations of compact stars.

The procedure to follow is now eminent, the true comparison between the present heavy-ion data and astrophysical data on compact stars is not located in the high-mass region but on the low-mass region of the mass-radius diagram of compact stars. The lightest neutron star known at present is the pulsar J1756-225 with a mass of $M = (1.18 \pm 0.02)M_\odot$.[20] Much lower values are probably not realized in nature as hot proto-neutron stars have a much larger minimum stable mass than cold neutron stars, for example a minimum mass of $0.86M_\odot$ has been found for an isothermal proto-neutron star.[42] Interestingly, a $1.2M_\odot$ neutron star has a maximum density of $n = 2n_0$ in our non-relativistic models,[29] so that exotic matter is likely to be not present. We find that the radius of such a low-mass neutron star is in fact highly sensitive to the nuclear equation of state (see also Ref. 43), in particular to the asymmetry energy at high densities which is well known.[31,44–46] There are several promising proposals for radii measurements of neutron stars, see Ref. 1 for a recent overview. The fascinating aspect is that heavy-ion experiments can address this density region and probe not only the equation of state but also the density dependence of the asymmetry energy. The ratio of the produced isospin partners K^+ and K^0 at subthreshold energies has been demonstrated to be sensitive to the isovector potential above saturation density.[47] The tantalising conclusion is that a direct comparison with heavy-ion data and compact star data seems to be feasible.

As always there are exceptions to the assumption, that the nuclear EoS just contains nucleons and leptons up to $1.2M_\odot$. In Ref. 48 strange quark matter is already present for only $0.3M_\odot$ which depends hugely on the choice of the MIT bag constant. In Ref. 49 hyperons appear already for a compact star mass of only $0.5M_\odot$ although the critical density for the onset of the hyperon population is around $2n_0$. The reason is that the equation of state is unphysically soft, so that the maximum mass is below the Hulse-Taylor mass limit. In any case, this provides another opportunity for the radius measurement of low-mass pulsars: if their radii turn out to be completely off the range predicted from our knowledge of the density dependence of the asymmetry energy, some exotic matter is present in the core of neutron stars!

4. Summary

The combined analysis of heavy-ion data on kaon production at subthreshold energies and neutron star mass measurements points towards a nuclear EoS that is soft at moderate densities and hard at high densities. A soft nuclear EoS as extracted from kaon production data is not in contradiction with heavy pulsars as mutually exclusive density regions are probed. The nuclear EoS above $n \approx 2n_0$ determines the maximum mass of neutron stars, which is controlled by unknown high-density physics (as hyperons and quark matter). Compact star matter constrained by the heavy-ion data seems to result in rather low maximum masses for compact stars when hyperons are included. A measurement of a heavy pulsar will make it quite difficult for having hyperons inside a neutron star and could exhibit an emerging conflict between hypernuclear and pulsar data. Properties of low-mass neutron stars ($M \leq 1.2 M_\odot$), however, are likely to be entirely determined by the EoS of nucleons and leptons only up to $n \approx 2n_0$, as hyperon and possibly quark matter could appear at larger densities. Thus, the measurement of the radii of low-mass pulsars provides the opportunity for a cross-check between heavy-ion and astrophysical data and possibly for the detection of an exotic phase in the interior of compact stars.

Acknowledgements

This work is supported in part by the Gesellschaft für Schwerionenforschung mbH, Darmstadt, Germany. Irina Sagert gratefully acknowledges support from the Helmholtz Research School for Quark Matter Studies.

References

1. J. M. Lattimer and M. Prakash, *Phys. Rep.* **442**, 109 (2007).
2. F. Özel, *Nature* **441**, 1115 (2006).
3. M. Alford, D. Blaschke, A. Drago, T. Klähn, G. Pagliara and J. Schaffner-Bielich, *Nature* **445**, E7 (2006).
4. J. Schaffner-Bielich, I. N. Mishustin and J. Bondorf, *Nucl. Phys. A* **625**, 325 (1997).
5. C. Sturm *et al.*, *Phys. Rev. Lett.* **86**, 39 (2001).
6. C. Fuchs, A. Faessler, E. Zabrodin and Y.-M. Zheng, *Phys. Rev. Lett.* **86**, 1974 (2001).
7. C. Hartnack, H. Oeschler and J. Aichelin, *Phys. Rev. Lett.* **96**, 012302 (2006).
8. A. Forster *et al.*, *Phys. Rev. C* **75**, 024906 (2007).
9. D. H. Rischke, *Prog. Part. Nucl. Phys.* **52**, 197 (2004).
10. M. Cheng, N. H. Christ, S. Datta, J. van der Heide, C. Jung, F. Karsch, O. Kaczmarek, E. Laermann, R. D. Mawhinney, C. Miao, P. Petreczky, K. Petrov, C. Schmidt, W. Soeldner and T. Umeda, *arXiv:0710.0354 [hep-lat]* (2007).
11. J. Schaffner-Bielich, *arXiv:0709.1043 [astro-ph]* (2007).
12. P. Senger, T. Galatyuk, D. Kresan, A. Kiseleva and E. Kryshen, *PoS* **CPOD2006**, 018 (2006).
13. H.-T. Janka, K. Langanke, A. Marek, G. Martínez-Pinedo and B. Müller, *Phys. Rep.* **442**, 38 (2007).
14. S. Rosswog and M. B. Davies, *Mon. Not. Roy. Astron. Soc.* **345**, 1077 (2003).
15. R. Oechslin, H.-T. Janka and A. Marek, *Astron. Astrophys.* **467**, 395 (2007).

16. J. A. Pons, S. Reddy, M. Prakash, J. M. Lattimer and J. A. Miralles, *Astrophys. J.* **513**, 780 (1999).
17. F. Weber, *Prog. Part. Nucl. Phys.* **54**, 193 (2005).
18. I. H. Stairs, *J. Phys. G* **32**, S259 (2006).
19. J. M. Weisberg and J. H. Taylor, The relativistic binary pulsar b1913+16: Thirty years of observations and analysis, in *Binary Radio Pulsars*, eds. F. A. Rasio and I. H. Stairs, Astronomical Society of the Pacific Conference Series **328**, 25 (2005).
20. A. J. Faulkner, M. Kramer, A. G. Lyne, R. N. Manchester, M. A. McLaughlin, I. H. Stairs, G. Hobbs, A. Possenti, D. R. Lorimer, N. D'Amico, F. Camilo and M. Burgay, *Astrophys. J.* **618**, L119 (2005).
21. D. J. Nice, E. M. Splaver, I. H. Stairs, O. Löhmer, A. Jessner, M. Kramer and J. M. Cordes, *Astrophys. J.* **634**, 1242 (2005).
22. David Nice and Ingrid Stairs, private communication.
23. Laura Kasian, David Nice and Ingrid Stairs, private communication.
24. S. M. Ransom, J. W. T. Hessels, I. H. Stairs, P. C. C. Freire, F. Camilo, V. M. Kaspi and D. L. Kaplan, *Science* **307**, 892 (2005).
25. Paolo Freire, talk given at the conference '40 Years of Pulsars', Montreal, August 12–17, 2007, see ns2007.org.
26. F. M. Walter and J. M. Lattimer, *Nature Physics* **2**, 443 (2006).
27. C. L. Korpa and M. F. M. Lutz, *Acta Phys. Hung. A* **22**, 21 (2005).
28. J. Schaffner-Bielich, *astro-ph/0703113* (2007).
29. I. Sagert, M. Wietoska, J. Schaffner-Bielich and C. Sturm, *arXiv:0708.2810 [astro-ph]* (2007).
30. L.-W. Chen, C. M. Ko and B.-A. Li, *Phys. Rev. Lett.* **94**, 032701 (2005).
31. B.-A. Li and A. W. Steiner, *Phys. Lett. B* **642**, 436 (2006).
32. L.-W. Chen, C. M. Ko, B.-A. Li and G.-C. Yong, *arXiv:0704.2340 [nucl-th]* (2007).
33. C. Fuchs, *arXiv:0706.0130 [nucl-th]* (2007).
34. B. M. Waldhauser, J. A. Maruhn, H. Stöcker and W. Greiner, *Phys. Rev. C* **38**, 1003 (1988).
35. N. K. Glendenning and S. A. Moszkowski, *Phys. Rev. Lett.* **67**, 2414 (1991).
36. Y. Sugahara and H. Toki, *Nucl. Phys. A* **579**, 557 (1994).
37. A. R. Bodmer, *Nucl. Phys. A* **526**, 703 (1991).
38. S. Gmuca, *Nucl. Phys. A* **547**, 447 (1992).
39. J. Schaffner and I. N. Mishustin, *Phys. Rev. C* **53**, 1416 (1996).
40. J. Schaffner-Bielich and A. Gal, *Phys. Rev. C* **62**, 034311 (2000).
41. H. Huber, F. Weber, M. K. Weigel and C. Schaab, *Int. J. Mod. Phys. E* **7**, 301 (1998).
42. D. Gondek, P. Haensel and J. L. Zdunik, *Astron. Astrophys.* **325**, 217 (1997).
43. J. M. Lattimer and M. Prakash, *Astrophys. J.* **550**, 426 (2001).
44. C. J. Horowitz and J. Piekarewicz, *Phys. Rev. Lett.* **86**, 5647 (2001).
45. C. J. Horowitz and J. Piekarewicz, *Phys. Rev. C* **64**, 062802 (2001).
46. A. W. Steiner, M. Prakash, J. M. Lattimer and P. J. Ellis, *Phys. Rep.* **411**, 325 (2005).
47. G. Ferini, T. Gaitanos, M. Colonna, M. Di Toro and H. H. Wolter, *Phys. Rev. Lett.* **97**, 202301 (2006).
48. K. Schertler, C. Greiner, J. Schaffner-Bielich and M. H. Thoma, *Nucl. Phys. A* **677**, 463 (2000).
49. H.-J. Schulze, A. Polls, A. Ramos and I. Vidana, *Phys. Rev. C* **73**, 058801 (2006).

SURFACE EMISSION FROM X-RAY DIM ISOLATED NEUTRON STARS

ROBERTO TUROLLA

Dept. of Physics, University of Padova, via Marzolo 8, 35131 Padova, Italy
E-mail: turolla@pd.infn.it

XDINSs (X-ray dim isolated neutron stars) are a group of seven soft X-ray sources originally discovered by ROSAT and characterized by purely thermal spectra ($kT \sim 40 - 100$ eV), lack of radio emission and rather long spin periods ($P \sim 3 - 11$ s). The absence of contamination from magnetospheric activity, a surrounding supernova remnant or a binary companion makes XDINSs the only sources in which we can have a clean view of the neutron star surface. As such they offer an unprecedented opportunity to confront theoretical models of surface emission with observations. This may ultimately lead to a direct measure of the star radius and thus probe the equation of state. XDINSs were long believed to be steady blackbody emitters. In the last few years new XMM and Chandra observations revelaled a much more faceted picture. The brightest source, RX J1856.5-3754, is the only one which indeed exhibits a purely blackbody spectrum. In all the other members of the class broad absorption features ($E \sim 200 - 700$ eV) have been discovered. The second brightest object, RX J0720.4-3125 was recently found to undergo (cyclic ?) spectral variations most probably produced by the star nutation. The increasing number of optical identifications confirms that the optical continuum lies above the Rayleigh-Jeans tail of the X-ray spectrum. All these facts challenge the standard picture of XDINSs as cooling neutron stars covered by an atmosphere and endowed with a dipolar magnetic field. In this talk I will review the current status of theoretical efforts in the modeling of XDINS surface emission, discuss the many still unsolved problems and address briefly the possible relationship between XDINSs and other isolated neutron star classes, like the recently discovered rotating radio transients (RRATs).

HIGH ENERGY NEUTRINO ASTRONOMY

E. MIGNECO

Istituto Nazionale di Fisica Nucleare, Laboratori Nazionali del Sud,
Via S. Sofia 62, 95123 Catania, Italy
E-mail: migneco@lns.infn.it

Neutrinos are very promising probes for high energy astrophysics. Indeed, many indications suggest that cosmic objects where acceleration of charged particles can takes place, e.g. GRBs and AGNs, are the sources of the detected UHECRs. Accelerated hadrons, interacting with ambient gas or radiation, can produce HE neutrinos. Contrary wise to charged particles and TeV gamma rays, neutrinos can reach the Earth from far cosmic accelerators, traveling in straight line, therefore carrying direct information on the source. Theoretical models indicate that a detection area of $\simeq$1 km^2 is required for the measurement of HE cosmic ν fluxes. The detection of Čerenkov light emitted by secondary leptons produced by neutrino interaction in large volume transparent natural media (water or ice) is today considered the most promising experimental approach to build high energy neutrino detectors. The experimental efforts towards the opening of the high energy neutrino astronomy are also reviewed.

1. Introduction

The only neutrinos ever detected from cosmos are the neutrinos from the supernova SN1987A and the Solar neutrinos. Indeed, up to know no significant excess of high energy neutrinos has been found to stem over the atmospheric neutrino background produced by the interaction of cosmic rays in the atmosphere that surrounds the Earth. The physics that can be addressed with high energy neutrino telescopes covers a very broad range of items, spanning from Dark Matter to Ultra High Energy Particle (UHECR) production. However, motivations for high energy neutrino astronomy and consequently for the construction of km^3-scale Čerenkov telescopes under-ice or underwater, mostly relay on the observation of high energy cosmic rays and on the rather recent discover of many unexpected γ-TeV emitter sources in our galaxy. Indeed, in spite of the continuous and remarkable progresses in cosmic ray physics, including the recent data of AUGER[1] on UHECR the problem of the origin of the cosmic rays is not completely solved. In the recent years, many theories and calculations about candidate sources, such as Supernova Remnants (SNR), Gamma Ray Bursts (GBR), Active Galactic Nuclei (AGN), have been developed and carried out. In particular, SNRs and GRBs seem to provide the environment and energy conditions needed to explain the galactic and the extragalactic cosmic rays respectively. Due to the presence of galactic and intergalactic magnetic field, charged particles with energy up to 10^{19} eV are scrambled thus not allowing to trace back towards the direction of the emitting sources. Moreover, is still much-discussed, on the basis of the recent AUGER data, a possible correlation between the observed particles on Earth and the closest potential cosmic accelerators. On the other hand, neutrinos, that are produced in hadronic interaction, have no charge and interact with matter only via the week force, represent a very promising probe. High energy neutrinos should reveal us which and where are the most powerful accelerators in the cosmos and how the acceleration mechanisms works in order to provide particles with the observed energy. A recent interesting experimental evidence concerns the observation of a rather large number of γ-TeV sources detected by the Čerenkov telescopes Hess[2] and Magic[3] in these last years. In particular, several of these sources show a power law spectrum with a spectral index around 2, consistently with a Fermi acceleration mechanism, and characteristics that strongly support the presence a proton acceleration mechanism. The detection of neutrinos from these sources would provide the smoking gun to disentangle between hadronic and leptonic acceleration processes. Calculations of fluxes for several neutrino candidate sources indicate, for both diffuse and point-like sources, that the opening of the high energy neutrino astronomy requires detector with effective areas of one km^2.

2. Detection Principle of Neutrino Čerencov Telescopes Under Water and Under Ice

Although a high energy neutrino telescopes aim at detecting neutrinos of all flavors, performance is optimized for muon detection. The energetic neutrino is detected indirectly through the muon produced in the neutrino interactions occurring nearby or inside the detector volume. Indeed, muons travel at a speed close to the light speed thus producing in sea-water Čerenkov light with an angle of about 42^0 with respect to the the muon track. A viable approach for a km^3-size neutrino telescope is to equip with optical sensors an adequate volume of a natural transparent medium such as deep sea water or the deep Antarctic ice, in fact about three thousands meters of water (or ice) reduce the flux of atmospheric muons by a factor 10^5. The ice or sea water acts as a shield for atmospheric muons, a target for the neutrino-lepton conversion and a Čerencov radiator for the lepton produced in the neutrino-medium interaction. A three-dimensional lattice made of several thousands of photomultipliers that measure the arrival time and the charge of the photons allows to reconstruct the direction and energy of the neutrino. Moreover, unlike conventional telescopes that are pointed towards the skies, high energy neutrino telescopes watch the sky upside down. Indeed, only up-going tracks from the opposite hemisphere can be unambiguously attributed to neutrinos that, for energy up to about 100 TeV, are the only particle that can pass the whole Earth. Neutrinos above 100 TeV are almost completely absorbed by the Earth. Another important source of background is the flux atmospheric neutrinos produced by the interaction of cosmic ray in the atmosphere. This flux, that cannot be suppressed, also provides a natural calibration source.

3. Detection Principle and International Context

Two telescope are needed in the two opposite hemispheres to cover the whole sky. The neutrino telescope in the Mediterranean Sea will survey the larger part of the Galactic disc, including the Galactic Center which is not visible from the South Pole. At the South Pole is currently under construction IceCube, the cubic kilometer neutrino telescope in the deep Antarctic ice. Also in Mediterranean Sea the efforts of the collaborations working in the field are eventually merging towards the construction of km^3 Čerenkov telescopes. In the following a will sketch the status of the experimental activity towards the km^3 Čerenkov telescopes.

3.1. *IceCube*

Following the successful experience of AMANDA, a prototype detector made of 677 OMs that established the best limits on neutrino diffuse and point-like sources, the IceCube collaboration started the construction of the km^3 neutrino telescope in ice. IceCube[4] consists of 80 strings and 4800 PMTS. 22 strings have been already deployed thus covering a volume that makes this detector the largest neutrino telescope actually operating in the world. The completion of the detector is expected in 2011. The analysis of IceCube data is in progress.

3.2. *The northern hemisphere: projects in the Mediterranean Sea*

The high energy neutrino telescope in the Mediterranean Sea requires very complex technologies to cope with the extreme conditions of the deep sea: corrosion, very high pressure. Moreover, the deployment of detector elements, maintenance and the remote handling of deep undersea connection represent hard technological challenges. Three different projects operate in the Mediterranean Sea: ANTARES in Toulon, Nestor in Pylos and NEMO Sicily. These projects will briefly described in the following. Since 2006 the three collaborations merged in the European KM3Net consortium which is supported by the European 6th Framework Program as a Design Study and in the 7th Framework Program as Preparatory Phase project both for the km^3 telescope.

ANTARES

The goal of ANTARES[5] is to build a prototype detector with effective area of about 0.1 km^2 40 km off-shore La Seyne-Toulon at a depth of about 2500 m. The whole detector is made of 12 lines (length 350 m) for a total of 900 PMTs. 5 lines are already in data taking while 4 more lines are in the sea waiting to be connected. The completion of the 12 lines detector is foreseen in early 2008. Several

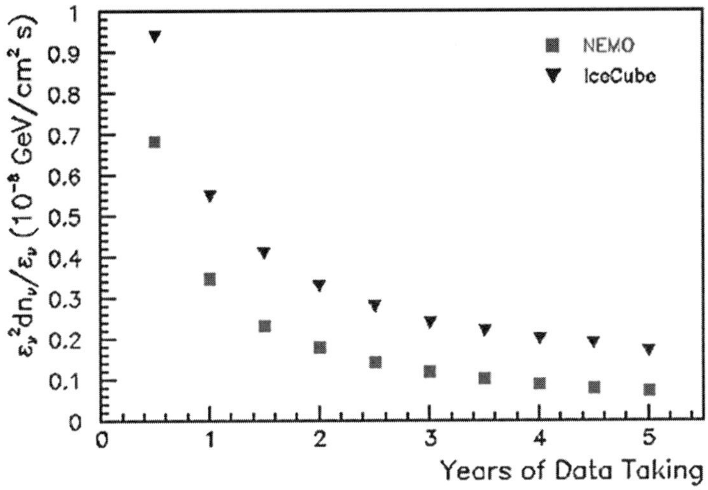

Fig. 1. Sensitivity of the NEMO km^3 detector to a neutrino point-like source with declination $\delta = -60°$ and a E^{-2} neutrino energy spectrum compared to the IceCube sensitivity.[9]

millions of down-going muon events have been collected thus allowing to study detector behavior in various bioluminescence conditions and several neutrino track candidates are reconstructed. The analysis for the comparison with the Monte Carlo expectation is in progress.

NESTOR

In March 2003 the NESTOR[6] collaboration deployed a test hexagonal floor module of a detector tower equipped with with 12 PMTs. The analysis of the data allowed the reconstruction of atmospheric muons and the comparison with the Monte Carlo estimations.

NEMO

NEMO[7] is an advanced R&D program aiming at the solution of the technological issues related to the realization of a km^3 neutrino telescope in the deep Mediterranean Sea. The project includes the characterization of an optimal site for detector installation and the construction and test of technological prototypes and their validation at depths up to 3500 m. With about 30 sea campaigns, the NEMO activity for the characterization of the Capo Passero site (3500 m), at about 80 km off-shore the SE coast of Sicily, provided a long term monitoring of the water properties such as the light transmission properties, the optical background noise, the sedimentation and the biological activity, the water current and the environmental properties (salinity, temperature) in situ. The measurements and site survey demonstrate that both the water optical properties and the oceanographic features are optimal for the installation of the km^3 Čerenkov telescope[8] so that Capo Passero was indicated

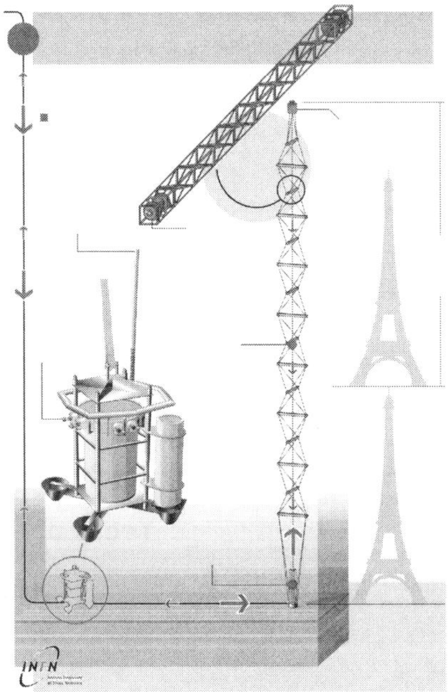

Fig. 2. The junction box and the tower of NEMO.

as a candidate site for the km^3 installation to ApPEC already in January 2003. One of the main concerns for the feasibility study was to reduce the number of structures that host the optical modules in order to minimize the number of underwater connections thus allowing the operations with Remotely Operated Vehicle (ROV) at depths of about 3500 m. The feasibility study carried out with the support of Monte Carlo simulations aiming at the optimization of detector performance, indicates that a km^3 neutrino telescope can be realized with about 100 structures (towers or strings) hosting a total number of 5000-6000 photomultipliers (PMT). The sensitivity of a detector made of 9×9 square lattice of towers (5832 PMT) for a generic point-like source is reported in Fig. 1 as a function of the integrated data taking time. The source position was chosen at a declination $\delta = -60°$ and a E^{-2} neutrino energy spectrum was considered. For a comparison the IceCube sensitivity[9] obtained for 1° search bin is also reported in the figure.

A study of the Moon shadow effect, due to the absorption of primary cosmic rays by the Moon, has also been undertaken with simulation. Indeed, this effects should provide a direct measurement of both the detector angular resolution and pointing accuracy. Preliminary results[10] show an angular resolution $\sigma = 0.19° \pm 0.02°$.

The NEMO-Phase1 project was aimed at the validation of the technological solution proposed for the realization and installation of the km^3 detector through the realization of a technological demonstrator including all the key elements of the km^3. The main elements of the telescope are the electro-optical cable that carries signals and power, the towers and a junction box that distributes power and data

Fig. 3. Preparation to the junction box deployment on the deck of the vessel Teliri.

from and to shore. A pictorial view of the key elements of the telescope is shown in Fig. 2.

The design of the junction box is innovative since the pressure-resistant steel vessels containing the electronics are hosted inside a corrosion-resistant fiberglass container filled with mineral oil. This solution decouples the two problems of pressure and corrosion resistance thus allowing to operate in the severe conditions of high-depth salt water, but with a significant cost decrease if compared to standard titanium containers. The tower consists of a three-dimensional semi-rigid structure composed by a sequence of bars, which host the optical modules and the instrumentation, interlinked by a system of ropes and anchored on the seabed. The structure is kept vertical by an appropriate buoyancy on the top. In the layout proposed for the km^3, each tower will consists of a sequence of 12 m long bars spaced vertically by 40 m, with the lowermost one placed at about 150 m above the sea bottom. NEMO-Phase1 went to completion in December 2006 with the successful deployment and connection of a junction box and a tower made of four 15 m long storeys at the underwater test site of the Laboratori Nazionali del Sud (LNS) of INFN at about 2000 m depth. Fig. 3 shows the preparation phase to the deployment of the junction box.

The transport of physics and control data occurs via an electro-optical cable that connects the LNS on-shore station in the Catania harbor to the off-shore station that is about 20 km far. The data to be recorded are selected with an onshore trigger procedure. Atmospheric muon events have been reconstructed (Fig. 4) and the analysis of the muon observed flux with a Monte Carlo comparison is ongoing. In order to validate the developed technologies at the depth of 3500 m and perform a long term site survey in Capo Passero, the NEMO Phase-2 started.

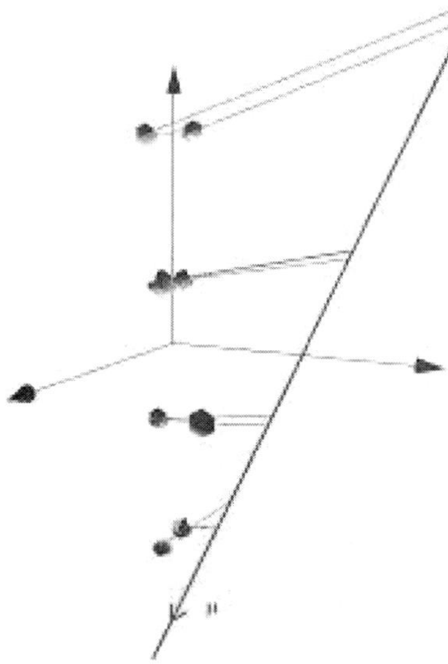

Fig. 4. A reconstructed down-going muon track where 8 PMTs were hit. The size of the optical module fired refers to the charge.

A 100 km electro-optical cable, with characteristics that are compatible with the needs of a km^3 detector, was laid in July 2007 to connect the deep sea infrastructures at 3500 m depth to a shore station. A complete 16 storey NEMO tower is under construction and will be deployed and connected in 2008. The realization of the NEMO project demonstrates that technological solutions exist for the realization of an underwater km^3 detector. The NEMO collaboration participates in the EU Design Study KM3NeT[11] and the following Preparatory Phase KM3NeT-PP. The NEMO experience will contribute to the advancement of the KM3NeT activities. Indeed, both the NEMO program and time schedule are well fitted to the Design Study and Preparatory Phase of KM3NeT.

References

1. The Pierre Auger collaboration, *Science* **318**, 938 (2007).
2. HESS web page at `http://www.mpi-hd.mpg.de/hfm/HESS`.
3. MAGIC web page at `http://wwwmagic.mppmu.mpg.de/physics/recent/index.html`.
4. A. Achterberg *et al.*, *Astrop. Phys.* **26**, 155 (2006); IceCube web page at `http://icecube.wisc.edu`.
5. ANTARES web page at `http://antares.in2p3.fr`.
6. S. E. Tzamarias, *Nucl. Instr. and Meth. A* **502**, 150 (2002); NESTOR web page at `http://nestor.org`.
7. NEMO web page at `http://nemoweb.lns.infn.it`.
8. G. Riccobene *et al.*, *Astropart. Phys.* **27**, 1 (2007).
9. J. Aharens *et al.*, *Astropart. Phys.* **20**, 507 (2004).
10. C. Distefano *et al.*, *Nucl. Instr. and Meth. A* **50**, 495 (2000).
11. KM3NeT web page at `http://www.km3net.org`.

WHAT GRAVITATIONAL WAVES SAY ABOUT THE INNER STRUCTURE OF NEUTRON STARS

V. FERRARI

Dipartimento di Fisica "G. Marconi," Sapienza Università di Roma and INFN, Sezione di Roma, I-00185 Roma, Italy

Neutron stars emit gravitational waves essentially by two mechanisms: because they rotate or because they pulsate. We will illustrate both mechanisms and discuss to what extent the detection of gravitational waves will allow to constrain the equation of state of matter at supranuclear densities.

Keywords: Gravitational waves; Neutron stars; Neutron star oscillations.

1. Introduction

In the past few years, a formidable effort has been done by the experimental groups working on interferometric detectors of gravitational waves, to improve the sensitivity of these instruments; both LIGO and VIRGO are now operating basically at the design sensitivity,[1] ultimately taking data in coincidence, and the scientific community is waiting for the results of these observational runs. Moreover, upgrading of the two experiments will start soon and much more sensitive, advanced detectors are planned for the future. Since neutron stars (NS) are sources of gravitational waves (GW) in many phases of their life, it is the right time to explore the possibility of using these waves to constrain the equation of state (EOS) of matter in their inner core. This is a relevant problem because the densities and pressures that prevail in a neutron star core are unreachable in terrestrial laboratories, and therefore the EOS of matter in such extreme conditions is only poorly constrained by empirical data. If gravitational waves carry an imprint of the EOS, they may provide an independent tool to explore the physics of matter at supranuclear densities.

In this paper we will consider two main mechanisms through which neutron stars emit gravitational waves: rotation and pulsations. We will discuss both mechanisms, also in connection with the possibility of detecting such waves.

2. Gravitational Waves from Rotating Stars

According to general relativity, a rotating neutron star emits gravitational waves if it has some degree of asymmetry with respect to the rotation axis. Such asymmetry may be generated by several physical processes occurring while the star is evolving, such as accretion, elastic deformations of the solid crust or core, or magnetic field effects. If the rotating star has a triaxial shape and rotates around one of the principal axes, say the z-axis, waves are emitted at twice the rotation frequency $\nu_{GW} = 2\nu$, with amplitude

$$h_{GW} = \frac{16\pi^2}{c^4} \frac{G I_{zz} \nu^2}{d} \epsilon \,, \tag{1}$$

where d is the source distance, I_{zz} is the moment of inertia and ϵ is the stellar *oblateness* defined as

$$\epsilon = 2\frac{a-b}{a+b} = \frac{I_{xx} - I_{yy}}{I_{zz}} \,, \tag{2}$$

where a and b are the equatorial semi-axes. It should be stressed that both the moment of inertia and the oblateness depend on the EOS of matter in the stellar interior. We also mention that if the star rotates around an axis which forms an angle θ (*wobble* angle) with the principal axes, waves are emitted *at* the rotation frequency, $\nu_{GW} = \nu$, and the amplitude is

$$h_{GW} = \frac{8\pi^2 G}{c^4} \frac{(I_{xx} - I_{zz})}{d} \nu^2 \theta \,. \tag{3}$$

The wave amplitudes (1) and (3) are derived on the assumption that the star is an isolated, rigid body.

We shall now discuss how gravitational waves emitted by rotating stars can be used to constrain the region where the oblateness ϵ and the moment of inertia of a neutron star can vary. In a recent paper of the LIGO collaboration,[2] the data of two observational runs, collected over about 3.5 months in 2004 and 2005, have been analyzed to look for the GW signal (1) emitted by 78 radio pulsars; since no signal has been detected, they have set upper limits on the GW emission of the different sources. It should be stressed that, at the time the data were collected, LIGO detectors where not working at their best sensitivity yet; therefore these upper limits are still quite high. In any event we shall use them to show how these data, combined with data obtained from astronomical observation, in the future will allow us to gain some insight on neutron stars structure. The upper limit set on the CRAB pulsar by LIGO is[2]

$$h_{GW\ LIGO} < 3.8 \cdot 10^{-24} \,. \tag{4}$$

An independent estimate of h_{GW} can be obtained from pulsars radio measurements. It is known that the CRAB pulsar spins down, and the same behaviour is observed in many pulsars; if the spin-down rate, $\dot{\nu}$, is known, an upper limit for the emitted gravitational signal can be found by assuming that the rotational energy lost per unit time $\frac{dE_{rot}}{dt} = 4\pi^2 I_{zz} \nu \dot{\nu}$, is entirely dissipated in gravitational waves, and therefore equates the gravitational wave luminosity $\frac{dE_{grav}}{dt} = \frac{2}{5}\frac{d^2 c^3}{G}\pi^2 \nu^2 h^2$. The resulting wave amplitude is:

$$h_{GW\ SD} = \left(\frac{2}{5} \frac{G I_{zz}}{c^3 d^2} \frac{|\dot{\nu}|}{\nu} \right)^{1/2} \,. \tag{5}$$

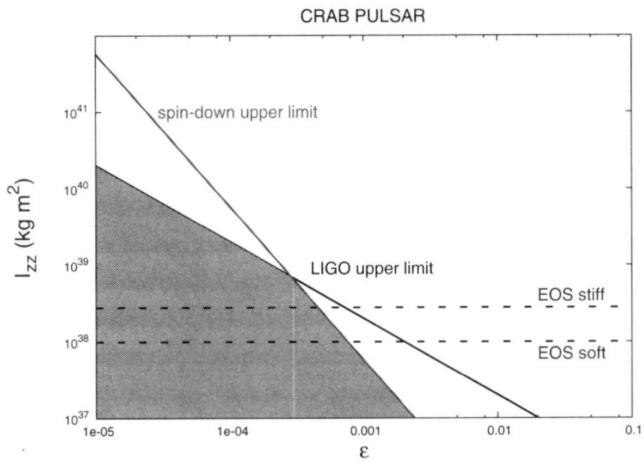

Fig. 1. The principal moment of inertia I_{zz} is plotted versus the stellar oblateness using the two independent relation (7) and (8) and the data obtained by LIGO and from radio measurements for the CRAB pulsar. The allowed region is shaded in green. The dashed lines refer to theoretical predictions of the moment of inertia based on a large set of modern EOS.

This is an upper limit because it is known that pulsars rotational energy is mainly dissipated through electromagnetic processes. Since for the CRAB $\nu = 29.6$ Hz and $\dot{\nu} = -3.73 \cdot 10^{-10}$ Hz/s, we find

$$h_{GW\ SD} < 1.4 \cdot 10^{-24}\ ; \tag{6}$$

this value is only three times smaller than the upper limit (4) set by LIGO measurements (obtained from a non-optimal data set). Thus, we can say that LIGO (and now also VIRGO) is approaching a regime of astrophysical interest.

Let us now see how the two independent estimates could be combined to gain information on the stellar structure. Assuming that the emitting star is a triaxial rotating body and that we know the wave amplitude from a direct detection, from Eq. (1) we can find the relation connecting the moment of inertia of the star and its oblateness

$$I_{zz} = \frac{c^4 d}{16\pi^2 G \nu^2 h_{GW\ detected}} \times \frac{1}{\epsilon}\ . \tag{7}$$

A second relation can be derived by formally equating the spin-down amplitude (5) and Eq. (1), finding

$$I_{zz} = \frac{5c^5|\dot{\nu}|}{512\pi^2 G \nu^5} \times \frac{1}{\epsilon^2}\ . \tag{8}$$

Eqs. (7) and (8) can be plotted as in Fig. 1, where the values of $h_{GW\ detected}$ used in Eq. (7) is the upper limit found by LIGO for the CRAB given in (4), whereas the values of and of ν and $\dot{\nu}$ are from radio measurements. The region allowed for I_{zz} and ϵ is that shaded region in the figure.

It is now interesting to compare these results with what we presently know about I_{zz} and ϵ from theoretical studies.

The maximal deformation that the crust of a neutron star can sustain without breaking, has been determined in several papers.[3–5] The main results of these studies are that if the star is composed by ordinary nuclear matter, then $\epsilon < 10^{-6}$; if it is composed by exotic matter (hyperons, deconfined quarks) it may be larger, $\epsilon < 5 \cdot 10^{-6}$ (or even as large as $\epsilon < 10^{-3}$ for very particular structures[5]). These limits refer to deformation from equilibrium of an old, cold NS. However, due to the effect of strong Lorentz forces induced by very intense magnetic fields, the star may be born with a highly non spherical shape; in this case large deformations would be possible, since the crust has not formed yet. As the star cools down and the crust forms, these deformations may be frozen so that the equilibrium shape of the final NS may have an oblateness much larger than the limits discussed above. Thus, depending on the mechanism which determines the stellar shape and on the evolutionary stage at which the star takes its shape, ϵ assumes values that can range in a wide interval, compatible with that shown in Fig. 1.

Theoretical studies on the moment of inertia of neutron stars have been performed in several papers;[6,7] these studies use a large set of EOS proposed to describe NS matter at supranuclear densities, and show that $I \sim [1-2.8] \cdot 10^{38}$ kg m^2, depending on the EOS and on the rotation rate. Again these results are compatible with Fig. 1. Thus, from Fig. 1 we see that the observational data actually provided by LIGO and by radio measurements are still too "loose" to constrain the region where the oblateness and the moment of inertia can vary, and so to give information on the EOS of matter in the NS inner core.

As mentioned above, observational runs by LIGO and VIRGO, operating at much higher sensitivity, will be available soon; thus, the black line referring to the LIGO upper limit in Fig. 1 will be considerably lowered, and will hopefully refer to a true detection rather than to an upper limit. In addition, the spin-down upper limit should also be improved in several respect. For instance, we found Eq. (5) assuming that the rotational energy lost by the star is dissipated entirely in gravitational waves. This is certainly not true, and to give a better estimate we need to know the magnetic field of the star, B. Unfortunately, at present B is estimated by assuming that the rotational energy is entirely dissipated by the magnetic field, thus we are clearly going around a circle. In order to correctly account for the electromagnetic and gravitational losses, we need an independent estimate of the magnetic field, which will hopefully be provided by future astronomical observations. Also, a measure of the wobble angle, certainly present in some pulsars, would be important to determine the expected gravitational wave amplitude.

From the discussion above, we understand that the detection of gravitational waves emitted by rotating stars will allow to restrict the interval where the moment of inertia and the oblateness of a NS could range, providing some information on the equation of state of matter in the stellar interior. However, this is only a piece of the puzzle. Further information could be ectracted from the detection of the gravitational signal emitted by a pulsating neutron star.

3. Stellar Pulsations

Stellar pulsations are a well known phenomenon in astronomy, since they underlay a variety of astrophysical processes. According to general relativity, non radial pulsations are associated to gravitational wave emission and, as we shall see, the mode frequencies carry interesting information on the inner structure of the emitting sources. Stars possess many different classes of oscillation modes. Fluid pulsation modes are classified following a scheme, introduced by Cowling in Newtonian gravity in 1942,[8] based on the restoring force which prevails when the generic fluid element is displaced from the equilibrium position. They are said g-modes, or gravity modes, if the restoring force is due to buoyancy, and p-modes if it is due to pressure gradients. The mode frequencies are ordered as follows

$$..\omega_{g_n} < .. < \omega_{g_1} < \omega_f < \omega_{p_1} < .. < \omega_{p_n}..$$

and are separated by the frequency of the fundamental mode (f-mode), which has an intermediate character between $g-$ and $p-$ modes. In addition, general relativity predicts the existence of spacetime modes, the w-modes, that do not exist in Newtonian gravity and have typically very high frequencies, larger than those of fluid modes.[9–12]

Pulsation modes are excited, for instance, when the star forms in a gravitational collapse, as a consequence of a binary coalescence, in a glitch, or due to a tidal interaction with a companion in a binary system. As an example, in Fig. 2 we show the Fourier transform of the gravitational signal emitted by a $1.4\,M_\odot$ neutron star inspiralling on a $10\,M_\odot$ Schwarzchild black hole. We plot the part of the signal which is due to the tidal deformation of the star, and which carries a signature of the NS's non-radial oscillation modes: they are manifested in the form of sharp peaks at the corresponding frequencies.[13]

Since numerical simulations of the most energetic astrophysical processes indicate that the most excited mode is the fundamental one, we will now focus on this mode. The issue we want to discuss is whether the identification of the peak corresponding to the excitation of the f-mode in a detected GW signal would allow to put constraints on the equation of state of matter in the stellar core. This problem has extensively been studied in the literature,[14–16] and the f-mode frequency, ν_f, has been computed for stars modeled using the EOS proposed to describe NS matter over the years. At densities exceeding the equilibrium density of nuclear matter, $\rho_0 = 2.67 \times 10^{14}$ g/cm^3, typical of NS inner core, modern EOS are derived within two main, different approaches: nonrelativistic nuclear many-body theory (NMBT) and relativistic mean field theory (RMFT); we will now show that, as discussed in Ref. 16, different ways of modeling hadronic interactions affect the pulsation properties of the star. To describe the inner and outer crust of the NS, we have used the Baym-Pethick-Sutherland[17] and the Pethick-Ravenhall-Lorenz EOS,[18] respectively. The EOS we choose to model the NS core are the following.

For the NMBT approach we select two groups of EOS: **Group I**, named (APRA, APRB, APRB200, APRB120), and **Group II**, named (BBS1, BBS2), respectively.

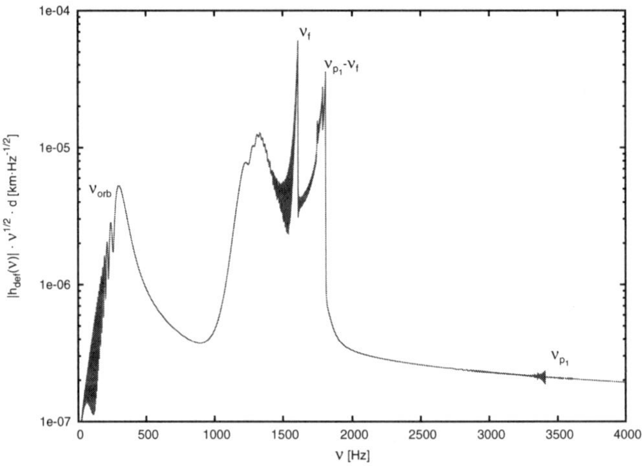

Fig. 2. The gravitational signal emitted by a neutron star which is tidally deformed by a $10\,M_\odot$ Schwarzchild black hole, on which it is inspiralling. Here we plot the strain amplitude $|h_+(\nu) + ih_\times(\nu)|\sqrt{\nu}$ multiplied by the distance d of the observer from the system. The sharp peaks indicates the excitation of the modes of oscillations of the neutron star, ν_f, ν_{p_1} and $\nu_{p_1} - \nu_f$.

In both cases matter is composed of neutrons, protons, electrons and muons in weak equilibrium, and the dynamics is described by a non-relativistic Hamiltonian which includes phenomenological potentials that describe two- and three-nucleon interactions. The potentials are obtained from fits of existing scattering data. For all EOS the two-body potential is v_{18}, whereas the three-body potential is Urbana IX for **Group I**, and Urbana VII for **Group II**. A first major difference between the two groups is that in **Group I** the ground state energy is calculated using variational techniques,[19,20] whereas in **Group II** it is calculated using G-matrix perturbation theory.[21] There are also differences among the EOS in each group:

Group I

APRB is an improved version of the APRA model. In APRA nucleon-nucleon potentials describe interactions between nucleons in their center of mass frame, in which the total momentum P vanishes. In the APRB the two-nucleon potential is modified including relativistic corrections which arise from the boost to a frame in which $P \neq 0$, up to order P^2/m^2. These corrections are necessary to use the nucleon-nucleon potential in a locally inertial frame associated to the star. As a consequence of this change, the three-body potential also needs to be modified in a consistent fashion.

The EOS APRB200 and APRB120 are the same as APRB up to $\sim 4\rho_0$, but at higher density there is a phase of deconfined quark matter described within the MIT bag model. The mass of the strange quark is assumed to be $m_s = 150$ MeV, the coupling constant describing quarks interaction is set to $\alpha_s = 0.5$, and the value of the bag constant is 200 MeV/fm^3 for APRB200 and 120 MeV/fm^3 for APRB120. The phase transition from nuclear matter to quark matter is described requiring

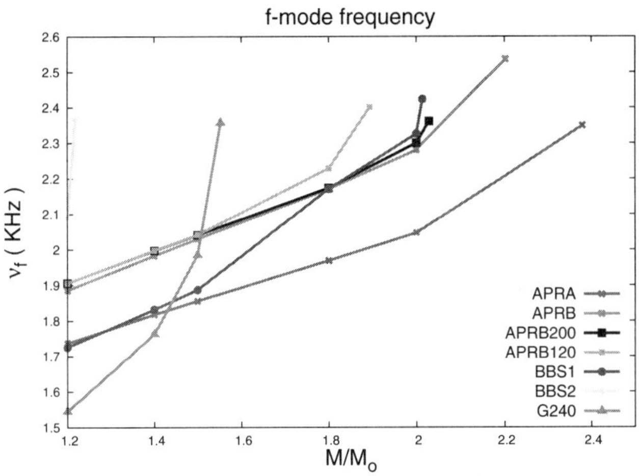

Fig. 3. The frequency of the fundamental mode is plotted as a function of the mass of the star, up to the maximum mass allowed by each EOS.

the fulfillment of Gibbs conditions, leading to the formation of a mixed phase, and neglecting surface and Coulomb effects.[19,22] Thus, these stars are *hybrid* stars.

Group II

The main difference between the equations of state BBS1 and BBS2 is that in BBS2 strange heavy baryons (Σ^- and Λ^0) are allowed to form in the core. Neither BBS1 nor BBS2 include relativistic corrections.

As representative of the RMFT, we choose the EOS named G240. Matter composition includes leptons and the complete octet of baryons (nucleons, $\Sigma^{0,\pm}$, Λ^0 and $\Xi^{\pm}$). Hadron dynamics is described in terms of exchange of one scalar and two vector mesons.[23]

In Fig. 3 we plot, for each EOS, the frequency of the fundamental mode as a function of the mass of the star, which is the only observable on which we might have reliable information. A comparison of the values of ν_f for APRA and APRB clearly shows that the relativistic corrections and the associated redefinition of the three-body potential, which improve the Hamiltonian of APRB with respect to APRA, lead to a systematic difference of about 150 Hz in the mode frequency. Conversely, the presence of quark matter in the inner core of the hybrid stars (EOS APRB200 and APRB120) produce changes in ν_f that are negligible with respect to those of the APRB stars. We also see that the frequencies corresponding to the BBS1 and APRA models, which are very close at $M \lesssim 1.4\ M_\odot$, diverge for larger masses. This behavior can be traced back to the different treatments of three-nucleon interactions, whose role in shaping the EOS becomes more and more important as the and central density and the mass of the star increase: while the variational approach of Ref. 19 used to derive the EOS APRA naturally allows for inclusion of the three-nucleon potential appearing in the Hamiltonian, in G-matrix perturbation theory used to derive the EOS BBS1 the three-body potential has to be replaced with an effective two-nucleon potential, obtained by averaging over the position of the third particle.[24] The f-mode frequency of the BBS2 model is significantly higher than that corresponding to the other EOS for the same mass. This is because the transition to hyperonic matter, predicted by the BBS2 model, produces a considerable softening of the EOS; thus, to a given mass corresponds a larger average density, and since ν_f scales linearly with the square root of the average density, ν_f increases. It should also be mentioned that the softening of the EOS leads to stable NS configurations of very low mass ($= 1.218\ M_\odot$).

It is also interesting to compare the f-mode frequencies corresponding to models BBS2, derived within the RMFT approach, and G240, derived using the NMBT approach, since they both predict the occurrence of heavy strange baryons. The behavior of ν_f shown in Fig. 3 directly reflects the relations between mass and central density: for a given mass, a larger central density corresponds to a smaller radius, and therefore to a larger average density. Consequently, higher frequencies correspond to larger central densities ρ_c. For example, the NS configurations of mass $1.2\ M_\odot$ correspond to $\rho_c \sim 7 \cdot 10^{14}\ \ \text{g/cm}^3$ for G240, and to a larger central density, $\rho_c \sim 2 \cdot 10^{15}\ \text{g/cm}^3$ for BBS2. On the other hand, the G240 model requires a central density of $\sim 2.5{\cdot}10^{15}\ \ \text{g/cm}^3$ to reach a mass of $\sim 1.55\ M_\odot$ and a value of ν_f equal to that of the BBS2 model.

In conclusion, from Fig. 3 and from the above discussion we clearly see that different ways of modeling hadronic interactions reflect into different pulsation properties of neutron stars.

Another interesting possibility which we would like to discuss is the following. In 1980 Witten suggested that strange stars - namely stars entirely made of a degenerate gas of up, down and strange quarks, except for a thin outer crust - may exist in nature.[25] However, a generally accepted observational evidence of the existence of such stars is still lacking and it is interesting to ask whether the detection of gravitational waves could provide any information in this respect. To this purpose, in Ref. 26 we have computed the f-mode frequency of strange stars modeled using the MIT bag model,[27] and we have compared the results with those shown in Fig. 3 for neutron stars. According to the bag model, quarks occur in color neutral clusters confined to a finite region of space – the bag – the volume of which is limited by the pressure of the QCD vacuum (the bag constant B); in addition, the residual interactions between quarks are assumed to be weak, and therefore are treated in low order perturbation theory in the color coupling constant α_s. Thus, the parameters of the model are the masses of up, down and strange quarks, α_s and the bag constant B. The mass of the up and down quarks, given in the Particle Data Book, are of the order of few MeV, negligible with respect to that of the strange quark, the value of which is in the ranges $(80 - 155)$ MeV. The value of the coupling constant α_s is constrained by the results of hadron collision experiments to range within $(0.4 - 0.6)$. In early applications of the MIT bag model B, α_s and m_s were adjusted to fit the measured properties of light hadrons (spectra, magnetic moments and charge radii). According to these studies B was shown to range from

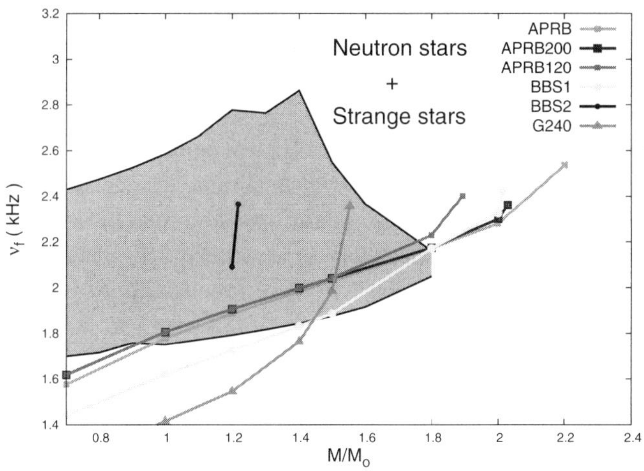

Fig. 4. The f-mode frequency is plotted as a function of the mass of the star, for neutron/hybrid stars (continuous lines), and for strange stars (shaded region), for the parameter range given in (9).

57.5 MeV/fm^3 [28] to 351.7 MeV/fm^3;[29] however, the requirement that strange quark matter be absolutely stable at zero temperature and pressure implies that B cannot exceed the maximum value $B_{max} \approx 95\,\text{MeV/fm}^3$.[30] For values of B exceeding B_{max}, a star entirely made of deconfined quarks is not stable, and quark matter can only occupy a fraction of the available volume, as in the models APRB200 and APRB120 considered above. Thus, if we want to study bare strange stars we need to restrict the values of B in the range $\in (57-95)\,\text{MeV/fm}^3$. In summary, the range of parameters we have considered to compute ν_f is

$$m_s \in (80-155)\ \text{MeV}, \qquad \alpha_s \in (0.4-0.6), \qquad B \in (57-95)\,\text{MeV/fm}^3. \tag{9}$$

The results are summarized in Fig. 4, where we plot ν_f as a function of the mass of the star, both for strange stars (the shaded region), and for the neutron/hybrid stars described above.

From Fig. 4 we can extract some overall features: according to the MIT bag model, strange stars cannot emit gravitational waves with $\nu_f \lesssim 1.7$ kHz, for any value of the mass in the range of stability. Note that 1.8 $M_\odot$ is the maximum mass above which no stable strange star can exist. There is a small range of frequency where neutron/hybrid stars are indistinguishable from strange stars; however, there is a large frequency region where only strange stars can emit. For instance if $M = 1.4\ M_\odot$, a signal with $\nu_f \gtrsim 2$ kHz would belong to a strange star. Even if we do not know the mass of the star (as it is often the case for isolated pulsars) the knowledge of ν_f allows to gain information about the source nature; indeed, if $\nu_f \gtrsim 2.2$ kHz, apart from a very narrow region of masses where stars with hyperons would emit (EOS BBS2 and G240), we can reasonably exclude that the signal is emitted by a neutron star. In addition, it is possible to show that if a signal emitted by an oscillating strange star would be detected, since ν_f is an increasing function of the bag constant it would be possible to set constraints on B much more stringent than those provided by the available experimental data.[26]

Since gravitational waves are emitted by pulsating neutron stars (or by strange stars) at frequencies tipically higher that 1 kHz, it is unlikely that LIGO and VIRGO in their present configuration will detect such signals, unless the oscillations are induced by an extremely energetic process and the source is in the Galaxy. However, upgrading of these detectors is already started and a 3rd generation gravitational wave detector is under design, which will have much higher sensitivity at the frequencies on interest. The detection of gravitational waves from pulsating stars will allow us to glance at their inner core and will shed some light on the secrets it conceals.

References

1. www.ligo.caltech.edu; www.virgo.infn.it.
2. B. Abbott *et al.*, *Phys. Rev. D* **76**, 042001 (2007).
3. G. Ushomirski, C. Cutler and L. Bildsten, *Mon. Not. R. Astron. Soc.* **319**, 902 (2000).
4. B. J. Owen, *Phys. Rev. Lett.* **95**, 211101 (2005).
5. B. Haskell, D. L. Jones and N. Andersson, *Mon. Not. R. Astron. Soc.* **373**, 1423 (2006).
6. M. Bejger, T. Bulik and P. Haensel, *Mon. Not. R. Astron. Soc.* **364**, 635 (2005).
7. O. Benhar, V. Ferrari, L. Gualtieri and S. Marassi, *Phys. Rev. D* **72**, 044028 (2005).
8. T. G. Cowling, *Mon. Not. R. Ast. Soc. Lond.* **101**, 367 (1942).
9. K. D. Kokkotas, *Mon. Not. R. Ast. Soc. Lond.* **268**, 1015 (1994).
10. S. Chandrasekhar and V. Ferrari, *Proc. R. Soc. Lond.* **434**, 449 (1991).
11. O. Benhar, E. Berti and V. Ferrari, *Mon. Not. Roy. Astron. Soc.* **310**, 797 (1999).
12. K. D. Kokkotas and B. F. Schutz, *Mon. Not. R. Ast. Soc. Lond.* **255**, 119 (1992).
13. V. Ferrari, L. Gualtieri and F. Pannarale, *Tidal imprint on the gravitational signal emitted by BH-NS coalescing binaries*, in preparation.
14. L. Lindblom and S. L. Detweiler, *Astrophys. J. Suppl.* **53**, 73 (1983).
15. N. Andersson and K. D. Kokkotas, *Mon. Not. R. Ast. Soc. Lond.* **299**, 1059 (1998).
16. O. Benhar, V. Ferrari and L. Gualtieri, *Phys. Rev. D* **70**, 124015 (2004).
17. G. Baym, C. J. Pethick and P. Sutherland, *Astrophys. J.* **170**, 299 (1971).
18. C. J. Pethick, B. G. Ravenhall and C. P. Lorenz, *Nucl. Phys. A* **584**, 675 (1995).
19. A. Akmal and V. R. Pandharipande, *Phys. Rev. C* **56**, 2261 (1997).
20. A. Akmal, V. R. Pandharipande and D. G. Ravenhall, *Phys. Rev. C* **58**, 1804 (1998).
21. M. Baldo, G. F. Burgio and H.-J. Schulze, *Phys. Rev. C* **61**, 055801 (2000).
22. O. Benhar and R. Rubino, *Astron. & Astroph.* **434**, 247 (2005).
23. N. K. Glendenning, *Compact Stars* (Springer, New York, 2000).
24. A. Lejeune, P. Grangé, P. Martzolff and J. Cugnon, *Nucl. Phys. A* **453**, 189 (1986).
25. E. Witten, *Phys. Rev. D* **30**, 272 (1984).
26. O. Benhar, V. Ferrari, L. Gualtieri and S. Marassi, *Gen. Rel. Grav.* **39**, 1323 (2007).
27. A. Chodos, R. L. Jaffe, K. Johnson, C. B. Thorne and W. F. Weiskopf, *Phys. Rev. D* **9**, 3471 (1974).
28. T. De Grand, R. L. Jaffe, K. Johnsson and J. Kiskis, *Phys. Rev. D* **12**, 2060 (1975).
29. C. E. Carlson, T. H. Hansson and C. Peterson, *Phys. Rev. D* **27**, 1556 (1983)
30. E. Farhi and R. L. Jaffe, *Phys. Rev. D* **30**, 2379 (1984).

RECONCILING 2 $M_\odot$ PULSARS AND SN1987A: TWO BRANCHES OF NEUTRON STARS

P. HAENSEL,[1,*] M. BEJGER[1,2] and J. L. ZDUNIK[1]

[1] *Nicholaus Copernicus Astronomical Center, Bartycka 18, 00-716 Warszawa, Poland*
[2] *LUTh, UMR 8102 du CNRS, Pl. Jules Janssen, 92195 Meudon, France*
[*] *E-mail: haensel@camk.edu.pl*

The analysis of SN1987A led Brown and Bethe[1,2] to conclude that the maximum mass of cold neutron stars is $M_{\rm max} \approx 1.5 M_\odot$. On the other hand, recent evaluation of the mass of PSR J0751+1807 implies $M_{\rm max} > 1.5 M_\odot$. This contradicts the original Bethe-Brown model but can be reconciled within evolutionary scenarios proposed in this talk.[3]

References

1. G. E. Brown and H. A. Bethe, *Astrophys. J.* **423**, 659 (1994).
2. H. A. Bethe and G. E. Brown, *Astrophys. J.* **445**, L129 (1995).
3. P. Haensel, M. Bejger and J. L. Zdunik, *Astron. Astrophys.*, to be submitted.

EOS OF DENSE MATTER AND FAST ROTATION OF NEUTRON STARS

J. L. ZDUNIK,* P. HAENSEL and M. BEJGER

Nicolaus Copernicus Astronomical Center, Polish Academy of Sciences,
Bartycka 18, PL-00-716 Warszawa, Poland
**E-mail: jlz@camk.edu.pl*

E. GOURGOULHON

LUTh, UMR 8102 du CNRS, Pl. Jules Janssen,
92195 Meudon, France

Recent observations of XTE J1739-285 suggest that it contains a neutron star rotating at 1122 Hz.[1] Such rotational frequency would be the first for which the effects of rotation are significant. We study the consequences of very fast rotating neutron stars for the potentially observable quantities as stellar mass and pulsar period.

1. Introduction

Neutron stars with their very strong gravity can be very fast rotators. Theoretical studies show that they could rotate at sub-millisecond periods, i.e., at frequency $f = 1/\text{period} > 1000$ Hz.[2,3] The first millisecond pulsar B1937+21, rotating at $f = 641$ Hz,[4] remained the most rapid one during 24 years after its detection. In January 2006, discovery of a more rapid pulsar J1748-2446ad rotating at $f = 716$ Hz was announced.[5] However, such sub-kHz frequencies are still too low to significantly affect the structure of neutron stars with $M > 1M_{\odot}$.[6] Actually, they belong to a *slow rotation* regime, because their f is significantly smaller than the mass shedding (Keplerian) frequency $f_{\rm K}$. Effects of rotation on neutron star structure are then $\propto (f/f_{\rm K})^2 \ll 1$. Rapid rotation regime for $M > 1M_{\odot}$ requires submillisecond pulsars with supra-kHz frequencies $f > 1000$ Hz.

Very recently Kaaret et al.[1] reported a discovery of oscillation frequency $f = 1122$ Hz in an X-ray burst from the X-ray transient, XTE J1739-285. According to Kaaret et al.[1] "this oscillation frequency suggests that XTE J1739-285 contains the fastest rotating neutron star yet found". If confirmed, this would be the first detection of a sub-millisecond pulsar (discovery of a 0.5 ms pulsar in SN1987A remnant announced in January 1989 was withdrawn one year later).

Rotation at $f > 1000$ Hz is sensitive to the stellar mass and to the equation of state (EOS). Hydrostatic, stationary configurations of neutron stars rotating at given rotation frequency f form a one-parameter family, labeled by the central

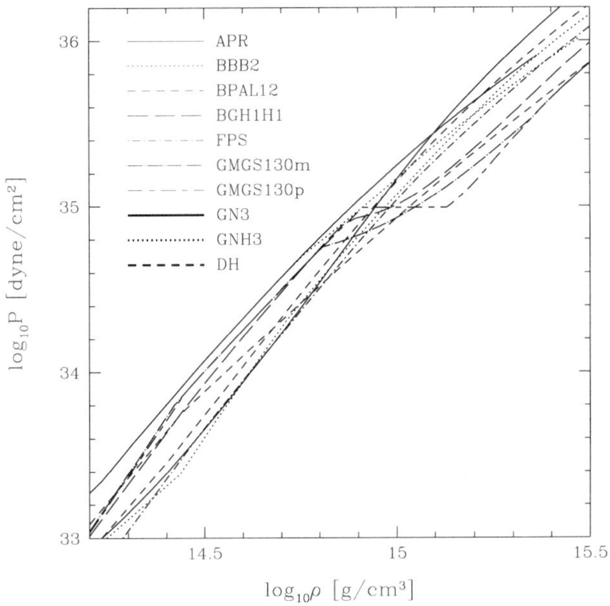

Fig. 1. The equations of state used in the paper.

density. This family - a curve in the mass - equatorial radius plane - is limited by two instabilities. On the high central density side, it is instability with respect to axi-symmetric perturbations, making the star collapse into a Kerr black hole. The low central density boundary results from the mass shedding from the equator. In the present paper we show how rotation at $f > 1000$ Hz is sensitive to the EOS, and what constraints on the EOS of neutron stars result from future observations of stably rotating sub-millisecond pulsars.

2. Method

We studied the properties of fast rotating neutron stars for a broad set of the models of dense matter. The set of equations of state (EOSs) considered in the paper is presented in Fig. 1.

Out of ten EOSs of neutron star matter, two were chosen to represent a soft (BPAL12) and stiff (GN3) extreme case. These two extreme EOSs should not be considered as "realistic", but they are used just to "bound" the neutron star models from the soft and the stiff side.

We consider four EOSs based on realistic models involving only nucleons (FPS, BBB2, DH, APR), and four EOSs softened at high density either by the appearance of hyperons (GNH3, BGN1H1), or a phase transition (GMGSm, GMGSp). The softening of the matter in latter case is clearly visible in Fig. 1 at pressure $P \sim 10^{35}$ dyn/cm^2.

EOSs GMGSp and GMGSm describe nucleon matter with a first order phase transition due to kaon condensation. In both cases the hadronic Lagrangian is the same. However, to get GMGSp we assumed that the phase transition takes place between two pure phases and is accompanied by a density jump. Assuming on the contrary that the transition occurs via a mixed state of two phases, we get EOS GMGSm. This last situation prevails when the surface tension between the two phases is below a certain critical value.

The stationary configurations of rigidly rotating neutron stars have been computed in the framework of general relativity by solving the Einstein equations for stationary axi-symmetric spacetime.[7,8] The numerical computations have been performed using the `Lorene/Codes/Rot_star/rotstar` code from the LORENE library (`http://www.lorene.obspm.fr`). One-parameter families of stationary 2-D configurations were calculated for ten EOSs of neutron-star matter, presented in Fig. 1.

Stability with respect to the mass-shedding from the equator implies that at a given gravitational mass M the circumferential equatorial radius $R_{\rm eq}$ should be smaller than $R_{\rm max}$ which corresponds to the mass shedding (Keplerian) limit. The value of $R_{\rm max}$ results from the condition that the frequency of a test particle at circular equatorial orbit of radius $R_{\rm max}$ just above the equator of the *actual rotating star* is equal to the rotational frequency of the star. This condition sets the bound on our rotating configurations from the right side on $M(R)$ plane (the highest radius and the lowest central density).

The limit for most compact stars (the lowest radius and the highest central density) is set by the onset of instability with respect to the axisymmetric oscillations defined by the condition:

$$\left(\frac{\partial M}{\partial \rho_{\rm c}}\right)_J = 0\ . \tag{1}$$

For stable configurations we have:

$$\left(\frac{\partial M}{\partial \rho_{\rm c}}\right)_J > 0\ . \tag{2}$$

3. Neutron Stars at 1122 Hz

In this section we present the parameters of the stellar configurations rotating at frequency 1122 Hz. For details and discussion see Bejger et al.[9]

It is interesting that the relation between the calculated values of M and $R_{\rm eq}$ at the "mass shedding point" is extremely well approximated by the formula for the orbital frequency for a test particle orbiting at $r = R_{\rm eq}$ in the Schwarzschild spacetime of a *spherical mass* M (which can be replaced by a point mass M at $r = 0$). We denote the orbital frequency of such a test particle by $f_{\rm orb}^{\rm Schw.}(M, R_{\rm eq})$. The formula giving the locus of points satisfying $f_{\rm orb}^{\rm Schw.}(M, R_{\rm eq}) = 1122$ Hz, represented by a dash line in Fig. 2, is

$$\frac{1}{2\pi}\left(\frac{GM}{{R_{\rm eq}}^3}\right)^{1/2} = 1122\ \text{Hz}\ . \tag{3}$$

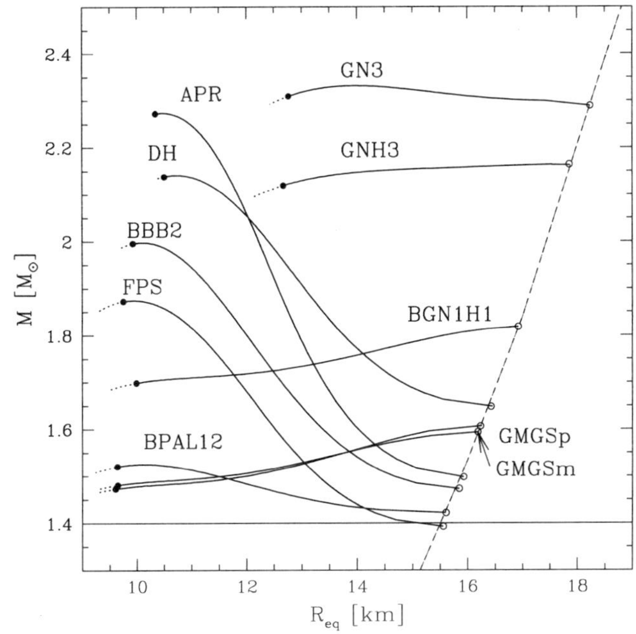

Fig. 2. Gravitational mass, M, vs. circumferential equatorial radius, $R_{\rm eq}$, for neutron stars stably rotating at $f = 1122$ Hz, for ten EOSs (Fig. 1). Small-radius termination by filled circle: setting-in of instability with respect to the axi-symmetric perturbations. Dotted segments to the left of the filled circles: configurations unstable with respect to those perturbations. Large-radius termination by an open circle: the mass-shedding instability. The mass-shedding points are very well fitted by the dashed curve $R_{\rm min} = 15.52\ (M/1.4M_{\odot})^{1/3}$ km. For further explanation see the text.

This formula for the Schwarzschild metric coincides with that obtained in Newtonian gravity for a point mass M. It passes through (or extremely close to) the open circles denoting the actual mass shedding (Keplerian) configurations. This is quite remarkable in view of rapid rotation and strong flattening of neutron star at the mass-shedding point. Equation (3) implies

$$R_{\rm max} = 15.52 \left(\frac{M}{1.4\ M_{\odot}}\right)^{1/3} \ \text{km} . \tag{4}$$

4. Submillisecond Pulsars

In this section we present results for neutron stars rotating at submillisecond periods for a broad range of frequencies (1000 - 1600 Hz)

In Fig. 3 we presented mass vs. radius relation for very fast rotating neutron stars. As we see the shape of the $M(R_{\rm eq})$ curves is more and more flat as rotational frequency increases. For $f_{\rm rot} = 1600$ Hz the curves $M(R_{\rm eq})$ are almost horizontal and the mass for each EOS is quite well defined and curves for different models of dense matter are typically well separated. In principle this property could be used

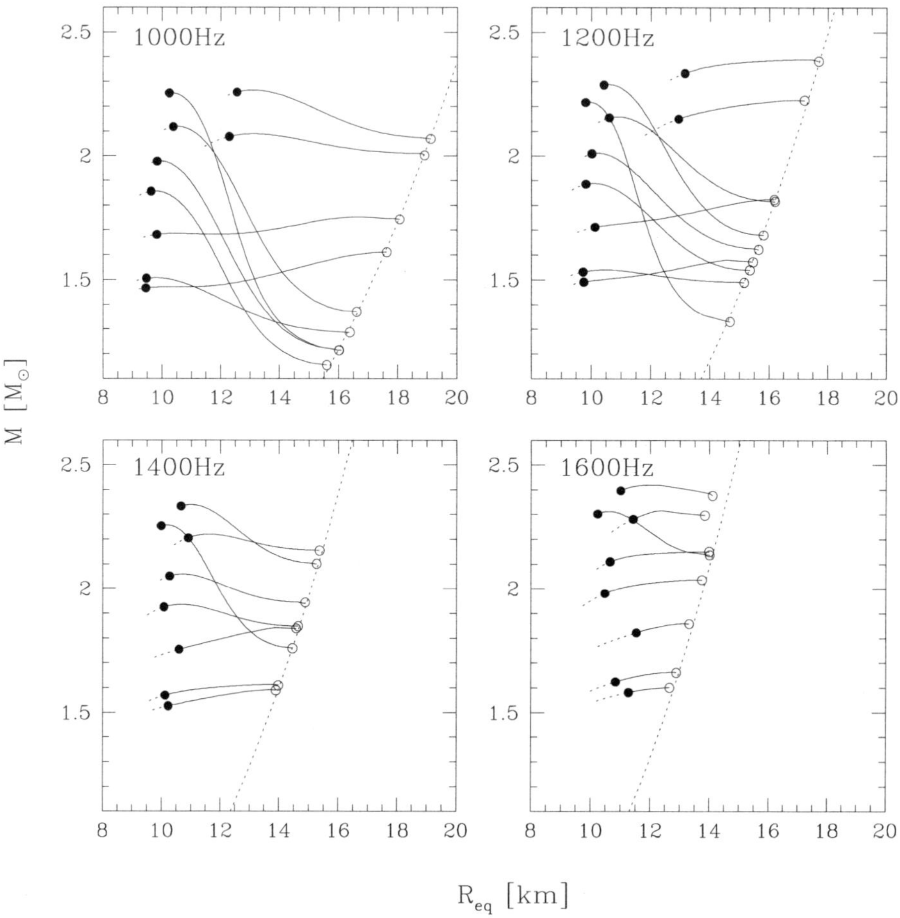

Fig. 3. Mass vs. radius relation for submillisecond neutron stars.

for selecting "true EOS" if we detect a very rapid pulsar and are able to estimate its mass.

In each panel the dotted curve corresponds to the formula (of which Eq. (3) is a special case for $f = 1122$ Hz)

$$M = \frac{4\pi^2 f^2}{G} R_{\rm eq}^3 \tag{5}$$

used for a given frequency (1000 Hz, 1200 Hz, 1400 Hz, 1600 Hz). As we see this formula works perfectly in very broad range of rotational frequencies (recently this formula has been tested by Krastev et al.[10] for the frequency 716 Hz.)

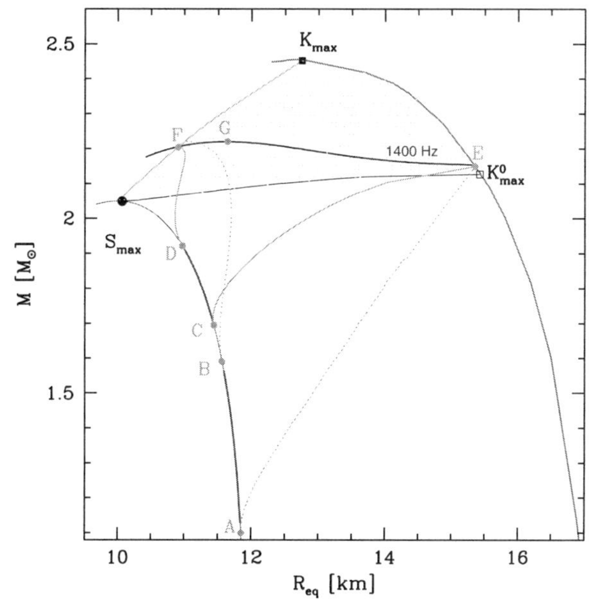

Fig. 4. Mass vs. radius relation for accreting NS with DH EOS. The example is for $f_{\rm rot} = 1400$ Hz.

5. Accretion

We studied the mechanism of the spin-up of neutron star due to the accretion from the last stable orbit (the innermost stable circular orbit - ISCO). As an example e discuss this subject for DH EOS.

We calculate the spin-up following the prescription given by Zdunik et al.[11,12] The value of specific angular momentum per unit baryon mass of a particle orbiting the neutron star at the ISCO, $l_{\rm IS}$, is calculated by solving exact equations of the orbital motion of a particle in the space-time produced by a rotating neutron star, given in Appendix A of Zdunik et al.[11]

Accretion of an infinitesimal amount of baryon mass $\mathrm{d}M_{\rm B}$ onto a rotating neutron star is assumed to lead to a new quasi stationary rigidly rotating configuration of mass $M_{\rm B} + \mathrm{d}M_{\rm B}$ and angular momentum $J + \mathrm{d}J$, with

$$\mathrm{d}J = x_l l_{\rm IS}\, \mathrm{d}M_{\rm B}\ , \tag{6}$$

where x_l denotes the fraction of the angular momentum of the matter element transferred to the star. The remaining fraction $1 - x_l$ is assumed to be lost via radiation or other dissipative processes.

We present results for two choices of x_l: $x_l = 1$ and $x_l = 0.5$ when all or half of the angular momentum of the accreting matter is transferred to the star from the ISCO. In Fig. 4 we plot the curve $M(R)$ corresponding to the frequency of rotation $f_{\rm rot} = 1400$ Hz for DH EOS. Point F on this curve corresponds to the onset of instability with respect to axi-symmetric oscillations (condition given by Eq. (1)).

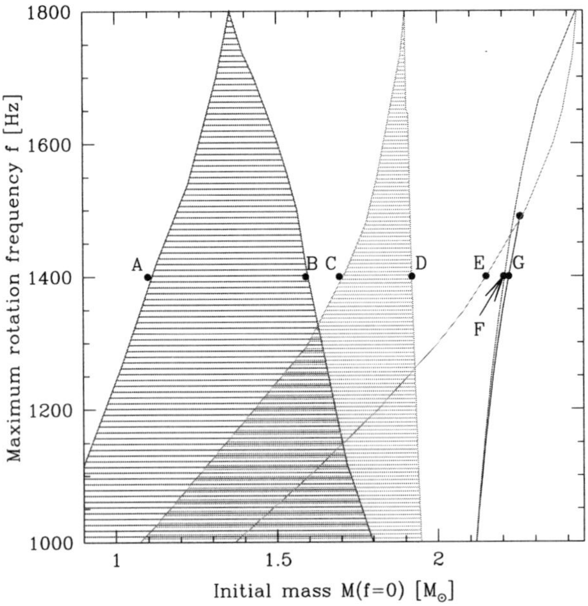

Fig. 5. The limits for the initial mass of accreting neutron star to be spun-up up to the given frequency. The labeled points correspond to the points in Fig. 3. The region to the left (red) corresponds to $x_l = 0.5$ and the central region (cyan) to $x_l = 1$. The non-shaded region (right) correspond to the allowable final masses.

Point E is the Keplerian configuration at frequency 1400 Hz, and G - corresponds to maximum mass along the curve with fixed frequency.

The curves starting at points A,B,C and D are the track of the accreting neutron star defined by the Eq. (6) for $x_l = 1$ (solid line) and $x_l = 0.5$ (dotted line) for cases C,D and A,B respectively. To reach the configuration rotating at frequency 1400 Hz we have to start with the nonrotating neutron star between the points C and D (if $x_l = 1$) of A and B (if $x_l = 0.5$) As we see the actual frequency of rotation sets the limits on the initial mass of nonrotating star, which can be spun-up to this frequency due to the accretion. For $f_{\rm rot} = 1400$ Hz these limits in the case $x_l = 1$ are: $1.7 M_\odot < M_i < 1.92 M_\odot$.

In Fig. 5 we plotted for DH EOS these limits for the initial mass (non-rotating) of accreting NS provided that this star could be spun-up to the given frequency. We presented results for two assumed values of the parameter x_l. The shaded area shows the allowed initial masses of nonrotating star to reach by the accretion a given rotational frequency. Left "triangle" corresponds to $x_l = 0.5$, the right shaded triangle (with points C and D) to $x_l = 1$

The curves on the right represent the limits on the actual mass of rotating NS. The three curves correspond to the location of the three points E,F,G in Fig. 3. The curve with the point E (green) is defined by the Keplerian frequency of rotating star, the point F (and magenta line) correspond to the boundary resulting from the instability with respect to axi-symmetric perturbations. The point G is the mass at the maximum point of the curve with fixed frequency (1400 Hz). As the frequency increases the mass at Keplerian point increases more rapidly than that defined by the onset of instability at the maximum mass (points F and G respectively). For high frequencies the maximum mass of the stars rotating at fixed frequency is given by the value for Keplerian configuration. For DH EOS at frequency $\simeq$ 1500 Hz the point G disappears and for faster rotation the curve $M(R_{\rm eq})$ monotonically increases. The region of the masses of the stars rotating at high frequency is very narrow. For $f_{\rm rot} > 1400$ Hz it is smaller than $0.1M_\odot$ (see also discussion of Fig. 3)

6. Discussion and Conclusions

The $M(R_{\rm eq})$ curve for $f \gtrsim 1400$ Hz is flat. Therefore, for given EOS the mass of NS is quite well defined. Conversely, measured mass of a NS rotating at $f \gtrsim 1400$ Hz will allow us to unveil the actual EOS. The "Newtonian" formula for the Keplerian frequency works surprisingly well for precise 2-D simulations and sets a firm upper limit on $R_{\rm eq}$ for a given f. Finally, observation of $f \gtrsim 1200$ Hz sets stringent limits on the initial mass of the nonrotating star which was spun up to this frequency by accretion.

Acknowledgments

This work was partially supported by the Polish MNiSW grant no. N203.006.32/0450 and by the LEA Astrophysics Poland-France (Astro-PF) program. MB was also partially supported by the Marie Curie Intra-european Fellowship MEIF-CT-2005-023644.

References

1. P. Kaaret, Z. Prieskorn, J. J. M. in't Zand, S. Brandt, N. Lund, S. Mereghetti, D. Goetz, E. Kuulkers and J. A. Tomsick, *Ap. J. Lett.* **657**, 97 (2007).
2. G. B. Cook, S. L. Shapiro and S. A. Teukolsky, *Ap. J.* **424**, 823 (1994).
3. M. Salgado, S. Bonazzola, E. Gourgoulhon and P. Haensel, *A&A* **108**, 455 (1994).
4. D. C. Backer *et al.*, *Nature* **300**, 615 (1982).
5. J. W. T. Hessels, S. M. Ransom, I. H. Stairs, P. C. C. Freire, V. M. Kaspi and F. Camilo, *Science* **311**, 1901 (2006).
6. S. L. Shapiro, S. A. Teukolsky and I. Wasserman, *Ap. J.* **272**, 702 (1983).
7. S. Bonazzola, E. Gourgoulhon, M. Salgado and J.-A. Marck, *A&A* **278**, 421 (1993).
8. E. Gourgoulhon, P. Haensel, R. Livine, E. Paluch, S. Bonazola and J.-A. Marck, *A&A* **349**, 851 (1999).
9. M. Bejger, P. Haensel and J. L. Zdunik, *A&A* **464**, 49 (2007).
10. Plamen G. Krastev, Bao-An Li and Aaron Worley, *astro-ph* 0707.3621.
11. J. L. Zdunik, P. Haensel and E. Gourgoulhon, *A&A* **381**, 933 (2002).
12. J. L. Zdunik, P. Haensel and M. Bejger, *A&A* **441**, 207 (2005).

PART E

Quark and Strange Matter in Neutron Stars

BULK VISCOSITY OF COLOR-SUPERCONDUCTING QUARK MATTER

MARK ALFORD

Washington University, St. Louis MO 63130, USA
E-mail: alford@wuphys.wustl.edu

At sufficiently high density, quark matter is expected to be in a color superconducting state. One way to detect the presence of such matter in a compact star is via the bulk and shear viscosities, which damp r-modes, preventing the star from spinning down quickly via an r-mode instability. I discuss recent calculations of the bulk viscosity of color-superconducting quark matter, and their application to observations of compact stars.

References

1. J. L. Friedman and K. H. Lockitch, gr-qc/0102114.
2. N. Andersson, *Class. Quant. Grav.* **20**, R105 (2003) [arXiv:astroph/0211057].
3. K. D. Kokkotas and N. Andersson, Proceedings of SIGRAV XIV, Genova 2000, Springer-Verlag (2001), arXiv:gr-qc/0109054.
4. J. Madsen, *Phys. Rev. Lett.* **85**, 10 (2000) [arXiv:astro-ph/9912418].

CHIRAL SYMMETRY RESTORATION AND QUARK DECONFINEMENT AT LARGE DENSITIES AND TEMPERATURE

A. DRAGO* and L. BONANNO

Dipartimento di Fisica, Università di Ferrara and INFN, Sez. Ferrara, Ferrara, 44100, Italy
**E-mail: drago@fe.infn.it*

A. LAVAGNO

Dipartimento di Fisica, Politecnico di Torino and INFN, Sez. Torino, Torino, 10125, Italy
E-mail: andrea.lavagno@polito.it

We study the nuclear equation state at large densities and intermediate temperatures by means of a purely hadronic chirally symmetric model and, from another point of view, by considering a mixed phase of hadrons and clusters of quarks. In both models a significant softening of the equation of state takes place due to the appearance of new degrees of freedom. However, in the first case the bulk modulus is mainly dependent on the density, while in the mixed phase model it also strongly depends on the temperature. We also show that the bulk modulus is not vanishing in the mixed phase due to the presence of two conserved charges, the baryon and the isospin one. Only in a small region of densities and temperatures the incompressibility becomes extremely small. Finally we compare our results with recent analysis of heavy ion collisions at intermediate energies.

1. Introduction

In last years a great deal of attention has been devoted to the study of the property of the Equation of State (EOS) of strongly interacting nuclear matter under extreme conditions such as large densities and/or high temperature. One of the main goals of high-energy heavy-ion research is to investigate the structure of nuclear phase and explore the existence of a deconfined phase of quarks and gluons. Such a new degree of freedom should generate a significant softening of the EOS.

The extraction of experimental information about the EOS of matter at large densities and temperatures from the data of intermediate and high energy Heavy Ion Collisions (HICs) is very complicated. Recently, possible indirect indications of a softening of the EOS at the energies reached at AGS have been discussed several times in the literature.[1–6] In particular, a recent analysis[7] based on a 3-fluid dynamics simulation suggests a progressive softening of the EOS tested through HICs at energies ranging from $2A$ GeV up to $8A$ GeV.

In a recent work[8] we studied two models for the softening: the first model is based on a purely hadronic chirally symmetric EOS[9–11] and the softening is due to partial restoration of the chiral symmetry. In that model the calculations are performed beyond mean field approximation and quantum fluctuations play a crucial role. In a second model the softening is due to the formation of clusters of quarks, which are the precursors of deconfinement. These clusters are at first metastable and they stabilize only at larger densities.

The aim of our calculation is to compare how the softening takes place in the two models and finally to relate our results to recent analysis of the experimental data. This may be helpful also in prevision of future experiments planned e.g. at Facility for Antiproton and Ion Research (FAIR) at GSI.[12]

2. The Chiral-Dilaton Model

Spontaneously chiral symmetry breaking has long been studied in several microscopic models because of such property is thought to be a fundamental feature of low-energy effective Lagrangians. At finite temperature the restoration of chiral symmetry in the linear and non-linear sigma model has been discussed but the attempt of describe nuclear dynamics fails due to the impossibility to reproduce basic properties of nuclei.[13] More sophisticated approaches have been proposed in the literature, both within a SU(2) chiral symmetric models[14,15] and also extending the symmetry to the strange sector.[16–18] Here we use the model introduced by the Minnesota group.[9–11] In that model chiral fields are present together with a dilaton field which reproduces at a mean field level the breaking of scale symmetry which takes place in QCD. In Ref. 9–11 it has been developed a formalism (which we adopt) allowing resummations beyond mean field approximation. This is important when studying a strongly non-perturbative problem as the restoration of chiral symmetry. By introducing the vector-isovector ρ meson, the lagrangian of the model can written as

$$
\begin{aligned}
\mathcal{L} = &\ \tfrac{1}{2}\partial_\mu\sigma\partial^\mu\sigma + \tfrac{1}{2}\partial_\mu\boldsymbol{\pi}\cdot\partial^\mu\boldsymbol{\pi} + \tfrac{1}{2}\partial_\mu\phi\partial^\mu\phi - \tfrac{1}{4}\omega_{\mu\nu}\omega^{\mu\nu} \\
&- \tfrac{1}{4}\boldsymbol{B}_{\mu\nu}\cdot\boldsymbol{B}^{\mu\nu} + \tfrac{1}{2}G_{\omega\phi}\phi^2\omega_\mu\omega^\mu + \tfrac{1}{2}G_{b\phi}\phi^2\boldsymbol{b}_\mu\cdot\boldsymbol{b}^\mu \\
&+ [(G_4)^2\omega_\mu\omega^\mu]^2 - \mathcal{V} \\
&+ \bar{N}\left[\gamma^\mu(i\partial_\mu - g_\omega\omega_\mu - \tfrac{1}{2}g_\rho\boldsymbol{b}_\mu\cdot\boldsymbol{\tau}) - g\sqrt{\sigma^2+\boldsymbol{\pi}^2}\right]N,
\end{aligned}
\tag{1}
$$

where

$$
\begin{aligned}
\mathcal{V} = &\ B\phi^4\left(\ln\frac{\phi}{\phi_0} - \frac{1}{4}\right) - \tfrac{1}{2}B\delta\phi^4\ln\frac{\sigma^2+\boldsymbol{\pi}^2}{\sigma_0^2} \\
&+ \tfrac{1}{2}B\delta\zeta^2\phi^2\left[\sigma^2+\boldsymbol{\pi}^2 - \frac{\phi^2}{2\zeta^2}\right] - \tfrac{3}{4}\epsilon_1' \\
&- \tfrac{1}{4}\epsilon_1'\left(\frac{\phi}{\phi_0}\right)^2\left[\frac{4\sigma}{\sigma_0} - 2\left(\frac{\sigma^2+\boldsymbol{\pi}^2}{\sigma_0^2}\right) - \left(\frac{\phi}{\phi_0}\right)^2\right].
\end{aligned}
\tag{2}
$$

Here σ and $\boldsymbol{\pi}$ are the chiral fields, ϕ the dilaton field, ω_μ the vector meson field and $\mathbf{b}_\mu$ the vector-isovector meson field, introduced in order to study asymmetric

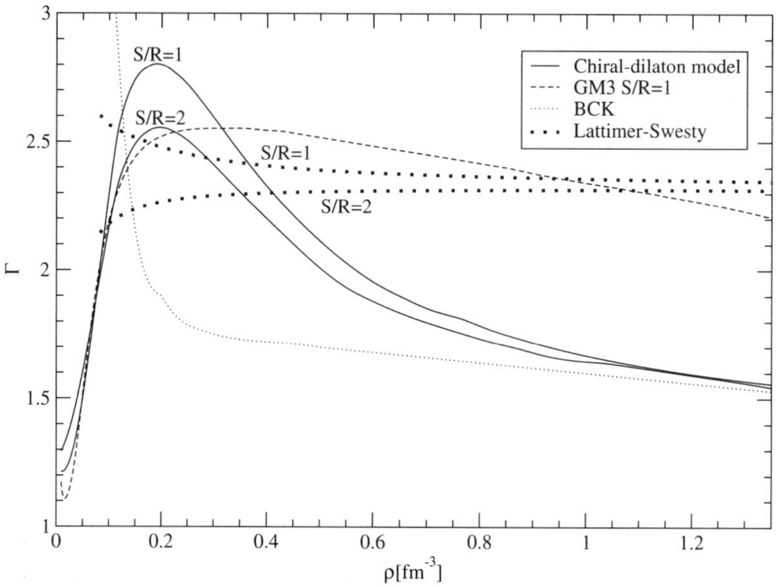

Fig. 1. Adiabatic index as a function of the baryon density for different isoentropic hadronic models.

nuclear matter. The field strength tensors are defined in the usual way $F_{\mu\nu} = \partial_\mu\omega_\nu - \partial_\nu\omega_\mu$, $\boldsymbol{B}_{\mu\nu} = \partial_\mu \boldsymbol{b}_\nu - \partial_\nu \boldsymbol{b}_\mu$. In the vacuum $\phi = \phi_0$, $\sigma = \sigma_0$ and $\boldsymbol{\pi} = 0$. The ω and ρ vacuum masses are generated by their couplings with the dilaton field so that $m_\omega = G^{1/2}_{\omega\phi}\phi_0$ and $m_\rho = G^{1/2}_{\rho\phi}\phi_0$. Moreover $\zeta = \phi_0/\sigma_0$, B and δ are constants and ϵ'_1 is a term that breaks explicitly the chiral invariance of the lagrangian. The potential $\mathcal{V}$ of Eq. (2) is responsible for the scale symmetry breaking. The choice of such a potential comes from the necessity to reproduce the same divergence of the scale current as in QCD. In our calculation we use the parameters set of Ref. 11 which was able to reproduce nuclear spectroscopy and also gives, within this model, the smallest value for the incompressibility at saturation density, $K^{-1} = 322$ MeV. Notice that this value is slightly larger than those traditionally used.

We have tested the model by reproducing all the relevant features of Ref. 11, in particular we obtain that at sufficiently high temperature the mean value of the σ field become small, signalling chiral restoration. As we will see below, the chiral-dilaton model implies important features in softening of the EOS at high densities and temperature, however, we would like to stress that such model can be very important in the framework of physics of compact star objects. In this context, in Fig. 1, we report the behavior of the adiabatic index for different isoentropic hadronic EOSs. We observe that adiabatic index is significantly reduced at large densities due to chiral symmetry restoration, even if such a reduction occurs at densities too large to have implication on supernova explosions.

In Fig. 2 we report the mass-radius relation for different hadronic equation of

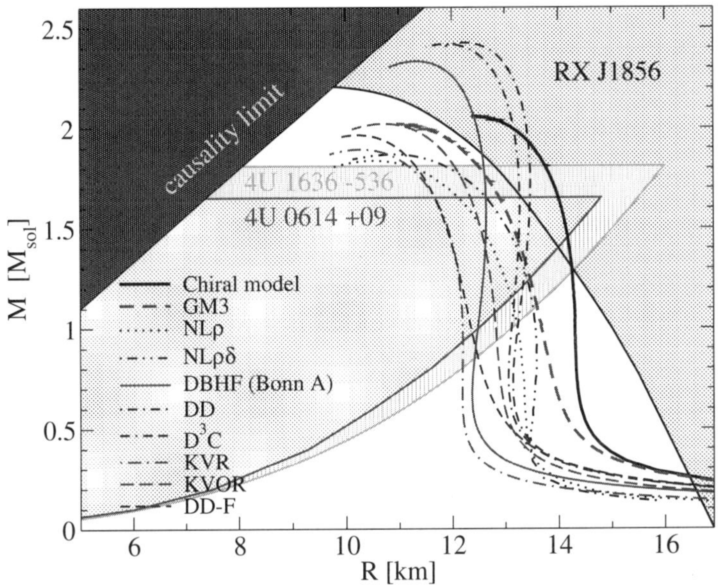

Fig. 2. Mass-radius relation for several hadronic models.

state with the recent experimental constraints. Notice that the chiral model provides good results, beeing in agreement with the experimental constraints and giving a maximum mass value larger than 2 solar masses.

3. The Mixed Hadron-Quark Phase

In this section we study the possibility that a mixed phase of hadrons and quark can form at high density and temperature. To describe the mixed phase we use the Gibbs formalism, which in Refs. 19–21 has been applied to systems where more than one conserved charge is present. In this contribution we are studying the formation of a mixed phase in which both baryon and isospin charge are preserved. The main result of this formalism is that, at variance with the so called Maxwell construction, the pressure is not constant in the mixed phase and therefore the incompressibility does not vanish. From the viewpoint of Ehrenfest's definition, the transition with two conserved charges is not of first, but of second order.[20,21]

Concerning the hadronic phase we use a relativistic field theoretical model, the NL$\rho\delta$,[22] taking into account also a scalar-isovector interaction, which increases the symmetry energy only at large densities. Qualitatively similar results can be obtained using other hadronic models.

For the quark phase we adopt an MIT bag like model. It is well known that, using the simplest version of the MIT bag model, if the bag pressure B is fixed to reproduce the critical temperature computed in lattice QCD, then at moderate temperatures the deconfinement transition takes place at very large densities. On the other hand

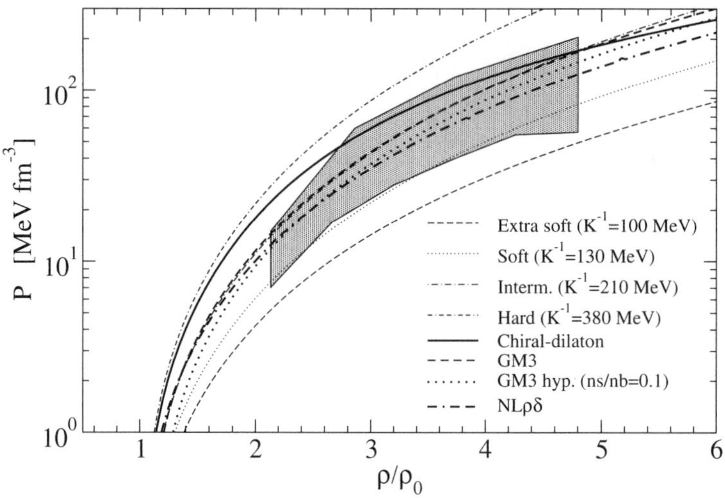

Fig. 3. Pressure at T=0 for symmetric nuclear matter, obtained using the chiral-dilaton model, the NL$\rho\delta$ model, the GM3 model and the set of simple parametric EOSs used in Ref. 7. Also the pressure obtained using the GM3 with a little hyperon fraction is shown.

there are strong theoretical indications that at moderate and large densities (and not too large temperatures) diquark condensates can form, whose effect can be approximately taken into account by reducing the value of the bag constant. A phenomenological approach can therefore be based on a density dependent bag constant, as proposed in Refs. 23,24. We have adopted a parametrization of the form

$$B_{\text{eff}} = B_0 - [\Delta(\mu)]^2\mu^2\,, \tag{3}$$

where $\Delta(\mu) = \bar{\Delta}\exp[-(\mu-\mu_0)^2/a^2]$. Here $B_0^{\frac{1}{4}} = 215$ MeV, $\bar{\Delta} = 100$ MeV, $\mu_0 = 300$ MeV and $a = 300$ MeV. One gluon exchange corrections are taken into account and we use $\alpha_s = 0.35$. One constraint on the parameter values is that at $\mu = 0$ the critical temperature is ~ 170 MeV, as suggested by lattice calculations, while the other constraint is the requirement that the mixed phase starts forming at a density slightly exceeding $3\,\rho_0$ for a temperature of the order of 90 MeV (as also suggested e.g. by Ref. 25).

An important issue concerns the effect of a surface tension at the interface between hadrons and quarks. The value of such a tension is poorly known, and in an MIT-bag-like model it is dominated by the effect of finite masses,[26] in particular of the strange quark. On the other hand in the system studied here strangeness plays a minor role, if any, and therefore the surface tension σ should be rather small. In the analysis we have used $\sigma = 10$ MeV/fm^2, but the results are qualitatively similar if a slightly larger value of σ is used.

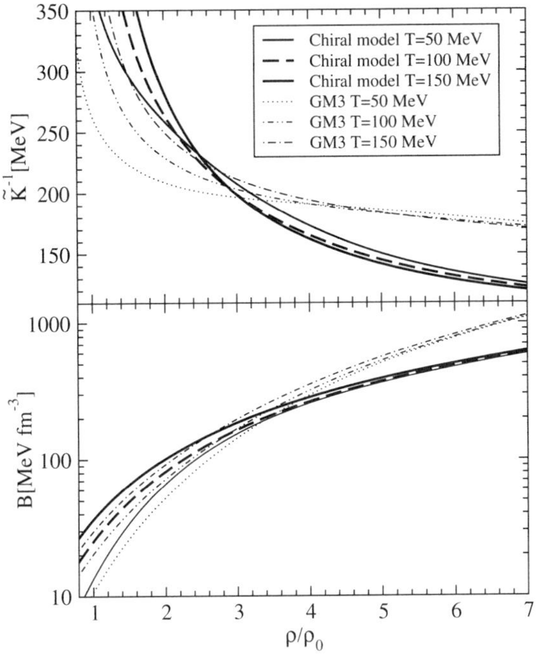

Fig. 4. The upper and lower panels show respectively the parameter $\widetilde{K}^{-1}$ and the bulk modulus at different temperatures as a function of density for the chiral-dilaton model and the GM3 parametrization ($Z/A = 0.4$).

4. Results and Conclusions

Let us start by presenting in Fig. 3 the pressure computed using the models here discussed with the limits obtained from the analysis of HICs at intermediate energies.[27] Because of in HICs a small number of hyperons is generated through associated production, we have also introduced an EOS with a fixed fraction of strangeness due to the presence of hyperons (using the GM3 parametrization). An estimate of the number of hyperons per participant can be obtained from the experimental ratio of the kaon yields per participant,[28] which indicates a strangeness fraction smaller than 10%. As a reference, we also show the parametric EOSs discussed in Ref. 7, which are characterized by different values of the incompressibility parameter K^{-1}. From Fig. 3 it is clear that only the extra soft and the hard EOSs are excluded by the experimental limit. The pressure computed using the chiral-dilaton model marginally exceeds the limit at low densities, due to the too large value of the incompressibility at saturation density. The effect of the partial restoration of chiral symmetry is clearly visible as a softening taking place at larger densities. Instead the small fraction of strangeness introduced using the GM3 parametrization does not produce a sizable softening of the EOS.

We now compare the compressibility computed in the various models discussed above with the one estimated in Ref. 7. In the following we always assume that

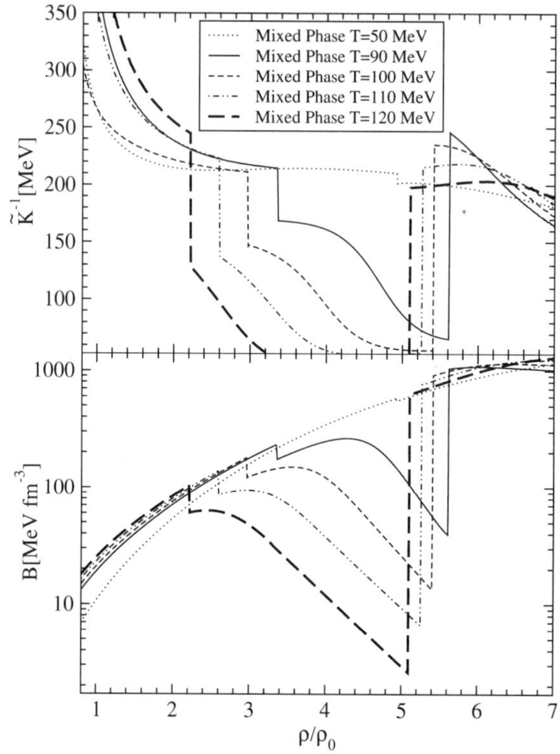

Fig. 5. Same as in Fig. 3, but using the mixed-phase model.

$Z/A = 0.4$. In the lower panels of Figs. 4 and 5 we plot the bulk modulus $B = \rho\,\partial P/\partial\rho$, respectively for the chiral dilaton model and for the mixed-phase model. Similarly, in the upper panels we show, as a function of the density and temperature, the value of the incompressibility parameter for which the bulk modulus of a parametric EOS has the same value of the bulk modulus computed in our models. This quantity, called $\widetilde{K}^{-1}$, is obtained by solving the equation:

$$\left.\frac{\partial P_{par}(\tilde{K})}{\partial\rho}\right|_{\rho,T} = \left.\frac{\partial P_{model}}{\partial\rho}\right|_{\rho,T}. \tag{4}$$

In this way we can directly compare with the analysis of Ref. 7 which explicitly indicates various values of the parameter $\widetilde{K}^{-1}$ as representative of the EOS tested at various energies. In the upper panel of Fig. 4 we show the $\widetilde{K}^{-1}$ parameter computed using the chiral-dilaton model and we see that it decreases significantly at large densities, while its dependence on the temperature is relevant only at small densities. By comparison, the GM3 parametrization provides a $\widetilde{K}^{-1}$ which is roughly constant at large densities. In Fig. 5, we show that the incompressibility computed using the mixed-phase model remains rather large above the lower critical density and it becomes really small only approaching the upper critical density.

In conclusion, in this contribution we have shown that a significant softening of the EOS can be obtained either via chiral symmetry restoration, if large densities are reached, or via the formation of a mixed phase of quarks and hadrons. In the chiral model the incompressibility strongly depends on the density but it is almost temperature independent at least up to $T \sim 150$ MeV. At the contrary a strong temperature dependence is found in the mixed phase model. Since in HICs at intermediate energies not too large densities are reached, the mixed-phase model seems to provide a better description of the results of the experiments analysis.

References

1. H. Liu, S. Panitkin and N. Xu, *Phys. Rev. C* **59**, 348 (1999).
2. S. Soff, S. A. Bass, M. Bleicher, H. Stoecker and W. Greiner (1999).
3. P. K. Sahu and W. Cassing, *Nucl. Phys. A* **712**, 357 (2002).
4. H. Stoecker, *Nucl. Phys. A* **750**, 121 (2005).
5. H. Stoecker, E. L. Bratkovskaya, M. Bleicher, S. Soff and X. Zhu, *J. Phys. G* **31**, S929 (2005).
6. M. Isse, A. Ohnishi, N. Otuka, P. K. Sahu and Y. Nara, *Phys. Rev. C* **72**, 064908 (2005).
7. V. N. Russkikh and Y. B. Ivanov, *Phys. Rev. C* **74**, 034904 (2006).
8. L. Bonanno, A. Drago and A. Lavagno (2007).
9. E. K. Heide, S. Rudaz and P. J. Ellis, *Nucl. Phys. A* **571**, 713 (1994).
10. G. W. Carter, P. J. Ellis and S. Rudaz, *Nucl. Phys. A* **603**, 367 (1996).
11. G. W. Carter and P. J. Ellis, *Nucl. Phys. A* **628**, 325 (1998).
12. P. Senger, *J. Phys. G* **30**, S1087 (2004).
13. R. J. Furnstahl, B. D. Serot and H.-B. Tang, *Nucl. Phys. A* **598**, 539 (1996).
14. I. Mishustin, J. Bondorf and M. Rho, *Nucl. Phys. A* **555**, 215 (1993).
15. R. J. Furnstahl, H.-B. Tang and B. D. Serot, *Phys. Rev. C* **52**, 1368 (1995).
16. P. Papazoglou, S. Schramm, J. Schaffner-Bielich, H. Stoecker and W. Greiner, *Phys. Rev. C* **57**, 2576 (1998).
17. P. Papazoglou *et al.*, *Phys. Rev. C* **59**, 411 (1999).
18. P. Wang, V. E. Lyubovitskij, T. Gutsche and A. Faessler, *Phys. Rev. C* **70**, 015202 (2004).
19. N. K. Glendenning, *Phys. Rev. D* **46**, 1274 (1992).
20. H. Muller and B. D. Serot, *Phys. Rev. C* **52**, 2072 (1995).
21. H. Muller, *Nucl. Phys. A* **618**, 349 (1997).
22. B. Liu, V. Greco, V. Baran, M. Colonna and M. Di Toro, *Phys. Rev. C* **65**, 045201 (2002).
23. G. F. Burgio, M. Baldo, P. K. Sahu and H.-J. Schulze, *Phys. Rev. C* **66**, 025802 (2002).
24. H. Grigorian, D. Blaschke and D. N. Aguilera, *Phys. Rev. C* **69**, 065802 (2004).
25. V. D. Toneev, E. G. Nikonov, B. Friman, W. Norenberg and K. Redlich, *Eur. Phys. J. C* **32**, 399 (2003).
26. D. N. Voskresensky, M. Yasuhira and T. Tatsumi, *Nucl. Phys. A* **723**, 291 (2003).
27. P. Danielewicz, R. Lacey and W. G. Lynch, *Science* **298**, 1592 (2002).
28. L. Ahle *et al.*, *Phys. Rev. C* **60**, 044904 (1999).

COLOR SUPERCONDUCTING QUARK MATTER IN COMPACT STARS

D. B. BLASCHKE

Institute for Theoretical Physics, University of Wroclaw,
50-204 Wroclaw, Poland
E-mail: blaschke@ift.uni.wroc.pl
www.ift.uni.wroc.pl/~blaschke

T. KLÄHN

Theory Division, Argonne National Laboratory,
Argonne IL 60439-4843, USA
E-mail: Thomas.Klaehn@googlemail.com

F. SANDIN

Department of Physics, Luleå University of Technology,
97187 Luleå, Sweden
E-mail: Fredrik.Sandin@gmail.com

Recent indications for high neutron star masses ($M \sim 2\,M_\odot$) and large radii ($R > 12$ km) could rule out soft equations of state and have provoked a debate whether the occurence of quark matter in compact stars can be excluded as well. We show that modern quantum field theoretical approaches to quark matter including color superconductivity and a vector meanfield allow a microscopic description of hybrid stars which fulfill the new, strong constraints. For these objects color superconductivity turns out to be essential for a successful description of the cooling phenomenology in accordance with recently developed tests. We discuss QCD phase diagrams for various conditions thus providing a basis for a synopsis for quark matter searches in astrophysics and in future generations of nucleus-nucleus collision experiments such as low-energy RHIC and CBM @ FAIR.

Keywords: Neutron stars; Chiral quark model; Color superconductivity.

1. Introduction

Recently, observations of compact stars have provided new data of high accuracy which put strong constraints on the high-density behaviour of the equation of state of strongly interacting matter otherwise not accessible in terrestrial laboratories. In particular, the high masses of $M = 1.96 + 0.09/-0.12\,M_\odot$ and $M = 2.73 \pm 0.25\,M_\odot$ obtained in very recent measurements on the millisecond pulsars PSR B1516+02B and PSR J1748-2021B, repectively,[1] together with the large radius of $R > 12$ km for the isolated neutron star RX J1856.5-3754 (shorthand: RX J1856)[2] point to a stiff equation of state at high densities. Measurements of high masses are also reported

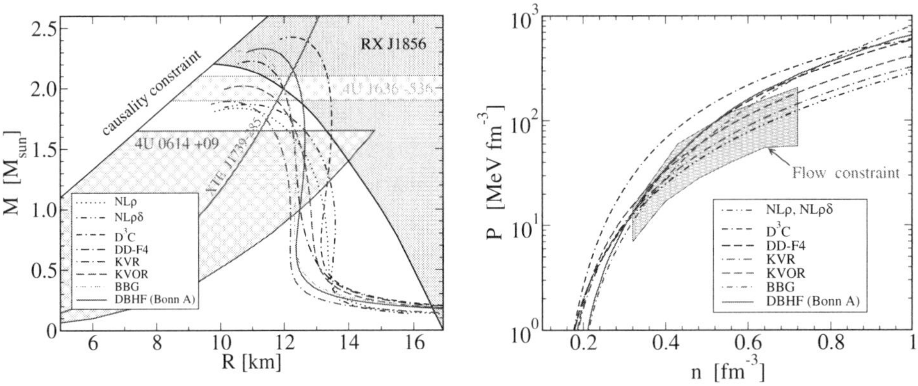

Fig. 1. Left panel: Mass and mass-radius constraints on compact star configurations from recent observations compared to solutions of TOV equations for nuclear EoS discussed in the text. Right panel: Flow constraint from heavy-ion collisions[13] compared to the same set of nuclear EoS used in the left panel.

for compact stars in low-mass X-ray binaries (LMXBs) as, e.g., $M = 2.0 \pm 0.1\, M_\odot$ for the compact object in 4U 1636-536.[3] For another LMXB, EXO 0748-676, constraints for the mass $M \geq 2.10 \pm 0.28\, M_\odot$ *and* the radius $R \geq 13.8 \pm 0.18$ km have been derived.[4] The status of these data is, however, unclear since the observation of a gravitational redshift $z = 0.35$ in the X-ray burst spectra[5] could not be confirmed thereafter despite numerous attempts.[6]

Measurements of rotation periods below ~ 1 ms as discussed for XTE J1739-285,[7] on the other hand, would disfavor too large objects corresponding to a stiff EoS and would thus leave only a tiny window of very massive stars in the mass-radius plane[8,9] for a theory of compact star matter to fulfill all above mentioned constraints.

In the left panel of Fig. 1 we show some of these modern observational constraints for masses and mass-radius relationships together with solutions of the Tolman-Oppenheimer-Volkoff (TOV) equations for a set of eight hadronic EoS classified in three groups: (i) *relativistic mean-field (RMF) approaches with non-linear (NL) self-interactions* of the σ meson.[10] In NLρ the isovector part of the interaction is described only by a ρ meson, while the set NL$\rho\delta$ also includes a scalar isovector meson δ that is usually neglected in RMF models;[11] (ii) *RMF models with density dependent couplings and masses* are represented here by four different models from two classes, where in the first one density dependent meson couplings are modeled so that a number of properties of finite nuclei (binding energies, charge and diffraction radii, surface thicknesses, neutron skin in ^{208}Pb, spin-orbit splittings) can be fitted.[12] D^{3}C has in addition a derivative coupling leading to momentum-dependent nucleon self-energies and DD-F4 is modeled such that the flow constraint[13] from heavy-ion collisions is fulfilled. The second class of these models is motivated by the Brown-Rho scaling assumption[14] that not only the nucleon mass but also the meson masses should decrease with increasing density. In the KVR and KVOR models[15] these dependences were related to a nonlinear scaling function of the σ- meson field such that the EoS of symmetric nuclear matter and pure neutron matter below four times the saturation density coincide with those of the Urbana-Argonne group.[16] In this way the latter approach builds a bridge between the phenomenological RMF models and (iii) *microscopic EoS* built on realistic nucleon-nucleon forces. Besides the variational approaches (APR,[16] WFF,[17] FPS[18]) such ab-initio approaches to nuclear matter are provided, e.g., by the relativistic Dirac-Brueckner-Hartree-Fock (DBHF)[19] and the nonrelativistic Brueckner-Bethe-Goldstone[20] approaches. Stiff EoS like D^3C, DD-F4, BBG and DBHF fulfill the demanding constraints for a large radius and mass, while the softer ones like NLρ don't. It is interesting to note that the flow constraint[13] shown in the right panel of Fig. 1 sets limits to the tolerable stiffness: it excludes the D^3C EoS and demonstrates that DD-F4, BBG and DBHF become too stiff at high densities above $\sim 0.55\ \mathrm{fm}^{-3}$. For a detailed discussion, see Ref. 34.

A key question asked in investigating the structure of matter at high densities is how the quark substructure of hadrons manifests itself in the EoS and whether the phase transition to quark matter can occur inside compact stars. In Ref. 4, Özel has debated that the new constraints reported above would exclude quark matter in compact star interiors reasoning that it would entail an intolerable softening of the EoS. Alford et al.[21] have given a few counter examples demonstrating that quark matter cannot be excluded. In the following section we discuss a recently developed chiral quark model[22] which is in accord with the modern constraints, see also Ref. 23.

2. Color Superconducting Quark Matter: Masquerade Revisited

We describe the thermodynamics of the deconfined quark matter phase within a three-flavor quark model of Nambu–Jona-Lasinio (NJL) type, with a mean-field thermodynamic potential given by

$$\Omega_{MF}(T,\mu) = \frac{1}{8G_S}\left[\sum_{i=u,d,s}(m_i^* - m_i)^2 - \frac{2}{\eta_V}(2\omega_0^2 + \phi_0^2) + \frac{2}{\eta_D}\sum_{A=2,5,7}|\Delta_{AA}|^2\right]$$
$$- \int\frac{d^3p}{(2\pi)^3}\sum_{a=1}^{18}\left[E_a + 2T\ln\left(1 + e^{-E_a/T}\right)\right] + \Omega_l - \Omega_0\,. \qquad (1)$$

Here, Ω_l is the thermodynamic potential for electrons and muons, and the divergent term Ω_0 is subtracted in order to assure zero pressure and energy density in vacuum ($T = \mu = 0$). The quasiparticle dispersion relations, $E_a(p)$, are the eigenvalues of the hermitean matrix

$$\mathcal{M} = \begin{bmatrix} \vec{\gamma}\ \ \vec{p}\ \ \ \hat{m}^*\ |\ \gamma^0\hat{\mu}^* & i\gamma_5 C\tau_A\lambda_A\Delta_{AA} \\ iC\gamma_5\tau_A\lambda_A\Delta_{AA}^* & -\vec{\gamma}^T\cdot\vec{p} + \hat{m}^* - \gamma^0\hat{\mu}^* \end{bmatrix}, \qquad (2)$$

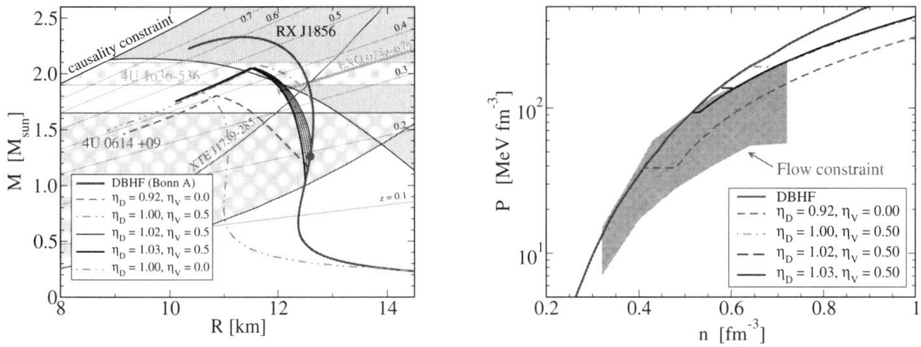

Fig. 2. Same as Fig. 1 for hybrid EoS with a low density hadronic branch described by the DBHF approach and a high density quark matter branch obtained from a three-flavor NJL model with color superconductivity (diquark coupling η_D) and isoscalar vector meson coupling (η_V).

in color, flavor, Dirac, and Nambu-Gorkov space. Here, Δ_{AA} are the diquark gaps. $\hat{m}^*$ is the diagonal renormalized mass matrix and $\hat{\mu}^*$ the renormalized chemical potential matrix, $\hat{\mu}^* = \mathrm{diag}_f(\mu_u - G_S\eta_V\omega_0, \mu_d - G_S\eta_V\omega_0, \mu_s - G_S\eta_V\phi_0)$. The gaps and the renormalized masses are determined by minimization of the mean-field thermodynamic potential (1). We have to obey constraints of charge neutrality which depend on the application we consider. In the (approximately) isospin symmetric situation of a heavy-ion collision, the color charges are neutralized, while the electric charge in general is non-zero. For matter in β-equilibrium in compact stars, also the global electric charge neutrality has to be fulfilled. For further details, see Ref. 24–27.

We consider η_D as a free parameter of the quark matter model, to be tuned with the present phenomenological constraints on the high-density EoS. Similarly, the relation between the coupling in the scalar and vector meson channels, η_V, is considered as a free parameter of the model. The remaining degrees of freedom are fixed according to the NJL model parameterization in table I of Ref. 28, where a fit to low-energy phenomenological results has been made.

As a unified description of quark-hadron matter, naturally including a description of the phase transition, is not available yet, although steps in this direction have been suggested within nonrelativistic[29] and relativistic[30] models. We apply here the so-called two-phase description, being aware of its limitations. The nuclear matter phase is described within the DBHF approach and the transition to the quark matter phase given above is obtained by a Maxwell construction. In the right panel of Fig. 2 it can be seen that the necessary softening of the high density EoS in accordance with the flow constraint is obtained for a vector coupling of $\eta_V = 0.5$ whereas an appropriate deconfinement density is obtained for a strong diquark coupling in the range $\eta_D = 1.02 - 1.03$. The resulting phase transition is weakly first order with an almost negligible density jump. Applying this hybrid EoS with so defined free parameters under compact star conditions a sequence of hybrid star configurations is obtained which fulfills all modern constraints, see the left panel of Fig. 2. In that figure we also indicate by a red dot the minimal mass M_{DU} for which the central density reaches a value allowing the fast direct Urca (DU) cooling process in DBHF neutron star matter to occur, leading to problems with cooling phenomenology.[31,32] Note that for a strong diquark coupling $\eta_D = 1.03$, the critical density for quark deconfinement is low enough to prevent the hadronic direct Urca (DU) cooling problem by an early onset of quark matter. For the given hybrid EoS, there is a long sequence of stable hybrid stars with two-flavor superconducting (2SC) quark matter, before the occurrence of the strange quark flavor and the simultaneous on set of the color-flavor-locking (CFL) phase renders the star gravitationally unstable. Comparing the hybrid star sequences with the purely hadronic DBHF ones one can conclude that the former 'masquerade' themselves as neutron stars[33] by having very similar mechanical properties.

Besides for mass-radius relationships, the masquerade effect can be discussed also for the moment of inertia,[22] which is becoming accessible to measurements in relativistic binary systems like the double pulsar J0737–3039.

As a new aspect of this discussion we add in these proceedings the energy release in the neutrino untrapping transition of a cooling protoneutron star (PNS). The process is pictured such that after the PNS formation in a supernova collapse it is hot enough ($T_{\rm PNS} \sim 30 \ldots 50$ MeV) to trap neutrinos since their mean free path is shorter than the size of the star so that the fast neutrino cooling cannot proceed from the volume but only from the surface. Until the neutrino opacity temperature ($T_{\rm opac} \sim 1$ MeV) is reached, the neutrinos take part in the β-equilibration processes and a finite lepton fraction is established. We will assume $Y_{le} = 0.4$ corresponding to a neutrino chemical potential $\mu_\nu \sim 200$ MeV in the PNS core. In the untrapping transition at $T \sim T_{\rm opac}$, the neutrinos decouple from the β-equilibrium, $Y_{le} \to 0$ and the equilibrium configuration gets readjusted with a new gravitational mass while naturally, the baryon number is conserved in this process. From Fig. 3 one can estimate the mass defect (energy release) to amount to $0.04\ M_\odot$ almost independent of the mass of the configuration and even its structure: neutron stars and hybrid stars result in the same energy release! This is another aspect of the masquerade effect. Note that the neutrino untrapped configurations fulfil the extended $M_G - M_B$ constraint.[34,35]

3. Unmasking Compact Star Interiors

To unmask the neutron star interior might therefore require observables based on transport properties, strongly modified from normal due to the color superconductivity. It has been suggested to base tests of the structure of matter at high densities on analyses of the cooling behavior[32,36,37] or the stability of fastly rotating stars against r-modes.[38,39] It has turned out that for these phenomena the fine tuning of color superconductivity in quark matter is an essential ingredient. The result of these studies suggests that on the one hand unpaired quark matter results in

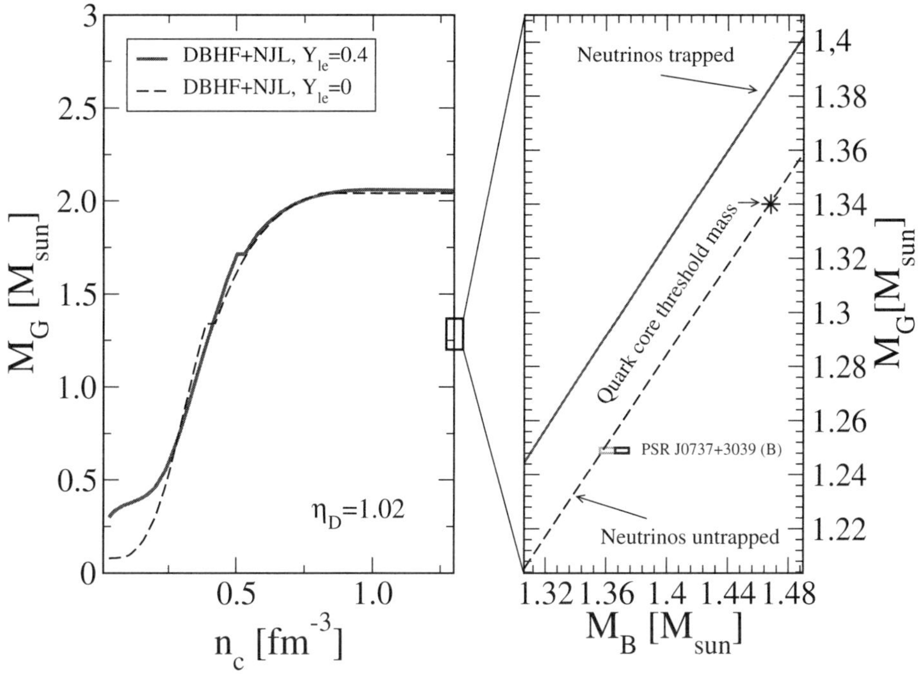

Fig. 3. The effect of neutrino untrapping on hybrid star configurations: a mass defect of 0.04 $M_\odot$ occurs in the transition from neutrino trapping case (solid red line, $Y_{le} = 0.4$) to untrapping (dashed black lines, $Y_{le} = 0$) at fixed baryon mass M_B, independent of the presence of a quark core for masses above the threshold indicated by an asterix. The blue rectangle corresponds to the constraint from the lighter companion (star B) in the double pulsar PSR J0737-3039.[35]

too fast cooling (same holds for pure two-flavor color superconductivity (2SC) due to one unpaired (blue) color) and on the other, complete pairing with large gaps (CFL phase, $\Delta_{\rm CFL} \sim 100$ MeV) would result in too slow cooling, r-mode instability and gravitational collapse. Viable alternatives for quark matter thus have to be color superconducting, with all quark modes paired, but a few modes (at least one) should have a small gap of $\Delta_X \sim 1$ MeV, wishfully with a decreasing density dependence, as suggested for the 2SC+X phase.[40,41] Unfortunately, there is no microscopic model supporting the 2SC+X pairing pattern yet. A suitable candidate, however, could be the recently suggested single flavor pairing in the isotropic spin-1 color superconductivity phase (iso-CSL).[42,43] It has recently been shown that the neutrino emissivity and bulk viscosity of the iso-CSL phase fulfills constraints from cooling and r-mode stability.[44]

4. Conclusions

In this contribution it is shown that modern quantum field theoretical approaches to quark matter including color superconductivity and a vector meanfield allow a microscopic description of hybrid stars which fulfill the new, strong constraints. The deconfinement transition in the resulting stiff hybrid equation of state is weakly first order so that signals of it have to be expected due to specific changes in transport properties governing the rotational and cooling evolution caused by the color superconductivity of quark matter. A similar conclusion holds for the investigation of quark deconfinement in future generations of nucleus-nucleus collision experiments at low temperatures and high baryon densities,[40] such as CBM @ FAIR.

Acknowledgements

We thank the organizers of the EXOCT07 conference for providing a perfect environment for discussios of the challenging questions in exotic matter and compact star research. D. B. is supported by the Polish Ministry of Science and Higher Education, T. K. is grateful for partial support from GSI Darmstadt and the Department of Energy, Office of Nuclear Physics, contract no. DE-AC02-06CH11357. F. S. acknowledges support from the Swedish Graduate School of Space Technology and the Royal Swedish Academy of Sciences.

References

1. P. C. C. Freire, S. M. Ransom, S. Begin, I. H. Stairs, J. W. T. Hessels, L. H. Frey and F. Camilo, arXiv:0711.0925 [astro-ph].
2. J. E. Trümper, V. Burwitz, F. Haberl and V. E. Zavlin, *Nucl. Phys. Proc. Suppl.* **132**, 560 (2004).
3. D. Barret, J. F. Olive and M. C. Miller, *Mon. Not. Roy. Astron. Soc.* **361**, 855 (2005).
4. F. Özel, *Nature* **441**, 1115 (2006).
5. J. Cottam, F. Paerels and M. Mendez, *Nature* **420**, 51 (2002).
6. J. Cottam, F. Paerels, M. Mendez, L. Boirin, W. H. G. Lewin, E. Kuulkers and J. M. Miller, arXiv:0709.4062 [astro-ph].
7. P. Kaaret *et al.*, *Astrophys. J.* **657**, L97 (2006).
8. G. Lavagetto, I. Bombaci, A. D'Ai', I. Vidana and N. R. Robba, arXiv:astro-ph/0612061.
9. M. Bejger, P. Haensel and J. L. Zdunik, *Astron. Astrophys.* **464**, L49 (2007).
10. T. Gaitanos, M. Di Toro, S. Typel, V. Baran, C. Fuchs, V. Greco and H. H. Wolter, *Nucl. Phys. A* **732**, 24 (2004).
11. B. Liu, V. Greco, V. Baran, M. Colonna and M. Di Toro, *Phys. Rev. C* **65**, 045201 (2002).
12. S. Typel, *Phys. Rev. C* **71**, 064301 (2005).
13. P. Danielewicz, R. Lacey and W. G. Lynch, *Science* **298**, 1592 (2002).
14. G. E. Brown and M. Rho, *Phys. Rev. Lett.* **66**, 2720 (1991).
15. E. E. Kolomeitsev and D. N. Voskresensky, *Nucl. Phys. A* **759**, 373 (2005).
16. A. Akmal, V. R. Pandharipande and D. G. Ravenhall, *Phys. Rev. C* **58**, 1804 (1998).
17. R. B. Wiringa, V. Fiks and A. Fabrocini, *Phys. Rev. C* **38**, 1010 (1988).
18. B. Friedman and V. R. Pandharipande, *Nucl. Phys. A* **361**, 502 (1981).
19. E. N. E. van Dalen, C. Fuchs and A. Faessler, *Nucl. Phys. A* **744**, 227 (2004); *Phys. Rev. C* **72**, 065803 (2005).
20. M. Baldo, G. F. Burgio and H.-J. Schulze, *Phys. Rev. C* **61**, 055801 (2000).
21. M. Alford, D. Blaschke, A. Drago, T. Klähn, G. Pagliara and J. Schaffner-Bielich, *Nature* **445**, E7 (2007).

22. T. Klähn *et al.*, *Phys. Lett. B* **567**, 170 (2007).
23. D. B. Blaschke, D. Gomez Dumm, A. G. Grunfeld, T. Klähn and N. N. Scoccola, *Phys. Rev. C* **75**, 065804 (2007).
24. D. Blaschke, S. Fredriksson, H. Grigorian, A. M. Öztas and F. Sandin, *Phys. Rev. D* **72**, 065020 (2005).
25. S. B. Rüster, V. Werth, M. Buballa, I. A. Shovkovy and D. H. Rischke, *Phys. Rev. D* **72**, 034004 (2005).
26. H. Abuki and T. Kunihiro, *Nucl. Phys. A* **768**, 118 (2006).
27. H. J. Warringa, D. Boer and J. O. Andersen, *Phys. Rev. D* **72**, 014015 (2005).
28. H. Grigorian, *Phys. Part. Nucl. Lett.* **4**, 382 (2007).
29. G. Röpke, D. Blaschke and H. Schulz, *Phys. Rev. D* **34**, 3499 (1986).
30. S. Lawley, W. Bentz and A. W. Thomas, *J. Phys. G* **32**, 667 (2006).
31. D. Blaschke, H. Grigorian and D. N. Voskresensky, *Astron. Astrophys.* **424**, 979 (2004).
32. D. Blaschke and H. Grigorian, *Prog. Part. Nucl. Phys.* **59**, 139 (2007).
33. M. Alford, M. Braby, M. W. Paris and S. Reddy, *Astrophys. J.* **629**, 969 (2005).
34. T. Klähn *et al.*, *Phys. Rev. C* **74**, 035802 (2006).
35. P. Podsiadlowski, J. D. M. Dewi, P. Lesaffre, J. C. Miller, W. G. Newton and J. R. Stone, *Mon. Not. Roy. Astron. Soc.* **361**, 1243 (2005).
36. S. Popov, H. Grigorian, R. Turolla and D. Blaschke, *Astron. Astrophys.* **448**, 327 (2006).
37. S. Popov, H. Grigorian and D. Blaschke, *Phys. Rev. C* **74**, 025803 (2006).
38. J. Madsen, *Phys. Rev. Lett.* **85**, 10 (1999).
39. A. Drago, G. Pagliara and I. Parenti, arXiv:0704.1510 [astro-ph].
40. H. Grigorian, D. Blaschke and T. Klähn, in: *Neutron Stars and Pulsars*, W. Becker and H. H. Huang (eds.), *MPE Report* **291**, 193 (2006); [arXiv:astro-ph/0611595].
41. H. Grigorian, D. Blaschke and D. Voskresensky, *Phys. Rev. C* **71**, 045801 (2005).
42. D. N. Aguilera, D. Blaschke, M. Buballa and V. L. Yudichev, *Phys. Rev. D* **72**, 034008 (2005).
43. D. N. Aguilera, D. Blaschke, H. Grigorian and N. N. Scoccola, *Phys. Rev. D* **74**, 114005 (2006).
44. D. B. Blaschke and J. Berdermann, arXiv:0710.5243 [hep-ph].

THERMAL HADRONIZATION, HAWKING-UNRUH RADIATION AND EVENT HORIZON IN QCD

P. CASTORINA

Physics Department, Catania University,
Via Santa Sofia 64, 95123 Catania, Italy
E-mail: paolo.castorina@ct.infn.it

Because of colour confinement, the physical vacuum forms an event horizon for quarks and gluons; this can be crossed only by quantum tunneling, i.e., through the QCD counterpart of Hawking radiation by black holes. Since such radiation cannot transmit information to the outside, it must be thermal, of a temperature determined by the strong force at the confinement surface, and it must maintain colour neutrality. The resulting process provides a common mechanism for thermal hadron production in high energy interactions, from e^+e^- annihilation to heavy ion collisions. The analogy with black-hole event horizon suggests a dependence of the hadronization temperature on the baryon density.

Keywords: Confinement; Thermal hadronization; Event horizon; Hawking radiation; Unruh effect.

1. Introduction

Over the years, hadron production studies in a variety of high energy collision experiments have shown a remarkably universal feature. From e^+e^- annihilation to $p-p$ and $p-\bar{p}$ interactions and further to collisions of heavy nuclei, covering an energy range from a few GeV up to the TeV range, the production pattern always shows striking thermal aspects, connected to an apparently quite universal temperature around $T_H \simeq 160 - 190$ MeV.[1,2]

What is the origin of this thermal behaviour? While high energy heavy ion collisions involve large numbers of incident partons and thus could allow invoking some "thermalisation" scheme through rescattering, in e^+e^- annihilation the predominant initial state is one energetic $q\bar{q}$ pair, and the number of hadronic secondaries per unit rapidity is too small to consider statistical averages.

A further piece in this puzzle is the observation that the value of the temperature determined in the mentioned collision studies is quite similar to the confinement/deconfinement transition temperature found in lattice studies of strong interaction thermodynamics.[3] While hadronization in high energy collisions deals with a dynamical situation, the energy loss of fast colour charges "traversing" the physical vacuum, lattice QCD addresses the equilibrium thermodynamics of unbound vs. bound colour charges. Why should the resulting critical temperatures be similar or even identical?

In Ref. 4, which is summarized in this contribution (see also Ref. 5), these hadronization phenomena are considered as the QCD counterpart of the Hawking radiation emitted by black holes (BH).[6] BHs provide a gravitational form of confinement that was quite soon compared to that of colour confinement in QCD,[7,8] where coloured constituents are confined to "white holes" (colourless from the outside, but coloured inside). The main results in Ref. 4 are:

- Colour confinement and the instability of the physical vacuum under pair production form an event horizon for quarks, allowing a transition only through quantum tunnelling; this leads to thermal radiation of a temperature T_H determined by the string tension.
- Hadron production in high energy collisions occurs through a succession of such tunnelling processes. The resulting cascade is a realization of the same partition process which leads to a limiting temperature in the statistical bootstrap and dual resonance models.
- In kinetic thermalization, the initial state information is successively lost through collisions, converging to a time-independent equilibrium state. In contrast, the stochastic QCD Hawking radiation is "born in equilibrium", since quantum tunnelling *a priori* does not allow information transfer.
- The temperature T_H of QCD Hawking radiation depends only on the baryon number and the angular momentum of the deconfined system. The former provides the dependence of T_H on the baryochemical potential μ, while the angular momentum pattern of the radiation allows a centrality-dependence of T_H and elliptic flow. In particular the μ dependence of T_H, will be discussed in more details in Sec. 4

2. Thermal Production Pattern

Let us first summarize the thermal production pattern in elementary collisions, $e^+e^-, pp, \bar{p}p, ...$ and in nucleus-nucleus scattering. The partition function of an ideal resonance gas is given by

$$\ln Z(T) = V \sum_i \frac{d_i}{(2\pi)^3} \phi(m_i, T), \tag{1}$$

where d_i is the degeneracy factor and $\phi(m_i, T)$ is the Boltzmann factor

$$\phi(m_i, T) = 4\pi m_i^2 T K_2(m_i/T). \tag{2}$$

Therefore the relative abundances of the species i and j turn out

$$\frac{N_i}{N_j} = \frac{d_i \phi(m_i, T)}{d_j \phi(m_j, T)} \tag{3}$$

and for transverse energy larger than T

$$\frac{dN}{dp_T^2} \sim \exp - \frac{1}{T} \sqrt{m_i^2 + p_T^2}. \tag{4}$$

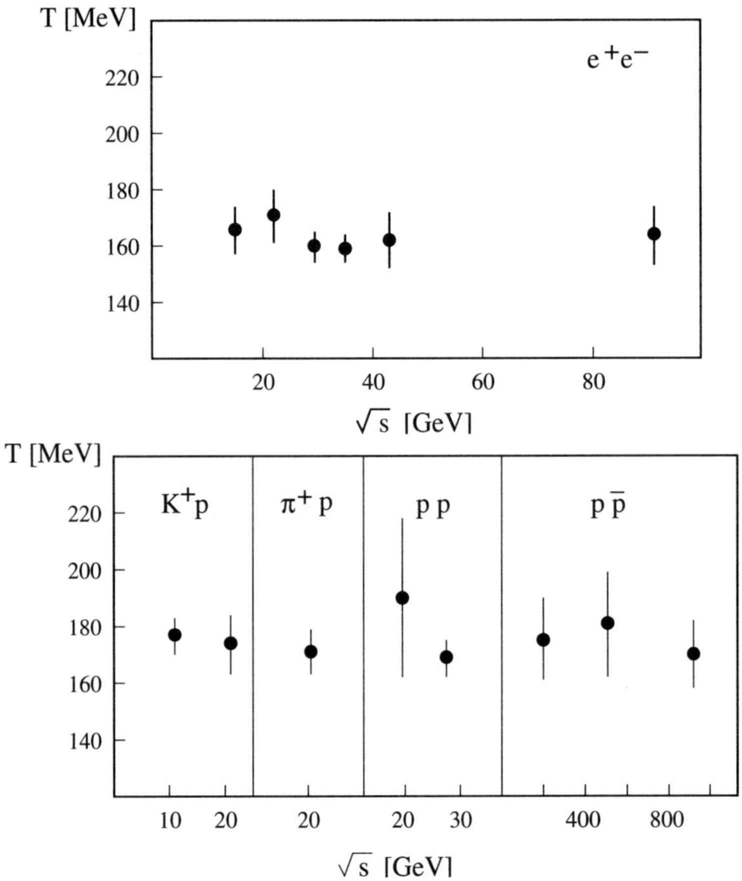

Fig. 1. Hadronization temperature at large energy for various elementary collisions.

In elementary collisions the statistical hadronization model[9] fits the data on the species abundances by two parameters: T and γ_s, that describes the strangeness suppression. For LEP data at $\sqrt{s} = 91.2$ GeV,[2] $T = 170 \pm 3 \pm 6$ MeV and $\gamma_s = 0.691 \pm 0.053$, where the systematic error is obtained by varying the resonance gas scheme and the contributing resonances. The PEP-PETRA data at different energies, $14 < \sqrt{s} < 45$ can be fitted[2] with an average temperature $T = 165 \pm 6$ MeV and $\gamma_s \simeq 0.7 \pm 0.05$. The pp SPS data at energies $\sqrt{s} = 19, 23.8, 26$ GeV give $T = 162.4 \pm 1.6$ Me and $\gamma_s \simeq 0 \pm 0.036$.[2] The other data for K^+p and π^+p scattering at energies close to SPS one and for $\bar{p}p$ at larger energy can be fitted by similar values. The fitted values of the temperature are depicted in Fig. 1.[10]

Therefore there is an universal hadronization temperatute $T_H = 170 \pm 10 - 20$ MeV which is independent on the species, on $\sqrt{s}$ and on the incident configuration. Moreover, also the transverse momentum spectra in elementary collisions can be fitted by the same value T_H.[2]

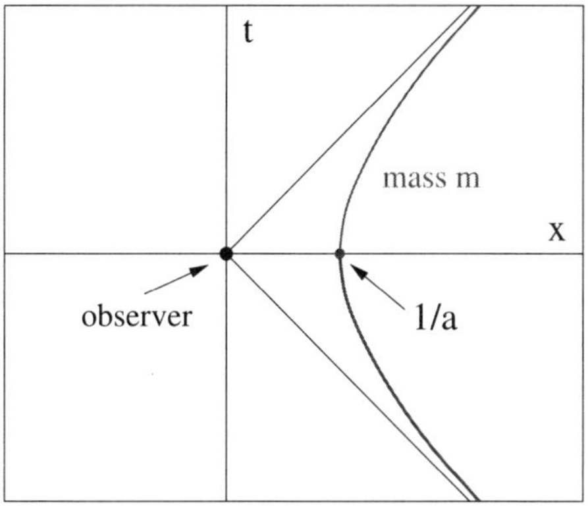

Fig. 2. Hyperbolic motion.

In heavy ion collisions there is a new parameter which describes the finite baryon density, i.e. the baryon chemical potential μ_B. The fits of the species abundances at high energy (peak SPS and RHIC) give[2] :

- $T_H = 168 \pm 2.4 \pm 10$ MeV; $\mu_b = 266 \pm 5 \pm 30$ MeV at $\sqrt{s} = 17$ GeV for (SPS) Pb-Pb,

- $T_H = 168 \pm 7$ MeV; $\mu_b = 38 \pm 11 \pm 5$ MeV at $\sqrt{s} = 130$ GeV for (RICH) Au-Au at y=0,

- $T_H = 161 \pm 2$ MeV; $\mu_b = 20 \pm 4$ MeV at $\sqrt{s} = 200$ GeV for (RICH) Au-Au.

In conclusion, hadron abundances in all high energy collisions (e^+e^- annihilation,hadron-hadron and heavy ion collisions) are those of an ideal resonance gas at universal temperature $T_H \simeq 170 \pm 10 - 20$ MeV.

3. Event Horizon and Hadronization

The idea that a thermal medium, with a kinetic thermalization by multiple partonic interaction, has been produced in the collisions could explain the previous phenomenon for a nucleus-nucleus scattering but does not work for e^+e^- and hadron-hadron scattering. One has to look for an universal, "non-kinetic" thermalization mechanism.

Indeed, in gravitation there is a well known example: the BH Hawking radiation has a thermal radiation spectrum due to tunnelling through the event horizon and the Hawking temperature is given by $T_{BH} = 1/8\pi GM$, where M is the (Schwarzschild) BH mass and G is the Newton constant.[6] Therefore the conjecture[4] is that colour confinement and hadronization mechanism are in strong analogy with BH physics and event horizon.

There are many reasons to believe that color confinement can be described by a color horizon in QCD because the theory is non linear and therefore it has an effective curved geometry.[11,12] However, since we discuss the hadronization mechanism, is better to consider the Unruh effect. As shown by Unruh,[13] systems with

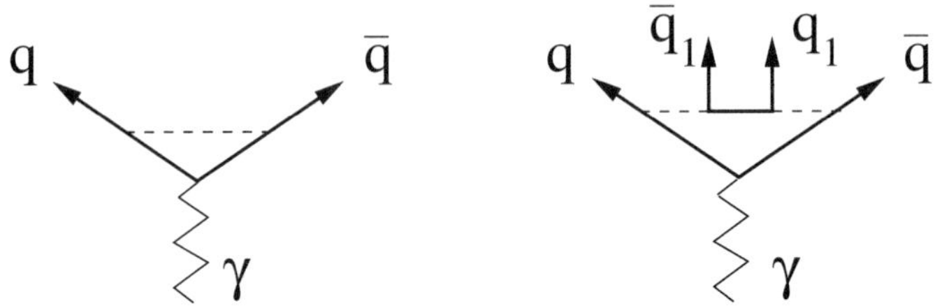

Fig. 3. Pair production in e^+e^-.

uniform acceleration, a, have an event horizon and see a thermal bath with temperature $T_U = a/2\pi$. For a particle of mass m, in uniform acceleration the equation of motion is solved by the parametric form

$$x = \frac{1}{a}\cosh a\tau \qquad t = \frac{1}{a}\sinh a\tau, \tag{5}$$

where $a = F/m$ denotes the acceleration in the instantaneous rest frame of m, and τ the proper time, with $d\tau = \sqrt{1-v^2}dt$. The resulting world line is shown in Fig. 2 with the event horizon beyond which m classically cannot pass. The only signal the observer can detect as consequence of the passage of m is thermal quantum radiation of temperature

$$T_U = \frac{a}{2\pi}. \tag{6}$$

In the case of gravity, a is the "surface gravity" (i.e. the acceleration at the horizon), $a = 1/(4GM)$, and hence one recovers the Hawking temperature.

In summary, the acceleration leads to a classical turning point and hence to an event horizon, which can be surpassed only by quantum tunnelling and at the expense of complete information loss, leading to thermal radiation as the only allowed signal.

On the other hand, at quantum level, it is well known that in a strong field the vacuum is unstable against pair production.[14] For example, in e^+e^- annihilation a $\bar{q}q$ pair is initially produced and when the linear potential is such that $\sigma x > \sigma x_Q \equiv 2m$ the string connecting $\bar{q}q$ breaks and the color neutralization induces an effective quantum event horizon (see Figs. 3 and 4)

The $\bar{q}q$ flux tube has a thickness given by[4]

$$r_T \simeq \sqrt{\frac{2}{\pi\sigma}} \tag{7}$$

and the $\bar{q}_1 q_1$ is produced at rest in e^+e^- cms but with a transverse momentum

$$k_T \simeq \frac{1}{r_T} \simeq \sqrt{\frac{\pi\sigma}{2}}. \tag{8}$$

The acceleration (or deceleration) associated with the string breaking and color neutralization mechanism turns out to be[4]

$$a \simeq 2\,k_T \simeq \sqrt{2\pi\sigma} \simeq 1.1 \text{ GeV}, \tag{9}$$

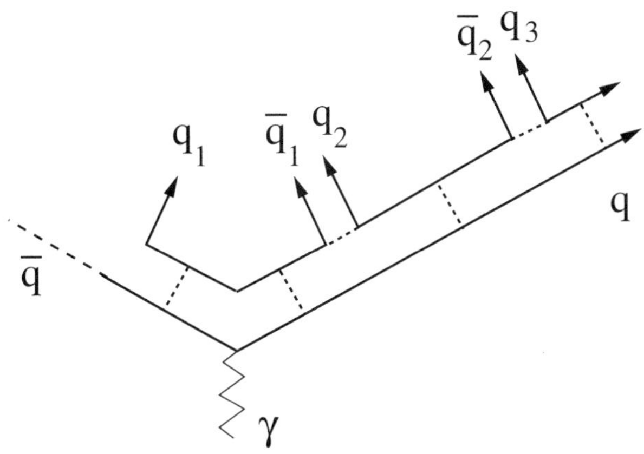

Fig. 4. Hadronization as a tunnelling process.

which leads to

$$T_q = \frac{a}{2\pi} \simeq \sqrt{\frac{\sigma}{2\pi}} \simeq 180 \text{ MeV} \tag{10}$$

for the hadronic Unruh temperature. It governs the momentum distribution and the relative species abundances of the emitted hadrons.

Notice that the previous hadronization mechanism can be described by saying that $\bar{q}_1$ reaches the $q_1\bar{q}_1$ event horizon and tunnels to become $\bar{q}_2$. The emission of hadrons $\bar{q}_1 q_2$ can be considered as Hawking radiation.

4. Vacuum Pressure and Baryon Density

It is interesting to consider the extension of the previous mechanism in the case of systems with a net baryon number,i.e with a new "charge" which can modify the tunnelling process and the Hawking-Unruh hadronization temperature.

In the BH case the effect of a total charge Q changes the Hawking temperature according to the formula (see for example Ref. 15)

$$T_{BH}(M,Q) = T_{BH}(M,0) \left\{ \frac{4\sqrt{1 - Q^2/GM^2}}{(1 + \sqrt{1 - Q^2/GM^2}\,)^2} \right\}. \tag{11}$$

Note that with increasing charge, the Coulomb repulsion weakens the gravitational field at the event horizon and hence decreases the temperature of the corresponding quantum excitations. As $Q^2 \to GM^2$, the gravitational force is fully compensated. The crucial quantity here is the ratio Q^2/GM^2 of the overall Coulomb energy, Q^2/R, to the overall gravitational energy, GM^2/R. Equivalently, $Q^2/GM^2 = P_Q/P_G$ measures the ratio of inward gravitational pressure P_G at the event horizon to the repulsive outward Coulomb pressure P_Q.

In QCD, we have a "white" hole containing coloured quarks, confined by chromodynamic forces or, equivalently, by the pressure B of the physical vacuum. If

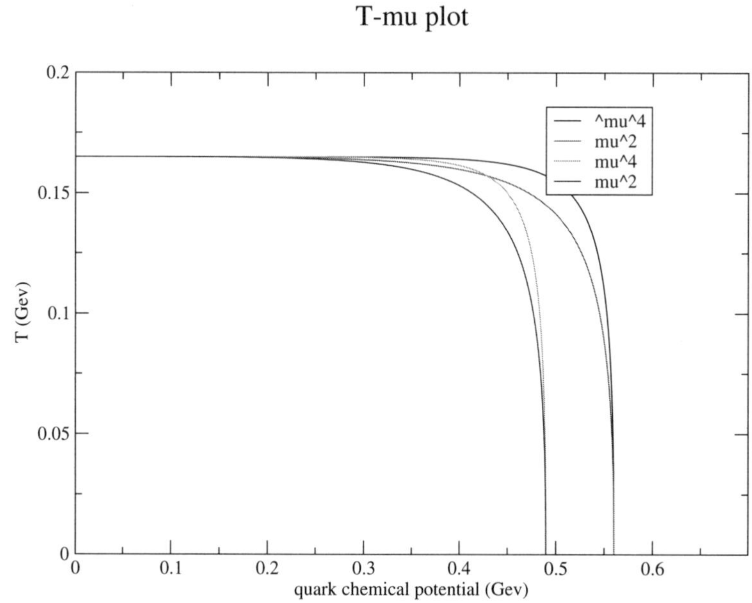

Fig. 5. T dependence on μ by Eq. (12) with $(\mu/\mu_0)^n$ for $n = 2, 4$.

the system has a non-vanishing overall baryon number, there will be a Fermi repulsion between the corresponding quarks, and this repulsion will provide a pressure $P(\mu)$ acting against B, with μ denoting the corresponding quark baryochemical potential. We thus expect a similar reduction of the hadronization temperature as function of μ. To quantify this aspect let us consider, at $T = 0$, the Fermi pressure, $P = d_g\mu^4/24\pi^2$, where d_g is the degeneracy factor, versus the QCD vacuum pressure due to the gluon condensate $B =< (\alpha_s/\pi)G^2_{\mu\nu} >$ which is a decreasing (largely unknown) function of the baryon density. By following the analysis of Ref. 16 for the dependence of the gluon condensate on the baryon number, the critical density, where B balances the Fermi pressure, turns out about $n_c = 5 - 6n_0$, with n_0 the nuclear saturation density. A slightly different result is obtained in the description of deconfinement by percolation:[17] $n_c \simeq 4n_0$.

In the intermediate region, where both T and μ are finite, we want to compare the effect of the Fermi repulsion to the vacuum pressure through the Hawking-Unruh form, i.e., we replace Q^2/GM^2 in Eq. (11) by $(\mu/\mu_0)^4$, giving

$$T(\mu)/T_0 = \frac{4\sqrt{1-(\mu/\mu_0)^4}}{(1+\sqrt{1-(\mu/\mu_0)^4})^2}. \tag{12}$$

The resulting behaviour of $T(\mu)$ is shown in Fig. 5 for μ_0 corresponding to the previous $n_c \simeq 4 - 6\ n_0$. In the same figure is shown the results obtained by using $(\mu/\mu_0)^2$ rather than $(\mu/\mu_0)^4$ in Eq. (12). In both cases the function remains rather flat up to large value of μ.

Clearly this approach is overly simplistic, since it reduces the effect of the additional quarks to only their Fermi repulsion. A more general way of addressing the problem would be to introduce an effective μ-dependence of the string tension.

References

1. R. Hagedorn, *Nuovo Cim. Suppl.* **3**, 147 (1965); *Nuovo Cim. A* **56**, 1027 (1968).
2. F. Becattini, *Z. Phys. C* **69**, 485 (1996) (e^+e^-);
   F. Becattini and U. Heinz, *Z. Phys. C* **76**, 268 (1997) ($pp/p\bar{p}$);
   J. Cleymans and H. Satz, *Z. Phys. C* **57**, 135 (1993) (heavy ions);
   F. Becattini *et al.*, *Phys. Rev. C* **64**, 024901 (2001) (heavy ions);
   F. Becattini and G. Passaleva, *Eur. Phys. J. C* **23**, 551 (2002);
   P. Braun-Munziger, K. Redlich and J. Stachel, (heavy ions).
3. See e.g., M. Cheng *et al.*, *Phys. Rev. D* **74**, 054507 (2006) for the latest state and references to earlier works.
4. P. Castorina, D. Kharzeev and H. Satz, *Eur. Phys. J. C* **52**, 187 (2007).
5. H. Satz, *Thermal Hadron Production and Hawking-Unruh Radiation in QCD*, CERN Particle Physics Seminar, 22 May 2007.
6. S. W. Hawking, *Comm. Math. Phys.* **43**, 199 (1975).
7. E. Recami and P. Castorina, *Lett. Nuovo Cim.* **15**, 347 (1976).
8. A. Salam and J. Strathdee, *Phys. Rev. D* **18** (1978).
9. E. Fermi, *Prog. Theor. Phys.* **5** (1950) 570; L. D. Landau, *Izv. Akad. SSSR Ser. Fiz.* **17**, 51 (1953); R. Hagedorn, *Nuovo Cim.* **15**, 434 (1960).
10. F. Becattini, *Nucl. Phys. A* **702**, 336 (2002), Proceedings of the Bielefeld Symposium "Statistical QCD," 26-30 August 2001.
11. M. Novello *et al.*, *Phys. Rev. D* **61**, 045001 (2000).
12. D. Kharzeev, E. Levin and K. Tuchin, *Phys. Lett. B* **547**, 21 (2002); *Phys. Rev. D* **70**, 054005 (2004).
13. W. G. Unruh, *Phys. Rev. D* **14**, 870 (1976).
14. J. Schwinger, *Phys. Rev.* **82**, 664 (1951).
15. See e.g., Li Zhi Fang and R. Ruffini, *Basic Concepts in Relativistic Astrophysics*, World Scientific, Singapore, 1983.
16. M. Baldo, P. Castorina and D. Zappalà, *Nucl. Phys. A* **743**, 13 (2004).
17. V. Magas and H. Satz, *Eur. Phys. J. C* **32**, 115 (2003).

FERROMAGNETISM IN THE QCD PHASE DIAGRAM

T. TATSUMI

Department of Physics, Kyoto University,
Kyoto 606-8502, Japan
E-mail: tatsumi@ruby.scphys.kyoto-u.ac.jp

A possibility and properties of spontaneous magnetization in quark matter are investigated. Magnetic susceptibility is evaluated within Fermi liquid theory, taking into account of screening effect of gluons. Spin wave in the polarized quark matter, as the Nambu-Goldstone mode, is formulated by way of the coherent-state path integral.

Keywords: Ferromagnetism; Fermi liquid; Screening; Spin wave; Quark matter.

1. Introduction

The phase diagram of QCD has been elaborately studied in density-temperature plane. Here we study the magnetic properties of QCD. Since the discovery of magnetars with super-strong magnetic field of $O(10^{14-15})$G the origin of strong magnetic field observed in compact stars has roused our interest again.[1] For the present there are three ideas about its origin: the first one reduces it to the fossil field, assuming the conservation of the magnetic flux during the evolution of stars. The second one applies the dynamo mechanism in the crust region. The third one seeks it at the core region, where hadron or quark matter develops. If spins of nucleons or quarks are aligned in some situations, they can provide the large magnetic field. Since elaborate studies about the spontaneous magnetization in nuclear matter have repeatedly shown the negative results, in the following, we explore the possibility of the spontaneous magnetization in quark matter.[2,3]

In the first paper[2] we have suggested a possibility of ferromagnetic phase in QCD, within the one-gluon-exchange (OGE) interaction. If such a phase is realized inside compact stars, we can see the magnetic field of $O(10^{15-17})$G is easily obtained. Directly evaluating the total energy of the polarized matter with a polarization parameter, we have shown that ferromagnetic phase is possibly realized at low densities, in analogy with the itinerant electrons by Bloch;[4] the Fock exchange interaction is responsible to ferromagnetism in QCD. We have also seen that the phase transition is weakly first order and one may also apply the technique for the second order phase transition to analyze it, while it is one of the specific features of the magnetic transition in gauge theories.

In the first half we discuss the magnetic susceptibility of quark matter within the

Fermi-liquid theory.[5] It is well known that we must properly take into account the screening effects in the gluon propagator to improve the IR behavior of the gauge interaction. For the longitudinal gluons we can see the static screening described in terms of the Debye screening mass. There is no static screening for the transverse gluons, while the dynamical screening appears instead due to the Landau damping.[6] We will figure out these screening effects in evaluating the magnetic susceptibility. It would be interesting to observe that there appears the non-Fermi-liquid effect to give an anomalous term in the finite temperature case as in the specific heat.[7]

In the second half we discuss how the spin wave, which is caused by the spontaneous magnetization, can be described within the coherent-state path integral.[8] Fist we map the quark matter to a spin system by assuming the spatial wave function for each quark and leaving the degree of freedom of the direction of the spin vector. This method is inspired by the spiral approach taken by Herring in old days to discuss the spin wave in the electron gas.[9] Introducing the collective variables $\boldsymbol{U}(\bar{\theta}, \bar{\phi}) \in S^2$, and integrating over the individual variables, we have an effective action to see the classical Landau-Lifshitz equation for the spin wave is naturally derived by the effective action for the collective variables. We also note that there are some geometrical aspects in the effective action, which may further quantize the classical spin wave to give magnons. Thus we have magnons in the ferromagnetic phase of quarks, which directly gives the $T^{2/3}$ dependence for the reduction of the magnetization.

2. Magnetic Susceptibility within the Fermi Liquid Theory

The magnetic susceptibility χ_M is defined as

$$\chi_M = \left.\frac{\partial \langle M \rangle}{\partial B}\right|_{N,T}, \tag{1}$$

with the magnetization $\langle M \rangle$. So we study the response of quark matter when a weak magnetic field is applied for a given quark number N and temperature T. Using the Gordon identity, the interaction Lagrangian may be written for the constant magnetic field, $\boldsymbol{A} = \boldsymbol{B} \times \boldsymbol{r}/2$;

$$\int d^4x \mathcal{L}_{ext} = \mu_q \int d^4x \bar{q} \left[-i\boldsymbol{r} \times \nabla + \boldsymbol{\Sigma}\right] \cdot \boldsymbol{B} q, \tag{2}$$

for quarks with electric charge e_q, where $\boldsymbol{\Sigma} = \begin{pmatrix} \boldsymbol{\sigma} & 0 \\ 0 & \boldsymbol{\sigma} \end{pmatrix}$, and μ_q denotes the Dirac magnetic moment $\mu_q = e_q/2m$. Then magnetization $\boldsymbol{M}$ may be written as

$$M_z = \langle \bar{q} \boldsymbol{\Sigma} q \rangle_z = \frac{\mu_q}{2} N_C \int \frac{d^3k}{(2\pi)^3} g_D(\boldsymbol{k})(n_{\boldsymbol{k},+} - n_{\boldsymbol{k},-}), \tag{3}$$

with the Fermi-Dirac distribution function $n_{\boldsymbol{k},\zeta}$ in the presence of $\boldsymbol{B}$. The gyromagnetic ratio g_D is defined as

$$g_D(\boldsymbol{k})\zeta = 2\mathrm{tr}\left[\Sigma_z \rho(k,\zeta)\right] = \left[1 - \frac{k_z^2}{E_k(E_k + m)}\right]\zeta, \tag{4}$$

in terms of the polarization density matrix $\rho(k,\zeta)$,

$$\rho(k,\zeta) = \frac{1}{2m}(\not{k} + m)P(a), \tag{5}$$

with the projection operator, $P(a) = 1/2 \cdot (1 + \gamma_5 \not{a})$.

The quasi-particle interaction consists of two terms,

$$f_{\boldsymbol{k}\zeta,\boldsymbol{q}\zeta'} = f^s_{\boldsymbol{k},\boldsymbol{q}} + \zeta\zeta' f^a_{\boldsymbol{k},\boldsymbol{q}}, \tag{6}$$

where $f^{s(a)}_{\boldsymbol{k},\boldsymbol{q}}$ is the spin-independent (dependent) interaction. Then we get the expression for the magnetic susceptibility at $T = 0$ written in terms of the Landau parameters:

$$\chi_M = \left(g^F_D \mu_q/2\right)^2 \left[\frac{\pi^2}{N_C k_F \mu} - \frac{1}{3} f^s_1 + \bar{f}^a\right]^{-1}, \tag{7}$$

where f^s_1 is a spin-averaged Landau parameter defined by

$$f^s_1 = \frac{3}{4} \sum_{\zeta,\zeta'} \int \frac{d\Omega_{\boldsymbol{kq}}}{4\pi} \cos\theta_{\boldsymbol{kq}} f^s_{\boldsymbol{k},\boldsymbol{q}} \bigg|_{|\boldsymbol{k}|=|\boldsymbol{q}|=k_F}, \tag{8}$$

with the relative angle $\theta_{\boldsymbol{kq}}$ of $\boldsymbol{k}$ and $\boldsymbol{q}$, and $\bar{f}^a$ the spin-dependent one,

$$\bar{f}^a \equiv \int \frac{d\Omega_{\boldsymbol{k}}}{4\pi} \int \frac{d\Omega_{\boldsymbol{q}}}{4\pi} f^a_{\boldsymbol{k},\boldsymbol{q}} \bigg|_{|\boldsymbol{k}|=|\boldsymbol{q}|=k_F}. \tag{9}$$

3. Static and Dynamic Screening Effects

When we consider the color-symmetric forward scattering amplitude of the two quarks around the Fermi surface by the one gluon exchange interaction (OGE), the direct term should be vanished due to the color neutrality of quark matter and the Fock exchange term gives a leading contribution. The color-symmetric and flavor-symmetric OGE interaction of quasi-particles on the Fermi surface may be written as

$$f_{\boldsymbol{k}\zeta,\boldsymbol{q}\zeta'}\big|_{|\boldsymbol{k}|=|\boldsymbol{q}|=k_F} = \frac{1}{N_C^2}\frac{1}{N_F^2} \sum_{a,b}\sum_{i,j} f_{\boldsymbol{k}\zeta ai,\boldsymbol{q}\zeta' bj} \bigg|_{|\boldsymbol{k}|=|\boldsymbol{q}|=k_F} = \frac{m}{E_k}\frac{m}{E_q} M_{\boldsymbol{k}\zeta,\boldsymbol{q}\zeta'} \bigg|_{|\boldsymbol{k}|=|\boldsymbol{q}|=k_F} \tag{10}$$

with the invariant matrix element, $M_{\boldsymbol{k}\zeta,\boldsymbol{q}\zeta'}$.[10] It has been well known that massless gluons often causes infrared (IR) divergences in the Landau parameters.[5]

Since the one gluon exchange interaction is a long-range force and we consider the small energy-momentum transfer between quasi-particles, we must improve the gluon propagator by taking into account the screening effect,

$$D_{\mu\nu}(k-q) = P^T_{\mu\nu} D_T(p) + P^L_{\mu\nu} D_L(p) - \xi \frac{p_\mu p_\nu}{p^4} \tag{11}$$

with $p = k - q$, where $D_{T(L)}(p) = (p^2 - \Pi_{T(L)})^{-1}$, and the last term represents the gauge dependence with a parameter ξ. $P_{\mu\nu}^{T(L)}$ is the standard projection operator onto the transverse (longitudinal) mode.

The self-energies for the transverse and longitudinal gluons are given as

$$\Pi_L(p_0, \boldsymbol{p}) = m_D^2 + i\frac{\pi m_D^2}{2v_F}\frac{p_0}{|\boldsymbol{p}|}\coth\frac{p_0}{2T},$$

$$\Pi_T(p_0, \boldsymbol{p}) = -i\frac{\pi p_F m_D^2}{4E_F}\frac{p_0}{|\boldsymbol{p}|}\coth\frac{p_0}{2T}, \tag{12}$$

in the limit $p_0/|\boldsymbol{p}| \to 0$, with the Debye screening mass, $m_D^2 \equiv (N_F/2\pi^2)g^2E_Fk_F$. Thus the longitudinal gluon is screened to have the Debye mass m_D, while the transverse gluon is not in the limit $p_0/|\boldsymbol{p}| \to 0$ and $T = 0$.[6] At finite temperature, however, the transverse gluons are also dynamically screened by the Landau damping. For quarks on the Fermi surface, the Lorentz invariant matrix element can be written as

$$M_{\boldsymbol{k}\zeta,\boldsymbol{q}\zeta'} = -\frac{N_C^2 - 1}{2N_C^2}g^2\left[-M^{00}D_L + M^{ij}P_T^{ij}D_T + \xi\frac{1}{|\boldsymbol{k}-\boldsymbol{q}|^4}(\boldsymbol{k}-\boldsymbol{q})_i(\boldsymbol{k}-\boldsymbol{q})_jM^{ij}\right], \tag{13}$$

with the coefficients

$$M^{\mu\nu} = \mathrm{tr}[\gamma^\mu\rho(k,\zeta)\gamma^\nu\rho(q,\zeta')]. \tag{14}$$

First of all, the matrix element is obviously independent of the gauge parameter ξ.

At temperature $T = 0$, there is no screening in the propagator of the transverse gluon, so that logarithmic divergence still remains in the Landau-Migdal parameters, f_1^s or $\bar{f}^a$. However, we can see that the divergences cancels each other to give a finite result for the susceptibility. The static screening effect gives a $g^2\ln g^{-2}$ contribution.[11]

In Fig. 1 we present an example of the magnetic susceptibility at $T = 0$. The ferromagnetic phase corresponds to the negative value of χ and the critical density is given by the divergence of χ. Compared with the OGE calculation, we can see that the static screening effect shifts the critical density to the lower value.

At finite temperature, the dynamical screening gives a $T^2\ln T$ contribution through the density of states near the Fermi surface, besides the standard T^2 dependence. This behavior is a kind of non-Fermi liquid effects, as in the specific heat.[7]

4. Spin Wave in the Polarized Quark Matter

When spontaneous magnetization occurs, the magnetization $\boldsymbol{M}$ is not vanished, so that rotation symmetry is violated in the ground state. Hence one can expect a Nambu-Goldstone mode, *spin wave* there. Different from the usual description of the spin wave, we must care about its realization in quark matter. Recalling a similar situation in electron gas, we take here an intuitive but correct framework

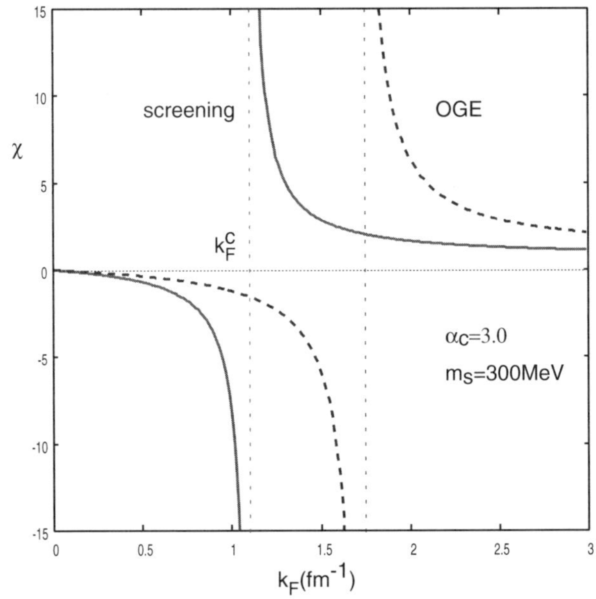

Fig. 1. Magnetic susceptibility as a function of the Fermi momentum. Divergence signals the onset of spontaneous spin polarization. The screening effect slightly shifts the critical density to lower densities.

called the spiral approach. Herring explicitly showed that the Bloch wall coefficient, which is closely related with the dispersion relation of the spin wave, is obtained for electron gas by the use of the spiral approach.[9]

Consider the fully polarized case. Then all the quarks have a definite spin state specified by $\boldsymbol{\zeta}$ (say, $\zeta = +1$). The single particle wave function with momentum k is simply given as

$$u^{(\zeta)}(k) = \frac{\not{k} + m}{\sqrt{2m(E_k + m)}} u^{(\zeta)}(m, \mathbf{0}) e^{-ikx} \tag{15}$$

with the spinor in the rest frame,

$$u^{(\zeta=+1)}(m, \mathbf{0}) = e^{ib} \begin{pmatrix} e^{-i\phi_k/2} \cos\theta_k/2 \\ e^{i\phi_k/2} \sin\theta_k/2 \\ 0 \end{pmatrix} \equiv \begin{pmatrix} g_k(\boldsymbol{r}) \\ 0 \end{pmatrix}, \tag{16}$$

taking the spin quantization axis specified by the polar angles (θ_k, ϕ_k). Thus the quark wave function is characterized by the momentum $\boldsymbol{k}$ and the polar variables θ_k, ϕ_k.

In the ground state, all the spins have the same direction, say $\theta_k = \bar{\theta}, \phi_k = \bar{\phi}$, so that the ground-state energy is degenerate in any value of them. In the following we allow them to be spatially dependent and introducing the small fluctuation fields,

$\xi_k(x), \eta_k(x)$, s.t.

$$\theta_k(x) = \bar{\theta}(x) + \xi_k(x), \phi_k(x) = \bar{\phi}(x) + \eta_k(x), \tag{17}$$

where $\bar{\theta}, \bar{\phi}$ or $\boldsymbol{U} = (U^1(\bar{\theta},\bar{\phi}), U^2(\bar{\theta},\bar{\phi}), U^3(\bar{\theta},\bar{\phi}))$ are the collective variables defined by

$$\cos\bar{\theta} \equiv \frac{1}{N_k}\sum_k \cos\theta_k, \bar{\phi} \equiv \frac{1}{N_k}\sum_k \phi_k, \tag{18}$$

and we shall see they describe the Nambu-Goldstone mode (spin wave).

In the spiral approach, we assume a spin configuration described by

$$g_k(\boldsymbol{r}) = \exp\left[it_z(d\bar{\phi}/dz)(\sigma_z/2)\right] g_k(\boldsymbol{r} - t) \tag{19}$$

with arbitrary displacement $\boldsymbol{t}$, which corresponds to a spin wave traveling along z axis with wave vector $d\bar{\phi}/dz$, so that $\bar{\phi}$ should be a linear function of z. Actually the mean value of the spin operator $\boldsymbol{\Sigma}$ is proportional to $\bar{\boldsymbol{\zeta}}$.

For a given Hamiltonian H we can evaluate the energy by putting (15), (16) into it,

$$H(\theta_k, \phi_k) = \int d^3x \langle H \rangle, \tag{20}$$

which may be regarded as a classical spin Hamiltonian for quark matter. Thus we mapped quark matter to assembly of spins.

5. Coherent-State Path Integral

One may reformulate the idea of the spiral approach in the framework of the path integral.[8] Consider a matrix element of the evolution operator

$$\langle \Omega'', t'' \mid \exp(-iT\hat{H}/\hbar) \mid \Omega', t' \rangle = \int \mathcal{D}\Omega \exp\left(i \sum_k \int_{t'}^{t''} dt \int d^3x \left[i\langle \Omega_k \mid \dot{\Omega}_k \rangle - H(\Omega) \right] \right), \tag{21}$$

where $|\Omega\rangle = |\Omega_1\rangle \otimes \cdots \otimes |\Omega_{N_k}\rangle$ is the spin coherent state

$$|\Omega_k\rangle = (\cos(\theta_k/2))^{2S} \exp\left[\tan(\theta_k/2)\, e^{i\phi_k} \hat{S}_-\right] |0\rangle, \tag{22}$$

with $S_z|0\rangle = S|0\rangle$.

Introducing collective variables $\bar{\Omega} = (\bar{\theta}, \bar{\phi})$, Eq. (21) may be rewritten as

$$\int \mathcal{D}\Omega\mathcal{D}\bar{\Omega}\delta\left(\bar{\Omega} - \frac{1}{N_k}\sum \Omega_k\right) e^{\left\{\frac{i}{\hbar}\sum_k \int_{t'}^{t''} dt \int d^3x \left[\frac{1}{2}(1-\cos\theta_k)\dot{\phi}_k - H(\theta_k,\phi_k)\right]\right\}} = \int \mathcal{D}\bar{\Omega} e^{iS_{\rm eff}(\bar{\Omega})}, \tag{23}$$

where the *effective action* is defined as

$$e^{iS_{\rm eff}(\bar{\Omega})} = \int \mathcal{D}\xi\mathcal{D}\eta\delta\left(\sum_k \xi_k\right)\delta\left(\sum_k \eta_k\right) \times e^{\left\{\frac{i}{\hbar}\sum_k \int_{t'}^{t''}\int d^3x\left[\frac{1}{2}(1-\cos(\bar{\theta}+\xi_k))(\dot{\bar{\phi}}+\dot{\eta}_k)-H(\bar{\theta},\bar{\phi},\xi_k,\eta_k)\right]dt\right\}}. \quad (24)$$

We expand the exponent with respect to ξ_k, η_k up to the second-order, discarding the higher-order terms in $\dot{\bar{\phi}}$ within the adiabatic approximation. Taking the stationary-phase approximation s.t.

$$\left.\frac{\delta H(\bar{\theta},\bar{\phi},\xi_k,\eta_k)}{\delta\xi_k}\right|_{\xi_k=\xi_k^c} = \left.\frac{\delta H(\bar{\theta},\bar{\phi},\xi_k,\eta_k)}{\delta\eta_k}\right|_{\eta_k=\eta_k^c} = 0, \quad (25)$$

we have

$$S_{\rm eff} \approx \frac{i}{\hbar}\int_{t'}^{t''} dt \int d^3x\left[\Sigma(1-\cos\bar{\theta})\dot{\bar{\phi}} - H(\bar{\theta},\bar{\phi},\xi_k^c,\eta_k^c)\right], \quad (26)$$

with $\Sigma = N_k/2V$. One may show that

$$H(\bar{\theta},\bar{\phi},\xi_k^c,\eta_k^c) = \frac{A}{2}\left(\nabla_r \boldsymbol{U}\right)^2, \quad (27)$$

as should be. Then, the Bloch wall coefficient reads

$$A = \frac{N_k/V}{8E_F} + O(g^2). \quad (28)$$

The classical equation of motion for $\boldsymbol{U}$ is given by $\delta S_{\rm eff} = 0$,

$$\Sigma\dot{\boldsymbol{U}} + 2A\Delta\boldsymbol{U}\times\boldsymbol{U} = 0, \quad (29)$$

which is the classical Landau-Lifshitz equation for the spin wave. Then, the dispersion relation for the spin wave is deduced,

$$\omega(\boldsymbol{q}) = (2A/\Sigma)q^2 = (1/2E_F)\,q^2 + O(g^2).$$

The effective action has some topological features, with which we can do the second quantization for the spin wave. The first term in the effective action can be written as the interaction with the dynamically induced vector potential $\boldsymbol{A}$:

$$(1-\cos\bar{\theta})\dot{\bar{\phi}} = \boldsymbol{A}(\bar{\boldsymbol{\zeta}})\cdot\dot{\bar{\boldsymbol{\zeta}}} \quad (30)$$

with

$$\boldsymbol{A} = \left(\frac{1-\cos\bar{\theta}}{\sin\bar{\theta}}\right)\hat{\boldsymbol{\phi}}. \quad (31)$$

Employing the Dirac quantization condition, we can see that Σ =integer or half-integer as should be.

Alternatively, the first term in terms of $\boldsymbol{U}$ can be rewritten as the line integral along the path on S^2, which is nothing but the Wess-Zumino term.

From the effective action (26), it is inferred that $\cos\bar{\theta}$ and $\bar{\phi}$ are canonical conjugate. Putting $\bar{\pi} \equiv \Sigma(1-\cos\bar{\theta})$, one can verify

$$\left\{\Sigma U^+(\boldsymbol{x},t), \Sigma U^-(\boldsymbol{y},t)\right\}_{PB} = 2i\Sigma U^3(\boldsymbol{x},t)\delta(\boldsymbol{x}-\boldsymbol{y}) \simeq 2i\Sigma\delta(\boldsymbol{x}-\boldsymbol{y}), \qquad (32)$$

where we have used the Kramers-Heller approximation for a large number of particles in the last step. Taking the Fourier transform, s.t.,

$$U^+(\boldsymbol{x},t) = \sqrt{\frac{2}{\Sigma}}\sum_{\boldsymbol{q}} a_{\boldsymbol{q}} e^{i\boldsymbol{q}\cdot\boldsymbol{x}}, \qquad (33)$$

and $U^-(\boldsymbol{x},t) = (U^+(\boldsymbol{x},t))^\dagger$, and assuming the quantum-mechanical commutation relation (*second quantization*),

$$[a_{\boldsymbol{q}}, a^\dagger_{\boldsymbol{q}'}] = \delta_{\boldsymbol{q},\boldsymbol{q}'}, \qquad (34)$$

we have an Hamiltonian as an assembly of magnons,

$$\int d^3x H(\bar{\theta}, \bar{\phi}, \xi^c_k, \eta^c_k) = \sum_{\boldsymbol{q}} \omega(\boldsymbol{q}) a^\dagger_{\boldsymbol{q}} a_{\boldsymbol{q}}. \qquad (35)$$

For low temperature the thermodynamical properties may be described by the excitation of the spin wave. We can easily see that magnetization is reduced by the excitation of the spin waves at finite temperature ($T^{3/2}$ Law), $(M(T)-M(0))/N_k g_D \mu_q = -\zeta(3/2)\left(\Sigma T/8\pi A\right)^{3/2}$, from which the Curie temperature reads, $T_c = 8\pi A/\Sigma(\rho/2\zeta(3/2))^{2/3}$. Thus we can roughly estimate the Curie temperature of several tens MeV for $2A/\Sigma \simeq 1/2E_F$.

6. Concluding Remarks

In this paper we have discussed the critical line of the spontaneous polarization on the density-temperature plane within the framework of the Fermi-liquid theory. We have evaluated the magnetic susceptibility by taking into account the screening effects for the gluon propagation, and figured out the important roles of the static and dynamic screening; the former gives the $g^2\ln(1/g^2)$ contribution, while the latter gives $T^2\ln T$ for finite temperature. Both effects surely reflect the specific feature of the gauge interaction. To get more realistic values for the critical density and temperature, we must consider some nonperturbative effects such as instantons as well as the systematic analysis of the higher-order terms in QCD.

We have presented a framework to deal with the spin wave as a Nambu-Goldstone mode. In the perspectives we can derive the magnon-quark coupling vertex, which may give a novel cooling mechanism and a novel type of the Cooper pairing.[12] They are not only theoretically interesting, but also phenomenologically important for the thermal evolution of compact stars bearing the ferromagnetic phase inside. It should be interesting if magnon effects could distinguish the microscopic origin from fossil field or the dynamo mechanism.

Acknowledgements

The author thanks K. Sato for his collaboration. This work has been partially supported by the Grant-in-Aid for the 21st Century COE "Center for the Diversity and Universality in Physics" and the Grant-in-Aid for Scientific Research Fund (C) of the Ministry of Education, Culture, Sports, Science and Technology of Japan (16540246).

References

1. A. K. Harding and D. Lai, *Rep. Prog. Phys.* **69**, 2631 (2006).
2. T. Tatsumi, *Phys. Lett. B* **489**, 280 (2000); T. Tatsumi, E. Nakano and K. Nawa, *Dark Matter* (Nova Science Pub., New York, 2006), p. 39.
3. E. Nakano, T. Maruyama and T. Tatsumi, *Phys. Rev. D* **68**, 105001 (2003); T. Tatsumi, E. Nakano and T. Maruyama, *Prog. Theor. Phys. Suppl.* **153**, 190 (2004); T. Tatsumi, T. Maruyama and E. Nakano, *Superdense QCD Matter and Compact Stars* (Springer, 2006), p. 241.
4. F. Bloch, *Z. Phys.* **57**, 545 (1929).
5. G. Baym and C. J. Pethick, *Landau Fermi-Liquid Theory* (WILEY-VCH, 2004); P. Noziéres, *Theory of Interacting Fermi Systems* (Westview Press, 1997); A. A. Abrikosov, L. P. Gorkov and I. Ye. Dzyaloshinskii, *Quantum Field Theoretical Methods in Statistical Physics* (Pergamon, Oxford, 1965).
6. M. Le Bellac, *Thermal Field Theory* (Cambridge U. Press, 1996); T. Schäfer and F. Wilczek, *Phys. Rev. D* **60**, 114033 (1999).
7. T. Holstein, R. E. Norton and P. Pincus, *Phys. Rev. B* **8**, 2649 (1973); S. Chakravarty, R. E. Norton and O. F. Syljuasen, *Phys. Rev. Lett.* **74**, 1423 (1995); A. Gerhold, A. Ipp and A. Rebhan, *Phys. Rev. D* **70**, 105015 (2004). A. Schäfer and K. Schwenzer, *Phys. Rev. D* **70**, 054007, 114037 (2004).
8. J. M. Radcliffe, *J. Phys. A* **4**, 313 (1971); J. R. Klauder, *Phys. Rev. D* **19**, 2349 (1979).
9. C. Herring, *Phys. Rev.* **85**, 1003 (1952); C. Herring, *Exchange Interactions among Itinerant Electrons: Magnetism IV* (Academic press, New York, 1966).
10. G. Baym and S. A. Chin, *Nucl. Phys. A* **262**, 527 (1976).
11. T. Tatsumi and K. Sato, in preparation.
12. N. Karchev, *J. Phys.: Condens. Matter* **15**, L385 (2003); D. Fay and J. Appel, *Phys. Rev. B* **22**, 3173 (1980).

ASYMMETRIC NEUTRINO EMISSION IN QUARK MATTER AND PULSAR KICKS

I. SAGERT and J. SCHAFFNER-BIELICH

Institute for Theoretical Physics, Goethe University, 60438 Frankfurt, Germany
E-mail: sagert@th.physik.uni-frankfurt.de

We derive conditions for an asymmetric neutrino emission of hot neutron stars with quark matter as a source for the observed large kick velocities of pulsars out of supernova remnants. We work out in detail the constraints for the initial temperature, the strength of the magnetic field and the electron chemical potential in the quark matter core. Also the neutrino mean free paths for quark matter and a possible hadronic mantle are considered for a successful kick mechanism. Heat capacities and neutrino emissivities in magnetised hot quark matter are investigated to delineate the maximum possible kick velocity. In addition, the influence of colour superconducting quark matter is taken into consideration as well. In a first study we find that ignoring neutrino quark scattering kick velocities of 1000 km/s can be reached very easily for quark phase radii smaller than 10 km and temperatures higher than 5 MeV. On the other hand, taking into account the small neutrino mean free paths it seems impossible to reach velocities higher than approximately 100 km/s, even when including effects from colour superconductivity where the neutrino quark interactions are suppressed.

EFFECTS OF THE TRANSITION OF NEUTRON STARS TO QUARK STARS ON THE COOLING

T. NODA* and M. HASHIMOTO

Department of Physics, Kyushu University,
Fukuoka-shi, Fukuoka 810-8560, Japan
**E-mail: tsune@gemini.rc.kyushu-u.ac.jp*

N. YASUTAKE

Science and Engineering, Waseda University,
Shinjuku-ku, Tokyo 169-8655, Japan

M. FUJIMOTO

Department of Physics, Hokkaido University,
Sapporo-shi, Hokkaido 060-0810, Japan

We investigate the thermal evolution of isolated neutron stars, including the transition of nuclear matter into quark matter at some time during the cooling stage. We show cooling curves by changing the transition periods in parametric manner. If there would be observational data which fit to our results, it suggests the existence of quark stars.

Keywords: Neutron Stars; Quark Stars; Cooling; Evolution.

1. Introduction

Neutron stars (NSs) are supported by the degenerated pressure against the gravity, and keep their masses about $1M_{\odot}$. NSs mainly consist neutron rich nuclear matter. Recently, studies of quark matter phase and/or quark gluon plasma are developed considerably. The quark matter phase may be realized in higher density and/or temperature region compared to normal nuclear matter of $\rho \sim 3 \times 10^{14}$ g cm^{-3}. It is believed that the existence of quark matter phase can be observed in high energy particle accelerators during the short period, in the early universe, or in the compact objects, such as NSs. Stars mainly composed of quark matter, are called quark stars (QSs), which properties are very similar to NSs. Therefore it is difficult to distinguish QS and NS. Method to find differences is to study the mass-radius relation, the differences of rotation properties, energy release at the phase transition, and cooling processes. We focus on the difference of cooling processes between compact stars, and investigate possible QS cooling scenarios.

2. Models

To calculate the cooling curves for the neutron star or quark star, we use the stellar evolutionary code which is assumed to be the hydrostatic equilibrium. This spherical symmetric code includes the general relativity.[1,2]

2.1. *Equation of state*

We consider two equations of state (EoSs) for each case of QS and NS. For NS, we choose Lattimer & Swesty EoS,[3] with the incompressibility $K = 220$ MeV for the high density region of $\rho > 10^8$ g cm^{-3}, and connect it to BPS EoS[4] for the low density region. For QS, we adopt the MIT bag model[5] with the bag constant $B^{1/4} = 174.5$ MeV which EoS is connected to NS EoS smoothly.

The maximum masses (radii) of our NS and QS models are $2.2M_{\odot}$ (10.7km) and $1.9M_{\odot}$ (11.6km), respectively. The central densities of these models are 2.3×10^{15}g cm^{-3} and 2.0×10^{15}g cm^{-3}, respectively.

2.2. *Compositions*

Surface compositions affect the NS cooling curves.[6] However, for simplicity, we choose ^{56}Fe as the surface composition. If we take into account ^{4}He, the opacity becomes smaller. Therefore the heat energy of the star can escape easily by the photon emission from the surface, and the cooling becomes slow at the beginning ($t < 10^4$ yr), but fast in the late phase.

2.3. *Neutrino emission processes*

The mean free path of neutrinos in NS ($\rho \sim 10^{14-15}$ g cm^{-3}) is about 10^5 km,[7] which is much larger than the NS or QS radius (~ 10 km). Therefore neutrinos easily escape from the star without any interactions with the NS/QS interior matter, and they carry internal heat energy off from the star.

We consider two models of neutrino emission process corresponding to the EoS of the star, the standard cooling for NS and the non-standard cooling for QS. The standard cooling model contains well-known modified URCA process and bremsstrahlung. The QS cooling model assumes simple quark β-decay process,[8] which includes u and d quarks in quark phase, and the standard cooling process in the baryonic phase. We plot the density dependence of each neutrino emission rate in Fig. 1. The quark cooling process becomes the most efficient cooling process beyond 10^{15} g cm^{-3}.

3. Scenarios

We consider two scenarios of QS cooling. In the first scenario, the star is assumed to be born as a QS, and cools through quark β-decay. In the second scenario, the star is born as NS after the supernova explosion, and evolves as NS with the standard

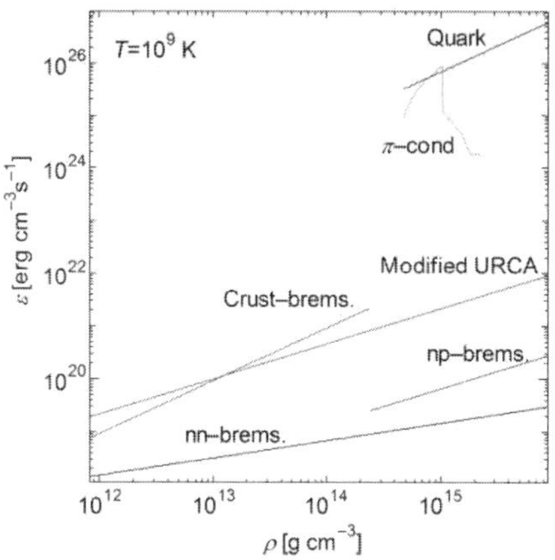

Fig. 1. Neutrino emission rates at $T = 10^9$ K. Quark: quark β-decay, π-cond: pion condensation, Crust-brems.: bremsstrahlung in crust, nn, np-brems.: bremsstrahlung of nucleons.

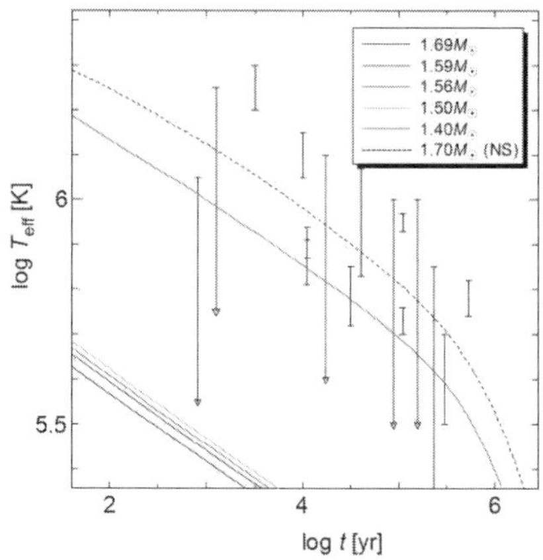

Fig. 2. Effective temperature measured from the time of the birth of NS/QS cooling curves of scenario 1. Observational data are indicated by the upper limits and/or lower limits of error bars.

cooling. Then the phase transition to QS occurs during some epoch of the evolution history.

For scenario 1, we use MIT bag model EoS connected to Lattimer & Swesty EoS, with quark β-decay process included. We take the mass of the star as a free parameter. We change the mass from $M = 1.40M_\odot$ to $M = 1.69M_\odot$, and calculate each cooling curve as seen in Fig. 2.

For scenario 2, we assume the mass to be $M = 1.7M_\odot$ and the choose parametric delay time to the phase transition (t_{QS}), which means the time between supernovae explosion and QS phase transition. We take t_{QS} artificially between 10 and 10^4 yr. t_{QS} depends on a mechanism of phase transition, which would be supercooling at the phase transition from baryonic to quark matter. only increases the mass of the star.

4. Results and Discussions

In scenario 1, we have calculated QS cooling curves. Fig. 2 compares the results with the observed effective temperatures of isolated neutron stars.[9,10] It is clear that cooling curves are significantly different at $1.4M_\odot < M < 1.5M_\odot$, due to the existence of quark core. Stars with the mass $M \geq 1.5M_\odot$ seem too cold compared with observations. Since simple quark β-decay we choose is too strong, we may need to include some suppression due to colour superconductivity.

In scenario 2, we have investigated cooling curves, where the transition occurs from NS to QS. In Fig. 3, we have compared the cooling curves with observed values.[9,10] Considering the transition delay time t_{QS}, we can make some variety of cooling curves, even if the mass is the same. In this scenario, we have the same prob-

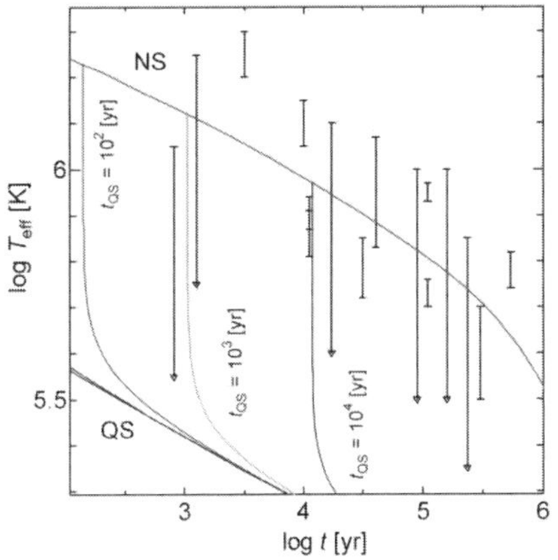

Fig. 3. Same as Fig. 2 except for the NS–QS transition cooling curves of scenario 2. t_{QS} is the beginning of the transition after the supercooling.

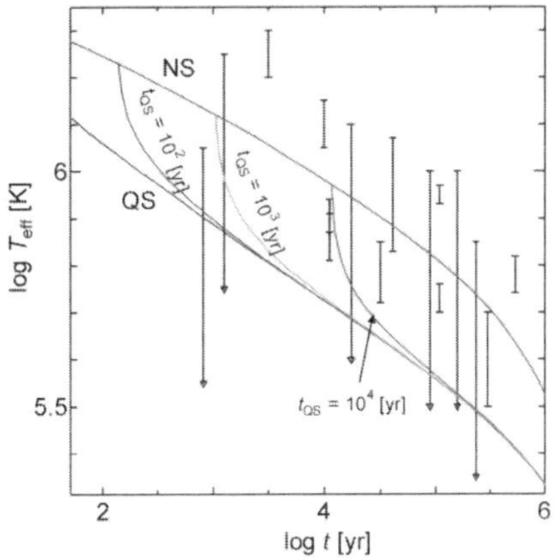

Fig. 4. Same as Fig. 3 except that the quark β-decay neutrino emission rate has been artificially reduced by a factor of 1/1000.

lem with the scenario 1: QS cooling is too strong. Thus, we need some suppression of the quark β-decay. There are also another problems, the triggering mechanism of transition and the energy release at the transition. The first one could be explained by supercooling, but it is not known whether the state can remain about thousands years after the formation of NS. For the second one, we have ignored the energy release, but it may not be realistic.

Also, we have calculated the cooling curves with the reduced quark β-decay. The neutrino emission rate is multiplied by 1/1000 artificially as seen in Fig. 4. If colour superconductivity reduces the rate up to this factor, the QS cooling can be consistent with the observation, and the existence of quark stars becomes more likely.

References

1. K. S. Thorne, *Astrophys. J.* **212**, 825 (1977).
2. M. Y. Fujimoto, T. Hanawa, I. Iben Jr. and M. R. Richardson, *Astrophys. J.* **278**, 813 (1984).
3. J. M. Lattimer and F. D. Swesty, *Nucl. Phys. A.* **535**, 331 (1992).
4. G. Baym, C. Pethick and P. Sutherland, *Astrophys. J.* **170**, 299 (1971).
5. N. Yasutake, K. Kotake, M. Hashimoto and S. Yamada, *Phys. Rev. D* **75**, 084012 (2007).
6. T. Noda, M. Hashimoto and M. Fujimoto, *Proceeding of Science* **PoS(NIC-IX)**, 153 (2006).
7. S. Shapiro and S. Teukolsky, *Black Holes, White Dwarfs and Neutron Stars* (Wiley-Interscience, 1983).
8. N. Iwamoto, *Phys. Rev. Lett.* **44**, 1637 (1980).
9. D. P. Page, `http://www.astroscu.unam.mx/neutrones/NS-Data/data_good.dat`.
10. P. O. Slane, D. J. Helfand and S. S. Murray, *Astrophys. J. Lett.* **571** L45 (2002).

THE ENERGY RELEASE – STELLAR ANGULAR MOMENTUM INDEPENDENCE IN ROTATING COMPACT STARS UNDERGOING FIRST-ORDER PHASE TRANSITIONS

M. BEJGER,[1,2] J. L. ZDUNIK,[2] P. HAENSEL[2] and E. GOURGOULHON[1]

[1] *LUTh, Observatoire de Paris, CNRS, Université Paris Diderot, 5 Place Jules Janssen, 92190 Meudon, France*

[2] *N. Copernicus Astronomical Center, Polish Academy of Sciences, Bartycka 18, PL-00-716 Warszawa, Poland*

E-mail: bejger@camk.edu.pl, jlz@camk.edu.pl, haensel@camk.edu.pl, eric.gourgoulhon@obspm.fr

We present the general relativistic calculation of the energy release associated with a first order phase transition (PT) at the center of a rotating neutron star (NS). The energy release, $E_{\rm rel}$, is equal to the difference in mass-energies between the initial (normal) phase configuration and the final configuration containing a superdense matter core, assuming constant total baryon number and the angular momentum. The calculations are performed with the use of precise pseudo-spectral 2-D numerical code; the polytropic equations of state (EOS) as well as realistic EOSs (Skyrme interactions, Mean Field Theory kaon condensate) are used. The results are obtained for a broad range of metastability of initial configuration and size of the new superdense phase core in the final configuration. For a fixed "overpressure", $\delta\overline{P}$, defined as the relative excess of central pressure of a collapsing metastable star over the pressure of the equilibrium first-order PT, the energy release up to numerical accuracy *does not depend* on the stellar angular momentum and coincides with that for nonrotating stars with the same $\delta\overline{P}$. When the equatorial radius of the superdense phase core is much smaller than the equatorial radius of the star, analytical expressions for the $E_{\rm rel}$ can be obtained: $E_{\rm rel}$ is proportional to $(\delta\overline{P})^{2.5}$ for small $\delta\overline{P}$. At higher $\delta\overline{P}$, the results of 1-D calculations of $E_{\rm rel}(\delta\overline{P})$ for non-rotating stars reproduce with very high precision exact 2-D results for fast-rotating stars. The energy release-angular momentum independence for a given overpressure holds also for the so-called "strong" PTs (that destabilise the star against the axi-symmetric perturbations), as well as for PTs with "jumping" over the energy barrier.

Keywords: Dense matter; Equation of state; Neutron stars; Rotation.

1. Introduction

Many theories of dense matter predict that at some density larger than the nuclear saturation density, a phase transition (PT) to some "exotic" state (i.e. boson condensate or quark deconfinement) occurs; for a review see e.g., Refs. 1–3. A first-order PTs are particularly interesting from the astrophysical and observational point of view, because are associated with a meta-stable state of dense matter. One can thus expect, in the case of PTs occurring in the interior of NSs, the release of non-negligible amount of energy. Here we will focus on the basic features of the

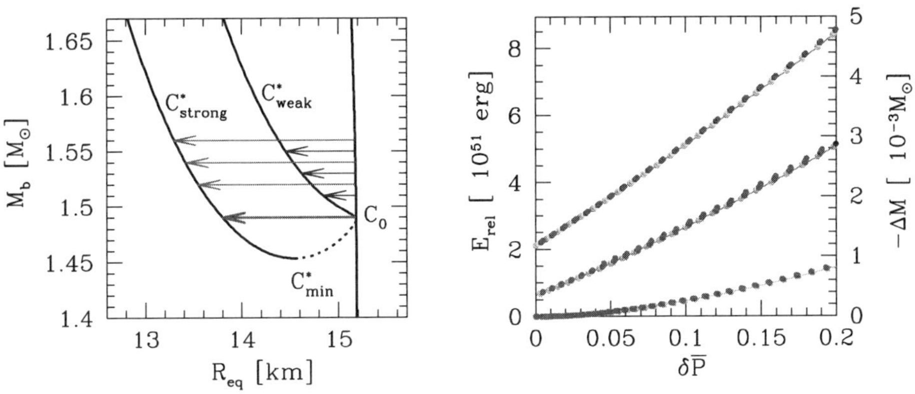

Fig. 1. Left panel: Strong and weak PTs on mass-radius diagram (dotted line marks unstable configurations). Right panel: the energy release $E_{\rm rel}$ vs the over-pressure parameter $\delta\bar{P}$ for strong (two upper curves) and weak (lowest curve) PTs. Points are colored differently for different total angular momenta $J = (0, 0.1, ..., 1.3) \times GM_\odot^2/c$ and follow the curve for the $J = 0$ (non-rotating) configurations.

energy release-angular momentum independence; for complete description we refer the reader to Refs. 4,5. The text is arranged as follows: in Sect. 2 we briefly describe the results of calculations. Sect. 3 contains conclusions and open questions.

2. Calculation of the Energy Release

The hydrostatic, axi-symmetric and rigidly rotating configurations of compact stars with and without PTs were obtained using the numerical GR library LORENE (`http://www.lorene.obspm.fr`). We assume that the baryon mass $M_{\rm b}$ and the total angular momentum J are constant during the PT from the metastable configuration $\mathcal{C}$ to stable configuration $\mathcal{C}^*$. From the microscopic point of view the first-order PT is characterised by the over-pressure parameter $\delta\bar{P} = (P_c - P_0)/P_0$, where P_c is the central pressure of the configuration $\mathcal{C}$ and P_0 is the equilibrium pressure, at which the PT occurs. We distinguish between the so-called *weak* and *strong* PTs, characterised by the density-jump parameter $\lambda = \rho^*/\rho$, where ρ and ρ^* are the densities of normal and condensed phase at P_0. PTs with $\lambda > \frac{3}{2}(1 + P_0/\rho c^2)$ (strong PTs) destabilise the stars with arbitrarily small cores against the axi-symmetric perturbations.[6–8] The energy release (energy difference) is defined as

$$E_{\rm rel} = c^2[M(\mathcal{C}) - M(\mathcal{C}^*)]_{M_{\rm b},J} \tag{1}$$

and it is quite remarkable that (up to numerical accuracy) it does not depend on the rotation state of the configuration i.e. on the total angular momentum J of the star and agrees well with the $E_{\rm rel}$ for non-rotating stars. In Fig. 1 we present the results for the polytropes (parametric EOSs in the form of $P = Kn^\Gamma$, where n is the number baryon density, K is is the pressure coefficient and Γ is called the adiabatic index; detailed description of parameters used can be found in Ref. 5.

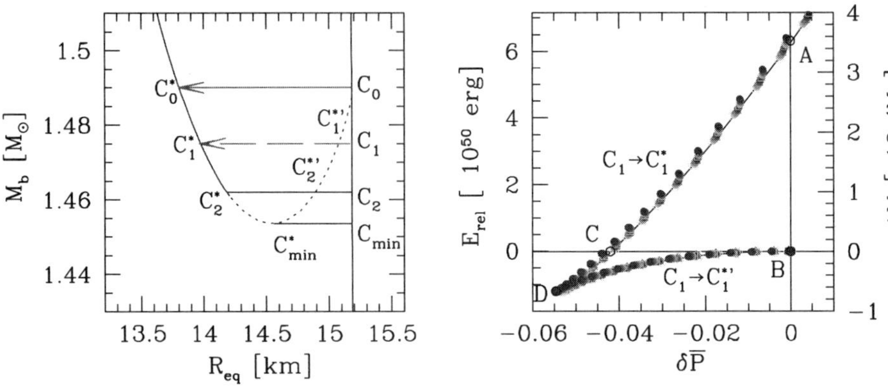

Fig. 2. Left panel: The strong PT with the "jumping" over the energy barrier on the mass-radius diagram. Right: Energy release $E_{\rm rel}$ vs the over-pressure parameter $\delta\bar{P}$ for the strong PT. Points with different colors correspond to PTs with different angular momentum $J = (0, 0.1, \ldots, 0.9) \times GM_{\odot}^2/c^2$ also for negative $\delta\bar{P}$. Solid line - result for $J = 0$ (non-rotating stars). Point A corresponds to $\mathcal{C}_0 \longrightarrow \mathcal{C}_0^\star$, point B to $\mathcal{C}_0 \longrightarrow \mathcal{C}_0^{\star\prime} = \mathcal{C}_0$, point C to $\mathcal{C}_2 \longrightarrow \mathcal{C}_2^\star$ and point D to $\mathcal{C}_{\rm min} \longrightarrow \mathcal{C}_{\rm min}^\star$.

In the case of realistic EOSs the results are qualitatively the same. We have checked the phenomenon for different realistic EOSs i.e. Skyrme interactions or the kaon condensate, presented as an example here. The crust obeys the EOS of Douchin & Haensel.[9] The matter below the PT is described using the relativistic mean-field theory.[10] The dense phase is the kaon condensate, with coupling of kaons to nucleons proposed by Glendenning & Schaffner-Bielich,[11] the optical potential $U_{\rm K}^{\rm lin}$ being equal to -115 MeV. The stability of configurations with P_c below that for the equilibrium PT is particularly interesting. If the star is excited initially, e.g. is pulsating, then the formation of a large dense phase core is possible, but it requires climbing ("jumping") over the energy barrier associated with formation of a small core. The results are shown in Fig. 2.

3. Conclusions and Open Questions

Our numerical calculations show the $E_{\rm rel}(J)$ independence during the weak as well as strong first-order PTs, for large rotation rates and large oblatnesses of the stars: it is therefore not an property of slow rotating stars only. The independence holds also for PTs with negative over-pressure $\delta\bar{P}$ i.e. when the configuration "jumps" over the energy barrier to reach another stable configuration. For small positive $\delta\bar{P}$ analytical relations were found: $E_{\rm rel} \propto (\delta\overline{P})^{2.5}$. The energy release $E_{\rm rel} \sim 10^{51} - 10^{52}$ erg is an absolute upper bound on the energy which can released in such PT. In astrophysical situation, the energy can be distributed between stellar pulsations, gravitational radiation, heating of stellar interior, X-ray emission from the neutron star surface, and even a gamma-ray burst.

Currently there is no mathematical proof of the J-independence of $E_{\rm rel}$. It may be interesting to look at the problem from the "thermodynamical" point of view. We write the total energy of the star (gravitational mass M) as

$$M = \frac{\mu}{u^t} A + 2\,\Omega J + 2\int_{\Sigma_t} P\, N \sqrt{\gamma}\, d^3x, \tag{2}$$

where μ is the baryon chemical potential, u^t the time component of the fluid 4-velocity, A the number of baryons in the star, Ω the angular velocity, P the fluid pressure, N the lapse function and $\sqrt{\gamma}\, d^3x$ the covariant volume element in the constant t hypersurface Σ_t. Eq. (2) is valid for any axi-symmetric stationary and rigidly rotating fluid star which obeys a barotropic EOS, as established by Bardeen & Wagoner in 1971.[12] Since $\mathcal{C}$ and $\mathcal{C}^*$ have the same baryon number A and the same angular momentum J, we get the energy release

$$E_{\rm rel} = \Delta E = A\,\Delta\Big(\frac{\mu}{u^t}\Big) + 2\,J\,\Delta\Omega + 2\left[\int_{\mathcal{C}} P\, N \sqrt{\gamma}\, d^3x - \int_{\mathcal{C}^*} P\, N \sqrt{\gamma}\, d^3x\right], \tag{3}$$

with

$$\Delta\Big(\frac{\mu}{u^t}\Big) := \frac{\mu}{u^t}\Big|_{\mathcal{C}} - \frac{\mu}{u^t}\Big|_{\mathcal{C}^*}, \qquad \Delta\Omega := \Omega|_{\mathcal{C}} - \Omega|_{\mathcal{C}^*}\,. \tag{4}$$

The terms on the right-hand-side of Eq. (3) are of comparable magnitude in their contribution to the energy release, so the possible "smallness" of some of them in relation to the others is not responsible for the $E_{\rm rel}(J)$ independence. We will address the problem of mathematically proving this property in the near future.

Acknowledgements

This work was partially supported by the Polish MNiI grant no. 1P03D.008.27, MNiSW grant no. N203.006.32/0450 and by the LEA Astro-PF programme. MB was also supported by the Marie Curie Fellowship no. MEIF-CT-2005-023644.

References

1. P. Haensel, A. Y. Potekhin and D. G. Yakovlev, *Neutron Stars 1: Equation of State and Structure* (Springer, New York, 2007).
2. N. K. Glendenning, *Compact Stars: Nuclear Physics, Particle Physics and General Relativity* (Springer, New York, 1997).
3. F. Weber, *Pulsars as Astrophysical Laboratories for Nuclear and Particle Physics* (IoP Publishing, Bristol & Philadelphia, 1999).
4. J. L. Zdunik, M. Bejger, P. Haensel and E. Gourgoulhon, *Astron. Astrophys.* **465**, 533 (2007).
5. J. L. Zdunik, M. Bejger, P. Haensel and E. Gourgoulhon, *ArXiv e-prints* **707** (2007).
6. Z. F. Seidov, *Soviet Astronomy* **15**, 347 (1971).
7. B. Kaempfer, *Phys. Lett. B* **101**, 366 (1981).
8. J. L. Zdunik, P. Haensel and R. Schaeffer, *Astron. Astrophys.* **172**, 95 (1987).
9. F. Douchin and P. Haensel, *Astron. Astrophys.* **380**, 151 (2001).
10. J. Zimanyi and S. A. Moszkowski, *Phys. Rev. C* **42**, 1416 (1990).
11. N. K. Glendenning and J. Schaffner-Bielich, *Phys. Rev. C* **60**, 025803 (1999).
12. J. M. Bardeen and R. V. Wagoner, *Astrophys. J.* **167**, 359 (1971).

HYPERON-QUARK MIXED PHASE IN DENSE MATTER

TOSHIKI MARUYAMA* and SATOSHI CHIBA

Advanced Science Research Center, Japan Atomic Energy Agency,
Tokai, Ibaraki 319-1195, Japan
**E-mail: maruyama.toshiki@jaea.go.jp*

HANS-JOSEF SCHULZE

INFN Sezione di Catania, Via Santa Sofia 64, Catania I-95123, Italy

TOSHITAKA TATSUMI

Department of Physics, Kyoto University, Kyoto 606-8502, Japan

We investigate the properties of the hadron-quark mixed phase in compact stars using a Brueckner-Hartree-Fock framework for hadronic matter and the MIT bag model for quark matter. We find that the equation of state of the mixed phase is similar to that given by the Maxwell construction. The composition of the mixed phase, however, is very different from that of the Maxwell construction; in particular, hyperons are completely suppressed.

Keywords: Neutron star; Mixed phase; Hyperon mixture; Quark matter; Pasta structure.

1. Introduction

Matter in neutron stars (NS) has a variety of density and chemical component due to the presence of gravity. At the crust of neutron stars, there exists a region where the density is lower than the normal nuclear density, $\rho_0 \approx 0.17 \text{ fm}^{-3}$ over a couple of hundreds meters. The pressure of such matter is mainly contributed by degenerate electrons, while baryons are clusterized and have little contribution. Due to the gravity pressure and density increase in the inner region (in fact, the density at the center amounts to several times ρ_0). Cold catalyzed matter consists of neutrons and the equal number of protons and electrons under chemical equilibrium. Since the kinetic energy of degenerate electrons is much higher than that of baryons, the electron fraction (or the proton one) decreases with increase of density and thus neutrons become the main component and drip out of the clusters. In this way baryons come to contribute to the pressure as well as electrons. At a certain density, other components such as hyperons and strange mesons may emerge. At even higher density, hadron-quark deconfinement transition may occur and quarks in hadrons are liberated.

It is well known that hyperons appear at several times ρ_0 and lead to a strong softening of the EOS with a consequent substantial reduction of the maximum neutron star mass. Actually the microscopic Brueckner-Hartree-Fock approach gives much lower masses than current observation values of $\sim 1.4 M_\odot$.

On the other hand, the hadron-quark deconfinement transition is believed to occur in hot and/or high-density matter. Taking EOS of quark matter within the MIT bag model, the maximum mass can increase to the Chandrasekhar limit once the deconfinement transition occurs in hyperon matter.[1,2]

The deconfinement transition from hadron to quark phase may occur as a first-order phase transition. There, a hadron-quark mixed phase should appear, where charge density as well as baryon number density is no more uniform. Owing to the interplay of the Coulomb interaction and the surface tension, the mixed phase can have exotic shapes called pasta structures.[3]

Generally, the appearance of mixed phase in matter results in a softening of the EOS. The bulk Gibbs calculation (BG) of the mixed phase, without the effects of the Coulomb interaction and surface tension, leads to a broad density region of the mixed phase (MP).[4] However, if one takes into account the geometrical structures in the mixed phase and applies the Gibbs conditions, one may find that MP is considerably limited and thereby the EOS approaches to the one given by the Maxwell construction (MC).[3,5,6]

In this report we explore the EOS and the structure of the mixed phase during the hyperon-quark transition, properly taking account of the Gibbs conditions together with the pasta structures.

2. Numerical Calculation

The numerical procedure to determine the EOS and the geometrical structure of the MP is similar to that explained in detail in Ref. 3. We employ a Wigner-Seitz approximation in which the whole space is divided into equivalent Wigner-Seitz cells with a given geometrical symmetry, sphere for three dimension (3D), cylinder for 2D, and slab for 1D. A lump portion made of one phase is embedded in the other phase and thus the quark and hadron phases are spatially separated in each cell. A sharp boundary is assumed between the two phases and the surface energy is taken into account in terms of a surface-tension parameter σ. The energy density of the mixed phase is thus written as

$$\epsilon = \frac{1}{V_W}\left[\int_{V_H} d^3r \epsilon_H(\boldsymbol{r}) + \int_{V_Q} d^3r \epsilon_Q(\boldsymbol{r}) + \int_{V_W} d^3r \left(\epsilon_e(\boldsymbol{r}) + \frac{(\nabla V_C(\boldsymbol{r}))^2}{8\pi e^2}\right) + \sigma S\right], \tag{1}$$

where the volume of the Wigner-Seitz cell V_W is the sum of those of hadron and quark phases V_H and V_Q, S the quark-hadron interface area. ϵ_H, ϵ_Q and ϵ_e are energy densities of hadrons, quarks and electrons, which are functions of local densities $\rho_a(\boldsymbol{r})$ $(a = n, p, \Lambda, \Sigma^-, u, d, s, e)$. The Coulomb potential V_C is obtained by solving the Poisson equation. For a given density ρ_B, the optimum dimensionality of the cell, the cell size R_W, the lump size R, and the density profile of each component

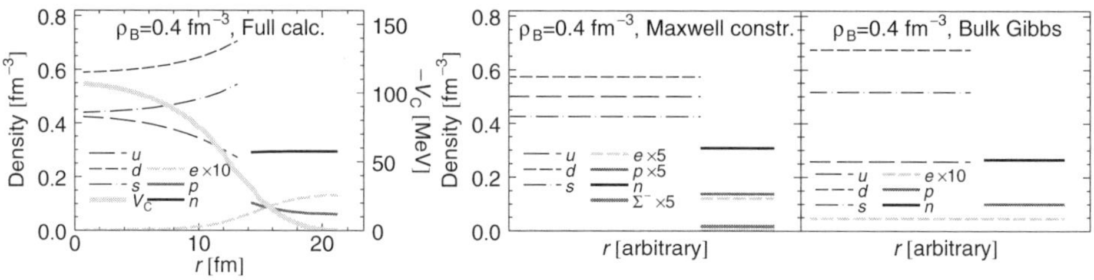

Fig. 1. Left: Density profiles and Coulomb potential V_C within a 3D (quark droplet) Wigner-Seitz cell of the MP at $\rho_B = 0.4$ fm^{-3}. The cell radius and the droplet radius are $R_W = 26.7$ fm and $R = 17.3$ fm, respectively. Center: Same as the left panel for the MC case. The radius r is in arbitrary units since there is no specific size. Right: The case of the BG calculation.

are searched for to give the minimum energy density. We employ $\sigma = 40$ MeV/fm^2 in the present study.

To calculate ϵ_H in the hadron phase, we use the Thomas-Fermi approximation for the kinetic energy density. The potential-energy density is calculated by the nonrelativistic BHF approach[1] based on microscopic NN and NY potentials.

$$\epsilon_H = \sum_{i=n,p,\Lambda,\Sigma^-} \sum_{k<k_F^{(i)}} \left[T_i(k) + \frac{1}{2} U_i(k) \right] , \tag{2}$$

$$U_i(k) = \sum_{j=n,p,\Lambda,\Sigma^-} U_i^{(j)}(k), \tag{3}$$

$$U_i^{(j)}(k) = \sum_{k'<k_F^{(j)}} \mathrm{Re}\langle kk'|G_{(ij)(ij)}\left[E_{(ij)}(k,k')\right]|kk'\rangle, \tag{4}$$

$$G_{ab}[W] = V_{ab} + \sum_c \sum_{p,p'} V_{ac}|pp'\rangle \frac{Q_c}{W - E_c + i\epsilon} \langle pp'|G_{cb}[W]. \tag{5}$$

The interaction parameters are chosen to reproduce the scattering phase shifts. Nucleonic three-body forces are included in order to (slightly) shift the saturation point of purely nucleonic matter to the empirical value.

For the quark phase, we use the MIT bag model with massless u and d quarks and massive s quark with $m_s = 150$ MeV. The energy density ϵ_Q consists of the kinetic term by the Thomas-Fermi approximation, the leading-order one-gluon-exchange term[7] proportional to the QCD fine structure constant α_s, and the bag constant B. Here we use $B = 100$ MeV/fm^3 and $\alpha_s = 0$ to get the quark EOS which crosses the hadronic one at an appropriate baryon density.

3. Hadron-Quark Mixed Phase

Figure 1 illustrates an example of the density profile in a 3D cell. One can see the non-uniform density distribution of each particle species together with the finite Coulomb potential; charged particle distributions are rearranged to screen the

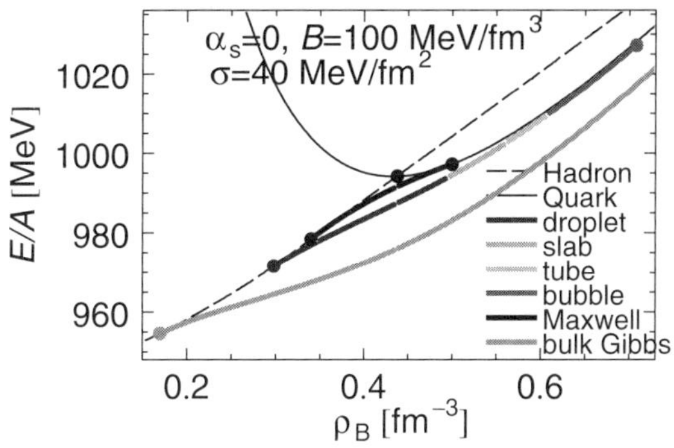

Fig. 2. EOS of the MP (thick curves) in comparison with pure hadron and quark phases (thin curves). Each segment of the MP is chosen by minimizing the energy.

Coulomb potential. The quark phase is negatively charged, so that d and s quarks are repelled to the phase boundary, while u quarks gather at the center. The protons in the hadron phase are attracted by the negatively charged quark phase, while the electrons are localized to the hadron phase. This density rearrangement of charged particles causes the screening of the Coulomb interaction between phases.

In the center and right panels, compared are the cases of MC and BG. MC assumes the local charge neutrality, while the BG does not. One can see that the local charge neutrality in the full calculation lies between two cases. The localization of electrons which is one of the charge screening effects, reduces the local charge density. But the local charge density remains still finite to some extent.

Figure 2 compares the resulting EOS with that of the pure hadron and quark phases. The thick black curve indicates the case of the MC, while the colored lines indicate the MP in its various geometric realizations starting with a quark droplet structure and terminating with a bubble structure. Note that the charge screening effect always tends to make matter locally charge neutral to save the Coulomb interaction energy. Hence, combined with the surface tension, it makes the non-uniform structures mechanically less stable and limits the density region of the MP.[3] Consequently the energy of the MP is only slightly lower than that of the MC.

However, the structure and the composition of the MP are very different from those of the MC, which is demonstrated in Fig. 3, where we compare the particle fractions as a function of baryon density in the full calculation (left panel) and the MC (right panel). One can see that the compositions are very different in two cases. In particular, a relevant hyperon (Σ^-) fraction is only present in the MC.

The suppression of hyperon mixture in the MP is due to the absence of the charge-neutrality condition in each phase. As shown in Fig. 4, hyperons (Σ^-) appear in charge-neutral hadronic matter at low density (0.34 fm^{-3}) to reduce the Fermi energies of electron and neutron. Without the charge-neutrality condition, on the other hand, there appears symmetric nuclear matter at lower density and

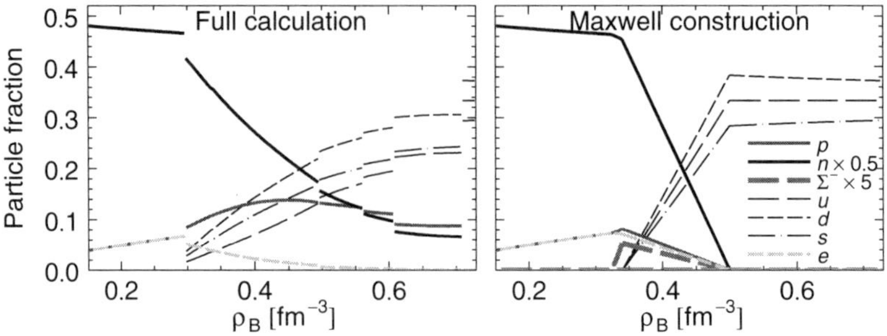

Fig. 3. Particle fractions in the MP by the full calculation (left panel) and the MC (right panel).

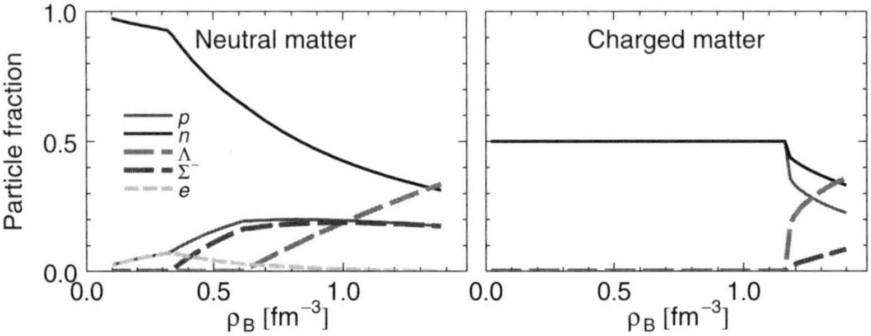

Fig. 4. Left panel: Particle fractions of neutral matter with electrons (corresponding to neutron star matter). Right panel: The same quantity for charged matter without electrons, the low-density part of which corresponds to symmetric nuclear matter. Both cases require chemical-equilibrium condition.

hyperons will be mixed above 1.15 fm^{-3} due to the large hyperon masses. The MP has positively charged hadron phase and negative quark phase though the Coulomb screening effect diminishes the local charge density. This brings the feature of symmetric nuclear matter into the hadron phase and, consequently, the mixture of hyperons is suppressed.[8]

4. Neutron Star Structure

Having the EOS comprising hadronic, mixed, and quark phase in the form $P(\epsilon)$, the equilibrium configurations of static NS are obtained in the standard way by solving the Tolman-Oppenheimer-Volkoff (TOV) equations[9] for the pressure $P(r)$ and the enclosed mass $m(r)$,

$$\frac{dP}{dr} = -\frac{Gm\epsilon}{r^2} \frac{(1+P/\epsilon)\left(1+4\pi r^3 P/m\right)}{1-2Gm/r}\,, \tag{6}$$

$$\frac{dm}{dr} = 4\pi r^2 \epsilon\,, \tag{7}$$

being G the gravitational constant. Starting with a central mass density $\epsilon(r=0) \equiv \epsilon_c$, one integrates out until the surface density equals the one of iron. This gives the

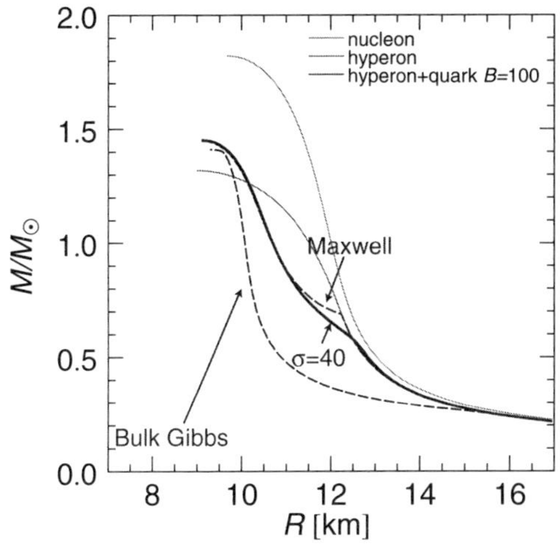

Fig. 5. Neutron star mass-radius relations for different EOS and three different hadron-quark phase transition constructions. For the hybrid stars (blue curves), the dashed lines indicate the Maxwell (upper curves) or bulk Gibbs (lower curves) constructions and the solid line the mixed phase of the full calculation.

stellar radius R and its gravitational mass $M = m(R)$. For the description of the NS crust, we have joined the hadronic EOS with the ones by Negele and Vautherin[10] in the medium-density regime, and the ones by Feynman-Metropolis-Teller[11] and Baym-Pethick-Sutherland[12] for the outer crust.

Fig. 5 compares the mass-radius relations obtained with the different models. The purely nucleonic EOS (green curve) yields a maximum NS mass of about $1.82\,M_\odot$, which is reduced to $1.32\,M_\odot$ when allowing for the presence of hyperons (red curve). This feature has been shown to be fairly independent of the nucleonic and hyperonic EOS that are used.[13] The canonical NS with mass of about $1.4\,M_\odot$ can therefore not be purely hadronic stars in our approach. In fact, the inclusion of quark matter augments the maximum mass of hybrid stars to about $1.5\,M_\odot$:

In general, the Maxwell construction leads to a kink in the $M(R)$ relation, because the transition from a hadronic to a hybrid star occurs suddenly, involving a discontinuous increase of the central density when the quark phase onsets in the core of a star. The bulk Gibbs calculation yields smooth mass-radius relations involving a continuous transition from a hadronic to a hybrid star beginning at rather low central density corresponding to very low NS mass.

The MP construction by the full calculation lies between the two extreme cases, and with our choice of $\sigma = 40$ MeV/fm^2 it is rather close to the Maxwell construction, smoothing out the kink of the hadron-hybrid star transition. This transition occurs generally at a fairly low NS mass, even below the natural minimum mass limit due to the formation via a protoneutron star[14] and is thus an unobservable feature.

Whereas the maximum masses are practically independent of the phase transi-

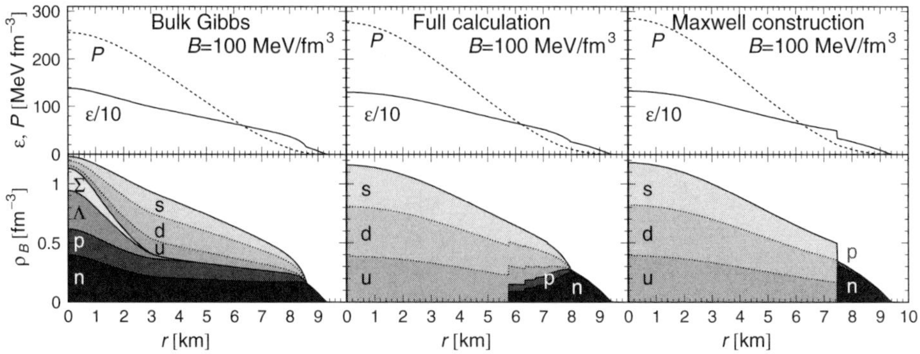

Fig. 6. Internal structure of a 1.4 $M_{\odot}$ neutron star obtained with three different phase transition constructions. The upper panels show total energy density and pressure and the lower panels the overall particle fractions as functions of the radial coordinate of the star, using the bulk Gibbs calculation (left panel), the mixed phase of the full calculation with $\sigma = 40$ MeV/fm^2 (central panel), and the Maxwell construction (right panel). In all cases $\alpha_s = 0$ and $B = 100$ MeV/fm^{-3} are used.

tion construction, there are evidently large differences for the internal composition of the star. This is illustrated in Fig. 6 which show the total energy density, pressure, and particle fractions as a function of the radial coordinate for a 1.4 $M_{\odot}$ NS. One observes with the bulk Gibbs construction (left panels) a coexistence of hadrons and quarks in a significant range of the star, whereas with the MC (right panels) an abrupt transition involving a discontinuous jump of energy and baryon density occurs at a distance $r \approx 7.5$ km from the center of the star. The small contamination with Σ^- hyperons in the hadronic phase is not visible on the scale chosen. The MP with the full calculation (central panels) lie between the two extreme cases, hadrons and quarks coexisting in a smaller range than in the bulk Gibbs cases.

5. Summary

In this article we have studied the properties of the mixed phase in the quark deconfinement transition in hyperonic matter, and their influence on compact star structure. The hyperonic EOS given by the BHF approach with realistic hadronic interactions is so soft that the transition density becomes very low if one uses the MIT bag model for the quark EOS.

The hyperon-quark mixed phase was consistently treated with the basic thermodynamical requirement due to the Gibbs conditions. We have seen that the resultant EOS is little different from the one given by the Maxwell construction. This is because the finite-size effects, the surface tension, and the Coulomb interaction tend to diminish the available density region through the mechanical instability, as has also been suggested in previous articles.[15,16]

For the bulk properties of compact stars, such as mass or radius, our EOS gives similar results as those given by the Maxwell construction. The maximum mass of a hybrid star is around $1.5\,M_\odot$, larger than that of the purely hyperonic star, $\approx 1.3\,M_\odot$. Hence we may conclude that a hybrid star is still consistent with the canonical NS mass of $1.4\,M_\odot$, while the masses of purely hyperonic stars lie below it.

On the other hand, the internal structure of the mixed phase is very different; e.g., the charge density as well as the baryon number density are nonuniform in the mixed phase. We have also seen that the hyperon number fraction is suppressed in the mixed phase due to the relaxation of the charge-neutrality condition, while it is always finite in the Maxwell construction. This has important consequences for the elementary processes inside compact stars. For example, coherent scattering of neutrinos off lumps in the mixed phase may enhance the neutrino opacity.[17] Also, the absence of hyperons prevents a fast cooling mechanism by way of the hyperon Urca processes.[18] These results directly modify the thermal evolution of compact stars.

References

1. M. Baldo, G. F. Burgio and H.-J. Schulze, *Phys. Rev. C* **58**, 3688 (1998); *Phys. Rev. C* **61**, 055801 (2000).
2. M. Baldo, G. F. Burgio and H.-J. Schulze, *Phys. Rev. C* **61**, 055801 (2000); O. E. Nicotra, M. Baldo, G. F. Burgio and H.-J. Schulze, *A & A* **451**, 213 (2006).
3. For a review, T. Maruyama, T. Tatsumi, T. Endo and S. Chiba, *Recent Res. Devel. Phys.* **7**, 1 (2006).
4. N. K. Glendenning, *Phys. Rev. D* **46**, 1274 (1992); *Phys. Rep.* **342**, 393 (2001).
5. T. Endo, T. Maruyama, S. Chiba and T. Tatsumi, *Prog. Theor. Phys.* **115**, 337 (2006).
6. T. Maruyama, T. Tatsumi, D. N. Voskresensky, T. Tanigawa and S. Chiba, *Phys. Rev. C* **73**, 035802 (2006).
7. E. Farhi and R. L. Jaffe, *Phys. Rev. D* **30**, 2379 (1984); M. S. Berger and R. L. Jaffe, *Phys. Rev. C* **35**, 213 (1987).
8. T. Maruyama, S. Chiba, H.-J. Schulze and T. Tatsumi, *Phys. Lett. B*, in press; *Phys. Rev. D*, in press.
9. S. L. Shapiro and S. A. Teukolsky, *Black Holes, White Dwarfs and Neutron Stars* (Wiley, New York, 1983).
10. J. W. Negele and D. Vautherin, *Nucl. Phys. A* **207**, 298 (1973).
11. R. P. Feynman, N. Metropolis and E. Teller, *Phys. Rev.* **75**, 1561 (1949).
12. G. Baym, C. Pethick and P. Sutherland, *Astrophys. J.* **170**, 299 (1971).
13. H.-J. Schulze, A. Polls, A. Ramos and I. Vidaña, *Phys. Rev. C* **73**, 058801 (2006).
14. O. E. Nicotra, M. Baldo, G. F. Burgio and H.-J. Schulze, *Astron. Astrophys.* **451**, 213 (2006); *Phys. Rev. D* **74**, 123001 (2006).
15. D. N. Voskresensky, M. Yasuhira and T. Tatsumi, *Phys. Lett. B* **541**, 93 (2002); D. N. Voskresensky, M. Yasuhira and T. Tatsumi, *Nucl. Phys. A* **723**, 291 (2003).
16. T. Tatsumi, M. Yasuhira and D. N. Voskresensky, *Nucl. Phys. A* **718**, 359 (2003); T. Endo, T. Maruyama, S. Chiba and T. Tatsumi, *Nucl. Phys. A* **749**, 333 (2005);
17. S. Reddy, G. Bertsch and M. Prakash, *Phys. Lett. B* **475**, 1 (2000).
18. M. Prakash, M. Prakash, J. M. Lattimer and C. J. Pethick, *Astrophys. J.* **390**, L77 (1992); T. Tatsumi, T. Takatsuka and R. Tamagaki, *Prog. Theor. Phys.* **110**, 179 (2003); T. Takatsuka, S. Nishizaki, Y. Yamamoto and R. Tamagaki, *Prog. Theor. Phys.* **115**, in press (2007).

NUCLEATION OF QUARK MATTER IN NEUTRON STARS: ROLE OF COLOR SUPERCONDUCTIVITY

IGNAZIO BOMBACI,[1] GERMÁN LUGONES[2] and ISAAC VIDAÑA[3]

[1] *Dipartimento di Fisica and INFN Sez. di Pisa, Italy*
[2] *CCNH, Universidade Federal do ABC, Santo André, SP, Brazil*
[3] *Dept. ECM, Universitat de Barcelona, Spain*

Pure hadronic compact stars ("neutron stars") above a critical mass $M_{\rm cr}$ are metastable[1,2] for the conversion to quark stars (hybrid or strange stars). This conversion process liberates an enormous amount of energy ($E_{\rm conv} \sim 10^{53}$ ergs), which could power some of the observed gamma ray bursts.[1–3] In cold deleptonized hadronic stars, the conversion process is triggered by the quantum nucleation of a quark matter drop in the stellar center. These drops can be made up of normal (i.e. unpaired) quark matter, or color superconducting quark matter, depending on the details of the equation of state of quark and hadronic matter.[4] In this talk, we present the results of recent calculations[5] of the effects of color superconductivity on the conversion of hadronic stars to quark stars. In particular, we study the dependence of the critical mass $M_{\rm cr}$ and conversion energy $E_{\rm conv}$ on the quark-quark pairing gap Δ, the bag constant B, and the surface tension σ of the quark-hadron interface.

References

1. Z. Berezhiani, I. Bombaci, A. Drago, F. Frontera and A. Lavagno, *Astrophys. J.* **586**, 1250 (2003).
2. I. Bombaci, I. Parenti and I. Vidaña, *Astrophys. J.* **614**, 314 (2004).
3. I. Bombaci and B. Datta, *Astrophys. J. Lett.* **530**, L72 (2000).
4. G. Lugones and I. Bombaci, *Phys. Rev. D* **72**, 065021 (2005).
5. I. Bombaci, G. Lugones and I. Vidaña, *Astron. & Astrophys.* **462**, 1017 (2007).

THE BULK VISCOSITY AND R-MODE INSTABILITY OF STRANGE QUARK MATTER

BASIL A. SA'D

Frankfurt International Graduate School for Science, J. W. Goethe-Universität, Max-von-Laue Strasse 1, D-60438 Frankfurt am Main, Germany
E-mail: sad@th.physik.uni-frankfurt.de

Compact stars are generically unstable against gravitational radiation emission from r-mode instabilities, but dissipative phenomena such as bulk and shear viscosities tend to dampen these instabilities, thus creating a window of stability. This may provide an opportunity to identify the phase of matter in such objects.[1,2] Here, bulk viscosity and r-mode instabilities of strange stars will be discussed for both normal and color-superconducting phases. The effect of Urca processes[3] will be emphasized.

References

1. N. Andersson and K. D. Kokkotas, "The r-mode instability in rotating neutron stars," *Int. J. Mod. Phys. D* **10**, 381 (2001) [arXiv:gr-qc/0010102].
2. J. Madsen, "Probing strange stars and color superconductivity by r-mode instabilities in millisecond pulsars," *Phys. Rev. Lett.* **85**, 10 (2000) [arXiv:astroph/9912418].
3. B. A. Sa'd, I. A. Shovkovy and D. H. Rischke, "Bulk viscosity of strange quark matter: Urca versus non-leptonic processes," submitted to *Phys. Rev. D* [arXiv:astro-ph/0703016].

NEUTRINO TRAPPING IN NEUTRON STARS IN THE PRESENCE OF KAON CONDENSATION

A. LI and W. ZUO

School of Physical Science and Technology, Lanzhou University, and Institute of Modern Physics, Chinese Academy of Sciences, Lanzhou 730000, P. R. China
E-mail: easylion@gmail.com, zuowei@impcas.ac.cn
www.impcas.ac.cn

G. F. BURGIO and U. LOMBARDO

INFN-LNS, Via Santa Sofia 44, I-95123 Catania, and Department of Physics and Astronomy, Catania University, Via Santa Sofia 64, I-95123, Italy
E-mail: fiorella.burgio@ct.infn.it, lombardo@lns.infn.it

We investigate the composition and the equation of state of the kaon condensed phase in neutrino-free and neutrino-trapped star matter within the framework of the Brueckner-Hartree-Fock approach with three-body forces. We find that neutrino trapping shifts the onset density of kaon condensation to a larger baryon density and reduces considerably the kaon abundance. As a consequence, when kaons are allowed, the equation of state of neutrino-trapped star matter becomes stiffer than the one of neutrino free matter. The effects of different three-body forces are compared and discussed. Neutrino trapping turns out to weaken the role played by the symmetry energy in determining the composition of stellar matter and thus reduces the difference between the results obtained by using different three-body forces.

Keywords: Dense nuclear matter; Nuclear symmetry energy; Kaon condensation; Neutrino trapping; Nuclear three-body force; Brueckner-Hartree-Fock approach.

1. Introduction

The effects of neutrino trapping in stellar matter and its implications for astrophysical phenomena were explored by several authors.[1,2] The equation of state (EOS) of the neutrino trapped stellar matter is particularly important for the evolution of a newly born neutron star (NS) and the mechanism of black hole formation.

The kaon condensation in neutrino-free NS matter has been studied[3] within the Brueckner-Hartree-Fock (BHF) approach,[4,5] and it was found that the composition of NS matter is sensitively dependent on the nuclear symmetry energy and the presence of kaons. In the present paper we shall extend the previous work to the study of cold neutrino trapped matter, using the BHF theory together with the nucleon three-body forces (TBF). Particularly, we shall mainly focus on the influence of different TBF on the global properties of proto-neutron stars (PNS) with kaons and the interplay between the roles played by the TBFs, neutrino trapping and

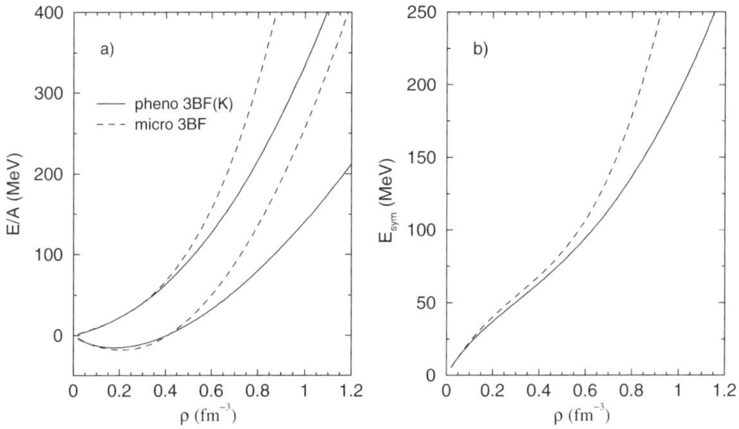

Fig. 1. The EoS is shown (panel (a)) for symmetric (lower curves) and pure neutron matter (upper curves). The symmetry energy as a function of the nucleon density is displayed in panel (b). See text for details.

kaon condensation. In the current work, thermal effects are neglected, although it has been found[2] that they could significantly affect the maximum mass of a PNS.

2. Results and Discussion

In Fig. 1, panel a), we show the equation of state for symmetric nuclear matter (lower curves) and pure neutron matter (upper curves). The solid lines represent the values obtained using the Urbana phenomenological TBF,[6,7] whereas the dashed lines are the values obtained by adopting the microscopic TBF.[8,9] The symmetry energy calculated in the BHF approach as a function of the nucleon density is also shown in panel b). We notice that the results of the two adopted TBFs show a similar behavior up to density $\rho \approx 0.4\ fm^{-3}$, but they differ a lot in the high density range. In particular, the microscopic TBF turns out to be more repulsive than the Urbana model at high densities, and the discrepancy between the two predictions becomes increasingly large as the density increases.

As to kaons, there is a well-known uncertainty about the attractive component of the kaon-nucleon interaction,[10] and we adopt the value of parameter $a_3 m_s$ in the range -310 MeV $< a_3 m_s <$ -134 MeV as done in Refs. 11,12, in order to investigate the sensitivity of our results to the variation. In Fig. 2 the particle population is plotted as a function of the baryon density for the neutrino-trapped matter. We notice that, because of trapped neutrinos, the onset of kaon condensation is shifted (about $0.2 fm^{-3}$) to larger densities than in neutrino free matter. Both TBF's produce very similar results, except for the less attractive case, i.e. $a_3 m_s =$ -134 MeV, where the onset for kaon condensation takes place at a lower density for the microscopic TBF. This is due to the fact that, at high density, the symmetry energy is larger, and this allows for a kaon condensed phase at baryon density lower than in the case with the Urbana TBF.

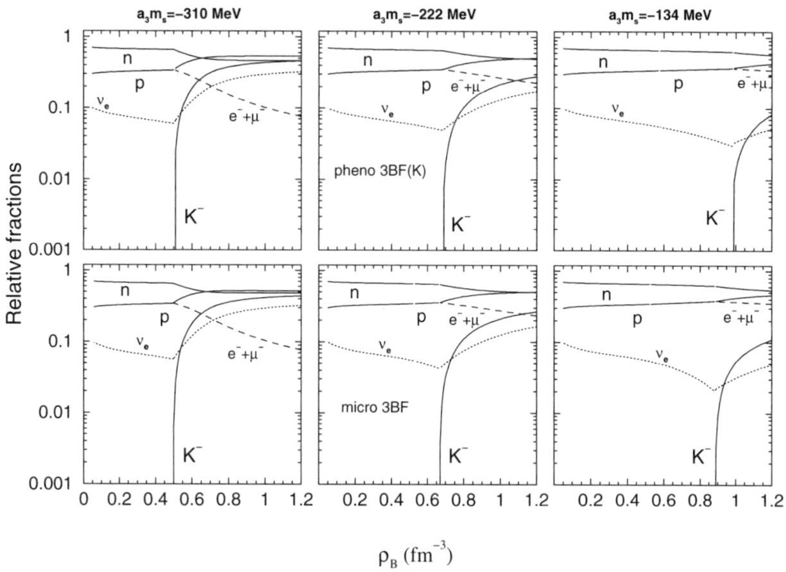

Fig. 2. The particle populations are shown as a function of the baryon density for neutrino-trapped neutron star matter with kaons, for three different values of $a_3 m_s$. In the upper (lower) panels results are displayed for the case when the phenomenological Urbana (microscopic) TBF is used.

Once the relative particle concentrations are known, we can calculate the equation of state, then finally the gravitational mass of the star can be obtained by solving the TOV equations.[13] This is shown in Fig. 3. In the purely nucleonic case, neutrino trapping generally produces a softer equation of state, because beta-stable matter turns out to be more symmetric in neutrons and protons. As a consequence, the maximum mass becomes smaller. However, the appearance of kaons changes this general picture and the equation of state becomes stiffer. This is due to the fact that the K^- onset depends on the lepton chemical potential, i.e., $\mu_e - \mu_{\nu_e}$, which stays at larger values in neutrino-trapped matter than in the neutrino-free case, thus delaying the appearance of K^- to higher baryon density. The resulting maximum mass increases, as shown in Fig. 3.

3. Summary and Conclusions

In summary, we have investigated the properties of kaon condensed PNS matter including the composition, the equation of state, and the radius-mass relation, in the framework of the BHF approach with both microscopic and phenomenological three-body forces. In particular, we have discussed the interplay among the effects of the different TBFs, kaon condensate, and neutrino trapping.

It is found that the contribution of trapped neutrinos makes the electron concentration keep high in β-equilibrated star matter and weakens the role played by the symmetry energy on the predicted composition of the star. As a result, the composition of the star matter becomes less sensitive to the two different three-body

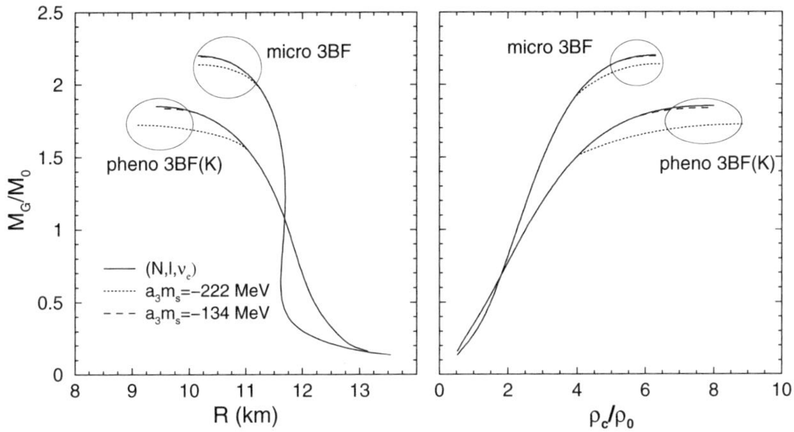

Fig. 3. The gravitational mass is shown as a function of the radius (left-hand side) and the normalized central density (right hand side), for the phenomenological (lower curves) and microscopic TBF's (upper curves) in neutrino trapped matter.

forces considered when neutrinos are trapped. In the star matter without kaons, neutrino trapping leads to a softening of the equation of state and consequently a lower maximum star mass. However when kaons are allowed, neutrino trapping makes the equation of state stiffer and therefore produces larger stars.

This general scenario could change if we include hyperons in our calculations, since hyperon onset might be a process in competition with kaon condensation.[1] We will present a detailed study in a forthcoming publication.

This work was supported by the EU grant CN/ASIA-LINK/008(94791).

References

1. M. Prakash *et al.*, *Phys. Rep.* **280**, 1 (1997) and references therein.
2. O. E. Nicotra, M. Baldo, G. F. Burgio and H.-J. Schulze, *Astron. Astrophys.* **451**, 213 (2006).
3. W. Zuo, A. Li, Z. H. Li and U. Lombardo, *Phys. Rev. C* **70**, 055802 (2004) and references therein.
4. I. Bombaci and U. Lombardo, *Phys. Rev. C* **44**, 1892 (1991).
5. W. Zuo, I. Bombaci and U. Lombardo, *Phys. Rev. C* **60**, 024605 (1999).
6. M. Baldo, I. Bombaci and G. F. Burgio, *Astron. Astrophys.* **328**, 274 (1997).
7. B. S. Pudliner, V. R. Pandharipande, J. Carlson and R. B. Wiringa, *Phys. Rev. Lett.* **74**, 4396 (1995); B. S. Pudliner, V. R. Pandharipande, J. Carlson, S. C. Pieper and R. B. Wiringa, *Phys. Rev. C* **56**, 1720 (1997).
8. P. Grangé, A. Lejeune, M. Martzolff and J.-F. Mathiot, *Phys. Rev. C* **40**, 1040 (1989).
9. W. Zuo, A. Lejeune, U. Lombardo and J.-F. Mathiot, *Nucl. Phys. A* **706**, 418 (2002).
10. D. B. Kaplan and A. E. Nelson, *Phys. Lett. B* **175**, 57 (1986).
11. V. Thorsson, M. Prakash and J. M. Lattimer, *Nucl. Phys. A* **572**, 693 (1994); **574**, 851 (1994), Erratum.
12. S. Kubis and M. Kutschera, *Nucl. Phys. A* **720**, 189 (2003).
13. S. L. Shapiro and S. A. Teukolsky, *Black Holes, White Dwarfs and Neutron Stars*, (John Wiley, New York, 1983).

P. AUGER OBSERVATORY: STATUS AND PRELIMINARY RESULTS

A. INSOLIA

Dipartimento di Fisica, Università di Catania, and INFN Catania, Via S. Sofia 64, I-95123 Catania, Italy
E-mail: antonio.insolia@ct.infn.it

The status of the P. Auger Observatory will be presented. The Fluorescence Detector has been completed and 1350 Cerenkov tanks have been already deployed and are currently taking data. The Observatory is expected to be completed by November 2007. Preliminary results on energy spectrum, small and large scale anisotropy search as well as photon flux and neutrino flux limits will be discussed.

PART F

Nuclear Structure from Laboratory to Stars

RECENT ADVANCES IN THE THEORY OF NUCLEAR FORCES AND ITS IMPACT ON MICROSCOPIC NUCLEAR STRUCTURE

R. MACHLEIDT

Department of Physics, University of Idaho,
Moscow, Idaho 83844, U.S.A.
E-mail: machleid@uidaho.edu

The theory of nuclear forces has made great progress since the turn of the millenium using the framework of chiral effective field theory (ChEFT). The advantage of this approach, which was originally proposed by Weinberg, is that it has a firm basis in quantum-chromodynamics and allows for quantitative calculations. Moreover, this theory generates two-nucleon forces (2NF) and many-body forces on an equal footing and provides an explanation for the empirically known fact that 2NF $\gg$ 3NF $\gg$ 4NF. I will present the recent advances in more detail and put them into historical context. In addition, I will also provide a critical evaluation of the progress made including a discussion of the limitations of the ChEFT approach.

Keywords: Nuclear forces; Nucleon-nucleon interaction; Effective field theory; Chiral perturbation theory; Nuclear matter.

1. Introduction and Historical Perspective

The theory of nuclear forces has a long history (cf. Table 1). Based upon the seminal idea by Yukawa,[1] first field-theoretic attempts to derive the nucleon-nucleon (NN) interaction focused on pion-exchange. While the one-pion exchange turned out to be very useful in explaining NN scattering data and the properties of the deuteron, multi-pion exchange was beset with serious ambiguities. Thus, the "pion theories" of the 1950s are generally judged as failures—for reasons we understand today: pion dynamics is constrained by chiral symmetry, a crucial point that was unknown in the 1950s.

Historically, the experimental discovery of heavy mesons in the early 1960s saved the situation. The one-boson-exchange (OBE) model[2] emerged which is still the most economical and quantitative phenomenology for describing the nuclear force.[3,4] The weak point of this model, however, is the scalar-isoscalar "sigma" or "epsilon" boson, for which the empirical evidence remains controversial. Since this boson is associated with the correlated (or resonant) exchange of two pions, a vast theoretical effort that occupied more than a decade was launched to derive the 2π-exchange contribution to the nuclear force, which creates the intermediate range attraction. For this, dispersion theory as well as field theory were invoked producing the Paris[5] and the Bonn[2,6] potentials.

Table 1. Seven Decades of Struggle: The Theory of Nuclear Forces

1935	**Yukawa: Meson Theory**
1950's	*The "Pion Theories"* One-Pion Exchange: o.k. Multi-Pion Exchange: disaster
1960's	Many pions ≡ multi-pion resonances: $\boldsymbol{\sigma}$, $\boldsymbol{\rho}$, $\boldsymbol{\omega}$, ... The One-Boson-Exchange Model
1970's	Refine meson theory: Sophisticated $\mathbf{2\pi}$ exchange models (Stony Brook, Paris, Bonn)
1980's	Nuclear physicists discover **QCD** Quark Cluster Models
1990's and beyond	Nuclear physicists discover **EFT** Weinberg, van Kolck **Back to Meson Theory!** *But, with Chiral Symmetry*

The nuclear force problem appeared to be solved; however, with the discovery of quantum chromo-dynamics (QCD), all "meson theories" were relegated to models and the attempts to derive the nuclear force started all over again.

The problem with a derivation from QCD is that this theory is non-perturbative in the low-energy regime characteristic of nuclear physics, which makes direct solutions impossible. Therefore, during the first round of new attempts, QCD-inspired quark models[7] became popular. These models are able to reproduce qualitatively and, in some cases, semi-quantitatively the gross features of the nuclear force.[8,9] However, on a critical note, it has been pointed out that these quark-based approaches are nothing but another set of models and, thus, do not represent any fundamental progress. Equally well, one may then stay with the simpler and much more quantitative meson models.

A major breakthrough occurred when the concept of an effective field theory (EFT) was introduced and applied to low-energy QCD.[10]

Note that the QCD Lagrangian for massless up and down quarks is chirally symmetric, i. e., it is invariant under global flavor $SU(2)_L \times SU(2)_R$ equivalent to $SU(2)_V \times SU(2)_A$ (vector and axial vector) transformations. The axial symmetry is spontaneously broken as evidenced in the absence of parity doublets in the low-mass hadron spectrum. This implies the existence of three massless Goldstone bosons which are identified with the three pions $(\pi^\pm, \pi^0)$. The non-zero, but small, pion mass is a consequence of the fact that the up and down quark masses are not exactly zero either (some small, but explicit symmetry breaking). Thus, we arrive at a low-energy scenario that consists of pions and nucleons interacting via a force governed by spontaneously broken approximate chiral symmetry.

To create an effective field theory describing this scenario, one has to write down the most general Lagrangian consistent with the assumed symmetry principles, particularly the (broken) chiral symmetry of QCD.[10] At low energy, the effective degrees of freedom are pions and nucleons rather than quarks and gluons; heavy mesons and nucleon resonances are "integrated out". So, the circle of history is closing and we are back to Yukawa's meson theory, except that we have learned to add one important refinement to the theory: broken chiral symmetry is a crucial constraint that generates and controls the dynamics and establishes a clear connection with the underlying theory, QCD.

It is the purpose of the remainder of this contribution to describe the EFT approach to nuclear forces in more detail.

2. Chiral Perturbation Theory and the Hierarchy of Nuclear Forces

The chiral effective Lagrangian is given by an infinite series of terms with increasing number of derivatives and/or nucleon fields, with the dependence of each term on the pion field prescribed by the rules of broken chiral symmetry. Applying this Lagrangian to NN scattering generates an unlimited number of Feynman diagrams. However, Weinberg showed[11] that a systematic expansion exists in terms of $(Q/\Lambda_\chi)^\nu$, where Q denotes a momentum or pion mass, $\Lambda_\chi \approx 1$ GeV is the chiral symmetry breaking scale, and $\nu \geq 0$ (cf. Fig. 1). This has become known as chiral perturbation theory (χPT). For a given order ν, the number of terms is finite and calculable; these terms are uniquely defined and the prediction at each order is model-independent. By going to higher orders, the amplitude can be calculated to any desired accuracy.

Following the first initiative by Weinberg,[11] pioneering work was performed by Ordóñez, Ray, and van Kolck[12,13] who constructed a NN potential in coordinate space based upon χPT at next-to-next-to-leading order (NNLO; $\nu = 3$). The results were encouraging and many researchers became attracted to the new field. Kaiser, Brockmann, and Weise[14] presented the first model-independent prediction for the NN amplitudes of peripheral partial waves at NNLO. Epelbaum *et al.*[15] developed the first momentum-space NN potential at NNLO, and Entem and Machleidt[16] presented the first potential at N^{3}LO ($\nu = 4$).

In χPT, the NN amplitude is uniquely determined by two classes of contributions: contact terms and pion-exchange diagrams. There are two contacts of order Q^0 [$\mathcal{O}(Q^0)$] represented by the four-nucleon graph with a small-dot vertex shown in the first row of Fig. 1. The corresponding graph in the second row, four nucleon legs and a solid square, represent the seven contact terms of $\mathcal{O}(Q^2)$. Finally, at $\mathcal{O}(Q^4)$, we have 15 contact contributions represented by a four-nucleon graph with a solid diamond.

Now, turning to the pion contributions: At leading order [LO, $\mathcal{O}(Q^0)$, $\nu = 0$], there is only the wellknown static one-pion exchange, second diagram in the first row of Fig. 1. Two-pion exchange (TPE) starts at next-to-leading order (NLO, $\nu = 2$) and all diagrams of this leading-order two-pion exchange are shown. Further TPE

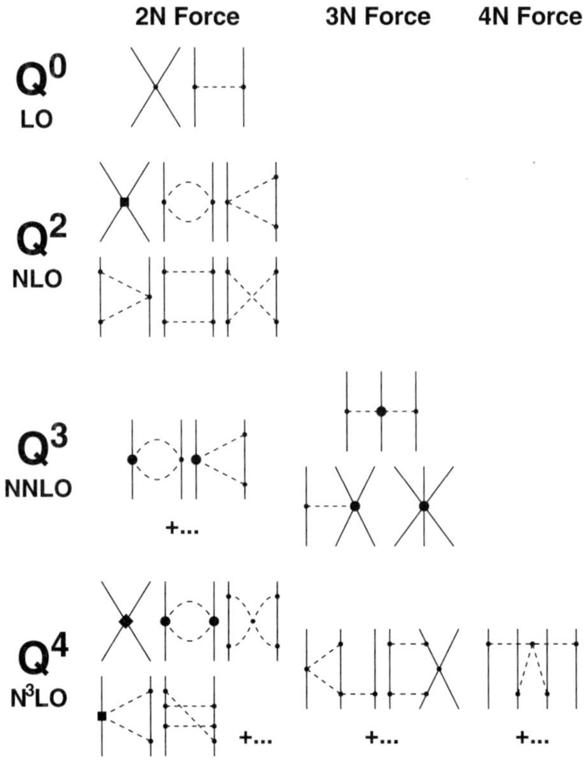

Fig. 1. Hierarchy of nuclear forces in χPT. Solid lines represent nucleons and dashed lines pions. Further explanations are given in the text.

contributions occur in any higher order. Of this sub-leading TPE, we show only two representative diagrams at NNLO and three diagrams at N^{3}LO. The TPE at N^{3}LO has been calculated first by Kaiser.[17] All 2π exchange diagrams/contributions up to N^{3}LO are summarized in a pedagogical and systematic fashion in Ref. 18 where the model-independent results for NN scattering in peripheral partial waves are also shown.

Finally, there is also three-pion exchange, which shows up for the first time at N^{3}LO (two loops; one representative 3π diagram is included in Fig. 1). In Ref. 19 it was demonstrated that the 3π contribution at this order is negligible.

One important advantage of χPT is that it makes specific predictions also for many-body forces. For a given order of χPT, two-nucleon forces (2NF), three-nucleon forces (3NF), ... are generated on the same footing (cf. Fig. 1). At LO, there are no 3NF, and at next-to-leading order (NLO), all 3NF terms cancel.[11,20] However, at NNLO and higher orders, well-defined, nonvanishing 3NF occur.[20,21] Since 3NF show up for the first time at NNLO, they are weak. Four-nucleon forces (4NF) occur first at N^{3}LO and, therefore, they are even weaker.

Table 2. χ^2/datum for the reproduction of the 1999 *np* database by families of *np* potentials at NLO and NNLO constructed by the Bochum/Juelich group.[22]

Bin (MeV)	# of *np* data	*Bochum/Juelich* NLO	NNLO
0–100	1058	4–5	1.4–1.9
100–190	501	77–121	12–32
190–290	843	140–220	25–69
0–290	2402	67–105	12–27

3. Chiral NN Potentials

The two-nucleon system is non-perturbative as evidenced by the presence of shallow bound states and large scattering lengths. Weinberg[11] showed that the strong enhancement of the scattering amplitude arises from purely nucleonic intermediate states. He therefore suggested to use perturbation theory to calculate the NN potential (Fig. 1) and to apply this potential in a scattering equation (Lippmann-Schwinger or Schrödinger equation) to obtain the NN amplitude. We follow this philosophy.

Chiral perturbation theory is a low-momentum expansion. It is valid only for momenta $Q \ll \Lambda_\chi \approx 1$ GeV. Therefore, when a potential is constructed, all expressions (contacts and irreducible pion exchanges) are multiplied with a regulator function,

$$\exp\left[-\left(\frac{p}{\Lambda}\right)^{2n}-\left(\frac{p'}{\Lambda}\right)^{2n}\right], \tag{1}$$

where p and p' denote, respectively, the magnitudes of the initial and final nucleon momenta in the center-of-mass system (CMS); and $\Lambda \ll \Lambda_\chi$. The exponent $2n$ is to be chosen such that the regulator generates powers which are beyond the order at which the calculation is conducted.

To what order in χPT do we have to go for sufficient accuracy? To discuss this issue on firm grounds, I show in Table 2 the χ^2/datum for the fit of the world *np* data below 290 MeV for a family of *np* potentials at NLO and NNLO. The NLO potentials produce the horrendous χ^2/datum between 67 and 105, and the NNLO are between 12 and 27. The rate of improvement from one order to the other is very impressive, but the quality of the reproduction of the *np* data at NLO and NNLO is obviously totally insufficient for reliable predictions.

Based upon these facts, it has been pointed out in 2002 by Entem and Machleidt[18,23] that NNLO is insufficient and one has to proceed to N^{3}LO. Consequently, the first N^{3}LO potential was created in 2003,[16] which showed that at this order a χ^2/datum comparable to the high-precision Argonne V_{18} potential can, indeed, be achieved, see Tables 3 and 4. This "Idaho" N^{3}LO potential[16] produces a χ^2/datum = 1.1 for the world *np* data below 290 MeV which compares well with the χ^2/datum = 1.04 by the Argonne potential (Table 3). In 2005, also the Bochum/Juelich group produced several N^{3}LO NN potentials,[25] the best of which fits the *np* data with

Table 3. χ^2/datum for the reproduction of the 1999 ***np* database** by various *np* potentials. Numbers in parentheses denote cutoff parameters in units of MeV.

Bin (MeV)	# of ***np*** data	*Idaho* N^3LO^{16} (500–600)	*Bochum/Juelich* N^3LO^{25} (600/700–450/500)	Argonne $V_{18}{}^{24}$
0–100	1058	1.0–1.1	1.0–1.1	0.95
100–190	501	1.1–1.2	1.3–1.8	1.10
190–290	843	1.2–1.4	2.8–20.0	1.11
0–290	2402	1.1–1.3	1.7–7.9	1.04

a χ^2/datum = 1.7 and the worse with a χ^2/datum = 7.9 (cf. Table 3). While 7.9 is clearly unacceptable for any meaningful application, a χ^2/datum of 1.7 may be sufficient for most purposes.

I turn now to the *pp* data, Table 4. Typically, χ^2 for *pp* data are larger than for *np* because of the higher precision of *pp* data. Thus, the Argonne V_{18} produces a χ^2/datum = 1.4 for the world *pp* data below 290 MeV and the best Idaho N^3LO *pp* potential obtains 1.5. The fit by the best Bochum/Juelich N^3LO *pp* potential results in a χ^2/datum = 2.9 and the worst is 22.3.

4. Limitations of χPT

Since χPT is a low-momentum expansion, we have to expect limitations concerning its applicability. This is demonstrated in Fig. 2, where phase shift predictions by various NN potentials are shown up to 1000 MeV lab. energy for the incident nucleon. The figure includes one representative of the family of the high-precision NN potentials, namely, the CD-Bonn potential[4] (solid line), which obviously predicts the phase shifts correctly up to the highest energies shown, even though it was adjusted only up to 350 MeV. The same is true for other high-precision potentials, like Argonne V_{18}[24] and the Nijmegen potentials.[3] On the other hand, the chiral NN potentials at order N^3LO (dashed line Ref. 16 and dotted line Ref. 25) do not make any reasonable predictions beyond about 300 MeV lab. energy. This is, of course, not unexpected since χPT applies only for momenta $Q \ll \Lambda_\chi \approx 1$ GeV, which is enforced by the regulator Eq. (1) where a typical choice for the cutoff parameter is $\Lambda \approx 500$ MeV. Thus, chiral potentials are reliable only for CMS momenta

Table 4. χ^2/datum for the reproduction of the 1999 ***pp* database** by various *pp* potentials. Numbers in parentheses denote cutoff parameters in units of MeV.

Bin (MeV)	# of ***pp*** data	*Idaho* N^3LO^{16} (500–600)	*Bochum/Juelich* N^3LO^{25} (600/700–450/500)	Argonne $V_{18}{}^{24}$
0–100	795	1.0–1.7	1.0–3.8	1.0
100–190	411	1.5–1.9	3.5–11.6	1.3
190–290	851	1.9–2.7	4.3–44.4	1.8
0–290	2057	1.5–2.1	2.9–22.3	1.4

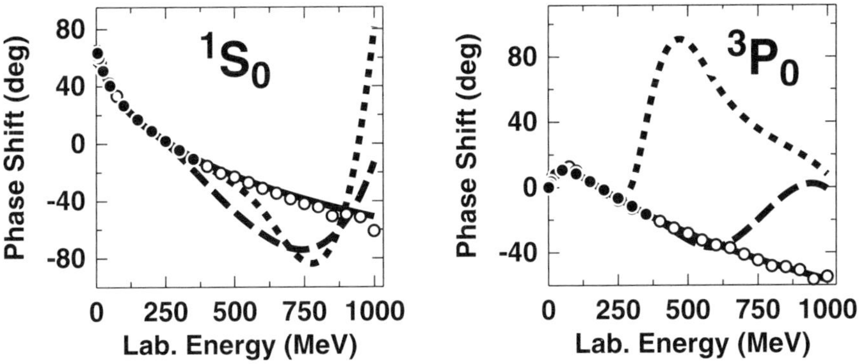

Fig. 2. *np* phase shifts of the $^1S_0$ and $^3P_0$ partial waves for lab. energies up to 1000 MeV. The solid curve shows the phase shifts predicted by the CD-Bonn potential.[4] Note that this curve is hardly visible because it agrees with the data and, thus, is buried under the symbols representing the data. The dashed and the dotted lines are the predictions by the N^3LO chiral potentials constructed by the Idaho[16] and the Bochum/Juelich[25] groups, respectively. Solid dots represent the Nijmegen multienergy *np* phase shift analysis[26] and open circles the GWU/VPI single-energy *np* analysis SM99.[27]

$p, p' \lesssim 2.2$ fm^{-1}. A Fermi momentum $k_F \approx 2.2$ fm^{-1} is equivalent to a nuclear matter density $\rho \approx 4\rho_0$ where ρ_0 denotes nuclear matter saturation density. Nuclear matter calculations in which chiral two-nucleon potentials are applied can be found in Refs. 28,29. Note, however, that for "complete" calculations also the chiral 3NF must be included. In any case, based upon the above arguments, one may trust the χPT approach up to densities around $4\rho_0$. In contrast, relativistic meson theory can be trusted to very high momenta (cf. the CD-Bonn curve in Fig. 2) and densities equivalent to those high momenta.

5. Conclusions

The theory of nuclear forces has made great strides since the turn of the millennium. Nucleon-nucleon potentials have been developed that are based on proper theory (EFT for low-energy QCD) and are of high-precision, at the same time. Moreover, the theory generates two- and many-body forces on an equal footing and provides a theoretical explanation for the empirically known fact that 2NF $\gg$ 3NF $\gg$ 4NF.

At N^3LO,[16,18] the accuracy can be achieved that is necessary and sufficient for microscopic nuclear structure. First calculations applying the N^3LO NN potential[16] in the (no-core) shell model,[30–33] the coupled cluster formalism,[34–38] and the unitary-model-operator approach[39] have produced promising results.

The 3NF at NNLO is known[20,21] and has been applied in few-nucleon reactions[21,40,41] as well as the structure of light nuclei.[42–44] However, the famous 'A_y puzzle' of nucleon-deuteron scattering is not resolved by the 3NF at NNLO. Thus, the most important outstanding issue is the 3NF at N^3LO, which is under construction.

Acknowledgments

This work was supported in part by the U.S. National Science Foundation under Grant No. PHY-0099444.

References

1. H. Yukawa, *Proc. Phys. Math. Soc. Japan* **17**, 48 (1935).
2. R. Machleidt, *Adv. Nucl. Phys.* **19**, 189 (1989).
3. V. G. J. Stoks *et al.*, *Phys. Rev. C* **49**, 2950 (1994).
4. R. Machleidt, *Phys. Rev. C* **63**, 024001 (2001).
5. M. Lacombe *et al.*, *Phys. Rev. C* **21**, 861 (1980).
6. R. Machleidt *et al.*, *Phys. Rep.* **149**, 1 (1987).
7. F. Myhrer *et al.*, *Rev. Mod. Phys.* **60**, 629 (1988).
8. D. R. Entem, F. Fernandez and A. Valcarce, *Phys. Rev. C* **62**, 034002 (2000).
9. G. H. Wu *et al.*, *Nucl. Phys. A* **673**, 273 (2000).
10. S. Weinberg, *Physica A* **96**, 327 (1979).
11. S. Weinberg, *Nucl. Phys. B* **363**, 3 (1991).
12. C. Ordóñez, L. Ray and U. van Kolck, *Phys. Rev. C* **53**, 2086 (1996).
13. U. van Kolck, *Prog. Part. Nucl. Phys.* **43**, 337 (1999).
14. N. Kaiser *et al.*, *Nucl. Phys. A* **625**, 758 (1997).
15. E. Epelbaum *et al.*, *Nucl. Phys. A* **671**, 295 (2000).
16. D. R. Entem and R. Machleidt, *Phys. Rev. C* **68**, 041001 (2003).
17. N. Kaiser, *Phys. Rev. C* **64**, 057001 (2001); **65**, 017001 (2002).
18. D. R. Entem and R. Machleidt, *Phys. Rev. C* **66**, 014002 (2002).
19. N. Kaiser, *Phys. Rev. C* **61**, 014003 (1999); *ibid.* **62**, 024001 (2000).
20. U. van Kolck, *Phys. Rev. C* **49**, 2932 (1994).
21. E. Epelbaum *et al.*, *Phys. Rev. C* **66**, 064001 (2002).
22. E. Epelbaum, W. Glöckle and U.-G. Meißner, *Eur. Phys. J. A* **19**, 401 (2004).
23. D. R. Entem and R. Machleidt, *Phys. Lett. B* **524**, 93 (2002).
24. R. B. Wiringa, V. G. J. Stoks and R. Schiavilla, *Phys. Rev. C* **51**, 38 (1995).
25. E. Epelbaum, W. Glöckle and U.G. Meissner, *Nucl. Phys. A* **747**, 362 (2005).
26. V. G. J. Stoks *et al.*, *Phys. Rev. C* **48**, 792 (1993).
27. R. A. Arndt *et al.*, SAID, solution of summer 1999 (SM99).
28. D. Alonso and F. Sammarruca, *Phys. Rev. C* **67**, 054301 (2003).
29. Z. H. Li *et al.*, *Phys. Rev. C* **74**, 047304 (2006).
30. L. Coraggio *et al.*, *Phys. Rev. C* **66**, 021303 (2002); *ibid.* **71**, 014307 (2005).
31. P. Navrátil and E. Caurier, *Phys. Rev. C* **69**, 014311 (2004).
32. C. Forssen *et al.*, *Phys. Rev. C* **71**, 044312 (2005).
33. J. P. Vary *et al.*, *Eur. Phys. J. A* **25**, 475 (2005).
34. K. Kowalski *et al.*, *Phys. Rev. Lett.* **92**, 132501 (2004).
35. D. J. Dean and M. Hjorth-Jensen, *Phys. Rev. C* **69**, 054320 (2004).
36. M. Wloch *et al.*, *J. Phys. G* **31**, S1291 (2005); *Phys. Rev. Lett.* **94**, 21250 (2005).
37. D. J. Dean *et al.*, *Nucl. Phys. A* **752**, 299 (2005).
38. J. R. Gour *et al.*, *Phys. Rev. C* **74**, 024310 (2006).
39. S. Fujii, R. Okamoto and K. Suzuki, *Phys. Rev. C* **69**, 034328 (2004).
40. K. Ermisch *et al.*, *Phys. Rev. C* **71**, 064004 (2005).
41. H. Witala *et al.*, *Phys. Rev. C* **73**, 044004 (2006).
42. A. Nogga *et al.*, *Nucl. Phys. A* **737**, 236 (2004).
43. A. Nogga, P. Navrátil, B. R. Barrett and J. P. Vary, *Phys. Rev. C* **73**, 064002 (2006).
44. P. Navrátil *et al.*, arXiv:nucl-th/0701038.

KOHN-SHAM DENSITY FUNCTIONAL APPROACH TO NUCLEAR BINDING

X. VIÑAS

Departament d'Estructura i Constituents de la Matèria,
Facultat de Física, Universitat de Barcelona,
Av. Diagonal 647, E-08028 Barcelona, Spain
E-mail: xavier@ecm.ub

M. BALDO

Istituto Nazionale di Fisica Nucleare, Sezione di Catania,
Via Santa Sofia 64, I-95123 Catania, Italy

P. SCHUCK

Institut de Physique Nucléaire, CNRS, UMR8608,
Orsay, F-91406, France

L. M. ROBLEDO

Departamento de Física Teórica C-XI, Universidad Autónoma de Madrid,
Cantoblanco, E-28049 Madrid, Spain

A non-relativisitic nuclear density functional theory is constructed, not as usual, from an effective density dependent nucleon-nucleon force but directly introducing in the functional results from microscopic nuclear and neutron matter Bruckner G-matrix calculations at various densities. A purely phenomenological finite range part to account for surface properties is added. The striking result is that only four to five adjustable parameters, spin-orbit included, suffice to reproduce nuclear binding energies and radii with the same quality as obtained with the most performant effective forces. For the pairing correlations, simply a density dependent zero range force is adopted from the literature. The ability of the proposed functional for describing deformed nuclei is also explored.

Keywords: Energy density functional; Nuclear matter; Many-body theory.

1. Introduction

A successful description of the properties of nuclei can be obtained quite often in the mean-field approximation using effective density dependent nucleon-nucleon interactions of Skyrme[1] or Gogny[2] type as well as in the relativistic mean field (RMF) approach.[3] In the non-relativistic case, Skyrme and Gogny forces has to be supplemented by an additional spin-orbit interaction which, however, is automatically included in the RMF approach. These effective forces or RMF parametrizations contain around ten parameters which are fitted to reproduce some ground state properties of some selected nuclei. With these ingredients nuclear mean field theories successfully describe not only ground state properties (e.g. binding energies and radii) but also excited states, fission barriers, and many things more.[3,4]

In this study we want to show that a description of nuclear properties of similar quality to that found using effective forces or RMF parametrizations can be obtained within the framework of the Density Functional Theory (DFT) widely and successfully used in other fields, like condensed matter, chemistry, atomic physics, etc.[8,9] The basis of this theory lies in the Hohenberg-Kohn theorem (HK),[10] which states that for a Fermi system, with a non-degenerate ground state, the total energy can be expressed as a functional of the local density $\rho(\mathbf{r})$ only. Such a functional reaches its variational minimum when evaluated with the exact ground state density. We propose a new energy density functional which combines very precise Bruckner G-matrix calculations with a phenomenological finite-range force for describing the surface properties and a spin-orbit interaction. Thus this new energy density functional contains a *previously determined* bulk part described by a microscopic calculation and a phenomenological part depending only on a few parameters which are fitted to experimental ground state properties.

To obtain a mean field description starting from an energy density functional the Kohn-Sham (KS)[7] scheme is used. In the KS-DFT method one introduces an auxiliary set of A orthonormal single particle wave functions $\psi_i(\mathbf{r})$, where A is the number of particles, and the density is assumed to be given by

$$\rho(\mathbf{r}) = \Sigma_{i,s,t}|\psi_i(\mathbf{r}, s, t)|^2, \tag{1}$$

where s and t stand for spin and iso-spin indices. The variational procedure to minimize the functional is performed in terms of the orbitals instead of the density. As it is done in condensed matter and atomic physics, the energy functional $E[\rho(\mathbf{r})]$ is split into two parts: $E = T_0[\rho] + W[\rho]$.[7] The first piece T_0 corresponds to the uncorrelated part of the kinetic energy which in the KS scheme is written as

$$T_0 = \frac{\hbar^2}{2m} \sum_{i,s,t} \int d^3r |\nabla \psi_i(\mathbf{r}, s, t)|^2. \tag{2}$$

The other piece $W[\rho]$ contains the potential energy as well as the correlated part of the kinetic energy. By performing variations respect to the single-particle orbitals it is obtained a closed set of A Hartree-like equations with an effective potential, which is the functional derivative of $W[\rho]$ with respect to the local density $\rho(\mathbf{r})$. Since the effective potential depends on the density, and therefore on the ψ_i's, a self-consistent procedure is necessary. Although such equations can in principle be exact, they are not useful unless a reliable approximation for the unknown part of the density functional $W[\rho]$ is found. In the KS-DFT formalism the exact ground state wave function is actually not known, the density being the basic quantity.[11]

The contribution of the spin-orbit interaction to the energy functional is very important in nuclear physics in contrast with the situation in condensed matter and atomic physics. To take into account the spin-orbit interaction the standard DFT has to be generalized to the non-local case.[12] Thus the spin-orbit contribution to the energy can be split in an uncorrelated part $E^{s.o.}$ plus a remainder. The form of the uncorrelated spin-orbit part is therefore taken exactly as in the Skyrme[1] or Gogny forces.[2]

We write the functional in the nuclear case as $E = T_0 + E^{s.o.} + E_{int} + E_C$, where we explicitly split off the Coulomb energy E_C. It is computed as the direct term plus the exchange contribution in the Slater approximation, that is $E_C^H = (1/2)\int\int d^3rd^3r'\rho_p(\mathbf{r})|\mathbf{r}-\mathbf{r}'|^{-1}\rho_p(\mathbf{r}')$, and $E_C^{ex} = -(3/4)(3/\pi)^{1/3}\int d^3r\rho_p(\mathbf{r})^{4/3}$ with $E_C = E_C^H + E_C^{ex}$ and $\rho_{p/n}$ the proton/neutron density.

The nuclear energy functional contribution $E_{int}[\rho_n, \rho_p]$, which contains the nuclear potential energy as well as additional correlations, can be split in a finite range term $E_{int}^{FR}[\rho_n, \rho_p]$ to account for surface properties and a bulk correlation part $E_{int}^{\infty}[\rho_n, \rho_p]$. This latest part is taken from a microscopic infinite nuclear matter calculation[13] as we will discuss below. Finally our KS-DFT functional reads:

$$E = T_0 + E^{s.o.} + E_{int}^{\infty} + E_{int}^{FR} + E_C. \tag{3}$$

For the finite range term we make the simplest phenomenological ansatz possible

$$\begin{aligned} E_{int}^{FR}[\rho_n, \rho_p] &= \frac{1}{2}\sum_{t,t'}\int\int d^3rd^3r'\rho_t(\mathbf{r})v_{t,t'}(\mathbf{r}-\mathbf{r}')\rho_{t'}(\mathbf{r}') \\ &\quad - \frac{1}{2}\sum_{t,t'}\gamma_{t,t'}\int d^3r\rho_t(\mathbf{r})\rho_{t'}(\mathbf{r}) \end{aligned} \tag{4}$$

with t = proton/neutron and $\gamma_{t,t'}$ the volume integral of $v_{t,t'}(r)$. The subtraction in (4) is made in order not to contaminate the bulk part, determined from the microscopic infinite matter calculation. For the finite range form factor $v_{t,t'}(r)$ we make a simple Gaussian ansatz: $v_{t,t'}(r) = V_{t,t'}e^{r^2/{r_0}^2}$. We choose a minimum of three open parameters: $V_{p,p} = V_{n,n} = V_L, V_{n,p} = V_{p,n} = V_U$, and r_0.

The only undetermined and most important piece in (3) is then the bulk contribution E_{int}^{∞}. As it was stated before, it is obtained from a microscopic infinite matter calculation, using a realistic bare force, together with a converged hole-line expansion.[13] We first reproduce very precisely by interpolating functions the *correlation* part of the ground state energy per particle of symmetric and pure neutron matters, and then make a quadratic interpolation for asymmetric matter. Finally the total correlation contribution to the energy functional in local density approximation reads:

$$E_{int}^{\infty}[\rho_p, \rho_n] = \int d^3r\big[P_s(\rho)(1-\beta^2) + P_n(\rho)\beta^2\big]\rho, \tag{5}$$

where P_s and P_n are two interpolating polynomials for symmetric and pure neutron matter, respectively, at the density $\rho = \rho_p + \rho_n$, and $\beta = (\rho_n - \rho_p)/\rho$ is the

Table 1. Coefficients of the polynomial fits P_s, Eq. (6) and P_n, Eq. (7) to the EOS of symmetric and neutron matter.

k	$b_k^{(s)}(MeV)$	$b_k^{(n)}(MeV)$	a_k(MeV)
1	-105.640069	-43.985736	-15.3563461
2	167.700968	49.784439	16.4197441
3	-181.762432	-42.400650	0.0
4	103.166047	21.894382	0.0
5	-22.4990207	-4.3071179	0.0

asymmetry parameter. For P_s we took ($x = \rho/\rho_0$, with ρ_0=0.17 fm^{-3}, see below)

$$P_s(\rho) = \begin{cases} \sum_{k=1}^{5} b_k^{(s)} x^k & x < 1, \\ P_s(\rho_0) + a_1 \cdot (x-1) + a_2 \cdot (x-1)^2 & x > 1, \end{cases} \tag{6}$$

where the coefficients (energy/particle in MeV) are given in Table 1. The two forms match at $x = 1$ ($\rho = \rho_0$) up to the second derivative. This functional form can be used up to $\rho = 0.24$ fm^{-3}, which is the interval where the independent fit of the microscopic calculation has been performed.

A similar expression holds for P_n,

$$P_n(\rho) = \sum_{k=1}^{5} b_k^{(n)} x^k, \tag{7}$$

where again the coefficients are displayed in Table 1, which is valid in the same density interval. The interpolating polynomial for symmetric matter has been constrained to allow a minimum exactly at the energy $E/A = -16MeV$ and Fermi momentum $k_F = 1.36fm^{-1}$, i.e. $\rho_0 = 0.17fm^{-3}$. Those values lie within the numerical uncertainty of the microscopic calculation. The present procedure resides in a more precise fit compared with the one given in Ref. 13, where more details of the microscopic EOS can be found. It is important to point out that the polynomial fits in Eqs. (6),(7) are done once and for ever before to include the bulk energy in the energy density functional and consequently they do not explicitly enter into the search of parameters in the fitting procedure to experimental data. The coefficients of P_s and P_n are therefore linked to the bare nuclear two and three body forces. To describe open shell nuclei pairing correlations have to be included. How to formally include pairing within DFT has recently been discussed in Ref. 14. In this work we simply add pairing through the simple BCS approach using the density dependent delta force defined in Ref. 15 for $m = m^*$. In our calculations the two-body center of mass correction has been included using the pocket formula, based on the harmonic oscillator, derived in Ref. 16 which nicely reproduces the exact correction as it has been shown in Ref. 12. Our functional is now fully defined and, henceforth, we call it **B**arcelona**C**atania**P**aris(BCP)-functional.

Table 2. Parameters of the Gaussian form factors and spin-orbit strength.

	r_0(fm)	V_L (MeV)	V_U (MeV)	W_0 (MeV)
BCP1	1.05	-93.520	-60.577	113.829
BCP2	1.25	-33.700	-32.483	110.812

In this exploratory investigation, we fitted two sets of parameters. We have considered only spherical nuclei which we choose as given below. The two fits were obtained in i) optimizing ground state energies only, ii) optimizing ground state energies together with rms radii. The open parameters, V_L, V_U, r_0 of Eq. (4) as well as the spin-orbit strength W_0, are fitted to reproduce the binding energies of the nuclei ^{16}O, ^{40}Ca, ^{48}Ca, ^{56}Ni, ^{78}Ni, ^{90}Zr, ^{116}Sn, ^{124}Sn, ^{132}Sn, ^{208}Pb and ^{214}Pb in the case of the parameter set BCP1 and additionally the charge rms radii ($r_c = \sqrt{r_p^2 + 0.64}$ fm) of the nuclei ^{16}O, ^{40}Ca, ^{48}Ca,^{90}Zr, ^{116}Sn, ^{124}Sn, ^{208}Pb and ^{214}Pb for the parameter set BCP2. Experimental binding energies and charge radii are taken from References 17 and 18 respectively. In Table 2 we give the obtained parameter sets of fits (i) and (ii).

Our intention is to compare our results from the BCP-functionals with the ones obtained from some of the most performant effective nucleon-nucleon forces available, namely D1S,[19] SkLy4[5] and NL3.[20] To this end, we have calculated the binding energies and charge radii of 161 even-even spherical nuclei. In Table 3 we report the energy and charge radii rms deviations between the corresponding experimental values and the theoretical ones obtained using the D1S force,[22] the NL3 parametrization,[21] the Skyrme force SkLy4[5] and our BCP1 and BCP2 functionals. We also display in Fig. 1 the differences between the theoretical and experimental energies and charge radii in the range between ^{16}Ne and ^{224}U calculated with the BCP1 functional in comparison with the same quantities obtained from the D1S effective force.[22]

We see that our theory with only four adjustable parameters, is at least as performant as the two other ones. One could say that, with BCP, radii are slightly worse, however, this is eventually compensated by slightly better energies. It may be a matter of taste to prefer functional BCP1 or BCP2. The common bulk prop-

Table 3. Energies (in MeV) and charge radii (in fm) rms deviations. The numerical values of energies and charge radii calculated with the BCP1 and BCP2 functionals used to obtain the rms values of this Table are given in Ref. 25

	BCP1	BCP2	D1S	NL3	SkLy4
rms_E	1.775	2.057	2.414	3.582	1.711
rms_R	0.031	0.028	0.020	0.020	0.024

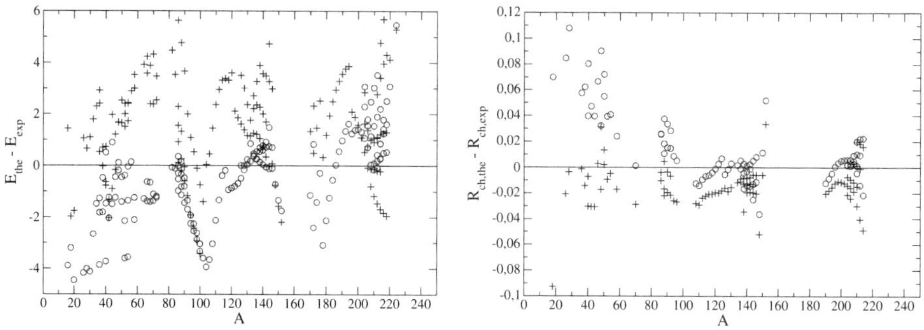

Fig. 1. Differences between the theoretical and experimental energies (left) and charge rms radii (right). Calculations are performed using the BCP1 functional (open circles) and the D1S parameter set (crosses).[22]

erties of both BCP-functionals are: Incompressibility modulus K_∞=240 MeV and symmetry energy J=33.55 MeV plus its derivative L=56.39 MeV, which are quite acceptable values. We have also computed the surface energies obtained with the BCP1 and BCP2 functionals which are 17.7 and 18.0 MeV respectively.[23] These values are close to the surface energy obtained with the Gogny D1S force which is 18.2 MeV. Our prescription for pairing is also reasonable, for instance we obtain an average gap of 1.5 MeV for the nucleus ^{116}Sn.

The fact that we need only 4 to 5 open parameters is in line with a recent analysis of Bertsch et al.[24] who pointed out that also in Skyrme forces implicitly only the same number of parameters are relevant. However, contrary to Skyrme functionals, here the bulk part of the functional comes from a very reliable microscopic calculation for all relevant proton and neutron densities what may be of crucial importance when extrapolating nuclei to the unknown regions.

As the BCP energy density functional has proved to work well in describing binding energies and charge rms radii of spherical nuclei, the next step is to study the performance of BCP in deformed nuclei. Spatial deformation is one of the most striking examples of the spontaneous symmetry breaking mechanism of the mean field approximation. The mechanism allows to incorporate short and long range correlations in an easy way into the simpler mean field picture. Long range correlations induce the breaking of rotational symmetry giving rise to the appearance of quadrupole deformed ground states. The concept of deformation also plays a relevant role in the description of the dynamics of the fission phenomena. There, the nucleus is assumed to proceed from its ground state shape up the scission in two fragments by changing continuously its shape. The energies of those configurations determine important parameters as the excitation energy of the fission isomers or the height of the fission barriers. In Figure 2 we display the fission barrier of the nucleus ^{240}Pu computed using the BCP1 and BCP2 functionals compared with the results obtained with the Gogny D1S force. As it can be seen from this figure the potential energy surface (PES) as a function of the quadrupole deformation

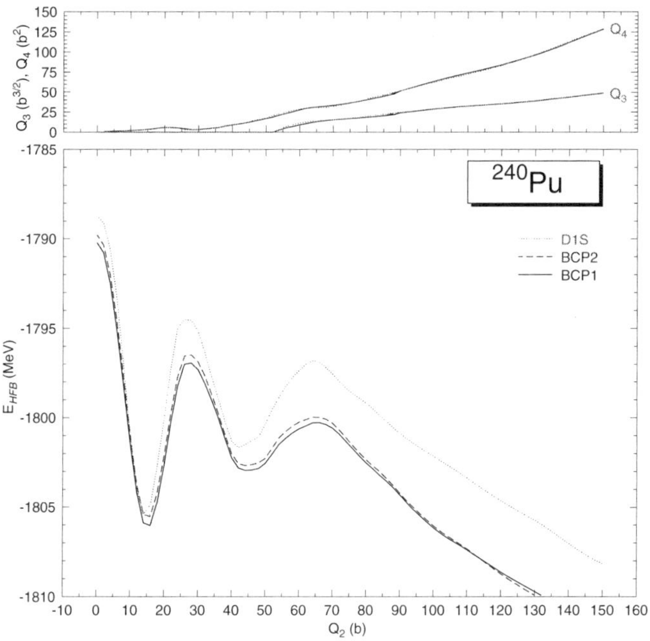

Fig. 2. Potential energy surface for ^{240}Pu as a function of the quadrupole deformation. Results are obtained using the BCP1 and BCP2 energy density functionals as well as the Gogny D1S force. In the upper panel the octupole and hexadecapole moments for the three cases are also depicted.

obtained with the BCP1 and BCP2 functionals are very similar and both closely follow the PES corresponding to the Gogny D1S force. The fission barriers provided by the BCP1 and BCP2 functionals are almost equal but slightly lower than the one calculated with D1S, as it is expected from the surface energy values reported previously. Although the shapes involved in fission are mainly characterized by its quadrupole moment, higher multipole deformations can be important for describing some fine details. From the top of Figure 2 we can see that the expectation values of the octupole and hexadecupole moments obtained with the BCP1 and BCP2 functionals are almost equal to the ones calculated with the D1S force.

In conclusion, we have applied a Kohn-Sham DFT approach to ground state properties of nuclei. The bulk part of the functional is provided by microscopic results using the converged hole line expansion based on a realistic bare force.[13] This allowed us to drastically reduce the number of open parameters in the non-pairing part of the functional. Three of those are contained in the surface part which was determined phenomenologically in convoluting the densities with a finite range Gaussian. A fourth adjustable parameter is the strength of the spin orbit potential for which we get values close to the usual.[5] In this pilot study we took for the pairing part of the functional the simplest possible procedure and adopted a previously adjusted density dependent δ-force.[15] The fit of open parameters was done only for spherical nuclei. In view of the surprisingly small number of adjustable parameters, the results are very encouraging and well compete with the most performant mean field theories presently in use. We have also checked that BCP functionals also yields excellent results in the deformed case similar to the ones obtained with the Gogny D1S force as is demonstrated here for the fission barrier of ^{240}Pu.

Acknowledgements

We are indebted to G. Bertsch and C. Maieron for valuable discussions. One of us (P.S.) is grateful to M. Casida for useful information. This work is partially supported by IN2P3-CICYT and INFN-CICYT. X.V. acknowledges Grant Nos. FIS2005-03142 (MEC, Spain and FEDER) and 2005SGR-00343 (Generalitat de Catalunya). L.M.R.'s work was supported in part by the DGI of MEC, Spain, under Project FIS2004-06697.

References

1. D. Vautherin and D. M. Brink, *Phys. Rev. C* **5**, 626 (1972).
2. J. Dechargé and D. Gogny, *Phys. Rev C* **21**, 1586 (1980).
3. P. Ring, *Prog. Part. Nucl. Phys.* **37** (1996) and references therein.
4. Li Guo-Quiang, *J. of Phys. G* **17**, 1 (1991).
5. E. Chabanat, P. Bonche, P. Haensel, J. Mayer and R. Schaeffer, *Nucl. Phys. A* **635**, 231 (1998).
6. B. Alex Brown, *Phys. Rev. C* **58**, 220 (1998).
7. W. Kohn and L. J. Sham, *Phys. Rev. A* **140**, 1133 (1965).
8. R. O. Jones and O. Gunnarsson, *Rev. Mod. Phys.* **61**, 689 (1989).
9. H. Eschrig, *The Fundamentals of Density Functional Theory* (Teubner, Stuttgart, 1996).
10. P. Hohenberg and W. Kohn, *Phys. Rev. B* **136**, 864 (1964).
11. In selfbound systems like nuclei the groundstate is always degenerate because of broken translational symmetry. This problem can, in principle, be solved with the approach advocated by J. Engel, *Phys. Rev. C* **75**, 014306 (2007). We therefore suppose that Eqs. (3-5) is a functional in the intrinsic system.
12. V. B. Soubbotin, V. I. Tselyaev and X. Viñas, *Phys. Rev. C* **67**, 014324 (2003).
13. M. Baldo, C. Maieron, P. Schuck and X. Viñas, *Nucl. Phys. A* **736**, 241 (2004) and references therein.
14. S. Krewald, V. B. Soubbotin, V. I. Tselyaev and X.Viñas, *Phys. Rev. C* **67**, 014324 (2003).
15. E. Garrido, P. Sarriguren, E. Moya de Guerra and P. Schuck, *Phys. Rev. C* **60**, 064312 (1999).
16. M. N. Butler, D. W. L. Sprung and J. Martorell, *Nucl. Phys. A* **422**, 157 (1984).
17. G. Audi and A. H. Waspra, *Nucl. Phys. A* **729**, 337 (2003).
18. I. Angeli, *Atomic Data and Nuclear Data Tables* **87**, 185 (2004).
19. J. F. Berger, M. Girod and D. Gogny, *Comput. Phys. Commun.* **63**, 365 (1991); *Nucl. Phys. A* **502**, 85c (1989).
20. G. A. Lalazissis, J. Köning and P. Ring, *Phys. Rev. C* **55**, 540 (1997).
21. G. A. Lalazissis and S. Raman, *Atomic Data and Nuclear Data Tables* **71**, 1 (1999).
22. `http://www-phys.cea.fr`; S. Hilaire and B. Nerlo-Pomorska, private communication.
23. M. Farine, private communication.
24. G. F. Bertsch, B. Sabbey and M. Uusnäkki, *Phys. Rev. C* **71**, 054311 (2005).
25. `http://www.ecm.ub.es/~assum/taula_xv`

STRUCTURE AND DECAY OF KAON-CONDENSED HYPERNUCLEI

T. MUTO

Department of Physics, Chiba Institute of Technology,
Narashino, Chiba 275-0023, Japan
E-mail: takumi.muto@it-chiba.ac.jp

The structure and decay modes of the K^--condensed hypernucleus, which may be produced in the laboratory as the strangeness-conserving system, is investigated on the basis of the effective chiral Lagrangian for the kaon-baryon interaction, combined with the nonrelativistic baryon-baryon interaction model. It is shown that a large number of negative strangeness is needed for the formation of highly dense and self-bound state with K^- condensates and that part of the strangeness should be carried by hyperons. Such a self-bound object may decay only through weak processes.

Keywords: Kaon condensation; Hyperons; Self-bound objects.

1. Introduction

Kaon dynamics in a hadronic medium is an important subject associated with role of strangeness in highly dense matter. In neutron stars, kaon condensation may exist as a macroscopic realization of strangeness (Bose-Einstein condensation).[1–4] In kaon-condensed matter, the net strangeness is produced through weak interaction processes, which play an decisive role on the ground state properties of the kaon-condensed phase in chemical equilibrium for weak processes such as $n \rightleftharpoons p + K^-$, $e^- \rightleftharpoons K^- + \nu_e$.[5] The existence of kaon condensation leads to softening of the equation of state (EOS) and has a sizable effect on the internal structure of neutron stars. Kaon condensation also important for thermal evolution of neutron stars since the neutrino emission processes are largely enhanced in the presence of kaon condensates.[6–9]

Meanwhile, possible formation of a deeply bound kaonic nucleus, in particular, $\bar{K}$ nuclear clusters, as a cold and highly dense system has been elaborated both theoretically and experimentally.[12–21] These studies of the kaonic nucleus may in turn provide us with information on kaon condensation in neutron stars.

Recently, we have discussed coexistence of kaon condensation with hyperonic matter, where hyperons (Y=Λ, Σ^-) are mixed in addition to the neutrons, protons and leptons in the ground state of neutron-star matter.[10,11] We pointed out that both the hyperon-mixing effect and the s-and p-wave (anti) kaon-baryon attractions soften the EOS of the kaon-condensed phase in hyperonic matter considerably, while the stiffness of the EOS is recovered at high densities due to the repulsive interaction between baryons. As a result, the energy of the system has a local minimum, which suggests the existence of the self-bound object with kaon condensates as a density isomer on any scale from an atomic nucleus to a neutron star. Based on the EOS, we have discussed the structure of the kaon-condensed neutron star as a self-bound star and its implications for observations of mass-radius relations and a possibility for baryonic dark matter.[10,11]

In this paper, we extend the idea of the self-bound object with kaon condensates to a strangeness-conserving system, and we consider a kaon-condensed hypernucleus (abbreviated as KCYN) being produced in laboratories. Role of hyperon-mixing on the ground state properties of the KCYN is clarified.

2. Formulation

2.1. *Overview of the KCYN*

An initial system is taken to be a target nucleus with mass number A and atomic number Z and the antikaons (K^-) , the number of which is $|S|$ (the total negative strangeness). The K^-'s are trapped in the nucleus, and the multi-antikaonic bound state (KCYN) with baryon number A, charge $Z-|S|$, and strangeness S is supposed to be ultimately formed. In the ground state of the KCYN, part of the strangeness S should be carried by hyperons, which are mixed in the nucleus through the strong processes, $K^-N \to Y$. In this paper, we take into account the Λ, Ξ^-, and Σ^- for hyperons as the constituents of the KCYN.

A spherical liquid-drop picture for the KCYN with radius R_A is adopted, and uniform distributions of both K^- and baryons are assumed inside the nucleus. In conformity with this assumption, the K^- field is taken to be uniform as a coherent state within the framework of chiral symmetry:

$$\langle K^- \rangle = \langle K^- | \hat{K}^- | K^- \rangle = \frac{f}{\sqrt{2}} \theta \exp(-i\mu_K t) \ , \tag{1}$$

where θ is the chiral angle, μ_K the chemical potential of the condensed K^-, and f ($\simeq$ 93 MeV) is the meson decay constant. The uniform baryon number density is denoted as $\rho_{\rm B}^0$. Then R_A and $\rho_{\rm B}^0$ are related by $\frac{4}{3}\pi R_A^3 \rho_{\rm B}^0 = A$.

The electromagnetic charge, strangeness number, and baryon number conservations are imposed on the KCYN:

$$\rho_p - \rho_{\Xi^-} - \rho_{\Sigma^-} - \rho_{K^-} = (Z - |S|)/A \cdot \rho_{\rm B}^0 \ , \tag{2a}$$

$$\rho_{K^-} + \rho_\Lambda + 2\rho_{\Xi^-} + \rho_{\Sigma^-} = |S|/A \cdot \rho_{\rm B}^0 \ , \tag{2b}$$

$$\rho_p + \rho_\Lambda + \rho_{\Xi^-} + \rho_{\Sigma^-} + \rho_n = \rho_{\rm B}^0 \ , \tag{2c}$$

in terms of the number density ρ_i ($i = K^-, p, \Lambda, \Xi^-, \Sigma^-, n$). We construct the effective Hamiltonian density by introducing the charge chemical potential μ, the strangeness chemical potential μ_s, and the baryon number chemical potential ν, as the Lagrange multipliers corresponding to these conditions. The resulting effective energy density is written in the form

$$\mathcal{E}_{\rm eff} = \mathcal{E}_{\rm vol} + \mu(\rho_p - \rho_{\Xi^-} - \rho_{\Sigma^-} - \rho_{K^-}) + \mu_s(\rho_{K^-} + \rho_\Lambda + 2\rho_{\Xi^-} + \rho_{\Sigma^-}) + \nu(\rho_p + \rho_\Lambda + \rho_{\Xi^-} + \rho_{\Sigma^-} + \rho_n) \ , \quad (3)$$

where $\mathcal{E}_{\rm vol}$ is the volume part of the energy density of the KCYN. From the extremum conditions for $\mathcal{E}_{\rm eff}$ with respect to variation of ρ_i, one obtains

$$\mu_K = \mu_\Lambda - \mu_p = \mu_{\Xi^-} - \mu_\Lambda = \mu_{\Sigma^-} - \mu_n = \mu - \mu_s \ , \quad (4a)$$

$$\mu_n = -\nu \ , \quad (4b)$$

where μ_i ($\mu_i = \partial\mathcal{E}_{\rm vol}/\partial\rho_i$) is the chemical potential of the particle i. Equations (4a) and (4b) show that the system is in chemical equilibrium for the strong processes, $K^-p \rightleftharpoons \Lambda$, $K^-\Lambda \rightleftharpoons \Xi^-$, $K^-n \rightleftharpoons \Sigma^-$.

The total energy of the KCYN at zero temperature is written as

$$E(\rho_{\rm B}^0, A, Z-|S|) = A\cdot\mathcal{E}_{\rm vol}/\rho_{\rm B}^0 + 4\pi R_A^2\sigma + \frac{3}{5}\frac{(Z-|S|)^2e^2}{R_A} \ , \quad (5)$$

where the second term on the r.h.s. of Eq. (5) is the surface energy with σ being the surface tension coefficient, and the third term is the bulk Coulomb energy. The surface energy coefficient is assumed to be similar to the value deduced from the empirical mass formula of the normal nucleus and is set to be σ = 1 MeV/fm^2.

2.2. *Volume part of the energy*

For the volume energy part $\mathcal{E}_{\rm vol}/\rho_{\rm B}^0$, the s-wave kaon-baryon interactions are incorporated from the effective chiral Lagrangian. In addition, we introduce the baryon potentials V_i ($i = p, \Lambda, \Xi^-, \Sigma^-, n$) in hyperonic matter as $V_i \equiv \partial\mathcal{E}_{\rm pot}/\partial\rho_i$ with the nonrelativistic potential energy density $\mathcal{E}_{\rm pot}$.[22] The parameters in the $\mathcal{E}_{\rm pot}$ are partly determined so as to satisfy the nuclear saturation properties and hyperon potential depths in symmetric nuclear matter, V_Λ= −27 MeV, V_{Ξ^-}=−16 MeV, and V_{Σ^-} = 23.5 MeV (repulsive case) at the standard nuclear density ρ_0 (=0.16 fm^{-3}) deduced from recent hypernuclear experiments.[23] The potential energy density $\mathcal{E}_{\rm pot}$ includes the many-body interaction terms. The exponent associated with those between nucleons (between hyperons and nucleons) is denoted as δ (γ). We choose the following two parameter sets: (A) $\delta = \gamma = 5/3$, for which the incompressibility in symmetric nuclear matter at ρ_0 is estimated to be K=306 MeV. (B) $\delta = 4/3$ and γ=2.0, for which K=236 MeV. We abbreviate the EOS model for hyperonic matter obtained with (A) and (B) as H-EOS (A) and H-EOS (B), respectively. The maby-body interactions between hyperons and nucleons become more repulsive for H-EOS (B) than for H-EOS (A) at high densities, so that the H-EOS (B) leads to stiffer EOS than H-EOS (A) at high densities.

The volume part of the energy density of the KCYN is written as

$$\mathcal{E}_{\rm vol} = \frac{3}{5}\frac{(3\pi^2)^{2/3}}{2M_N}\left(\rho_p^{5/3} + \rho_n^{5/3} + \rho_\Lambda^{5/3} + \rho_{\Xi^-}^{5/3} + \rho_{\Sigma^-}^{5/3}\right)$$

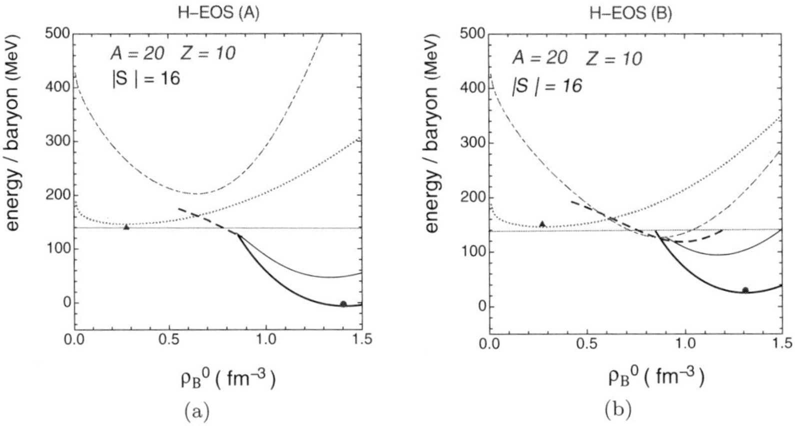

Fig. 1. (a) The energy per baryon of the KCYN as functions of the baryon number density inside the nucleus $\rho_{\rm B}^0$ for H-EOS (A) and Σ_{Kn}=305 MeV. The case for A=20, Z=10, and $|S|$=16 is shown. (b) The same as in (a), but for H-EOS (B).

$$
\begin{aligned}
&+ (\rho_\Lambda \delta M_{\Lambda N} + \rho_{\Xi^-}\delta M_{\Xi^- N} + \rho_{\Sigma^-}\delta M_{\Sigma^- N}) + \mathcal{E}_{\rm pot} \\
&- \left(\rho_p \Sigma_{Kp} + \rho_n \Sigma_{Kn} + \rho_\Lambda \Sigma_{K\Lambda} + \rho_{\Xi^-}\Sigma_{K\Xi^-} + \rho_{\Sigma^-}\Sigma_{K\Sigma^-}\right)(1-\cos\theta) \\
&+ \frac{1}{2} f^2 \mu_K^2 \sin^2\theta + f^2 m_K^2 (1-\cos\theta) \ , \qquad (6)
\end{aligned}
$$

where the first term on the right hand side of Eq. (6) denotes the baryon kinetic energy density with the free nucleon mass M_N, the second term comes from the mass difference between the hyperon and nucleon, δM_{iN}, the third term the baryon potential energy density, the fourth term the s-wave kaon-baryon scalar interaction brought about by the kaon-baryon sigma terms Σ_{Ki}, and the last two terms stand for the free parts of the condensed kaon energy density (coming from kinetic energy and free mass terms). The Kn sigma term Σ_{Kn}, simulating the s-wave Kn scalar attractive interaction, is taken to be 305 MeV throughout this paper. The energy of the ground state is obtained in the mean-field approximation with use of Eqs. (5) and (6), together with the classical field equation for the kaon, $\partial\mathcal{E}_{\rm vol}/\partial\theta = 0$, the electromagnetic charge, strangeness number, and baryon number conservations (2a)$\sim$(2c), and the chemical equilibrium conditions for the strong interaction processes (4).

3. Numerical Results

3.1. *Ground state energy*

The curves for the energy per baryon of the KCYN with A=20, Z=10, and $|S|$=16 are shown as functions of the baryon number density inside the nucleus, $\rho_{\rm B}^0$, in Fig. 1. The H-EOS (A) and the H-EOS (B) are used as the noncondensed hyperonic matter EOS in (a) and (b), respectively. Figures 1(a) and (b) show qualitatively similar features for transition of the ground state with increase in density $\rho_{\rm B}^0$: As

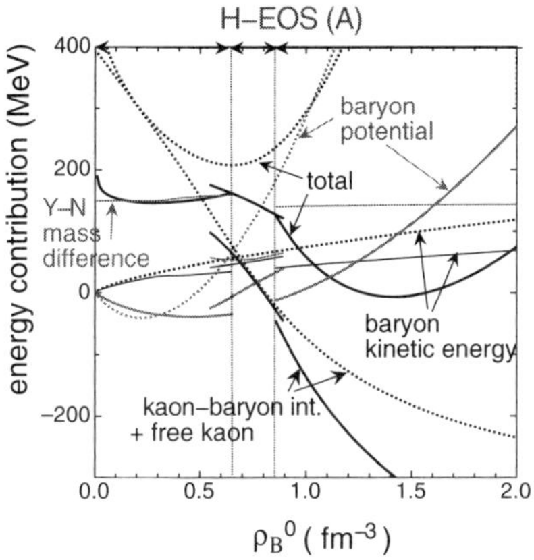

Fig. 2. The contributions to the total energy per baryon for the KCYN (solid lines) and those for the kaon-condensed nucleus without hyperon-mixing (dotted lines) as functions of $\rho_{\rm B}^0$ for A=20, Z=10, and $|S|$=16. The case for H-EOS (A) is shown.

$\rho_{\rm B}^0$ increases, the ground state changes from the noncondensed hypernuclear state (dotted line) to the K^--condensed state with n, p, Λ without mixing of Ξ^- and Σ^- (dashed line), and to the K^--condensed state with n, p, Λ, Ξ^-, and Σ^- (bold solid line) at higher densities. It is to be noted that these transitions of the ground states occur discontinuously as density $\rho_{\rm B}^0$ increases. There is a local energy minimum with respect to $\rho_{\rm B}^0$ for the K^--condensed state with full hyperon-mixing (the bold solid line). The equilibrium density $\rho_{\rm B,min}^0$ is read as 8.8 ρ_0 (8.2 ρ_0) for H-EOS (A) [H-EOS (B)]. For comparison, the K^--condensed state with n, p, Λ, Σ^- without mixing of Ξ^- is shown by the thin solid line. One can see that mixing of Ξ^- further reduces the energy and pushes up the equilibrium density. The binding energy and saturation density at the local minimum for H-EOS (B) are smaller than those for H-EOS (A), which reflects the fact that the repulsive interactions between hyperons and nucleons from many-body terms at high densities are more marked for H-EOS (B) than for H-EOS (A).[11]

The energy at the local minimum is lower than that of the free Λ-nucleon system where the energy per baryon is given by $|S|/A \cdot \delta M_{\Lambda N} \sim$ 140 MeV (shown by the thin straight line in parallel with the $\rho_{\rm B}^0$ axis). Therefore, the energy minimum state is stable against the strong processes, and it is expected to have a long lifetime, decaying only through weak processes such as $N + \langle K^- \rangle \to N + e^- + \bar{\nu}_e$ $(N = p, n)$.

3.2. *Effects of hyperon-mixing*

In Figs. 1(a) and (b), we also show the energy per baryon of the K^--condensed state without including hyperons as a function of $\rho_{\rm B}^0$, where the strangeness $|S|$ is totally carried by the K^- condensates (the dash-dotted line). From comparison of the dash-dotted line and bold solid line, one finds that mixing of hyperons leads to a more deeply bound and denser equilibrium state of the KCYN. In Fig. 2, the

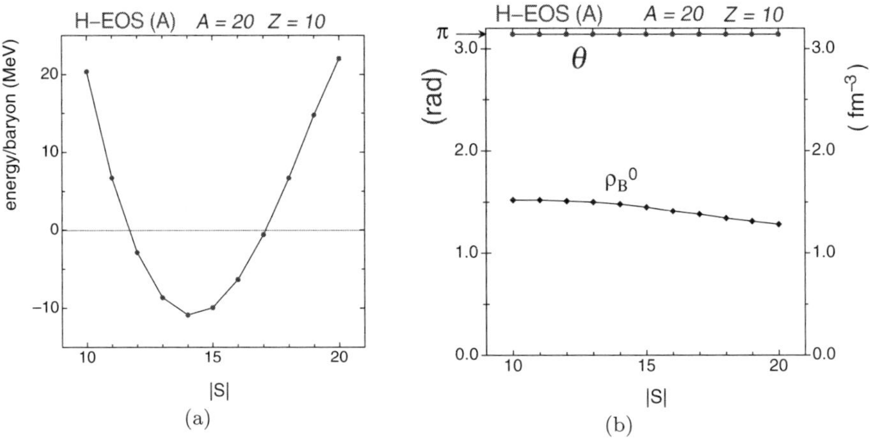

Fig. 3. (a) The energy per baryon of the energy minimum state of the KCYN with A=20, Z=10 as a function of the strangeness $|S|$ for H-EOS (A) and Σ_{Kn}=305 MeV. (b) The baryon number density ρ_B^0 and the chiral angle θ for the energy minimum state of the KCYN with A=20, Z=10 as functions of the strangeness $|S|$ for H-EOS (A) and Σ_{Kn}=305 MeV.

contributions to the total energy per baryon of the K^--condensed state with and without mixing of hyperons (solid lines and dotted lines, respectively) as functions of ρ_B^0 for A=20, Z=10, and $|S|$=16. The result by the use of H-EOS (A) is shown. [The case for H-EOS (B) is qualitatively the same as that for H-EOS (A).]

The dependence of the total energy per baryon on ρ_B^0 is mainly determined from (i) the sum of the s-wave kaon-baryon interaction and free kaon energy and (ii) the baryon potential energy $\mathcal{E}_{\rm pot}/\rho_B^0$. The attractive effect in (i), coming from the kaon-baryon interaction, gets remarkable for the hyperon-mixing case than for the case without hyperons at high densities. In addition, the repulsive effect from (ii) is much reduced for the hyperon-mixing case, since the nucleon-nucleon repulsion is avoided by mixing of hyperons at high densities. As a result, the KCYN becomes a very dense and compact object due to mixing of hyperons.

3.3. *Decay of the KCYN*

Here we discuss the properties of the equilibrium state of the KCYN systematically by changing the strangeness number $|S|$. The energy per baryon, the baryon number density inside the nucleus ρ_B^0, and the chiral angle θ, for the energy minimum state with A=20, Z=10 for H-EOS (A) are shown as functions of the strangeness number $|S|$ in Figs. 3(a) and (b).

From Figs. 3(a) and (b), one finds that the KCYN with A=20, Z=10 has the lowest energy for $|S|$=14, while the baryon number density inside the nucleus is saturated around $\rho_B^0 \sim 9\rho_0$, and the chiral angle θ has a maximum value π irrespective of $|S|$. The KCYN may decay into the lowest energy state with $|S|$=14 sequentially through weak processes. However, the typical weak processes,

$N + \langle K^- \rangle \to N + e^- + \bar{\nu}_e$ ($N = p, n$) cannot occur, because the relevant matrix elements for these processes are proportional to $\sin^2\theta$,[7] so that they vanish for $\theta = \pi$. Other weak reactions such as $\langle K^- \rangle + n \to \Xi^-$, $\langle K^- \rangle + \Lambda \to \Sigma^-, \cdots$, must be considered as possible decay modes of the KCYN.

4. Summary and Concluding Remarks

We have considered the structure and decay modes of the K^--condensed hypernucleus, aiming at producing in the laboratory as the strangeness conserving system. The effective chiral Lagrangian for the kaon dynamics including the kaon-baryon interactions has been based, combined with the nonrelativistic baryon-baryon interaction model. It has been shown that hyperon-mixing effect is important for the formation of highly dense and deeply bound state with K^- condensates. A large number of negative strangeness, $|S| = O(A)$, is needed in order to form a density isomer state, which may decay only through weak processes. The KCYN may have a close connection to a strangelet as a dense self-bound object where asymptotically free u, d, s quarks are almost equally distributed.[24–27] It is also interesting to make clear a continuity between the KCYN and the experimentally proposed kaonic nuclei systematically by varying the number of the trapped antikaons in nuclei. Related work is under way with a relativistic mean-field theory.[28]

The underlying EOS should be consistent with neutron star mass observations. Our results imply the considerably soft EOS at some density intervals as a result of coexistence of kaon condensates and hyperons. There are several observational candidates which have large gravitational masses ($\sim 2\ M_\odot$ with $M_\odot$ being the solar mass) for neutron stars.[29–31] Some attempts reconciling the observations of large neutron star masses and the soft EOS associated with phase transitions have been discussed.[32] If it is confirmed that some neutron stars have indeed such large masses, one should consider additional repulsive effects making the EOS stiffer at very high densities. With regard to this problem, the ambiguity of the kaon-baryon and baryon-baryon interactions at high densities should be examined. It is pointed out that three-body repulsive interactions should work universally between hyperons and nucleons at high densities in hyperonic matter in order to prevent too softening of the EOS due to mixing of hyperons.[33]

For future issues, one should consider the validity of the uniform distribution of the kaon condensates and baryons inside the nucleus, relativistic effects on the structure of the KCYN and the EOS. Formation mechanisms of the KCYN should be explored to settle technical problems to make the multi-antikaons trapped inside the nucleus.

Acknowledgements

The author is grateful to the organizers of the EXOCT 2007, the international symposium on Exotic States of Nuclear Matter, held at University of Catania, Italy. The work is supported in part by the Grant-in-Aid for Scientific Research Fund (C) of the Ministry of Education, Science, Sports, and Culture (No. 18540288).

References

1. D. B. Kaplan and A. E. Nelson, *Phys. Lett. B* **175**, 57 (1986).
2. T. Muto, R. Tamagaki and T. Tatsumi, *Prog. Theor. Phys. Suppl.* **112**, 159 (1993); T. Muto, T. Takatsuka, R. Tamagaki and T. Tatsumi, *Prog. Theor. Phys. Suppl.* **112**, 221 (1993).
3. C. -H. Lee, *Phys. Rep.* **275**, 197 (1996).
4. M. Prakash, I. Bombaci, M. Prakash, P. J. Ellis, J. M. Lattimer and R. Knorren, *Phys. Rep.* **280**, 1 (1997).
5. T. Muto and T. Tatsumi, *Phys. Lett. B* **283**, 165 (1992).
6. G. E. Brown, K. Kubodera, D. Page and P. Pizzochero, *Phys. Rev. D* **37**, 2042 (1988).
7. T. Tatsumi, *Prog. Theor. Phys.* **80**, 22 (1988).
8. D. Page and E. Baron, *Astrophys. J.* **254**, L17 (1990).
9. H. Fujii, T. Muto, T. Tatsumi and R. Tamagaki, *Nucl. Phys. A* **571**, 758 (1994); *Phys. Rev. C* **50**, 3140 (1994).
10. T. Muto, *Nucl. Phys. A* **697**, 225 (2002); *Nucl. Phys. A* **754**, 305c (2005); *AIP Conference Proceedings* **847**, 439 (2006).
11. T. Muto, arXiv: nucl-th/0702027.
12. Y. Akaishi and T. Yamazaki, *Phys. Rev. C* **65**, 044005 (2002); A. Dote *et al.*, *Phys. Lett. B* **590**, 51 (2004); *Phys. Rev. C* **70**, 044313 (2004).
13. T. Yamazaki, A. Dote and Y. Akaishi, *Phys. Lett. B* **587**, 167 (2004).
14. T. Kishimoto *et al.*, *Nucl. Phys. A* **754**, 383c (2005); *Prog. Theor. Phys.* **118**, 181 (2007).
15. M. Iwasaki *et al.*, nucl-ex/0310018; T. Suzuki *et al.*, *Phys. Lett. B* **597**, 263 (2004).
16. M. Agnello *et al.*, *Phys. Rev. Lett.* **94**, 212303 (2005).
17. J. Mareš, E. Friedman and A. Gal, *Phys. Lett. B* **606**, 295 (2005); *Nucl. Phys. A* **770**, 84 (2006).
18. J. Yamagata, H. Nagahiro, Y. Okumura and S. Hirenzaki, *Prog. Theor. Phys.* **114**, 301 (2005); J. Yamagata, H. Nagahiro and S. Hirenzaki, *Phys. Rev. C* **74**, 014604 (2002).
19. E. Oset and H. Toki, *Phys. Rev. C* **74**, 015207 (2006); V. K. Magas, E. Oset, A. Ramos and H. Toki, *Phys. Rev. C* **74**, 025206 (2006).
20. T. Koike and T. Harada, *Phys. Lett. B* **652**, 262 (2007).
21. Y. Ikeda and T. Sato, *Phys. Rev. C* **76**, 035203 (2007).
22. S. Balberg and A. Gal, *Nucl. Phys. A* **625**, 435 (1997).
23. A. Gal, *Prog. Theor. Phys. Suppl.* **156**, 1 (2004), and references therein.
24. N. Itoh, *Prog. Theor. Phys.* **44**, 291 (1970).
25. S. A. Chin and A. K. Kerman, *Phys. Rev. Lett.* **43**, 1292 (1979).
26. E. Witten, *Phys. Rev. D* **30**, 272 (1984).
27. E. Farhi and R. L. Jaffe, *Phys. Rev. D* **30**, 2379 (1984).
28. T. Muto, T. Maruyama and T. Tatsumi, in progress.
29. D. J. Nice *et al.*, *Astrophys. J.* **634**, 1242 (2005).
30. J. M. Lattimer, this volume.
31. D. G. Yakovlev, this volume.
32. P. Haensel, M. Bejger and J. L. Zdunik, arXiv: 0705.4594v1 [astro-ph]; P. Haensel, this volume.
33. S. Nishizaki, Y. Yamamoto and T. Takatsuka, *Prog. Theor. Phys.* **108**, 703 (2002); T. Takatsuka, *Prog. Theor. Phys. Suppl.* **156**, 84 (2004).

ISOSCALAR AND ISOVECTOR NUCLEAR MATTER PROPERTIES AND GIANT RESONANCES

H. SAGAWA

Center for Mathematical Sciences, The University of Aizu,
Aizu-Wakamatsu, Fukushima 965-8580, Japan
E-mail: sagawa@u-aizu.ac.jp

S. YOSHIDA

Science Research Center, Hosei University,
2-17-1 Fujimi, Chiyoda, Tokyo 102-8160, Japan
E-mail: s_yoshi@hosei.ac.jp

The isoscalar and isovector nuclear matter properties are investigated in the Skyrme Hartree-Fock (SHF) and relativistic mean field (RMF) models. The correlations between the nuclear matter incompressibility and the isospin dependent term of the finite nucleus incompressibility is elucidated by using various different Skyrme Hamiltonians and RMF Lagrangians. Microscopic HF+random phase approximation (RPA) calculations are performed with Skyrme interactions for Sn isotopes to study the strength distributions of isoscalar giant monopole resonances (ISGMR). The symmetry term of nuclear incompressibility is extracted to be $K_\tau = -(500 \pm 50)$ MeV from the recent experimental data of ISGMR in Sn isotopes.

Keywords: Nuclear matter properties; Incompressibility; Giant monopole resonance.

1. Introduction

Nuclear mean field models have been successful in providing a microscopic description of many properties of the nuclear ground states and of collective excitations including giant resonances. The Skyrme Hartree–Fock (SHF) model is one commonly used non-relativistic mean field model. Relativistic mean field (RMF) models, based on an effective Lagrangian for the interacting many-body system, have been used also as a successful model. The nuclear binding energy and the saturation density are precisely known experimental quantities that can be used to determine the parameter sets of all Skyrme interactions and RMF Lagrangians. Other nuclear matter properties such as the incompressibility and the symmetry energy are less well determined from experiments and are still open questions in the determination of precise values.

In this talk, we study relations among nuclear matter properties of the Hamiltonian density for a large number of different Skyrme and RMF parameter sets. Firstly, we will show correlations among nuclear matter properties in both the SHF and RMF models. Secondly, we study the correlations between the isoscalar term K_∞ and the isovector term K_τ of the incompressibility. Then, we will find out relations between the isovector nuclear matter properties and the isospin dependence of giant monopole resonances. Finally, we extract the values of K_∞ and K_τ from experimental isoscalar giant monopole resonances (ISGMR) recently observed in RCNP, Osaka University, and also in Texas A&M University. Skyrme HF+RPA (random phase approximation) calculations are also performed to study the detailed structure of ISGMR. In Section II, we study the correlations among the quantities which characterize the isoscalar and the isovector nuclear matter properties. Section III is devoted to describe the nuclear matter and finite nucleus incompressibilities by using various Skyrme and RMF parameter sets. HF+RPA results of ISGMR will be shown for Sn isotopes in Section III. The summary is given in section IV.

2. Isoscalar and Isovector Nuclear Matter Properties

The Skyrme interactions used in many recent studies have nine parameters $(t_0, t_1, t_2, t_3, x_0, x_1, x_2, x_3, \alpha)$ in addition to the parameters of the spin-orbit interaction. The energy density functionals of the Skyrme interaction are expressed by using the Thomas-Fermi approximation for the kinetic energy density. The nuclear matter properties are defined by using these energy density functionals. Various correlations among nuclear matter properties have been discussed in the cases of the SHF and RMF models recently.[1–5] It was shown that the Skyrme parameters can be expressed analytically in terms of the isoscalar and the isovector nuclear matter properties of the Hamiltonian density in Ref. 2.

In this section, we will study correlations between the isovector nuclear matter properties. The isoscalar part $h(\rho)$ and the isovector part $\varepsilon_\delta(\rho)$ of the Hamiltonian density H_{nm} are defined by

$$h(\rho) = \lim_{I\to 0} H_{nm}, \tag{1}$$

$$\varepsilon_\delta(\rho) = \frac{1}{2} \lim_{I\to 0} \frac{\partial^2}{\partial I^2}\left(\frac{H_{nm}}{\rho}\right), \tag{2}$$

where I is the asymmetric parameter $I = (\rho_n - \rho_p)/\rho$. The derivative terms and the Coulomb term in the Hamiltonian density do not give any contribution in the infinite nuclear matter calculations. The explicit forms of the Hamiltonian densities H_{nm} are found in Ref. 2.

The physical properties of infinite symmetric nuclear matter with $\rho_n = \rho_p$ can be obtained from the following six equations:

$$0 = \frac{\partial}{\partial \rho}\left(\frac{h}{\rho}\right)\Big|_{\rho=\rho_{nm}}, \tag{3}$$

$$-E_0 = \frac{h(\rho_{nm})}{\rho_{nm}} \qquad (\text{ in SHF }),$$

$$-E_0 = \frac{h(\rho_{nm})}{\rho_{nm}} - M \qquad (\text{ in RMF }), \tag{4}$$

$$K_\infty = 9\rho^2 \frac{\partial^2}{\partial \rho^2}\left(\frac{h}{\rho}\right)\Big|_{\rho=\rho_{nm}}, \tag{5}$$

$$J = \varepsilon_\delta(\rho_{nm}), \tag{6}$$

$$L = 3\rho \frac{\partial}{\partial \rho}\varepsilon_\delta(\rho)\Big|_{\rho=\rho_{nm}}, \tag{7}$$

$$K_{sym} = 9\rho^2 \frac{\partial^2}{\partial \rho^2}\varepsilon_\delta(\rho)\Big|_{\rho=\rho_{nm}}, \tag{8}$$

where ρ_{nm}, E_0, K_∞ and J are the nuclear saturation density, the binding energy per nucleon, the incompressibility of symmetric nuclear matter and the symmetry energy, respectively.

In Figs. 1(a)-(d), we study the range of values of the four quantities K_∞, J, L and K_{sym} that occur in SHF and RMF models investigated here, as well as correlations between these quantities. Clear linear correlations in both Skyrme and RMF parameter sets are found between J and L, and between J and K_{sym} as shown in Figs. 1(a) and (b), respectively. We should also notice that the values of not only L but also K_{sym} are very close in the Skyrme interactions denoted by 6, 7, 8 and 13 with $\alpha = \frac{1}{6}$ which is the power of the nuclear density-dependent term in the Skyrme interaction. While the incompressibilities K_∞ of RMF model vary over a large range, the isovector nuclear matter values J, L and K_{sym} of RMF lie in a narrow range, and are much larger than those of SHF except recently proposed parameter sets DD-ME1 and DD-ME2 with the density dependent meson−nucleon couplings. The two parameter sets DD-ME1 and DD-ME2 show similar isovector nuclear matter values J, L and K_{sym} to those of the Skyrme interactions.

We do not see any clear correlations between the incompressibility K_∞ and other properties in Figs. 1(c) and (d) at first glance. However we can see a kind of correlation when we arrange the results according to the value of α. Namely there is a singular point for K_∞ at 306MeV, i.e., α=2/3. The analytic formulas of this singularity was discussed in detail in Ref. 2. Before the singular point, the values J and L increase as a function of K_∞, while the J and L decrease after the singularity.

3. Isospin Dependence of ISGMR and Isovector Nuclear Matter Properties

The value of K_∞ is the most fundamental quantity used in determining the nuclear matter equation of states (EOS). The important experimental information on K_∞

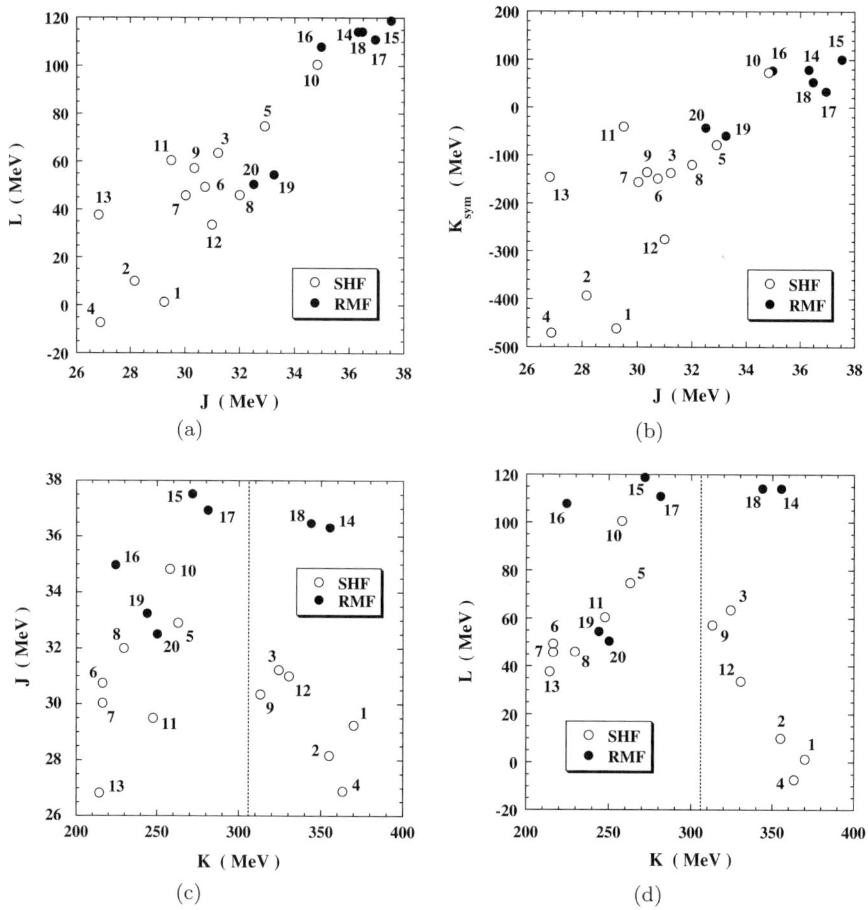

Fig. 1. Correlations among nuclear-matter properties calculated by using 13 parameter sets of the SHF (open circles) and 7 parameter sets of RMF (filled circles). Correlations between (a) J and L, (b) J and K_{sym}, (c) $K(=K_{\infty})$ and J, (d) $K(=K_{\infty})$ and L are shown. The numbers denote the different parameter sets: 1 for SI, 2 for SIII, 3 for SIV, 4 for SVI, 5 for Skya, 6 for SkM, 7 for SkM*, 8 for SLy4, 9 for MSkA, 10 for SkI3, 11 for SkI4, 12 for SkX, 13 for SGII, 14 for NLSH, 15 for NL3, 16 for NLC, 17 for TM1, 18 for TM2, 19 for DD-ME1 and 20 for DD-ME2.

is provided by ISGMR in finite nuclei. The Hartree-Fock(HF)+random phase approximation (RPA) calculations with Skyrme interactions were performed to obtain the response functions of ISGMR.[6,7] The RMF model was also applied to calculate ISGMR in the time-dependent Hartree(TDRMF) approximation and also in RPA. The calculated results were then compared with experimental data in order to extract the nuclear incompressibility.[8,9]

The nuclear matter incompressibility K_{∞} is determined by the second derivative of the energy per particle E/A with respect to the density ρ at the saturation point in Eq. (5) . The nuclear matter incompressibility K_{∞} is not a directly measurable quantity. Instead, the energy of ISGMR E_{ISGMR} is expressed in terms of the finite nucleus incompressibility K_A as[6]

$$E_{ISGMR} = \sqrt{\frac{\hbar^2 K_A}{m < r^2 >_m}}, \tag{9}$$

where m is the nucleon mass and $< r^2 >_m$ is the mean square mass radius of the ground state. The finite nucleus incompressibility can be parameterized by means of a similar expansion to the liquid drop mass formula with the volume, surface, symmetry and Coulomb terms,[6]

$$K_A = K_\infty + K_{surf} A^{-1/3} + K_\tau \delta^2 + K_{Coul} \frac{Z^2}{A^{4/3}}, \tag{10}$$

where $\delta = (N - Z)/A$. We denote K_τ the symmetry term of the finite nucleus incompressibility because the symbol K_{sym} has been already used as one of the isovector nuclear matter properties defined by Eq. (8). The volume term of the finite nucleus incompressibility K_A is identified as the nuclear matter incompressibility K_∞. The symmetry term K_τ is related to nuclear matter properties as[6,10]

$$K_\tau = K_{sym} + 3L - LB, \tag{11}$$

where B is proportional to the third derivative of the hamiltonian density with respect to the density at the saturation point,

$$B = \frac{27\rho_{nm}^2}{K_\infty} \left. \frac{d^3 h}{d\rho^3} \right|_{\rho=\rho_{nm}}. \tag{12}$$

The analytic formulas for K_{surf} and K_{Coul} are given by

$$K_{surf} = 4\pi\, r_0{}^2 \left[4\sigma(\rho_{nm}) + 9\rho_{nm} \left. \frac{d^2\sigma}{d\rho^2} \right|_{\rho=\rho_{nm}} + 2\sigma(\rho_{nm})B \right], \tag{13}$$

$$K_{Coul} = \frac{3}{5} \frac{e^2}{r_0} (1 - B), \tag{14}$$

where r_0 is the radius constant. In Eq. (13), σ is a surface tension in symmetric semi-infinite nuclear matter. K_{surf} can be evaluated by the extended Thomas-Fermi approximation and the scaled HF calculations on semi-infinite nuclear matter in the SHF model. These evaluations show an approximate relation $K_{surf} \sim -K_\infty$ within an accuracy of a few % in the SHF model. In RMF, the study of an extended Thomas-Fermi approximation gives a slightly larger surface contribution, for example, $K_{surf} \sim -1.16K_\infty$ in the case of NL3.

The values of K_τ and K_{Coul} calculated by using various Skyrme Hamiltonians and RMF Lagrangians are shown in Figs. 2(a) and (b). K_τ is largely negative and shows anti-correlation with the nuclear matter incompressibility K_∞ in both SHF and RMF models. Correlation lines for the two models are drawn in Fig. 2(a). The correlation coefficients r are evaluated to be $r = 0.827$ for SHF and $r = 0.759$ for RMF, respectively. It was pointed out in Ref. 2 that there are no clear correlations between L and K_∞, and between K_{sym} and K_∞ (see Fig. 1(d)). On the other hand

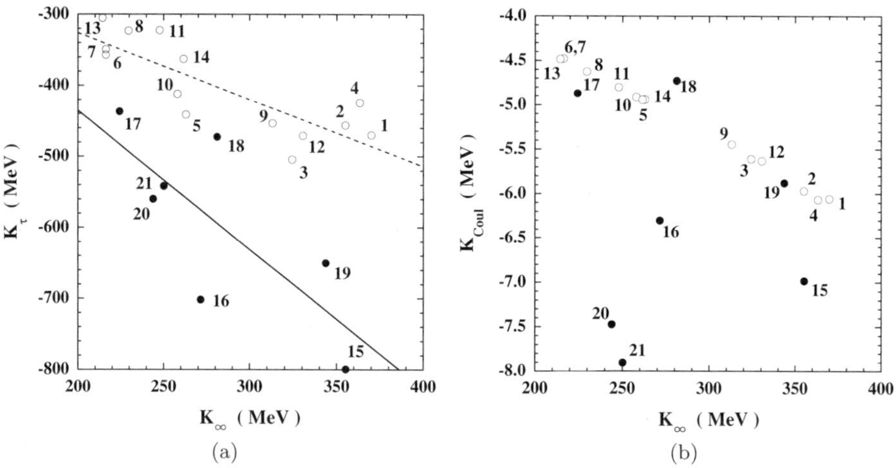

Fig. 2. (a) Correlation between the nuclear matter incompressibility K_∞ and the symmetry term K_τ in the finite nucleus incompressibility; (b) Correlation between the nuclear matter incompressibility K_∞ and the Coulomb contribution K_{Coul} in the finite nucleus incompressibility calculated by using 14 parameter sets of the SHF (open circles) and seven parameter sets of RMF (filled circles). In the left panel, the solid line is drawn for RMF model with the correlation coefficient r=0.759, while the dashed line is for SHF model with r=0.827. The numbers denote the different parameter sets: 1 for SI, 2 for SIII, 3 for SIV, 4 for SVI, 5 for Skya, 6 for SkM, 7 for SkM*, 8 for SLy4, 9 for MSkA, 10 for SkI3, 11 for SkI4, 12 for SkX, 13 for SGII, 14 for SGI, 15 for NLSH, 16 for NL3, 17 for NLC, 18 for TM1, 19 for TM2, 20 for DD-ME1 and 21 for DD-ME2.

it is remarkable that K_τ, as a linear combination of K_{sym}, L and B, show a clear correlation with K_∞ having a large correlation coefficient, especially in SHF.

As one can see in Fig. 2(a), any Hamiltonian which has a larger K_∞ gives a smaller K_τ. The variations of K_τ for the Skyrme interactions are

$$K_\tau = (-400 \pm 100) \text{ MeV}. \tag{15}$$

On the other hand, the values of RMF are largely negative and have more variation among the seven effective Lagrangians,

$$K_\tau = (-620 \pm 180) \text{ MeV}. \tag{16}$$

In principle, the value K_{Coul} should be model-independent. Nevertheless, we can see a weak correlation between K_∞ and K_{Coul} in Fig. 2(b). The correlation between K_∞ and K_{Coul} can be expressed analytically by using the isoscalar nuclear matter properties as given in the Appendix in Ref. 10. Among the 13 parameter sets of Skyrme interactions, the variation of K_{Coul} is rather small,

$$K_{Coul} = (-5.2 \pm 0.7) \text{ MeV},$$

compared with that of K_τ. The values of K_{Coul} in RMF show essentially the same trend, but have a larger variation.

Recently, the ISGMR strength distributions in the Sn isotopes from ^{112}Sn to ^{124}Sn have been measured by using inelastic α scatterings at RCNP, Osaka Univer-

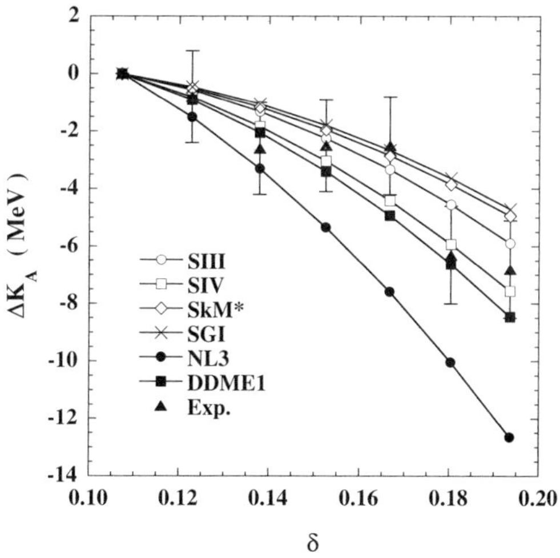

Fig. 3. The difference of incompressibility $\Delta K = K_A - K_{A=112}$ as a function of $\delta = \frac{N-Z}{A}$. Experimental data are determined by using the excitation energies of ISGMR in Ref. 11 and the HF mass radii. See the text for details.

sity.[11] The ISGMR of Sn isotopes were also studied in Texas A & M University.[13,14] We examine the K_τ dependence of the E_{ISGMR} by using the formula (10). In Fig. 3, the difference of the compressibility $\Delta K_A = K_A - K_{A=112}$ is plotted as a function of $\delta = (N-Z)/A$. We adopt four Skyrme interactions and two RMF Lagrangians. In the analysis, the surface term is taken to be $K_{surf} = -K_\infty$ for the Skyrme model and $K_{surf} = -1.18K_\infty$ for the RMF model. The adopted interactions vary from a smaller K_τ value of -350 MeV for SkM* to a larger value of -700 MeV for NL3. The empirical isospin dependence of ΔK_A is close to the results of SIII, SIV and DD-ME1, which have $K_\tau = -(500 \pm 50)$ MeV.

As seen in Fig. 4, the RPA result gives single collective peak in each Sn isotope being consistent with the experimental data obtained by Texas A&M[13,14] and also by RCNP.[11] In general, the calculated average energies by RPA with SkM* are few hundreds keV higher than the empirical ones. Recently, a relativistic RPA model[15] and SHF+Quasi-Particle RPA model[16] are applied to calculate the ISGMR in Sn isotopes. These models predict also higher ISGMR peaks in Sn isotopes close to the present results. The calculated excitation energies decrease from ^{112}Sn and ^{124}Sn by about 1 MeV which is consistent with the observed data. This decrease is expected from a large negative symmetry term K_τ in the finite nucleus incompressibility as was pointed out in Fig. 2(a).

4. Summary

We first studied the correlations between the isovector nuclear matter properties J, L and K_{sym} and found clear linear correlations among the three quantities in both SHF and RMF models. In the SHF model the correlations between the isovec-

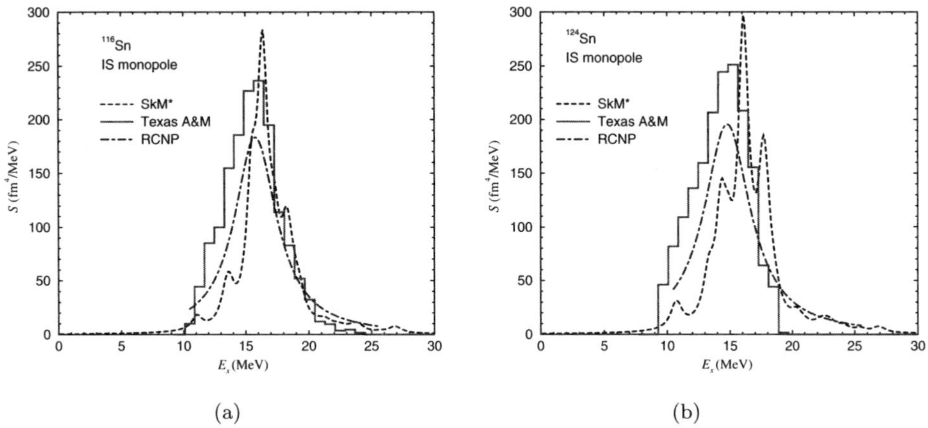

Fig. 4. HF+RPA response functions of (a) ^{116}Sn and (b) ^{124}Sn nuclei with Skyrme interaction SkM*. The experimental data are taken from Refs. 12 (Texas A&M) and 11 (RCNP).

tor quantities and the incompressibility K have a singularity at $K_c = 306$ MeV caused by a particular value $\alpha = \frac{2}{3}$ which is the power of the density-dependent term of the Skyrme force. We studied also the various terms of the finite nucleus incompressibility K_A in the SHF and RMF models. The correlations between K_∞ and K_τ and also between K_∞ and K_{Coul} are elucidated by using analytic formulas for Skyrme Hamiltonians and RMF Lagrangians. We extracted empirically the symmetry term to be $K_\tau = -(500 \pm 50)$ MeV from the analysis of the isospin dependence of the excitation energies of ISGMR in Sn isotopes.

References

1. S. Yoshida and H. Sagawa, *Phys. Rev. C* **69**, 024318 (2004).
2. S. Yoshida and H. Sagawa, *Phys. Rev. C* **73**, 044320(2006).
3. B. A. Brown, *Phys. Rev. Lett.* **85**, 5296 (2000).
4. S. Typel and B. A. Brown, *Phys. Rev. C* **64**, 027302 (2001).
5. R. J. Furnstahl, *Nucl. Phys. A* **706**, 85 (2002).
6. J. P. Blaizot, *Phys. Rep.* **64**, 171 (1980).
7. I. Hamamoto, H. Sagawa and X. Z. Zhang, *Phys. Rev. C* **56**, 3121 (1997).
8. J. Piekarewicz, *Phys. Rev. C* **66**, 041301 (2002).
9. D. Vretenar, T. Nikšić and P. Ring, *Phys. Rev. C* **68**, 024310 (2003).
10. H. Sagawa, S. Yoshida, Guo-Mo Zeng, Jian-Zhong Gu and Xi-Zhen Zhang, *Phys. Rev. C*, in press (2007).
11. U. Garg, Proc. of 2nd COMEX Meeting (2006) and private communications; T. Li *et al.*, *Phys. Rev. Lett.* in press (2007).
12. D. H. Youngblood, Y.-W. Lui, H. L. Clark, B. John, Y. Tokimoto and X. Chen, *Phys. Rev. C* **69**, 034315 (2004); Y.-W. Lui, private communications.
13. Y.-W. Lui, D. H. Youngblood, Y. Tokimoto, H. L. Clark and B. John, *Phys. Rev. C* **70**, 014307 (2004).
14. D. H. Youngblood, H. L. Clark and Y.-W. Lui, *Phys. Rev. Lett.* **82**, 691 (1999).
15. J. Piekarewicz, *Phys. Rev. C*, in press (2007).
16. G. Colò, private communications (2007).

THE SKYRME INTERACTION AND ITS TENSOR COMPONENT

GIANLUCA COLÒ and PIER FRANCESCO BORTIGNON

Dipartimento di Fisica, Università degli Studi and INFN, Sezione di Milano, 20133 Milano, Italy

HIROYUKI SAGAWA

Center for Mathematical Sciences, University of Aizu, Aizu-Wakamatsu, Fukushima 965-8560, Japan

In the present volume, several contributions are devoted to the study of the properties of exotic nuclear matter (nuclei at the drip lines or neutron stars) by using implementations of the Density Functional Theory (DFT) in atomic nuclei, either in the nonrelativistic or covariant formalism. In our contribution, we argue that tensor correlations are important to be included in the functionals; the necessity to include other kind of correlations, namely those associated with the particle-vibration coupling, is discussed by using the specific example of ^{132}Sn.

Keywords: Hartree-Fock; Tensor force; Single-particle states; Particle-vibration coupling.

1. Introduction

The atomic nucleus is a strongly correlated many-body system. Medium-heavy nuclei can be studied if an "effective" interaction is employed, which takes implicitly into account the effect of correlations. We define effective interaction every interaction which is meant to be used in connection with a given many-body approximation scheme. One of the most widely used approaches in nuclear physics is the so-called self-consistent mean-field (SCMF) method based on an effective interaction $V_{\rm eff}$.[1]

In the nuclear structure community, both zero-range interactions like the Skyrme force,[2] or finite-range interactions like the Gogny force,[3] are quite popular and lead to satisfactory descriptions of many nuclear properties. This fact is an indication that the self-consistent mean field based on these interactions is, among exisiting approaches, the one which is closest to a realization of the Density Functional Theory (DFT) for nuclei.

Recently, much attention has been paid to the importance of including the tensor correlations in the existing energy functionals. Within the shell-model framework, theoretical studies have elucidated the specific effects of the proton-neutron tensor force: in particular, in Ref. 4 it has been pointed out that this tensor force provides a specific attraction (repulsion) between protons and neutrons lying in $j_>$ and $j_<$ ($j_>$ and $j_>$, or $j_<$ and $j_<$) orbitals. In the framework of mean-field calculations employing effective interactions, the issue has been addressed in Ref. 5 using the Gogny interaction and in Ref. 6 using the Skyrme interaction. At the same time, works about the relevance of the tensor force when one considers the evolution of single-particle states have been carried out by other groups as well.[7–10]

In the present paper, part of the results already reported in Ref. 10 are discussed, together with some new result aimed at demonstrating that the inclusion of the tensor force should be carried out on equal footing with other kind of correlations like those associated with the particle-vibration coupling.

2. Theoretical Framework

In our work, we employ a Skyrme parameter set plus the triplet-even and triplet-odd tensor zero-range tensor terms, which read

$$\begin{aligned} V_{\rm eff}^{(tensor)} &= \frac{T}{2}\{[(\sigma_1\cdot k')(\sigma_2\cdot k') - \frac{1}{3}(\sigma_1\cdot\sigma_2)k'^2]\delta(r_1-r_2) \\ &+ \delta(r_1-r_2)[(\sigma_1\cdot k)(\sigma_2\cdot k) - \frac{1}{3}(\sigma_1\cdot\sigma_2)k^2]\} \\ &+ U\{(\sigma_1\cdot k')\delta(r_1-r_2)(\sigma_2\cdot k) - \frac{1}{3}(\sigma_1\cdot\sigma_2)[k'\cdot\delta(r_1-r_2)k]\}. \end{aligned} \tag{1}$$

In the above expression, the operator $k = (\nabla_1 - \nabla_2)/2i$ acts on the right and $k' = -(\nabla_1 - \nabla_2)/2i$ acts on the left. The coupling constants T and U denote the strength of the triplet-even and triplet-odd tensor interactions, respectively. The tensor interactions (1) give contributions both to the binding energy and to the spin-orbit splitting, which are, respectively, quadratic and linear in the proton and neutron spin-orbit densities,

$$J_q(r) = \frac{1}{4\pi r^3}\sum_i v_i^2(2j_i+1)\left[j_i(j_i+1) - l_i(l_i+1) - \frac{3}{4}\right]R_i^2(r). \tag{2}$$

In this expression $q = 0(1)$ labels neutrons (protons), while where $i = n, l, j$ runs over all states having the given q. The v_i^2 is the BCS occupation probability of each orbital and $R_i(r)$ is the radial part of the wavefunction. It should be noticed that the exchange part of the central Skyrme interaction gives the same kind of contributions to the total energy density and spin-orbit splitting. The spin-orbit potential is given by

$$U_{s.o.}^{(q)} = \frac{W_0}{2r}\left(2\frac{d\rho_q}{dr} + \frac{d\rho_{q'}}{dr}\right) + \left(\alpha\frac{J_q}{r} + \beta\frac{J_{q'}}{r}\right), \tag{3}$$

where the first term on the r.h.s comes from the Skyrme spin-orbit interaction whereas the second term includes both the central exchange and the tensor contributions, that is, $\alpha = \alpha_C + \alpha_T$ and $\beta = \beta_C + \beta_T$. The central exchange contributions are written in terms of the usual Skyrme parameters, $\alpha_C = \frac{1}{8}(t_1 - t_2) - \frac{1}{8}(t_1x_1 + t_2x_2)$, $\beta_C = -\frac{1}{8}(t_1x_1 + t_2x_2)$, while the tensor contributions are expressed as $\alpha_T = \frac{5}{12}U$ and $\beta_T = \frac{5}{24}(T+U)$. In this work, we employ the SLy5 parameter set.[11]

Except for the double-magic systems, we perform HF-BCS in order to take into account the pairing correlations. Our pairing force is the zero-range, density-dependent one, already employed along the Z=50 isotopic chain in connection with the SLy4 set in Ref. 12.

One should remark that since the Skyrme spin-orbit contribution (proportional to W_0) gives a value of the spin-orbit splitting which is linear in the derivatives of the proton and neutron densities, the associated mass number and isospin dependence is very moderate in heavy nuclei. On the other hand, the second term in Eq. (3) depends on the spin density J_q which has a more peculiar behavior. J_q gives essentially no contribution in the spin-saturated cases, but it increases linearly with the number of particles if only one of the spin-orbit partners is filled. The sign of the J_q will change depending upon the quantum numbers of the orbitals which are progressively filled: that is, the orbital with $j_>$ gives a positive contribution to J_q while the orbital with $j_<$ gives a negative contribution to J_q.

3. Results for the Z=50 Isotopes and N=82 Isotones

In Fig. 1, the energy differences of the proton single-particle states, $\varepsilon(h_{11/2})-\varepsilon(g_{7/2})$, along the Z=50 isotopes are shown as a function of the neutron excess N-Z. The original SLy5 interaction fails to reproduce the experimental trend both qualitatively and quantitatively. On the other hand, we can see a substantial improvement due to the introduction of the tensor interaction with (α_T, β_T)=(-170,100) MeV·fm^5. This choice gives a very nice agreement with the experimental data in the range $20 \leq$ (N-Z) ≤ 32, both quantitatively and qualitatively. The HF+tensor results can be qualitatively understood by simple arguments, looking at Eq. (3). In the Z=50 core, only the proton $g_{9/2}$ orbital dominates the proton spin density J_p (cf. Eq. (2)); consequently, with a negative value of α_T, the spin-orbit potential (3) is enlarged in absolute value (notice that W_0 is positive and the radial derivatives of the densities are negative), the values of the proton spin-orbit splittings are increased, and the energy difference $\varepsilon(h_{11/2}) - \varepsilon(g_{7/2})$ is reduced with respect to HF-BCS without tensor. This reduction is seen better around N-Z=20: in fact, ^{120}Sn is, to a good extent, spin-saturated as far as the neutrons are concerned so that one gets no contribution from J_n. It should be noticed that the term in α does not give any isospin dependence to the spin-orbit potential for a fixed proton number, but only the term in β can be responsible for the isospin dependence. In a pure HF description, from N-Z=6 to 14, the $g_{7/2}$ neutron orbit is gradually filled and J_n is reduced. Then, the positive value of β_T enlarges in absolute value the spin-orbit potential and increases the spin-orbit splitting, so that the energy difference $\varepsilon(h_{11/2}) - \varepsilon(g_{7/2})$ becomes smaller. Because of pairing, this decrease is not so pronounced in the results of Fig. 1. Moreover, from N-Z=14 to 20, the $s_{1/2}$ and $d_{3/2}$ neutron orbits are occupied and in this region the spin density is not so much changed since the $s_{1/2}$ orbital does not provide any contribution. Instead, for N-Z=20 to 32, the $h_{11/2}$ orbital is gradually filled. This gives a positive contribution to the spin-orbit potential (3)

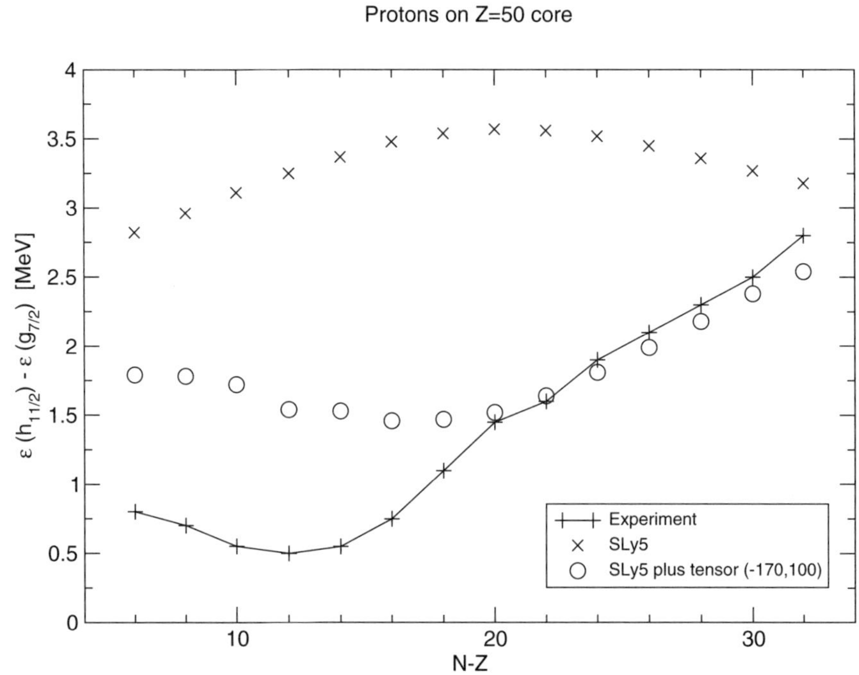

Fig. 1. Energy differences between the $1h_{11/2}$ and $1g_{7/2}$ single-proton states along the Z=50 isotopes. The calculations are performed without (crosses) and with (circles) the tensor term in the spin-orbit potential (3), on top of SLy5 (which includes the central exchange, or J^2, terms). The experimental data are taken from Ref. 13. See the text for details.

and the the spin-orbit splitting becomes smaller. $\varepsilon(h_{11/2}) - \varepsilon(g_{7/2})$ consequently increases, and this effect is well pronounced in our theoretical results.

4. The Effect of Particle-Vibration Coupling

As discussed at length in Ref. 14, the coupling of single-particle states to vibrational states has the net effect of increasing the level density around the Fermi surface by about 30%, by shifting occupied and unoccupied states in opposite directions. In our work, we have taken into account this coupling in the case of the nucleus ^{132}Sn (i.e., without pairing correlations), by using the formalism which is described in detail in Ref. 15. We have calculated the single-particle states first within simple Skyrme-HF, and then by including the tensor terms with the parameters mentioned above. We have finally added perturbatively the energy shift due to the coupling with phonons calculated by means of Random Phase Approximation (RPA). The results are depicted in Fig. 2, where we display the valence neutron hole states.

One can notice that the tensor force is quite important for the shift it can gives to the $h_{11/2}$ state. Both tensor correlations and the particle-vibration coupling cooperate in giving the correct ordering and energy separation between the $g_{7/2}$

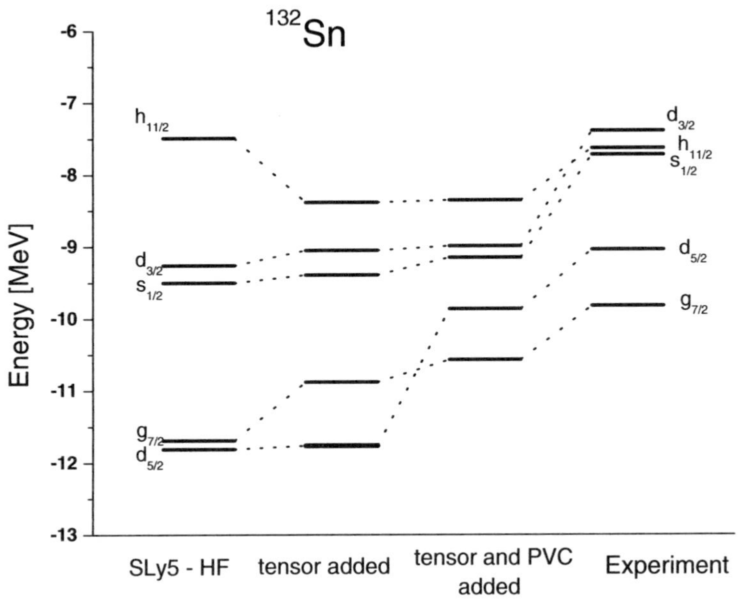

Fig. 2. Results for the neutron hole states in the nucleus ^{132}Sn. The four columns, from left to right, correspond respectively to: simple HF, HF including the tensor terms, HF including the tensor terms plus the energy shift due to coupling with vibrations and experimental energies.

and $d_{5/2}$ states. Moreover, the large compression of levels provided by the particle-vibration coupling is clearly evident in Fig. 2. The final result, as compared with the experimental finding, is extremely better than pure HF, in keeping with the fact that it provides the correct magnitude of the energy range of the valence shell. The only feature which is still not reproduced is the ordering of the two last occupied levels, $d_{3/2}$ and $h_{11/2}$. Apparently, no mean field model can provide the correct ordering of these two levels.[1]

5. Conclusions

In conclusion, we have shown that by including the tensor terms on top of existing Skyrme functionals, it is possible to explain the isotope trend of single-particle states. We have discussed only the case of the Z=50 isotopes, but in another contribution included in the present volume,[16] it is shown that the same parameters provide good results also for lighter systems. By combining together the mean field results which include the tensor correlations, and the coupling of single-particle states with nuclear shape vibrations, we have tried to reproduce in detail the spectrum of ^{132}Sn. Our results are preliminary, and more work along this line is in progress. However, it looks quite promising to try to extend this strategy to a number of different systems, in keeping with what has been shown in Fig. 2.

Acknowledgments

We acknowledge the financial support from the Asia-Link project [CN/ASIA-LINK/008(94791)] of the European Commission.

References

1. M. Bender, P.-H. Heenen and P.-G. Reinhard, *Rev. Mod. Phys.* **75**, 121 (2003).
2. D. Vautherin and D. M. Brink, *Phys. Rev. C* **5**, 626 (1972); M. Beiner, H. Flocard, N. Van Giai and Ph. Quentin, *Nucl. Phys. A* **238**, 29 (1975).
3. J. Dechargé and D. Gogny, *Phys. Rev. C* **21**, 1568 (1980); J. F. Berger, M. Girod and D. Gogny, *Comp. Phys. Comm.* **63**, 365 (1991).
4. T. Otsuka, T. Suzuki, R. Fujimoto, H. Grawe and Y. Akaishi, *Phys. Rev. Lett.* **95**, 232502 (2005).
5. T. Otsuka, T. Matsuo and D. Abe, *Phys. Rev. Lett.* **97**, 162501 (2006).
6. B. A. Brown, T. Duguet, T. Otsuka, D. Abe and T. Suzuki, *Phys. Rev. C* **74**, 061303(R) (2006).
7. D. M. Brink and Fl. Stancu, *Phys. Rev. C* **75**, 064311 (2007).
8. J. Dobaczewski, in: T. Duguet, H. Esbensen, K. M. Nollett and C. D. Roberts (eds.), *Proceedings of the Third ANL/MSU/JINA/INT RIA Workshop*, (World Scientific, Singapore, 2006).
9. T. Lesinski *et al.*, *Phys. Rev. C* **76**, 014312 (2007).
10. G. Colò, H. Sagawa, S. Fracasso and P. F. Bortignon, *Phys. Lett. B* **646**, 227 (2007).
11. E. Chabanat *et al.*, *Nucl. Phys. A* **635**, 231 (1998).
12. S. Fracasso and G. Colò, *Phys. Rev. C* **72**, 064310 (2005).
13. J. P. Schiffer *et al.*, *Phys. Rev. Lett.* **92**, 162501 (2004).
14. C. Mahaux, P. F. Bortignon, R. A. Broglia and C. H. Dasso, *Phys. Rep.* **120**, 1 (1985).
15. G. Colò, T. Suzuki and H. Sagawa, *Nucl. Phys. A* **695**, 167 (2001).
16. Cf. the contribution of Wei Zou *et al.* in this volume.

SPIN-ISOSPIN PHYSICS AND ICHOR PROJECT

H. SAKAI for the ICHOR collaboration

Department of Physics, University of Tokyo,
7-3-1 Hongo, Bunkyo, Tokyo 113-0033, Japan
E-mail: sakai@phys.s.u-tokyo.ac.jp
www.nucl.phys.s.u-tokyo.ac.jp

After a brief introduction of general aspects on the spin-isospin responses of nuclei, the ICHOR project is introduced. The ICHOR Project aims to explore the unexplored spin-isospin responses of nuclei by using the exothermic heavy-Ion charge-exchange reactions employing radioactive beams.

Keywords: Spin giant resonances; Landau-Migdal parameters; IVSMR; IVSDR.

1. Introduction

Spin-isospin interactions produce various interesting responses in nuclei such as the Gamow-Teller (GT) β-decay, GT giant resonance (GTGR), isovector-spin monopole giant resonance (IVSM), spin dipole resonance (SDR), double GT giant resonance (DGTGR) etc. depending on momentum and/or energy of nuclear system. The knowledge of those responses often plays an essential role in understanding the astrophysical phenomena.

One good example is the extraction of the Landau-Migdal parameter $g'_{N\Delta}$ which provides us the information on the critical density of pion condensation ρ_c. The pion condensation might be realized in neutron stars. If it were realized, the equation-of-state (EOS) is largely modified resulting in radius and/or mass changes as well as in the cooling speed of neutron stars. The $g'_{N\Delta}$ value can be estimated from the quenching value of the GT responses. The quenching value is a reduction factor from the GT spin sum-rule (occasionally called Ikeda sum-rule) value of $S(\beta^-) - S(\beta^+)$ =3(N-Z) which is the model independent quantity. Thus one needs to measure full GT responses for both β^+ and β^- directions. Such measurements have been performed employing the (p, n) and (n, p) reactions on ^{90}Zr target at 300 MeV. The GT strengths were very reliably extracted, which were used to determine the quenching value of 0.86±0.07. The $g'_{N\Delta}$ (g'_{NN}) value was estimated to be 0.2-0.4 ($\sim$ 0.6) and this small value leads to $\rho_c \sim 2\rho_0$, assuming $g'_{\Delta\Delta} = 0.5$. Detailed description on how to reduce these values can be found in the recent review article "Spin-isospin responses via (p, n) and (n, p) reactions" by Ichimura *et al.*[1]

Another example is the use of the model independent spin-dipole sum-rule

$$S_- - S_+ = \frac{9}{4\pi}(N < r^2 >_n - Z < r^2 >_p), \qquad (1)$$

from which the neutron skin thickness $\delta_{np} = \sqrt{< r^2 >_n} - \sqrt{< r^2 >_p}$ can be extracted. The δ_{np} value gives a strong constraint on the neutron matter EOS and consequently on neutron stars. It was reported in this conference by Yako[2] how to derive the total spin-dipole strengths for β^- and β^+ directions from the experiments. The δ_{np} value is successfully extracted.[3,4]

2. The ICHOR Project

So far the spin-isospin responses have been studied mainly by using the charge-exchange (CE) reactions with stable beams. Note that the CE reactions with stable beams lead to endothermic reactions. Generally the spin-isospin responses under consideration ($\Delta L = 0, \Delta S = 1$) are most strongly excited at small momentum transfer (preferably $q = 0$) by the reaction. However, the reaction kinematics due to the endothermic reaction is such that the excitation of nucleus is always associated with the finite momentum transfer ($\omega < q$ (space-like region)). Therefore it becomes extremely difficult to search for spin-isospin responses in the highly excited region due to the increasing momentum transfer. The exothermic Heavy-Ion Charge-Exchange (HICE) reaction employing radioactive projectiles having a large positive reaction Q-value allows us to access an unexplored $\omega > q$ region (time-like region).

We have initiated the ICHOR project[a] (Isospin-spin responses in CHarge-exchange exOthermic Reactions) in 2005. The ICHOR project consists of mainly two subjects related with spin-isospin responses.

The first subject concerns with the study of the intermediate states of the double beta decay nucleus through the (p, n) and (n, p) reactions. The 2ν double beta decay proceeds through the GT states in the intermediate nucleus in two steps, the GT transitions from mother nucleus to daughter nucleus and subsequent GT transitions from daughter nucleus to grand-daughter nucleus. The first step can be studied by the (p, n) reaction on the mother nucleus, while the second step by the (n, p) reaction on the grand-daughter nucleus. We have measured the (p, n) reaction on ^{48}Ca, ^{76}Ge, ^{100}Mo and ^{116}Cd and the (n, p) reaction on ^{48}Ti and ^{116}Sn, respectively. Data analyses are under way.

The second subject aims to explore the spin-isospin modes in the highly excited region with q ~small by using the exothermic HICE reactions. To realize such research with radio-isotope (RI) secondary beams, we are presently constructing the magnetic spectrometer SHARAQ in the RIKEN Nishina Center.

In the next section, we show the basic principle of the exothermic HICE reaction how to realize a condition $\omega \gg 0$ with $q = 0$. Then we describe the magnetic

[a]Grant-in-Aid for Specially Promoted Research of MEXT, Japan. Spokespaerson : H. Sakai.

spectrometer SHARAQ which is under construction dedicated to the radio active secondary beam experiments.

3. Exothermic HICE Reaction

For the nuclear reaction of $A(a,b)B$, momentum transfer q, excitation energy ω of the particle B and the mass difference $\Delta m(\Delta M)$ between the particle $a(A)$ and $b(B)$ are expressed by

$$\begin{aligned}\vec{q} &= \vec{p}_B - \vec{p}_A = \vec{p}_a - \vec{p}_b,\\ \omega &= M'_B - M_B,\\ \Delta m &= M_a - M_b,\\ \Delta M &= M_A - M_B,\end{aligned}$$

where $\vec{p_i}$ and M_i denote the momentum and the rest mass of particle i, respectively. It is often defined $Q_{gg} = \Delta m + \Delta M$ in a heavy ion reaction. For charge exchange reaction, $|\Delta m| \ll M_a$, with small energy and momentum transfer compared to the rest mass of the relevant particle at small scattering angle ($q \ll M_A$, $q \ll \sqrt{M_a^2 + p_a^2}$, and $\theta \ll 1$), the relationship between ω and q is expressed as:

$$\begin{aligned}\omega &\simeq \frac{\Delta m}{\gamma} + \Delta M \pm \beta\sqrt{q^2 - (p_a\theta)^2},\\ (\beta q)^2 &\simeq \left(\omega - \frac{\Delta m}{\gamma} - \Delta M\right)^2 + (\beta p_a \theta)^2 \geq \left(\omega - \frac{\Delta m}{\gamma} - \Delta M\right)^2,\end{aligned} \tag{2}$$

where β and γ denote the velocity and the γ factor of the incident particle a. Because of $\Delta m < 0$ and $\Delta M < 0$ for the charge exchange reaction with a stable beam and stable target combination, the momentum transfer q is always greater than ω/β, i.e. restriction to the space-like region (see Eq. (2)). However, if we use a charge exchange reaction from an unstable nuclear projectile with a large mass excess to a stabler ejectile, we can access to the time-like region ($q < \omega$) due to the offset from $\Delta m/\gamma + \Delta M > 0$ in the right hand side of Eq. (2).

4. Excitations of IVSMR and DGTR by Exothermic HICE Reactions

To show the merit of the exothermic HICE reactions, two examples of research are briefly introduced

4.1. *IVSMR*

The first example is the study of the isovector spin monopole resonances (IVSMR) in both β^- and β^+ directions. There are some experimental hints on IVSMR(β^-)[5,6] but none exists on IVSMR(β^+). Excitation energies and widths provide us very important information on the nuclear matter compressibility with spin degrees of

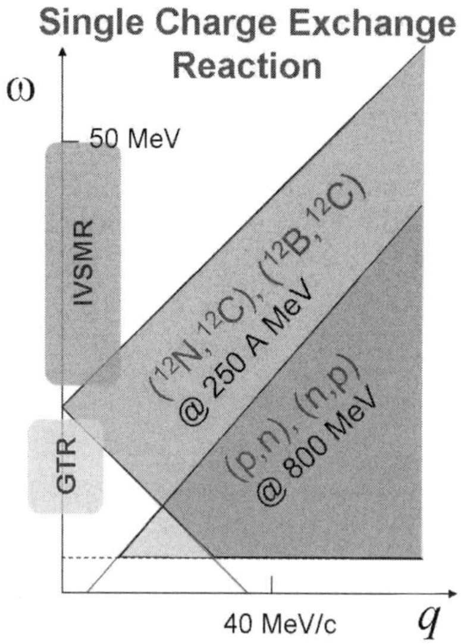

Fig. 1. Kinematical condition for the studies of the IVSMR using unstable and stable nuclear beams indicated in the figure.

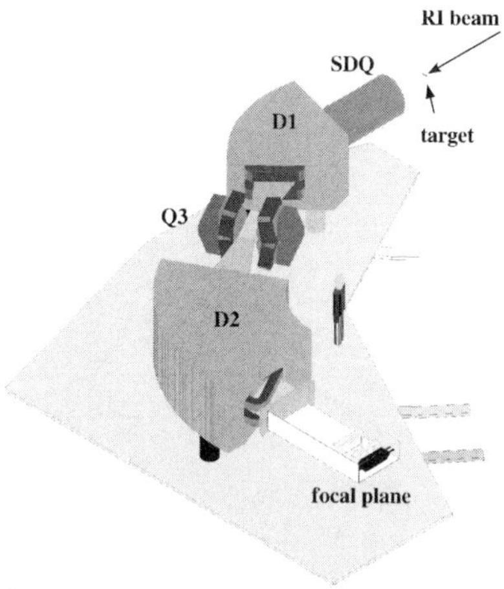

Fig. 2. The SHARAQ spectrometer.

freedom. In addition the precise neutron skin thickness can be extract by employing the model independent non-energy weighted sum rule which is

$$S_- - S_+ = 3[N < r^4 >_n - Z < r^4 >_p], \tag{3}$$

where $< r^4 >_n$ and $< r^4 >_p$ are the mean quadruple radii (mqr) of the neutron and proton distributions, respectively. Because of the r^4 dependence, it is highly sensitive to the neutron/proton distribution. Please be aware it is the r^2 dependence for SDR (see Eq. 1). Thus one can probably deduce more reliably the $\Delta_n p = \sqrt[4]{< r^4 >_n} - \sqrt[4]{< r^4 >_p}$ value from the measured S_- and S_+ values. Note that the precise charge radius value $\sqrt[4]{< r^4 >_p}$ can be obtained from the electron scatterings.

For the study of IVSMR, the exothermic (^{12}N,^{12}C) and (^{12}B,^{12}C) reactions should be a good choice for β^- and β^+ directions, since they have large positive Δm values, namely +16.83 MeV and +13.88 MeV, respectively. A kinematic relation between ω and q is schematically plotted in Fig. 1.

4.2. *DGTR*

The double Gamow-Teller giant resonance (DGTR) has never observed. It is a two phonon state, the GT giant resonance built on the GT giant resonance. There are several attempts to search for DGTR using the endothermic (^{18}O,^{18}Ne) or (^{11}B,^{11}Li) reactions, which were not successful. However the exothermic (^{20}Mg,^{20}Ne) reaction, for example, with a large mass excess of Δm =25.6 MeV can achieve easily an excitation energy of ~20 MeV with $q = 0$, which could be an essential condition to excite DGTR.

Table 1. Specifications of Q1–3 magnets.

	Q1	Q2	Q3
Bore [mm]	$340^W \times 230^H$		$\phi 270$
Max. gradient [T/m]	14.1	14.1	7.4
Effective length [mm]	530	1020	840

Table 2. Specifications of D1 and D2 magnets.

	D1	D2
Bending radius [m]	4.4	4.4
Bending angle [deg]	32.7	60
Gap [mm]	230	200
Max. field strength [T]	1.55	1.55
Weight [t]	102	290

5. Magnetic Spectrometer SHARAQ

The SHARAQ spectrometer is a high resolution magnetic spectrometer designed for radioactive isotope (RI) beam experiments at the RI Beam Factory (RIBF). The spectrometer will be used in nuclear physics experiments in combination with a variety of radioactive isotope (RI) beams from BigRIPS.[7]

The SHARAQ spectrometer can analyze particles with magnetic rigidity of 1.8–6.8 Tm, corresponding to an energy of 40–440 MeV/A for $A/Z = 2$ particles. It is designed for momentum and angular resolutions of $\delta p/p \sim 1/15000$ and $\delta\theta \sim 1$ mrad, respectively. The SHARAQ can be rotated around its target position in the range from $-2°$ to $15°$ to allow 0 degree and forward angle measurements. Figure 2 gives an overview of the SHARAQ spectrometer.

RI beams, produced by the primary heavy-ion beams in BigRIPS, are transported to the SHARAQ target position through the BigRIPS beam-line and a SHARAQ beam-line which is in preparation.[8] The beam-line and the spectrometer are designed to fulfill dispersion matching conditions both in both lateral and angular dispersions simultaneously which are crucial to achieve the high spacial and angular resolving powers of the spectrometer because of the large intrinsic momentum spread of RI beams. The beam-line is also designed to operate in an achromatic mode to provide a small beam spot at the target position for diagnostic and experimental purposes.

5.1. *Ion-optical design*

The SHARAQ spectrometer consists of three quadrupole (Q) and two dipole (D) magnets in the order of QQDQD.[9] The first quadrupole doublet (SDQ) consists of two superconducting quadrupoles (Q1 and Q2). The following normal-conducting dipole (D1) and quadrupole (Q3) magnets are recycled from the decommissioned spectrograph SMART. The second dipole magnet, D2, is a new magnet specially designed to achieve the spectrometer specifications.[10] The D2 magnet is a 60° bending magnet with a pole gap of 200 mm. The specifications of the quadrupole magnets (Q1, Q2 and Q3) and the dipole magnets (D1 and D2) are listed in Tables 1 and 2, respectively.

6. Summary

The ICHOR project which has started in 2005 is briefly introduced. It is an abbreviation of the Isospin-spin responses in CHarge-exchange exOthermic Reactions and it consists mainly of two parts: study of the intermediate states of double beta decay nucleus by the (p, n) and $(n.p)$ reactions and investigation of IVSMR and/or DGTR by the exothermic HICE reactions. The merit of using the exothermic HICE reactions is stressed. For the latter purpose, the high-resolution SHARAQ spectrometer dedicated to the secondary RI beam is now under construction at RI Beam Factory (RIBF) of RIKEN. It is designed to achieve a high momentum resolution of $\delta p/p$=1/14700 and a high angular resolution of $\delta\theta \sim 1$ mrad. The spectrometer is scheduled to be operational in 2008.

Acknowledgements

The author thanks Profs. Shimoura and Uesaka for allowing to use some materials from Refs. 11,12. This work is supported by the Grant-in-Aid of Specially Promoted Research (Grant No. 17002003) of the Ministry of Education, Culture, Sports, Science, and Technology of Japan.

References

1. M. Ichimura, H. Sakai and T. Wakasa, *Prog. Part. Nucl. Phys.* **56**, 446 (2006).
2. K. Yako, these conference proceedings.
3. K. Yako, H. Sagawa and H. Sakai, *Phys. Rev. C* **74**, 051303(R) (2006).
4. H. Sagawa *et al.*, *Phys. Rev. C* **76**, 024301 (2007).
5. D. L. Prout *et al.*, *Phys. Rev. C* **63**, 014603 (2000).
6. R. G. T. Zegers *et al.*, *Phys. Rev. Lett.* **90**, 202501 (2003).
7. T. Kubo, *Nucl. Instrum. Methods Phys. Res. B* **204**, 97 (2003).
8. T. Kawabata *et al.*, in XIth International Conference on Electromagnetic Isotope Separators and Techniques Related to their Applications (EMIS2007), Deauville, 2007.
9. T. Uesaka *et al.*, CNS Annual Report 2005.
10. G. P. Berg, "Design Study of a New Dipole Magnet D2 for the SHARAQ Spectrometer," *unpublished.*
11. S. Shimoura, *ibid.* in EMIS2007.
12. T. Uesaka, *ibid.* in EMIS2007.

NEUTRON SKIN THICKNESS OF ^{90}ZR DETERMINED BY (P,N) AND (N,P) REACTIONS

K. YAKO* and H. SAKAI

Department of Physics, University of Tokyo,
Bunkyo, Tokyo 113-0033, Japan
**E-mail: yakou@phys.s.u-tokyo.ac.jp*

H. SAGAWA

Center for Mathematical Sciences, The University of Aizu,
Aizu-Wakamatsu, Fukushima 965-8580, Japan

Charge exchange spin-dipole (SD) excitations of ^{90}Zr are studied by the ^{90}Zr(p, n) and ^{90}Zr(n, p) reactions at 300 MeV. A multipole decomposition technique is employed to obtain the SD strength distributions in the cross section spectra. A model-independent SD sum rule value is obtained: 148 ± 12 fm^2. The neutron skin thickness of ^{90}Zr is determined to be 0.07 ± 0.04 fm from the SD sum rule.

Keywords: Charge exchange reactions; Charge exchange spin-dipole sum rule; Neutron skin thickness.

1. Introduction

Proton and neutron distributions in nuclei are among the most fundamental issues in nuclear physics. The neutron skin thickness, defined as the difference between the root mean square (rms) radii of the proton and neutron distributions, is known to be correlated with the nuclear symmetry energy and to be related to the equation of state of the neutron matter.[1–5] Several attempts have been made to determine neutron distributions experimentally.[6–10] Nevertheless the results are model-dependent and they should be checked against one another.

We present a method for determining the neutron rms radius by using the model-independent sum rule strength of charge exchange spin-dipole (SD) excitations.[11] The operators for SD transitions are defined by $\hat{S}_\pm = \sum_{im\mu} t^i_\pm \sigma^i_m r_i Y^\mu_1(\hat{r}_i)$ with the isospin operators $t_3 = t_z$, $t_\pm = t_x \pm i t_y$. The model-independent sum rule is derived as

$$S_- - S_+ = \frac{9}{4\pi}\left(N\langle r^2\rangle_{\rm n} - Z\langle r^2\rangle_{\rm p}\right), \tag{1}$$

where $S_\pm$ are the total SD strengths. The mean square radii of the neutron and proton distributions are denoted as $\langle r^2\rangle_{\rm n}$ and $\langle r^2\rangle_{\rm p}$, respectively. In this work the model-independent sum rule value is extracted from both (p, n) and (n, p) spectra on ^{90}Zr.

2. Analysis

The dipole components of the cross section spectra are identified by multipole decomposition (MD) analysis of the ^{90}Zr(p, n)[12] and ^{90}Zr(n, p)[13] data at 300 MeV. Then the SD strengths are obtained by assuming a proportionality relation between B(SD) and the $\Delta L = 1$ component of the cross section $\sigma_{\Delta L=1,\pm}(q,\omega)$, given by

$$\sigma_{\Delta L=1,\pm}(q,\omega) = \hat{\sigma}_{\mathrm{SD}_\pm}(q,\omega)B(\mathrm{SD}_\pm), \tag{2}$$

where $\hat{\sigma}_{\mathrm{SD}_\pm}(q,\omega)$ is the SD cross section per $B(\mathrm{SD}_\pm)$, which depends on the momentum transfer q and the energy transfer ω. The $\sigma_{\Delta L=1}(q,\omega)$ data of the (p, n) and (n, p) channels were taken from the result of MD analysis at 4.6° and 4–5°, respectively.[13] The $\sigma_{\Delta L=1}(q,\omega)$ spectra at these angles are most sensitive to the SD cross sections since the $\Delta L = 1$ cross sections take on maximum values.

Distorted wave impulse approximation (DWIA) calculations are performed to obtain $\hat{\sigma}_{\mathrm{SD}_\pm}(q,\omega)$ for the 0^-, 1^-, and 2^- transitions using the computer code DW81.[14] The one-body transition densities are calculated from the simplest 1p1h configurations considering all configurations within the $1\hbar\omega$ excitation. The optical model potential parameters are taken from Ref. 15. The effective NN interaction is taken from the t-matrix parameterization of the free NN interaction by Franey and Love at 325 MeV.[16] We use the averaged unit cross section of $\hat{\sigma}_{\mathrm{SD}_-}(4.6°, 0\ \mathrm{MeV}) = 0.27$ mb/sr/fm^2 and $\hat{\sigma}_{\mathrm{SD}_+}(4\text{–}5°, 0\ \mathrm{MeV}) = 0.26$ mb/sr/fm^2 in the analysis. The unit cross sections are calculated at each energy bin of the cross section histogram. Details of the analysis are given in Ref. 17. The systematic uncertainty is estimated to be 14%.

3. Results

The $dB(\mathrm{SD})/dE$ distributions obtained by using Eq. (2) are shown in Fig. 1. The horizontal axis in the $dB(\mathrm{SD}_-)/dE$ spectrum is the excitation energy of the residual ^{90}Nb nucleus. The SD$_-$ strength spectrum shows a dominant resonance structure centered at $E_x = 20$ MeV and the strength extends to ~ 50 MeV excitation. The $dB(\mathrm{SD}_+)/dE$ distribution is shifted by +17 MeV to account for the Coulomb displacement energy and the nuclear mass difference. A comparison between the data and some random-phase approximation calculations has been presented by H. Sagawa.[18]

The integrated SD strengths, $S_\pm \equiv \int_0^{E_x} \frac{dB(\mathrm{SD}_\pm)}{dE} dE$, and the sum rule value, $S_- - S_+$, are plotted in Fig. 2. The sum rule value is almost constant in the excitation energy range of 30–50 MeV.[17] The sum rule value up to 40 MeV yields $S_- - S_+ = 148 \pm 6(\mathrm{stat.}) \pm 7(\mathrm{syst.}) \pm 7(\mathrm{MD})$ fm^2, where the statistical uncertainty, the systematic uncertainty of the normalization in the cross section data, and the uncertainty in the MD analysis are given. The corresponding SD strengths are $S_- = 247 \pm 4(\mathrm{stat.}) \pm$

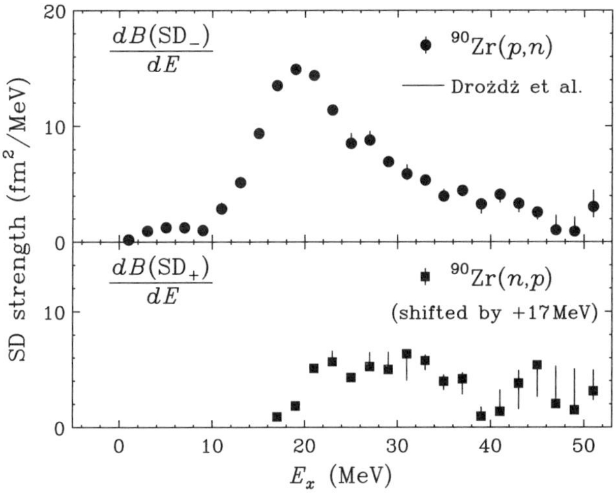

Fig. 1. Charge exchange SD strength $dB(SD_-)/dE$ (upper panel) and $dB(SD_+)/dE$ (lower panel). The circles and squares are the experimental data. The $dB(SD_+)/dE$ spectrum is shifted by +17 MeV.

13(syst.) $\pm$ 12(MD) fm^2 and $S_+ = 98 \pm 4$(stat.) $\pm$ 5(syst.) $\pm$ 5(MD) fm^2. The rms radius of the proton matter in ^{90}Zr is estimated to be 4.19 fm[18] after correcting for the effect of the proton form factor from the charge radius.[19] The neutron skin thickness calculated from Eq. (1) is then 0.07 ± 0.04 fm, where the statistical and systematic uncertainties are combined in quadrature with the uncertainty in MD analysis. The value obtained in the present study is consistent with that obtained from the analysis of the proton elastic scattering (0.09 ± 0.07 fm)[6] but with a smaller uncertainty.

4. Summary

A consistent MD analysis of the (p, n) and (n, p) reaction data from ^{90}Zr has been performed and the SD strength distributions of both channels are obtained experimentally for the first time. By using the two experimentally obtained integrated SD strengths, the model-independent formula (1) yields a SD sum rule value of 148 ± 12 fm^2, which corresponds to a neutron skin thickness of 0.07 ± 0.04 fm. This method is applicable to heavy and medium heavy nuclei on which both the (p, n) and (n, p) measurements are possible.

References

1. S. Typel and B. A. Brown, *Phys. Rev. C* **64**, 027302 (2001).
2. R. J. Furnstahl, *Nucl. Phys. A* **706**, 85 (2002).
3. J. M. Lattimer and M. Prakash, *Science* **304**, 536 (2004).
4. P. Danielewicz, *Nucl. Phys. A* **727**, 233 (2003).
5. S. Yoshida and H. Sagawa, *Phys. Rev. C* **73**, 044320 (2006).

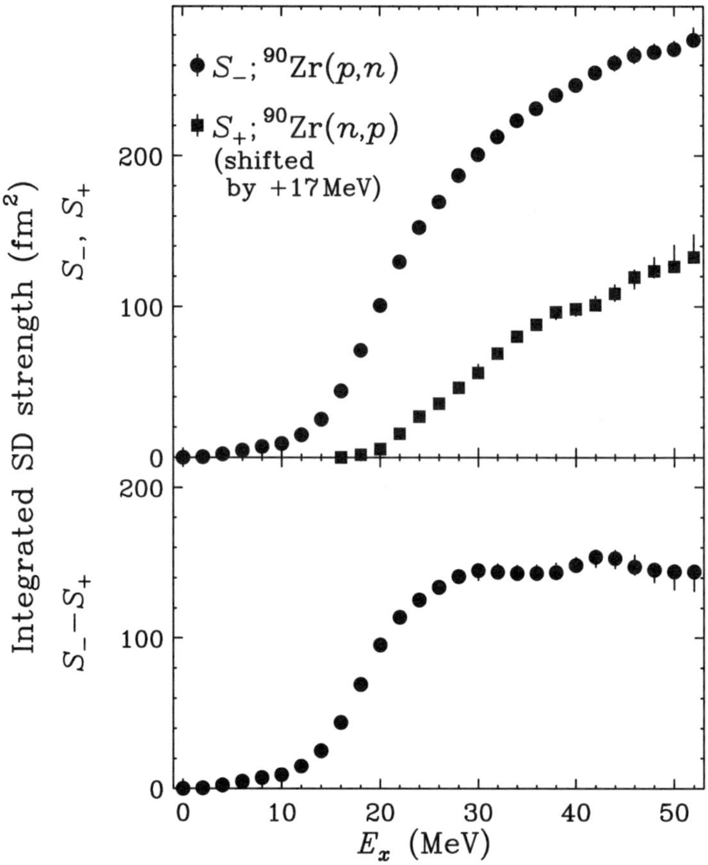

Fig. 2. Integrated charge exchange SD strengths. The upper panel shows the S_- and S_+ spectra. The lower panel shows the $S_- - S_+$ spectrum.

6. L. Ray *et al.*, *Phys. Rev. C* **18**, 1756 (1978).
7. A. Krasznahorkay *et al.*, *Phys. Rev. Lett.* **66**, 1287 (1991).
8. A. Krasznahorkay *et al.*, *Phys. Rev. Lett.* **82**, 3216 (1999).
9. A. Trzcińska *et al.*, *Phys. Rev. Lett.* **87**, 082501 (2001).
10. C. J. Horowitz *et al.*, *Phys. Rev. C* **63**, 025501 (2001).
11. C. Gaarde *et al.*, *Nucl. Phys. A* **369**, 258 (1981).
12. T. Wakasa *et al.*, *Phys. Rev. C* **55**, 2909 (1997).
13. K. Yako *et al.*, *Phys. Lett. B* **615**, 193 (2005).
14. M. A Schaeffer and J. Raynal, *Program DWBA70* (unpublished); J. R. Comfort, Extended version *DW81*.
15. E. D. Cooper, S. Hama, B. C. Clark and R. L. Mercer, *Phys. Rev. C* **47**, 297 (1993).
16. M. A. Franey and W. G. Love, *Phys. Rev. C* **31**, 488 (1985).
17. K. Yako, H. Sagawa and H. Sakai, *Phys. Rev. C* **74**, 051303(R) (2006).
18. H. Sagawa *et al.*, *Phys. Rev. C* **76**, 024301 (2007).
19. H. de Vries, C. W. de Jager and C. de Vries, *At. Data and Nucl. Data Tables* **36**, 495 (1987).

SYNTHESIS OF SUPER-HEAVY NUCLEI IN A MODIFIED DI-NUCLEAR SYSTEM MODEL

E. G. ZHAO,[1] J. Q. LI,[2] S. G. ZHOU,[1] W. ZUO[2] and W. SCHEID[3]

[1] *Institute of Theoretical Physics, Chinese Academy of Sciences, Beijing 100080, China*
[2] *Institute of Modern Physics, Chinese Academy of Sciences, Lanzhou 730000, China*
[3] *Institute of Theoretical Physics, Justus-Liebig-University Giessen, 35392 Giessen, Germany*

Based on a modified di-nuclear system model in which the nucleon transfer is coupled with dissipation of energy and angular momentum, the evaporation residue cross sections of some cold and hot fusion reactions to synthesize super-heavy nuclei are calculated. The influences of deformation and relative orientation of projectile and target are discussed. Our results show that the waist to waist orientation is in favor of the formation of super-heavy compound nuclei. The isotopic dependence of the maximal evaporation residue cross section is also investigated. The calculated results show that the cross sections do not change much with increasing neutron number of the target.

Keywords: Super-heavy nuclei; Heavy-ion fusion reaction; Di-nuclear system model.

The synthesis of super-heavy elements (SHE) has been a very hot field in nuclear physics for decades. Till now, many isotopes of SHE from $Z = 103$ to 118 have been produced by heavy-ion fusion reactions in experiments.[1] However, the production cross sections and the corresponding life time of SHE decrease rapidly as the charge number Z increases. To understand the mechanism of the heavy-ion fusion reaction and to guide future experiments, many theoretical models have been developed.[2] Among them the di-nuclear system (DNS) model has made great progresses in reproducing the available experimental data. Based on the DNS model, some modifications to it were made.[3] With this modified DNS model, the evaporation residue cross sections of some cold and hot fusion reactions to synthesize super-heavy nuclei are calculated. The influences of deformation and relative orientation of the projectile and the target are discussed.

In the DNS model the evaporation residue cross section is written as

$$\sigma_{ER}(E_{cm}) = \sum_{J=0}^{J_f} \sigma_c(E_{cm}, J) P_{CN}(E_{cm}, J) W_{sur}(E_{cm}, J) \ , \tag{1}$$

where σ_c, P_{CN} and W_{sur} are the capture cross section, the probability of compound nucleus formation by nucleon transfer and the survival probability of the compound nucleus respectively.[4] The nucleon transfer can be described with the

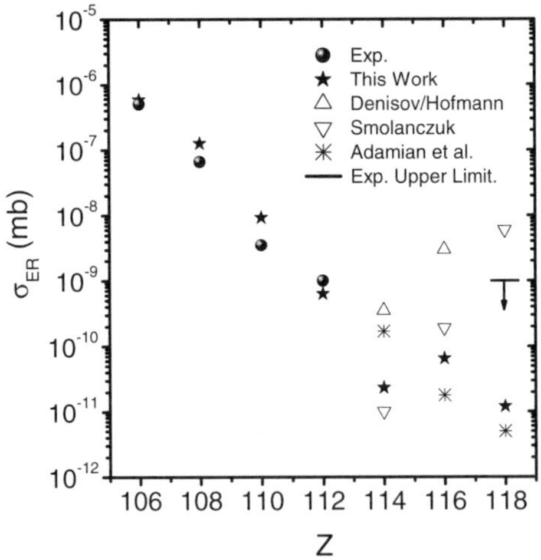

Fig. 1. The evaporation residue cross sections for one neutron emission in Pb-based reactions.[3] Our calculated results are indicated by solid stars, the experimental values by solid dots. Some estimated data for elements 114, 116 and 118 are indicated with different symbols.

master equation

$$\frac{dP(A_1,E_1,t)}{dt} = \sum_{A_1'} W_{A_1A_1'}\left[d_{A_1}P(A_1',E_1',t) - d_{A_1'}P(A_1,E_1,t)\right] - \Lambda^{qf}_{A_1,E_1,t}(\Theta)P(A_1,E_1,t)\,, \tag{2}$$

where the meaning of all the notations here can be found in Ref. 3. In the original DNS model, the nucleon transfer starts after all the relative kinetic energy is converted into the intrinsic excitation energies of the target and projectile. In our treatments, however, the nucleon transfer is coupled with the dissipation of the relative kinetic energy and angular momentum, and the transition probability $W_{A_1A_1'}$ is calculated microscopically.[3]

Based on the above modifications to the DNS model, the evaporation residue cross sections for some cold fusion reactions to synthesize super-heavy nuclei are calculated. The results are shown in Fig. 1. It can be seen that our calculation can reproduce the existing experimental data quite well. For the case of Z equal to or larger than 114, different models gave rather different results which need to be examined by future experiments.

Some hot fusion reactions were also investigated.[5] In Fig. 2 the results for reaction of ^{48}Ca + ^{208}Pb are given. We can see that the capture cross sections are reproduced very well though it has nothing to do with DNS. The experimental evaporation residue cross sections for evaporating different number of neutrons are reproduced qualitatively.

In the left panel of Fig. 3, the calculated P_{CN} for the reaction ^{86}Kr + ^{208}Pb in two different orientations are displayed. The full and open circles are for the pole to pole and the waist to waist orientations respectively. We can see that the P_{CN} for

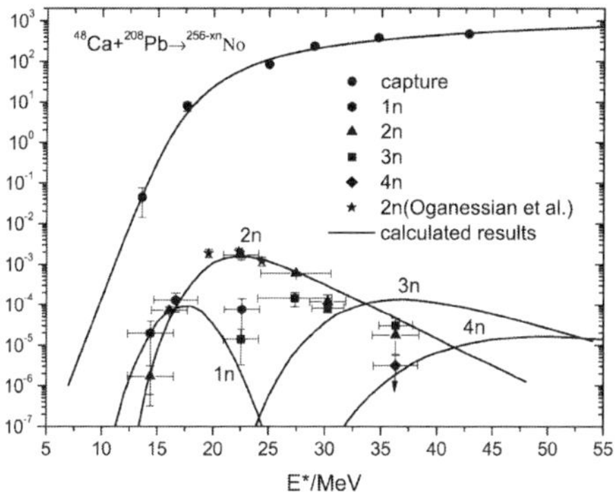

Fig. 2. Comparison of measured and calculated excitation functions for capture and xn evaporation channels in the reaction ^{48}Ca + ^{208}Pb.[5] The experimental values are represented by solid circles for capture cross sections and hexagons, triangles, squares and diamonds for evaporation residue cross sections from Refs. 6,7. Pentagrams denote experimental data from Dubna.[8]

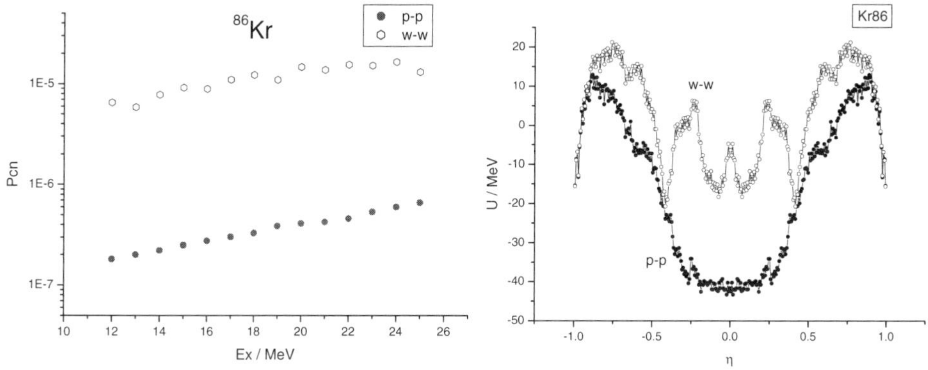

Fig. 3. Left: The calculated P_{CN} for reaction of ^{86}Kr + ^{208}Pb in different orientations. The open and full circles are for waist to waist (ww) and pole to pole (pp) orientations respectively; Right: The driving potentials for reaction ^{86}Kr + ^{208}Pb in different orientations. The open and full circles are for waist to waist (ww) and pole to pole (pp) orientations respectively.

the waist to waist orientation are much larger than that for the pole to pole. The reason for these results is related to the difference between the driving potentials for different orientations which is shown in the right panel of Fig. 3. It can be seen that the quasi-fission barrier for the waist to waist orientation is much larger than the one for the pole to pole.

Since the life times for all the super-heavy nuclei produced till now are very short and they are getting shorter when the proton number increases. The main reason for this situation seems to be the neutron number of these super-heavy nuclei being too small. Here we made some investigations to see what happens if one increases the neutron number of the target. In Fig. 4, the isotopic dependence of the maximal evaporation residue cross section of the reaction ^{48}Ca + ^{A}Pu for 3n and 4n channels is shown.[5] The evaporation residue cross section does not change much with increasing the neutron number of target.

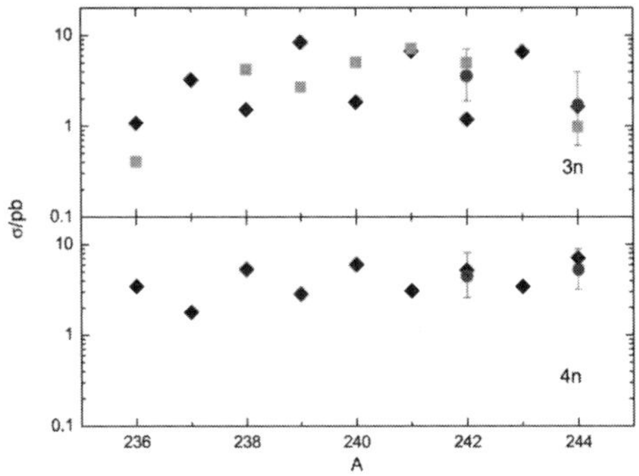

Fig. 4. Isotopic dependence of the maximal production cross-sections of the reaction $^{48}Ca + {}^{A}Pu \to 114$ for 3n (upper panel) and 4n (lower panel) evaporation channels.[5] The diamonds are our calculation. The solid circles with error bars are experimental data from Ref. 9. The full squares are calculated results from Ref. 10.

In summary, a modified di-nuclear system model in which the nucleon transfer is coupled with dissipation of energy and angular momentum is used to calculate the evaporation residue cross sections of some cold and hot fusion reactions to synthesize super-heavy nuclei. The effects of deformation and relative orientation of the projectile and the target are discussed. It's shown that the waist to waist orientation is in favor of formation of super-heavy compound nuclei. An investigation of the isotopic dependence of the evaporation residue cross section indicates that the cross sections do not change much with increasing neutron number of the target.

Acknowledgments

This work was partly supported by NSFC (10435010, 10475003 and 10575036), the Major State Basic Research Development Program of China (2007CB815000) and Knowledge Innovation Projects of CAS (KJCX3-SYW-N02 and KJCX2-SW-N17) and DFG of Germany. Two of us (EGZ and WZ) also acknowledge the support from University of Catania, the Asia-Link project [CN/ASIA-LINK/008(94791)] of the European Commission, and the helpful discussions with Prof. U. Lombardo.

References

1. S. Hofmann and G. Muezenberg, *Rev. Mod. Phys.* **72**, 733 (2000); Yu. Ts. Oganessian *et al.*, *Phys. Rev. C* **69**, R021601 (2004); K. Morita *et al.*, *J. Phys. Soc. Jpn.* **73**, 2593 (2004).
2. V. V. Volkov, *Physics of Particles and Nuclei* **35**, 425 (2004), and references therein.
3. W. F. Li *et al.*, *J. Phys. G* **32**, 1143 (2006).
4. G. G. Adamian *et al.*, *Nucl. Phys. A* **633**, 409 (1998).
5. Z. Q. Feng, G. M. Jin, F. Fu and J. Q. Li, *Nucl. Phys. A* **771**, 50 (2006).
6. E. V. Prokhorova *et al.*, arXiv: nucl-ex/0309021.
7. A. V. Belozerov *et al.*, *Eur. Phys. J. A* **16**, 447 (2003).
8. Yu. Ts. Oganessian *et al.*, *Phys. Rev. C* **64**, 054606 (2001).
9. Yu. Ts. Oganessian *et al.*, *Phys. Rev. C* **70**, 064609 (2004); **69**, 054607 (2004).
10. G. G. Adamian, N. V. Antonenko and W. Scheid, *Phys. Rev. C* **69**, 014607 (2004).

PART G

Nuclear Superfluidity

MESOSCOPIC TREATMENT OF SUPERFLUID NEUTRON CURRENT IN SOLID STAR CRUST

B. CARTER

Observatoire de Paris-Meudon, France
E-mail: brandon.carter@obspm.fr

The flow velocities in a rotating neutron star crust are low compared with the speed of light, so that a Newtonian description can be used without serious loss of accuracy at a local level. However, on a larger scale it will be necessary to take account of the curvature due to the very the strong gravitational field, so that for a global treatment it will be necessary to use a fully General Relativistic description.

EQUATION OF STATE IN THE INNER CRUST OF NEUTRON STARS: DISCUSSION OF THE UNBOUND NEUTRONS STATES

J. MARGUERON and N. VAN GIAI
Institut de Physique Nucléaire, Université Paris-Sud, F-91406 Orsay CEDEX, France

N. SANDULESCU
Institute of Physics and Nuclear Engineering, R-76900 Bucharest, Romania

In this paper, we calculate the stable Wigner-Seitz (W-S) cells in the inner crust of neutron stars and we discuss the nuclear shell effects. A distinction is done between the shell effects due to the bound states and those induced by the unbound states, which are shown to be spurious. We then estimate the effects of the spurious shells on the total energy and decompose it into a smooth and a residual part. We propose a correction to the Hartree-Fock binding energy in Wigner-Seitz cell (HF-WS).

Keywords: Equation of state; Self-consistent mean-field model; Pairing correlations.

1. Introduction

From the very first nuclear models based on extrapolations of a mass formula,[1] several models have been developed to provide the equation of state in the inner crust, like the BBP[2] and BPS[3] semi-classical models or fully quantal self-consistent mean-field models.[4–7,9,10] The latter models are based on the Wigner-Seitz (W-S) approximation and the continuum states are discretized in a box (spherical or not). In semi-classical models, a clear limitation is coming from the absence of nuclear shell effects which is known to play an important role in determining the composition of the ground state prior to the neutron drip,[4] in the inner crust[6] as well as in the transition zone between the crust and the core.[11] Since the seminal work of Negele & Vautherin,[4] many efforts have been invested in the developpement of self-consistent mean-field modelization of the inner crust, by introducing deformation of the nuclear cluster in the high density part of the crust,[5,10] or pairing correlations,[7–9] or more recently by improving of the lattice description within the band theory.[12] Band theory takes into account the proper symmetries of the system, but the equations are numerically very complicated to solve and have not yet been solved in a self-consistent framework. Instead, it has been coupled to a self-consistent mean-field model at the W-S approximation.[13] It has then been shown that the W-S approximation is justified if the temperature is larger than about

Table 1. Comparison of the proton number Z and the radius of the W-S cell R_{WS} obtained in different calculations. See the text for more explanations.

N_{zone}	k_F	Z				R_{WS} (fm)			
	fm^{-1}	N-V[4]	HF-BCS[a]	HF	HFB	N-V[4]	HF-BCS[a]	HF	HFB
2	1.12	40	20	22	22	19.6	15	16.6	16.6
3	0.84	50	36	22	20	27.6	24	20.6	20.6
4	0.64	50	56	24	24	33.2	36	26.2	26.6
5	0.55	50	58	28	20	35.8	40	30.4	27.0
6	0.48	50		28	20	39.4		32.0	29.2
7	0.36	40		22	20	42.4		36.0	33.4
8	0.30	40		22	22	44.4		38.2	37.8
9	0.26	40		22	22	46.4		38.0	38.8
10	0.23	40		36	36	49.4		46.6	47.0
11	0.20	40	52	36	28	53.8	57	48.4	43.8

Note: [a] The numbers shown in this column interpolate the results presented in Ref. 9.

100 keV, or if the quantity of interest averages the density of states on a typical scale of about 100 keV around the Fermi surface. The typical scale which has been found, 100 keV, is in fact related to the average inter-distance energy between the unbound states which are discretized in the W-S box. In this paper, we clarify the effect of the discretization of the unbound states on the ground state energy as well as the role of pairing correlations. Consequences for the equation of state are presented.

2. Equation of State in the Inner Crust

The inner crust matter is divided in 11 zones shown in Tab. 1, as in Ref. 4 (notice that the denser zone has been removed since it is in the deformed pasta region). Each W-S cell is supposed to contain in its center a spherical neutron-rich nucleus surrounded by unbound neutrons and immersed in a relativistic electron gas uniformly distributed inside the cell. For a given baryonic density $\rho_B = A/V$, neutron number N and proton number Z, the total energy of a W-S cell, E_{tot} has contributions coming from the rest mass of the particles, $E_m(N,Z) = Z(m_p + m_e) + N m_n$, from the nuclear components (including Coulomb interaction between protons) $E_n(\rho_B, N, Z)$, from the lattice energy,

$$E_l(\rho_B, N, Z, \{\rho_p\}) = \frac{1}{2}\int d^3r_1 d^3r_2 \rho_e(r_1)\frac{e^2}{|r_{12}|}\left(\rho_e(r_2) - \rho_p(r_2)\right) , \qquad (1)$$

which is induced by the difference between the uniform electron density and proton density, and finaly from the kinetic energy T_e of the relativistic electron gas.[4] The Coulomb exchange energy as well as the screening correction for the electrons gas has been neglected.

The self-consistent Hartree-Fock-Bogoliubov (HFB) approach has been presented in detail in Ref. 14. It has been extended to describe the dense part of the inner crust at finite temperature in Ref. 15. With minor modifications, it has

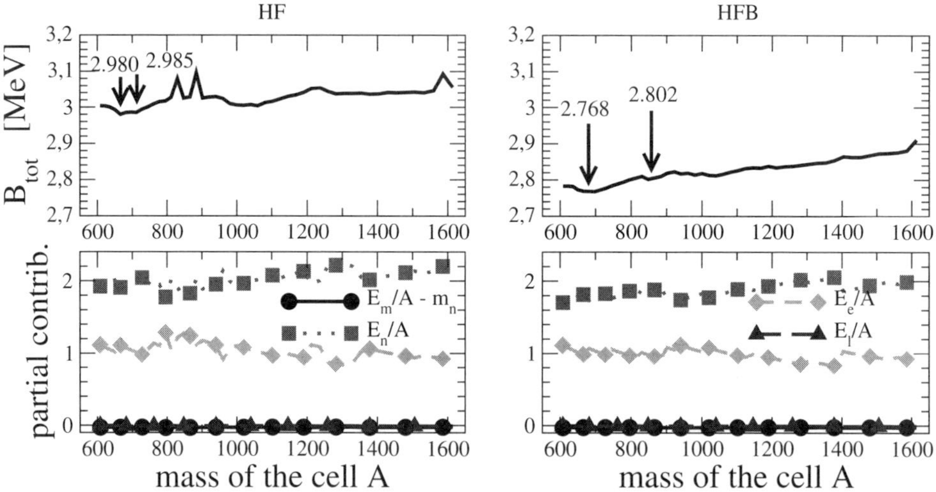

Fig. 1. The binding energy ϵ/ρ_B of the W-S cells in zone N_{zone}=4 as well as the partial contributions of the rest mass energy $E_m/A - m_n$, the nucleons, the electrons and the lattice. HF and HFB results are shown on the left and right panels, respectively. See the text for more details.

been successfully applied to calculate the specific heat of the whole part of the inner crust.[16] The effective N-N interaction should provide a reasonable description of both the nuclear clusters and the neutron gas. We use the Skyrme force SLy4[17] which was adjusted to describe properly the mean field properties of neutron-rich nuclei and infinite neutron matter. The choice of the pairing force is more problematic since at present it is not yet clear what is the strength of pairing correlations in low density neutron matter. However, in the present paper we have chosen to adjust the effective pairing interaction on the results of the pairing gap obtained with the bare interaction. We then have a maximum gap in neutron matter of about 3 MeV located at $k_F = 0.85$ fm^{-1}.

The minimization is performed in two steps: first, at a given density ρ_B and for a given mass number A, we look for the cell which satisfies the beta-stability criterion, $\mu_p + \mu_e = \mu_n$. The neutron and proton chemical potentials are provided by the HFB calculation while the electron chemical potential is deduced from $\mu_e = dE_{tot}/dN_e$. Note that, due to shell effects we may find several cells which satisfy this criterion. We then choose the one which has a minimal energy. In the second step, we minimize the total energy B_{tot} with respect to the mass number A. It should be noted that, in the second step all the cells must be calculated at the same density while the minimum energy is searched over the variable A. As the volume is modified by unit steps (the step of the mesh is 0.2 fm), the mass number A can take only discrete values so that $A = \rho_B V$ is always satisfied. The step between two values of A is not constant and vary as the square of the WS radius R_{WS}. We have approximatively $dA \sim \rho_B 4\pi R_{WS}^2 dR_{WS}$.

For N_{zone}=4, we show in Fig. 1 a typical example of step 2 in the search for the

W-S cell which minimizes the energy. On the l.h.s. is represented the total binding energy E_{tot}/A of the W-S cells obtained in Hartree-Fock (HF) mean-field model (without pairing). The position of the lowest minimum is indicated as well as two other local minima. In this case, there is a difference of 5 keV per nucleon between the two first minimum. On the r.h.s., we plot the results of HFB calculations which show the effects of the pairing correlations: the first minimum is deeper (about 34 keV per nucleons). On the lower part of the graph, we show the partial contributions of the rest mass energy $E_m/A - m_n$, the nuclear binding energy E_n/A, the electron kinetic energy T_e/A and lattice energie E_l/A. One sees that the inner crust is a typical example of frustrated system: when the nuclear energy is minimum, the electron energy is maximum and vice-versa. The minimum in energy is a compromise between these two opposite energies. The consequence is that it gives rise to numerous local minima which are not very far in energy from the absolute minimum.

The equation of state obtained at the HF and HFB approximations is given in Table 1. In total, 19714 cells configurations has been calculated. The pairing correlations do not strongly modify the HF results, it may slightly decrease the proton number and the WS radius. We also show previous calculations of equations of state. We have obtained proton numbers Z different from those of Refs. 4,9. The results of Ref. 4 have been obtained in a density matrix expansion framework interpolating microscopic calculations for homogeneous matter. The difference may be due to the fact that the Skyrme interaction SLy4 is based on more recent microscopic parametrizations. It is difficult to compare with the results of Ref. 9 since those calculations have not been done at the same densities and they have essentially been carried out in the dense part of the crust. We then show an interpolation of their results. The comparison shows a similar trend in the dense cells.

3. Discussion of the Shell Effects

In ordinary nuclei, the bound states play an essential role in determining the ground state properties and the role of the continuum states is usually negligeable. The coupling between bound states and continuum states can be more important for nuclei near the drip lines, as the Fermi energy is increasing. However, even in these most extreme situations continuum states are only virtually populated and the occupation numbers associated to continuum states are small compared to those of the bound states. In the inner crust of neutron stars, the situation is completely different: the Fermi energy is positive and most of the mass is due to the unbound neutrons while bound states account only for a small fraction. Thus, the continuum states contribute in an essential way to physical properties of the crust like its density, total energy or specific heat. The presence of bound states, or nuclear clusters, is not negligible but it induces only corrections to the ground-state properties.[16] It is then very important to have an accurate model of the unbound states.

It should also be noted that, as the continuum states are embedded in a Coulomb lattice, their asymptotic behavior is different from that occurring in nuclei. The continuum states are indeed periodically distorted within a distance equal to the lattice step (20-100 fm). The proper theory for the description of the continuum states is the band theory which requires to solve the Schrödinger equation,[12,13] $(h_0 + h_{\mathbf{k}})u_{\alpha,\mathbf{k}}(\mathbf{r}) = \epsilon_{\alpha,\mathbf{k}}u_{\alpha,\mathbf{k}}(\mathbf{r})$, where h_0 is the usual single-particle Hamiltonian of mean-field models. The second term $h_{\mathbf{k}}$ is induced by the Floquet-Bloch boundary conditions for the single-particle wave function $\varphi_{\alpha,\mathbf{k}}$: $\varphi_{\alpha,\mathbf{k}}(\mathbf{r}) = u_{\alpha,\mathbf{k}}(\mathbf{r})e^{i\mathbf{k}\cdot\mathbf{r}}$. The spherical W-S approximation is obtained by setting $\mathbf{k} = 0$ and replacing the W-S elementary polyhedron by a sphere. Then, the remaining index α is discrete, like in usual mean-field models, and runs over the eigenstates of the Hamiltonian h_0. The main difference between mean-field models and band theory is indeed due to the term $h_{\mathbf{k}}$. In band theory, this term introduces a new index $\mathbf{k}$ which is a continuous variable ranging between 0 and π/a fm^{-1} where a is the lattice step. As a consequence, the density of unbound states is continuous (notice that it may also have structures and gaps). It is then clear that the discrete distribution of unbound states present in usual mean-field models is a consequence of the dropping of $h_{\mathbf{k}}$ in the W-S approximation.

We can characterize the distribution of unbound states by the average distance between the unbound states energies, which can be related to the radius of the W-S cell as $\Delta\epsilon \sim \hbar/2mR_{WS}^2$: the smaller the W-S radius, the larger the average distance $\Delta\epsilon$. One could notice that the W-S radius is a function of the density and it is fixed by the equation of state, as shown in Table 1. At the W-S approximation, the distribution of unbound states is then also fixed by the density. The non-continuous distribution of unbound states could be interpreted as a spurious effect, induced by the W-S approximation, since in band theory calculations, the density of unbound states varies in a continuous way.[13] One can then distinguish between the physical shell effects due to the bound states from the spurious shell effects due to the unbound states and the W-S approximation.

4. Spurious Shell Effects

In order to estimate the spurious shell effects due to the unbound neutrons, we simulate an homogeneous gas of neutron matter in W-S cell by removing all the proton states. In homogeneous matter, the energy density should be independent of the volume, then of the W-S radius, and it should depend only on the density. In our simulation of homogeneous matter, we then vary the W-S radius from 10 to 50 fm, at constant neutron density. We call $B_{WS-hom.}(\rho_{unb.}, R_{WS})$ the binding energy per particle obtained for a given W-S radius R_{WS} and for a given density of unbound neutrons $\rho_{unb.}$. We show the binding energy versus the W-S radius in Fig. 2(a) for several densities corresponding to N_{zone}=6, 7 and 8. It is clearly seen that the binding energy is not independent of the W-S radius, except for large W-S radii where the binding energy converge to the energy of homogeneous matter, $B_{hom.}(\rho_{unb.})$. The binding energy of homogeneous matter $B_{hom.}(\rho_{unb.})$ is calculated

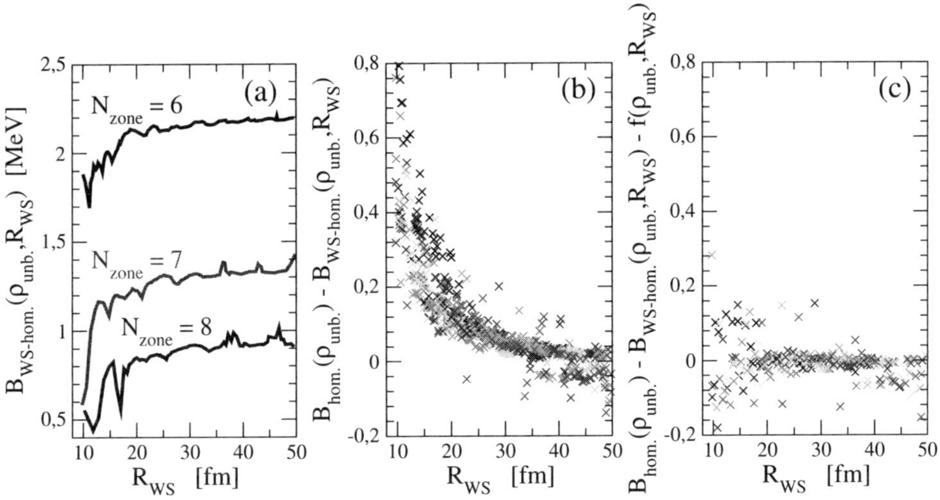

Fig. 2. We represent several quantities versus the W-S radius: (a) the binding energy $B_{WS-hom.}(\rho_{unb.}, R_{WS})$ corresponding to N_{zone}=6, 7 and 8 (b) the difference $B_{hom.}(\rho_{unb.}) - B_{WS-hom.}(\rho_{unb.}, R_{WS})$ (dots) and the function $f(\rho_{unb.}, R_{WS})$ which fit the smooth component (solid lines), and (c) the difference between the dots presented in (b) and the function $f(\rho_{unb.}, R_{WS})$.

from the average value of the Skyrme two-body interaction on a plane wave basis. The difference $B_{hom.}(\rho_{unb.}) - B_{WS-hom.}(\rho_{unb.}, R_{WS})$ is represented on Fig. 2(b) (dots). As expected, this quantity is converging to zero for large W-S radii, but for small radii, around 20 fm for instance, the difference can be as large as 300 keV per nucleons.

From the dots represented on Fig. 2(b), we obtain an universal function which fits the smooth radial dependence of the difference $B_{hom.}(\rho_{unb.}) - B_{WS-hom.}(\rho_{unb.}, R_{WS})$. As this effect is due to the discretization of the unbound states, it is proportional to the average distance $\Delta\epsilon$ between the box states. This difference should then decrease with the W-S radius as R_{WS}^{-2}. It should also be proportional to the number of unbound neutrons, while the density dependence is not known. We then represent it by a power law $\rho_{unb.}^{\alpha}$ where the power α is adjusted to reproduce the dots presented in Fig. 2(b). In this figure we show the fitted function,

$$f(\rho_{unb.}, R_{WS}) = 89.05(\rho_{unb.}/\rho_0)^{0.1425} R_{WS}^{-2} \,. \tag{2}$$

It interpolates the smooth part of the shell effects on the binding energy. There is, however, a residual difference which cannot be fitted. Finally, to estimate the residual difference between $B_{hom.}(\rho_{unb.}) - B_{WS-hom.}(\rho_{unb.}, R_{WS})$ and the adjusted function $f(\rho_{unb.}, R_{WS})$, shown on the r.h.s of Fig. 2(c). The average residual fluctuations are now of the order of 50 keV.

5. Modified HF Binding Energy in W-S Cells (HF-WS)

In Sec. 4, it has been shown that the fluctuations in the binding energy induced by the spurious shell effect can be decomposed into a smooth and a residual term and we have obtained a universal function for the smooth term. We then propose a systematic correction to the HF binding energy in the W-S cell:

$$\begin{aligned} B_{HF-WS}(\rho_B, \rho_{unb.}, R_{WS}) &\equiv B_{WS-inhom.}(\rho_B, R_{WS}) \\ &\quad + \Big(B_{hom.}(\rho_{unb.}) - B_{WS-hom.}(\rho_{unb.}, R_{WS})\Big)_{smooth} \\ &= B_{WS-inhom.}(\rho_B, R_{WS}) + f(\rho_{unb.}, R_{WS}) \,. \end{aligned} \quad (3)$$

In the calculation of the total energy, the nuclear binding energy E_n/A should then be replaced by $B_{HF-WS}(\rho_B, \rho_{unb.}, R_{WS})$. The minimum energy can be unambiguously identified if the difference in energy between the first and the second minimum is larger that the residual fluctuation, about 50 keV. In the HF calculation, the difference in energy between different cell configurations are of the order of few keV to few tens of keV. The residual fluctuation of the proposed correction is then too large. However, as seen above, the pairing correlations help in producing a deeper first minimum. It may also help in reducing the size of the residual fluctuations. In a future investigation, we plan to obtain an improved binding energy for W-S cells including pairing correlations. It must be remarked that the effects of the coupling between the protons and the unbound neutrons have been neglected in B_{HF-WS}. This is a good approximation since the protons are deeply bound.

6. Conclusions

In the search for the W-S cells which minimize the energy, it is important to look not only at the lowest minimum energy, but also at the other minima. The differences between the first and other minima show how stable is the lowest minimum with respect, for instance, to thermal fluctuations or spurious shell effects. In HF calculations, the differences in total energy between the first minimum and the others are about few keV per nucleon (few MeV in total energy). It is shown that the spurious shell effects, induced by the W-S approximation, introduce a fluctuation in the calculation of the total energy up to about 300 keV per nucleon. Interpolating the smooth part of the spurious shell effects, we have then proposed a simple method to reduce the fluctuations down to about 50 keV per nucleon (HF-WS). The residual fluctuations are still too large, and the removal of the smooth part is not enough to improve the results obtained within the W-S approximation. We have also shown that the pairing correlations help in producing a deeper first minimum. We have then compared the equation of state obtained from HF and HFB binding energies. Several issues should still be addressed in future studies: spurious shell correction with pairing correlations as well as effects of the temperature. These two effects should give lower residual fluctuations than the one we obtained at the $T = 0$ HF approximation.

References

1. H. A. Bethe, G. Börner and K. Sato, *Astron. and Astrophys.* **7**, 279 (1970).
2. G. A. Baym, H. A. Bethe and C. J. Pethick, *Nucl. Phys. A* **175**, 225 (1971).
3. G. Baym, C. Pethick and P. Sutherland, *Astrophys. J.* **170**, 299 (1971).
4. J. W. Negele and D. Vautherin, *Nucl. Phys. A* **207**, 298 (1973).
5. P. Magierski, A. Bulgac and P.-H. Heenen, *Nucl. Phys. A* **719**, 217 (2003).
6. F. Montani, C. May and H. Müther, *Phys. Rev. C* **69**, 065801 (2004).
7. F. Barranco, R. A. Broglia, H. Esbensen and E. Vigezzi, *Phys. Lett. B* **390**, 13 (1997).
8. N. Sandulescu, N. Van Giai and R. J. Liotta, *Phys. Rev. C* **69**, 045802 (2004).
9. M. Baldo, E. E. Saperstein and S. V. Tolokonnikov, *Nucl. Phys. A* **775**, 235 (2006).
10. P. Gögelein and H. Müther, arXiv:0704.1984.
11. A. Bulgac and P. Magierski, *Nucl. Phys. A* **683**, 695 (2001); **703**, 892 (2002).
12. B. Carter, N. Chamel and P. Haensel, *Nucl. Phys. A* **748**, 675 (2005).
13. N. Chamel, S. Naimi, E. Khan and J. Margueron, *Phys. Rev. C*, 055806 (2007).
14. J. Dobaczewski, H. Flocard and J. Treiner, *Nucl. Phys. A* **422**, 103 (1984).
15. N. Sandulescu, *Phys. Rev. C* **70**, 025801 (2004).
16. C. Monrozeau, J. Margueron and N. Sandulescu, *Phys. Rev. C* **75**, 065807 (2007); C. Monrozeau, PhD thesis, Université Paris-Sud (2007), unpublished.
17. E. Chabanat *et al.*, *Nucl. Phys. A* **627**, 710 (1997).

PAIRING AND BOUND STATES IN NUCLEAR MATTER

JOHN W. CLARK

Department of Physics & McDonnell Center for the Space Sciences,
Washington University, St. Louis, Missouri 63130, USA
E-mail: jwc@wustl.edu

ARMEN SEDRAKIAN

Institute for Theoretical Physics, J. W. Goethe-Universität,
D-60054 Frankfurt am Main, Germany
E-mail: sedrakian@th.physik.uni-frankfurt.de

We map out the temperature-density ($T-n$) phase diagram of an infinitely extended fermion system with pairwise interactions that support two-body (dimer) and three-body (trimer) bound states in free space. Adopting interactions representative of nuclear systems, we determine the critical temperature T_{cs} for the superfluid phase transition and the limiting temperature T_{ce3} for the extinction of trimers. The phase diagram at subnuclear densities features a Cooper-pair condensate in the higher-density, low-temperature domain; with decreasing density there is a crossover to a Bose condensate of strongly bound dimers. The high-temperature, low-density domain is populated by trimers. The trimer binding energy decreases as the point (T, n) moves toward the domain occupied by the superfluid and vanishes at a critical temperature $T_{ce3} > T_{cs}$. The ratio of the trimer and dimer binding energies is found to be a constant independent of temperature.

Keywords: Nuclear matter; Pairing; BCS-BEC crossover; In-medium bound states; Phase diagram.

1. Introduction

Pairing correlations and in-medium bound states of two, three, and more particles are generic phenomena occurring in systems of fermions exerting mutual attraction. Specific examples of physical contexts in which these phenomena are observed or expected include the following.

Dilute gases of cold fermionic atoms. Dramatic progress in trapping and manipulating cold atoms obeying Fermi statistics, including the ability to tune the strengths of their interactions, has opened the way to observation of the BCS-BEC crossover[1] and exploration of the phase diagrams of these fascinating and fundamental systems. Three-body bound states can be created and studied if three different atomic species or three different hyperfine states of the same atom are trapped.

Quark matter. Pairing of quarks in a variety of exotic phases has been predicted to exist in the high-density QCD of deconfined quark matter,[2] which could be realized in the interiors of compact, massive stars. Three-quark states dominate the low-energy (confined) limit of the theory, whereas higher quark-number states are extremely rare.

Dilute nucleonic matter. In different isospin mixtures, this substance plays key roles in supernova and neutron-star physics, notably through its equation of state at relevant temperatures and pressures.[3] Information on few-body correlations in infinite nuclear matter is also valuable for the study of nuclei far from stability.[4]

Since the binding force is different in these examples, the dependence of binding energy on the number of bound fermions is non-universal, as is immediately apparent when one compares nuclear and quark systems.

In this contribution we shall focus entirely on the third example, dilute nuclear matter. Our primary goal is to identify, in a simple but relevant model of symmetrical nuclear matter, the regions of the temperature-density phase diagram where pair condensation (precipitated by Cooper pairing), two-body bound states (dimers), and three-body bound states (trimers) are important. Certainly the actual phase diagram will be more complicated, since one must anticipate the formation, in the nuclear medium, of higher-A clusters, prominently α particles. However, it will be seen that the physics of pairing, two- and three-body clustering, and their interplay, is interesting in its own right.

The model to be studied, based on the Malfliet-Tjon nucleon-nucleon (NN) interaction, is specified in Sec. 2. Two-body states, including Cooper pairs and deuteron-like dimers, are considered in Sec. 3. A finite-temperature pairing formalism is implemented, with results that permit qualitative discussion of the BCS-BEC crossover from a condensate of $^3S_1$ Cooper pairs in the high-density limit to a Bose condensate of deuterons at asymptotically low density. The T-matrix formalism for describing bound pairs in the normal medium (dimers) at elevated temperatures is also introduced. Three-body states in the medium (trimers) are considered in Sec. 4; there we outline the T-matrix formalism leading to the three coupled nonsingular integral equations used to solve for the three-body bound-state wave functions and energies. Results from numerical calculations of dimer/trimer energies and trimer wave functions are reported and discussed in Sec. 5. The combined results for pairing and in-medium bound states are assembled into a temperature-density phase diagram of the nuclear-matter model at subnuclear densities.

2. Model of Dilute Nuclear Matter in the (T, n) Plane

We study a schematic model of nuclear matter that will give ready quantitative access to the parametric domain characterized by competition of Cooper pairing and formation of two- and three-body in-medium bound states. This model involves a number of simplifications and specifications.

We consider only symmetrical nuclear matter, with equal populations of neutrons and protons. The model could quite naturally be extended to explore behavior in the asymmetry variable $(N - Z)/2$, with the expectation of interesting behavior.

As is the usual practice, we do not include the effects of Coulomb repulsion between protons. For the NN interaction, we assume a central two-body potential of the Malfliet-Tjon class,[5] a linear combination

$$V(r) = g_1 \frac{e^{-\lambda_1 r}}{r} + g_2 \frac{e^{-\lambda_2 r}}{r} \qquad (g_1 > 0\,, \quad g_2 < 0) \tag{1}$$

of repulsive and attractive Yukawas in each spin-isospin channel. The range and strength parameters have been adjusted to fit basic properties of few-nucleon systems, including the S-wave phase shifts and binding energies of the deuteron and triton. The MT-III parametrization is chosen. Since the MT interactions contain no tensor component, S and D states are not coupled and the deuteron quadrupole moment is not reproduced.

We focus on relatively low densities, where the most attractive $T = 0$ pairing interaction, as indicated by NN phase shifts, is in the $^3S_1$–$^3D_1$ partial-wave channel.[6] The pairing problem at finite temperature is treated within a Green's function formulation[7] in an approximation leading to the usual temperature-dependent version of BCS theory, with a gap equation driven by the *bare* two-body interaction. Solving the gap equation enables us to follow the crossover from BCS pairing of protons and neutrons at the higher densities to clustering into deuterons and their Bose-Einstein condensation at the lowest densities.

We use the scattering-matrix formalism to deal with bound states (dimers) above the superfluid critical temperature. In the same spirit, we adapt the Faddeev–Skornyakov–Ter-Martirosian[8–10] formalism to solve the three-body problem in the presence of the background nuclear medium.

3. Two-Body States: Cooper Pairs, BCS-BEC Crossover, Dimers

To describe pairing in our model of dilute nuclear matter, we work within the real-time Green's function formalism.[7] Since a time-local NN interaction has been assumed, the pairing interaction and pairing gap are energy independent. We impose the quasiparticle approximation, in which only the pole parts of the (retarded) normal and anomalous propagators are retained. The anomalous propagator is thus expressed as

$$F^R = F^\dagger = u_p v_p \left[(\omega - \omega_+ + i\eta)^{-1} - (\omega - \omega_- + i\eta)^{-1} \right] \tag{2}$$

in terms of the quasiparticle spectrum $\omega_\pm = \pm\sqrt{E^2(p) + \Delta^2(p)}$ and Bogolyubov amplitudes u_p and v_p, with $u_p^2 = 1/2 + E(p)/2\omega_+$ and $u_p^2 + v_p^2 = 1$. The assumed mean-field approximation to the anomalous self-energy (the gap function) is then

$$\Delta^R(\mathbf{p}) = 2 \int \frac{d\omega d\mathbf{p}'}{(2\pi)^4} V(\mathbf{p}, \mathbf{p}') \mathrm{Im} F^R(\omega, \mathbf{p}') f_\beta(\omega)\,, \tag{3}$$

with f_β denoting the Fermi distribution.

Substituting for F^R, and doing partial-wave expansions, the two coupled integral equations for the gap in the $^3S_1$–$^3D_1$ channel are (with $l, l' = 0, 2$)

$$\Delta_l(p) = -\int \frac{dp'p'^2}{(2\pi)^2} \sum_l V^{3SD1}_{ll'}(p,p') \frac{\Delta_{l'}(p')}{\sqrt{E^2(p') + D^2(p')}} \left[f_\beta(\omega_+) - f_\beta(\omega_-)\right] , \quad (4)$$

where $D^2(k) \equiv (3/8\pi)[\Delta_0^2(k) + \Delta_2^2(k)]$ is the angle-averaged neutron-proton gap function and $V^{3SD1}(p,p')$ is the NN interaction in the ${}^3S_1$–${}^3D_1$ channel.

We study the problem at constant density and temperature. Accordingly, the chemical potential is to be determined self-consistently from the gap equation and the formula for the density,

$$n = -8\int \frac{d\mathbf{p}\, d\omega}{(2\pi)^4} \mathrm{Im}\, G^R_+(\omega, \mathbf{p}) f(\omega) = 4\int \frac{d^3p}{(2\pi)^3} \left[u_p^2 f_\beta(\omega_+) + v_p^2 f_\beta(\omega_-)\right] , \quad (5)$$

having substituted the retarded normal propagator in quasiparticle approximation,

$$G^R_\pm = u_p^2\,(\omega - \omega_\pm + i\eta)^{-1} + v_p^2(\omega - \omega_\mp + i\eta)^{-1} . \quad (6)$$

It is well known that the presence of a strong short-range repulsion in the pairing force creates problems for straightforward iterative solution of the gap equation and invalidates the popular weak-coupling formula for the gap.[4] As an alterative to the very effective separation approach introduced by Khodel et al.[11] for solution of the gap equation, we implement a modified iterative procedure[12] in which the long-range momentum dependence associated with the inner repulsion is dealt with by introducing an ultraviolet cutoff at $p' = \Lambda$, starting with $\Lambda \ll \Lambda_P$ = natural (soft) cutoff. The routine runs as follows:

1. For fixed Λ, initialize an internal loop by solving for $D(p_F)$ assuming the gap function to be a constant. Then take the initial input gap on the RHS of the gap equation as $\Delta^{(i=0)} = V(p_F, p)D(p_F)$. The internal loop leads to rapid convergence to a Λ-dependent gap function.
2. An external iteration loop increments the cutoff Λ until it exceeds Λ_P and the gap becomes insensitive to the cutoff, i.e., $d\Delta(p,\Lambda)/d\Lambda \approx 0$.
3. A third loop seeks convergence between the output gap function and the chemical potential, such that the starting density is reproduced.

Numerical results for the pairing gap $\Delta = \Delta(p = p_F)$ at the Fermi surface and the consistently determined chemical potential μ are shown as functions of temperature in Fig. 1. In representing the density dependence of these quantities, it is convenient to introduce (i) a normalized volume per particle defined as $f = n_0/n$, where n_0 the saturation density of symmetrical nuclear matter, taken as $n_0 = 0.16$ fm^{-3}, and (ii) a diluteness parameter $n|a^3|$, the scattering length a being set to 5.4 fm for neutron-proton scattering.

The low- and high-temperature asymptotics of the calculated gap are well described by the BCS relations

$$\Delta(T \to 0) = \Delta(0) - [2\pi\alpha_1 \Delta(0) T]^{1/2} \exp(-\Delta(0)/T) , \quad (7)$$

$$\Delta(T \to T_{cs}) = 3.06\alpha_2 [T_{cs}(T_{cs} - T)]^{1/2} , \quad (8)$$

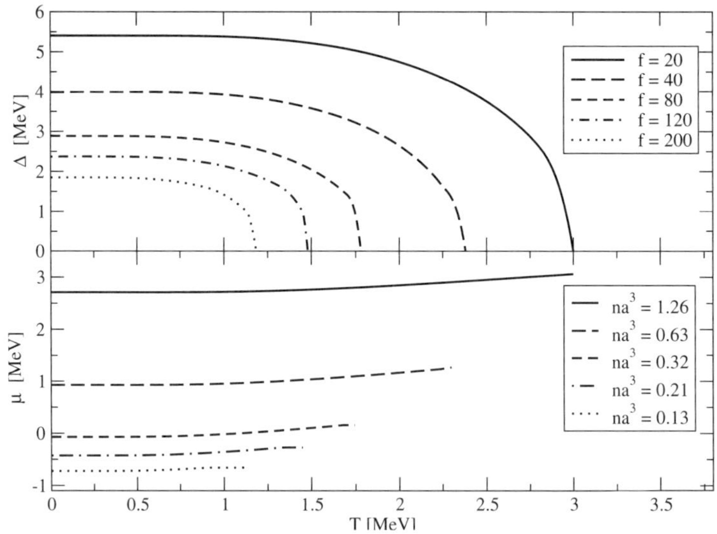

Fig. 1. Dependence of the pairing gap (upper panel) and chemical potential (lower panel) on temperature for fixed values of the ratio $f = n_0/n$ and the diluteness parameter $n|a|^3$, respectively, where $n_0 = 0.16$ fm^{-3} and $a = 5.4$ fm.

where T_{cs} is the critical temperature for the superfluid transition. However, the BCS weak-coupling values $\alpha = 1 = \beta$ are replaced by $\alpha_1 \approx 0.2$ and $\alpha_2 \approx 0.9$, so the familiar BCS result $\Delta(0)/T_{cs} = 1.76$ no longer holds. These deviations are understandable, since weak coupling cannot be assumed and the pairing is spin-triplet rather than spin-singlet.

The ratio $\Delta(T = 0)/|\mu|$ provides one useful measure of the coupling strength as the system makes the transition between the weak-coupling and strong-coupling limits associated with the BCS-BEC crossover. Examining Fig. 1, one sees that within the parametric domain studied numerically, the strong-coupling regime is attained for $f \geq 40$ (where $\Delta \gg \mu$). At $f = 20$, the system is in an intermediate regime ($\Delta \sim \mu$). The diluteness parameter $n|a|^3$ furnishes an alternative measure of coupling; by this measure, the system is in the strong-coupling regime for $f \geq 40$. We note that the change in sign of the chemical potential from positive to negative under increasing diluteness, occurring near $f = 80$, may also be taken as a signature of the crossover between weak- and strong-coupling.

In the limit of vanishing density, $f \to \infty$, the chemical potential $\mu(T = 0)$ goes to -1.1 MeV, half the deuteron binding energy. In this limit, the gap equation reduces to the Schrödinger equation for the two-body bound state, with the chemical potential playing the role of the energy eigenvalue. Thus, within our description, the BCS condensate of $^3S_1$ Cooper pairs is seen to evolve into a Bose-Einstein condensate of deuterons as the system moves from high density to low, i.e., from weak coupling to strong coupling in a smooth crossover taking place without change of symmetry of the many-body wave function (cf. Ref. 1).

Now let us consider two-body bound states *above* the critical temperature T_{cs} of the superfluid transition. As is well known, the two-body T-matrix sums particle-particle ladders via the Lippmann-Schwinger equation, or

$$\mathrm{T} = \mathrm{V} + \mathrm{V}\mathrm{G}_0\mathrm{T} = \mathrm{V} + \mathrm{T}\mathrm{G}_0\mathrm{V}\,. \tag{9}$$

Denoting four-momenta in the CM system by $P = (E, \mathbf{P})$ and $p = (\epsilon, \mathbf{p})$, and in the Lab system by $k_{1,2} = (\omega_{1,2},\, \mathbf{k}_{1,2})$, the retarded part T^R of the T-matrix, expressed in the momentum representation, is given by

$$T^R(\mathbf{p}, \mathbf{p}'; \mathbf{P}, E) = V(\mathbf{p}, \mathbf{p}') + \int \frac{d\mathbf{p}''}{(2\pi)^3} V(\mathbf{p}, \mathbf{p}'') G_0^R(\mathbf{p}'', \mathbf{P}, E) T^R(\mathbf{p}'', \mathbf{p}'; \mathbf{P}, E)\,. \tag{10}$$

We are of course concerned with *in-medium* two-body bound states. Working again in the quasiparticle approximation, the relevant two-body Green's function is

$$G_0^R(\mathbf{k}_1, \mathbf{k}_2, E) = \frac{Q_2(\mathbf{k}_1, \mathbf{k}_2)}{E - \epsilon(\mathbf{k}_1) - \epsilon(\mathbf{k}_2) + i\eta}\,. \tag{11}$$

The two-body phase-space occupation factor

$$Q_2(\mathbf{k}_1, \mathbf{k}_2) = 1 - f_\beta(\mathbf{k}_1) - f_\beta(\mathbf{k}_2)\,, \tag{12}$$

operating in intermediate states, allows for propagation of particles and holes, thereby incorporating time-reversal invariance.

The two-body T^R-matrix has a pole at the energy of the two-body bound state. If $Q_2 = 1$, as in free space, the pole lies at minus the deuteron binding energy. At finite density and temperature, the pole, shifted to the right in energy, corresponds to a dimer with deuteron quantum numbers swimming in the background nuclear medium. As the temperature drops at fixed density, the dimer binding energy $-E_d$ keeps decreasing until it reaches zero at a critical extinction temperature T_{ce2}, where the dimer dissolves into the medium. Numerical results for the dimer energy $E_d(T, n)$ will be presented in Sec. 5 along with results for the trimer energy $E_t(d, n)$.

4. Three-Body States in the Medium

The in-medium three-body problem is approached within the Faddeev framework,[9,10] suitably adapted to the finite temperature and density of the ambient medium.[10,12] A fermionic system supports a three-body bound state when there exists a negative-energy solution of the homogeneous counterpart of the three-body T-matrix equation

$$\mathcal{T} = \mathcal{V} + \mathcal{V}\mathcal{G}\mathcal{V} = \mathcal{V} + \mathcal{V}\mathcal{G}_0\mathcal{T}\,, \tag{13}$$

where $\mathcal{V} = V_{12} + V_{23} + V_{31}$ is the full interaction in the three-body system and $\mathcal{G}$ and $\mathcal{G}_0$ are respectively the full and free three-body Green's functions. It is easily seen that the kernel in this problem is not square-integrable. This obstacle is overcome by complete resummation of the ladder series in each two-body ij channel, in terms of in-medium transition operators

$$\mathcal{T}_{ij} = V_{ij} + V_{ij}\mathcal{G}_0\mathcal{T}_{ij}\,. \tag{14}$$

The resummations lead to a system of three coupled, nonsingular integral equations of Fredholm type,

$$\mathcal{T}^{(k)} = T_{ij} + T_{ij}\mathcal{G}_0\left(\mathcal{T}^{(i)} + \mathcal{T}^{(j)}\right), \tag{15}$$

to be solved for the components $\mathcal{T}^{(k)}$ of the three-body T-matrix in the Faddeev decomposition $\mathcal{T} = \mathcal{T}^{(1)} + \mathcal{T}^{(2)} + \mathcal{T}^{(3)}$. Analysis of the bound-state problem is based on the homogeneous version of these equations.

It is more convenient to work with the components $\psi^{(k)}$ of the wave function $\Psi = \psi^{(1)} + \psi^{(2)} + \psi^{(3)}$ of the three-body state, solving the homogeneous analogs

$$\psi^{(k)} = \mathcal{G}_0 T_{ij}(\psi^{(i)} + \psi^{(j)}) \tag{16}$$

of the three equations (15) for the $\mathcal{T}$-matrix components $\mathcal{T}^{(k)}$. For identical particles, the equations (16) reduce to

$$\psi^{(1)} = \mathcal{G}_0 T_{23} P \psi^{(1)}, \tag{17}$$

where the operator P is given by $P = P_{12}P_{23} + P_{31}P_{23}$ in terms of particle interchange operators, and $\Psi = (1 + P)\psi^{(1)}$. In the momentum representation, the relevant three-body propagator, treated in quasiparticle approximation, is

$$\mathcal{G}_0(\mathbf{k}_1, \mathbf{k}_2, \mathbf{k}_3, \Omega) = \frac{Q_3(\mathbf{k}_1, \mathbf{k}_2, \mathbf{k}_3)}{\Omega - \epsilon(\mathbf{k}_1) - \epsilon(\mathbf{k}_2) - \epsilon(\mathbf{k}_3) + i\eta}, \tag{18}$$

where

$$Q_3(\mathbf{k}_1, \mathbf{k}_2, \mathbf{k}_3) = [1 - f_\beta(\mathbf{k}_1)][1 - f_\beta(\mathbf{k}_2)][1 - f_\beta(\mathbf{k}_3)] - f_\beta(\mathbf{k}_1)f_\beta(\mathbf{k}_2)f_\beta(\mathbf{k}_3) \tag{19}$$

is the intermediate-state phase-space occupation factor for three-particle propagation. Inspecting Eqs. (11), (12), (18), and (19), it is seen that the background medium exerts its presence on the two- and three-body problems through the temperature-dependent suppression of the phase space available for intermediate two-body and three-body states and through renormalization of the single-particle energies. (However, the latter effect is of little importance at the low densities involved.)

Further analysis and reduction of the three-body problem involves the following steps. First, recalling that the momentum space for this problem is spanned by Jakobi four-momenta

$$K = k_i + k_j + k_k\,, \quad p_{ij} = (k_i - k_j)/2\,, \quad q_k = (k_i + k_j)/3 - 2k_k/3\,, \tag{20}$$

we substitute for k_1, k_2, and k_3 in terms of these variables. Second, we work in an angular-momentum basis of states $|pq\alpha\rangle_i \equiv |pq(l\lambda)LM(s\tfrac{1}{2})SM_S\rangle_i$, where p and q are the magnitudes of the relative momenta of the pair $\{kj\}$ (indicated by complementary index i), l and λ are their associated relative angular-momentum quantum numbers, s is their total spin, and $LMSM_S$ are the orbital and spin quantum numbers of the three-body system. This strategy enables reduction of the problem to solution of an integral equation in two continuous variables. Finally, the phase-space occupation factors Q_2 and Q_3 and single particle energies $\epsilon_{+,-}$ are treated in

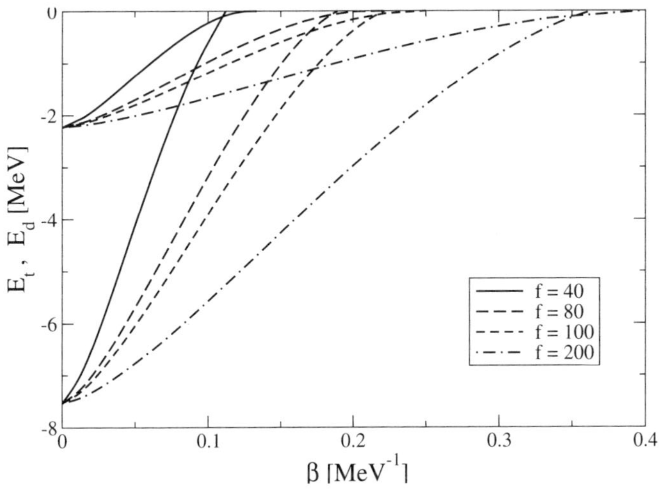

Fig. 2. Dependence of the two-body (E_d) and three-body (E_t) binding energies on inverse temperature, for fixed values of the ratio $f = n_0/n$.

angle-average approximation. The propagator is assumed to be independent of the momentum of the three-body system with respect to the background.

5. Numerical Results and their Interpretation

In this concluding section we discuss numerical results on three-body in-medium bound states and the evolution of the trimer and dimer binding energies with temperature and density. These results, together with those on the critical temperature of the superfluid transition, are assembled into a (T, n) phase diagram of dilute symmetrical nuclear matter. Figure 2 shows how the dimer and trimer energies evolve with temperature at selected values of the normalized specific volume f. The two- and three-body bound states enter the continuum at respective critical extinction temperatures T_{ce2} and T_{ce3} determined by $E_d(\beta) = 0$ and $E_t(\beta) = 0$. The essential message of this figure is that the binding of dimers and trimers decreases with rising density and with declining temperature. In addition, the solutions obtained show a remarkable feature: the ratio $\eta = E_t(\beta)/E_d(\beta)$ is a constant *independent of temperature.* Due to this feature, virtually equivalent results are obtained with an alternative definition of extinction temperature that accounts for the breakup channel $t \to d + n$. From the free-space binding energies $E_d(0) - 2.23$ MeV and $E_t(0) = -7.53$ MeV, we have $\eta = 3.38$. (It should be noted that the deviations from constancy of the ratio η visible near the points of extinction are artifacts of the numerics.)

We next consider the results for the in-medium three-body wave function $\Psi(p, q)$. Contour plots[12] in the Jacobi variables p and q show that as the temperature drops toward zero, the wave function becomes increasingly concentrated around the ori-

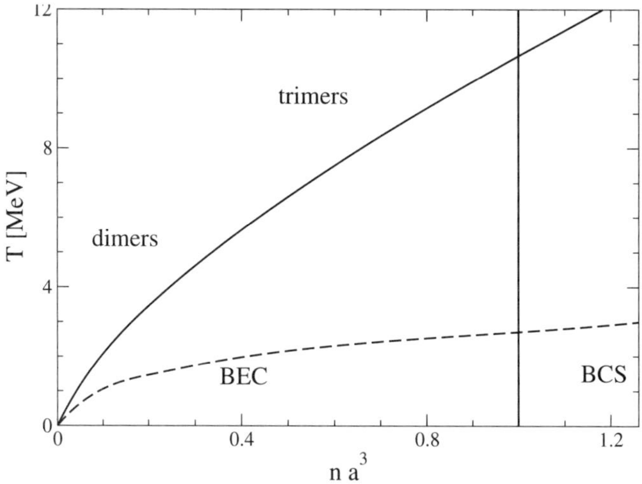

Fig. 3. Schematic temperature-density ($T - na^3$) phase diagram of dilute, isospin-symmetric nuclear matter. The dependence of the critical temperature T_{ce3} of extinction of three-body bound-states on density is traced by the solid curve. Trimers exist above this critical line. The dependence of the critical temperature T_{cs} of the superfluid phase transition is shown by the dashed curve. The condensate is a weakly-coupled BCS superfluid far to the right of the vertical line and is a BE condensate of tightly-bound pairs far to the left, with a smooth crossover transition in between.

gin in momentum space. The radius of the bound state increases correspondingly, tending to infinity at the extinction transition. The wave function develops oscillations near the transition temperature. This behavior is a precursor of the transition to the continuum; in the absence of a trimer-trimer interaction one ends up with plane-wave states.

Finally, we may combine the results on pairing and bound states to draw a phase diagram (Fig. 3), which shows several distinct regions in the $T - na^3$ plane.

A. The low-density, high-temperature domain (upper left corner) is populated by trimers, which enter the continuum when the critical line $T_{ce3}(n)$ is crossed from above.
B. The low-temperature and low-density domain (lower left: $na^3 \ll 1$, $f < 40$) contains a Bose condensate of tightly-bound deuterons.
C. The low-temperature, high-density domain (lower right: $na^3 \gg 1$) features a BCS condensate of weakly-bound Cooper pairs.
D. The domain between the two critical lines $T_{ce3}(n)$ and $T_{cs}(n)$ contains nucleonic $\{p, n\}$ liquid.

Commenting on crossings, phase transitions, and symmetry breaking, we note first that the phases B and C are characterized by broken symmetry associated with the $\langle\psi\psi\rangle$ superfluid condensate. The transition C $\rightarrow$ B does not involve any symmetry changes – it is a smooth crossover from a BCS to a BEC condensate. On the other hand, B $\rightarrow$ D and C $\rightarrow$ D are second-order phase transitions related to the vanishing of the condensate along the line $T_{cs}(na^3)$. The transition A $\rightarrow$ D (e.g., along the solid vertical line, tentatively of second order) is characterized by an order parameter given by the fraction of trimers, which goes to zero at T_{ce3}. Some caution is needed in interpreting this schematic phase diagram. For example, at zero temperature the system is in equilibrium only at the saturation density of symmetrical nuclear matter.

To conclude, we observe that this study serves in part to illustrate a two-fold complexity of dilute nuclear matter. On the one hand, the system supports liquid-gas and superfluid phase transitions: depending on the temperature and density, it can be in a gaseous, liquid, or superfluid state. These features are in fact *generic* to systems of fermions interacting with forces of van der Waals type that are repulsive at short range and attractive at larger separations. On the other hand, the attractive component leads to clustering, i.e., formation of dimers, trimers, etc. The clustering pattern depends on the actual form of the attractive force and is therefore *not* universal. Naturally, the true phase diagram of nuclear matter at finite density and temperature must be considerably richer than that depicted here, with modifications due to quartetting and isospin asymmetry, but the general outlines sketched in the present analysis should persist.

Acknowledgements

This research was supported in part under NSF Grant No. PHY-0140316 and by the McDonnell Center for the Space Sciences.

References

1. P. Nozières and S. Schmitt-Rink, *J. Low Temp. Phys.* **59**, 195 (1985).
2. M. Alford and K. Rajagopal, in *Pairing in Fermionic Systems: Basic Concepts and Modern Applications*, edited by A. Sedrakian, J. W. Clark and M. Alford (World Scientific, Singapore, 2006), pp. 1-36 [arXiv:nucl-th/0607028].
3. J. M. Lattimer and D. Swesty, *Nucl. Phys. A* **535**, 331 (2001).
4. D. J. Dean and M. Hjorth-Jensen, *Rev. Mod. Phys.* **75**, 607 (2003).
5. R. A. Malfliet and J. A. Tjon, *Nucl. Phys. A* **127**, 161 (1969).
6. A. Sedrakian and J. W. Clark, in *Pairing in Fermionic Systems: Basic Concepts and Modern Applications*, edited by A. Sedrakian, J. W. Clark and M. Alford (World Scientific, Singapore, 2006), pp. 145-184 [arXiv:nucl-th/0607028].
7. J. W. Serene and D. Reiner, *Phys. Rep.* **101**, 221 (1983).
8. G. V. Skorniakov and K. A. Ter-Martirosian, *Sov. Phys. JETP* **4**, 648 (1957).
9. L. D. Faddeev, *Sov. Phys. JETP* **12**, 1014 (1961).
10. A. Sedrakian and G. Röpke, *Ann. Phys. (NY)* **266**, 524 (1998).
11. V. A. Khodel, V. V. Khodel and J. W. Clark, *Nucl. Phys. A* **598**, 390 (1996).
12. A. Sedrakian and J. W. Clark, *Phys. Rev. C* **73**, 035803 (2006) [arXiv:nucl-th/0511076].
13. T. Alm, B. L. Friman, G. Röpke and H. Schulz, *Nucl. Phys. A* **551**, 45 (1993); U. Lombardo, P. Nozières, P. Schuck, H.-J. Schulze and A. Sedrakian, *Phys. Rev. C* **64**, 064314 (2001) [arXiv:nucl-th/0109024].

PAIRING IN BCS THEORY AND BEYOND

L. G. CAO,[1,2,*] U. LOMBARDO[3,4] and P. SCHUCK[5]

[1] *Institute of Modern Physics, Chinese Academy of Science, Lanzhou 730000, China*
[2] *Center of Theoretical Nuclear Physics, National Laboratory of Heavy Ion Accelerator, Lanzhou 730000, China*
[3] *Laboratori Nazionali del Sud, INFN, Via Santa Sofia 62, I-95123 Catania, Italy*
[4] *Dipartimento di Fisica dell'Università, Viale Andrea Doria 62, I-95123 Catania, Italy*
[5] *Institut de Physique Nucléaire, Université Paris-Sud, F-91406 Orsay Cedex, France*
**E-mail: caolg@impcas.ac.cn*

Medium effects on $^1S_0$ pairing in neutron and nuclear matter are studied in a microscopic approach. The screening potential is calculated in the RPA limit, suitably renormalized to cure the low density mechanical instability of nuclear matter. The selfenergy corrections are consistently included resulting in a strong depletion of the Fermi surface. The selfenergy corrections always lead to a quenching of the gap, which is enhanced by the screening effect of the pairing potential in neutron matter, whereas it is almost completely compensated by the antiscreening effect in nuclear matter.

Keywords: Pairing in nuclear and neutron matter; Self-energy effect; Medium effect; Induced interaction.

1. Introduction

Superfluidity in nuclear matter has been payed more attention since the first application of the BCS theory to nuclear systems,[1] but up to now it has not yet been achieved satisfactorily. The attractive components of the bare nuclear interaction have led to the investigation of several pairing configurations, e.g. neutron-neutron or proton-proton pairing in the $^1S_0$ channel in neutron stars[2] disregarding possible repulsive effect exerted by screening of the force via the medium. A pairing suppression has in fact been found by most calculations of pairing in neutron matter (see Ref. 3 and references therein). On the contrary, other pairing configurations have not yet been explored since the repulsive components of the direct nuclear interaction cannot support the formation of Cooper pairs. But there are strong indications that, in a nuclear rather than neutron matter environment, the medium polarization of the interaction can favor the formation of Cooper pairs similar to the lattice vibrations in ordinary superconductors. In nuclear matter the medium enhancement of neutron-neutron $^1S_0$ pairing is to be traced back to the proton particle-hole excitations,[4] and in finite nuclei to the surface vibrations.[5]

Another distinctive feature of the nuclear environment is the presence of strong short range correlations that induce two effects relevant for the pairing: one is the depletion of the Fermi surface; the other one is the strong mass renormalization caused by short-range particle-particle correlations.[6] The two effects conspire against the pairing formation.[7]

A complete microscopic treatment of the very subtle pairing problem requires vertex and selfenergy corrections to be treated on the same footing. In the present work we use a realistic two body force (V18,[8] see below) in the Born term, and use as vertices in the induced interaction a force which is based on a more modern G-matrix calculation[9] as this was the case for the Gogny force.[4,10] We now get reasonable renormalization effects of the pairing force and we calculate the corresponding gaps as a function of density in pure neutron matter as well as in symmetric nuclear matter.

Firstly the screening interaction is discussed in the RPA limit, and then the summation of bubble diagrams and the resummation of dressed bubble diagrams both using the Landau parameters are derived. Then the results are presented. Last part is devoted to the comparison with other calculations and to the conclusions.

2. Screening Interaction

2.1. *One-bubble screening interaction*

In the present calculation we adopt the G-matrix itself and we try to reduce its complexity with reasonable approximations. For the sake of application to the pairing in the $^1S_0$ channel we select the two particle state with total spin S=0 and isospin T=1. Then the one-bubble interaction can be written as

$$< 1\bar{1}|\mathcal{V}_1|1'\bar{1}' >= \frac{1}{4}\sum_{2,2'}\sum_{ST}(-)^S(2S+1) < 12|G^{ph}_{ST}|1'2' >_A< 2'\bar{1}|G^{ph}_{ST}|2\bar{1}' >_A \Lambda^0(22'), \tag{1}$$

where $1 \equiv (\vec{k}_1, \sigma_1, \tau_1)$ ($1' \equiv (\vec{k}_{1'}, \sigma_{1'}, \tau_{1'})$) and $\bar{1} \equiv (-\vec{k}_1, \sigma_1, \tau_1)$ ($\bar{1}' \equiv (-\vec{k}_{1'}, \sigma_{1'}, \tau_{1'})$) are the momenta of the pair in the entrance (exit) channel. Λ is the static polarization part. The G-matrix is converted into the ph sector, as it is required to solve the Bethe-Salpeter equation and to sum up the bubble series $\tilde{V}_2$. The standard recoupling procedure from pp sector to ph sector yields

$$G^{ph}_{ST} = \sum_c (2S_c+1)(2T_c+1)(-1)^{S_c+T_c}\begin{Bmatrix} \frac{1}{2} & \frac{1}{2} & S_c \\ \frac{1}{2} & \frac{1}{2} & S \end{Bmatrix}\begin{Bmatrix} \frac{1}{2} & \frac{1}{2} & T_c \\ \frac{1}{2} & \frac{1}{2} & T \end{Bmatrix} G_{S_cT_c}, \tag{2}$$

where the brackets are the 6j symbols. The sum runs over the spin S_c and isospin T_c of the pp channels included in the calculation. Since the G-matrix incorporates short range pp correlations, its momentum range is shrunk remarkably in comparison with the bare interaction. At variance with the bare interaction, the G-matrix cannot sustain large momentum transfers $\vec{q} = \vec{k} - \vec{k}'$, that justifies the approximation to average it around the Fermi surface, in the limit q=0. As a consequence the q dependence is only located in the integral of the polarization part, giving the Lindhard function.

The problems with the calculation of the ph multi bubble contribution are the following. First, the bubble series with G-matrix insertions have to be previously summed up. But, since the interaction vertices in the ph channel involve particle excitations around the Fermi surface, they can be approximated by the Landau parameters. Second, even replacing the bare interaction vertices by G-matrices , there appears the low density singularity in the RPA in nuclear matter ($F_0 = -1$). This problem, discussed in Ref. 3 (see also references therein) is remedied by dressing the vertex insertions according to the Babu-Brown induced interaction theory.[11]

2.2. *Landau parameters from the BHF approximation*

The microscopic basis of the ph effective interaction can be set in terms of the energy functional of symmetric nuclear matter

$$\begin{aligned} N(0) f_{\sigma\tau,\sigma'\tau'}(\vec{k},\vec{k'}) &= \frac{\delta^2 E}{\delta n_{\sigma\tau}(\vec{k})\delta n_{\sigma'\tau'}(\vec{k'})} \\ &= F + F'(\tau\cdot\tau') + G(\sigma\cdot\sigma') + G'(\sigma\cdot\sigma')(\tau\cdot\tau'), \end{aligned} \tag{3}$$

where the density of states $N(0)$ is introduced to make the Landau parameters dimensionless. In the BHF[6] approximation the energy functional is given by

$$E = \sum_k \frac{\hbar^2 k^2}{2m} + \frac{1}{2} \sum_{k_1,k_2} \langle k_1, k_2 | G(\omega) | k_1, k_2 \rangle_A . \tag{4}$$

One can determine the Landau parameters from the microscopic Brueckner theory, performing the double variational derivative, Eq. (3), of the energy per particle, Eq. (4). So doing, a number of contributions are generated that can be calculated one by one[12] in some approximation due to the complex structure of G-matrix. A simple and powerful way to calculate the Landau parameters is to suitably fit the BHF energy and the corresponding sp spectrum with a functional of the occupation numbers and then to perform the double derivative. A Skyrme-like functional has proved to reproduce accurately the equation of state (EoS) of symmetric as well as spin and isospin asymmetric nuclear matter.[13] Therefore we determine the Landau parameters in that way. The latter are plotted in Fig. 1 as a function of the Fermi momentum. As expected F_0 exhibits the well known instability below the saturation point, which makes the RPA series difficult to handle. As in previous papers[4,14] this drawback can be overcome by the induced interaction theory of Babu and Brown.[11] The numerical results are depicted in Fig. 1. The salient feature of the induced interaction is that the renormalization of F_0 prevents any singular behavior to occur below the saturation density.

Therefore we dressed the residual interaction first with the short range correlations (G-matrix instead of bare interaction) and then by the renormalized long range correlations V_{ph} replacing the G-matrix in the RPA series. Since the calculation of the induced interaction with the G-matrix is a quite complex job, we have simplified the problem replacing the G-matrix with the Landau parameters. The way we determine the Landau parameters, the approximation turns out to be better than starting from the G-matrix itself.

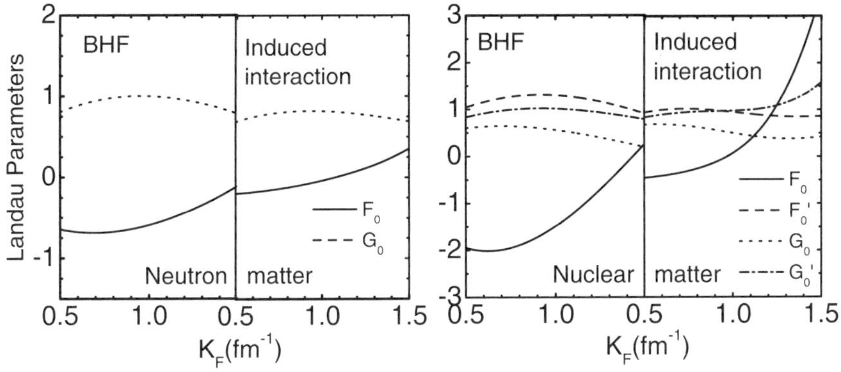

Fig. 1. Landau parameters of pure neutron matter and nuclear matter.

2.3. *Bubble series*

The vertex insertions dressing the bubbles must be treated on different footing than the external ones. Using the Landau parameters corresponds to the Landau limit, which is a quite reasonable approximation. In this case the RPA summation of the ph interaction turns out to be algebraic and, expressed in term of the dressed bubble, it is written as

$$\Lambda(q)_{ST} = \frac{\Lambda^0(q)}{1 + \Lambda^0(q)\mathcal{L}_{ST}}, \tag{5}$$

where $\mathcal{L}_{ST}$ are the Landau parameters, whose components are commonly denoted by: $\mathcal{L}_{00} = F$, $\mathcal{L}_{01} = F'$, $\mathcal{L}_{10} = G$, $\mathcal{L}_{11} = G'$. In this expression we clearly see how the induced interaction prevents any divergence to occur since $|\Lambda\mathcal{L}| \leq |\mathcal{L}| \leq 1$. Replacing in Eq. (1) the bare bubble Λ^0 with the dressed bubble Λ we get the full screening interaction used in the calculation.

3. Results

3.1. *Screening interaction*

In this paper we only focus on the $^1S_0$ pairing interaction in the two extreme situations of pure neutron matter and symmetric nuclear matter. The G-matrix is generated from a selfconsistent BHF calculation.[6] The Argonne V18 two body force[8] is adopted as the input bare interaction plus a microscopic model for the three body force.[15] The calculation also provides the self-energy from which we extract the sp spectrum and the Z-factors.

Let us start with neutron matter. In this case the screening interaction can be decomposed in two terms: S=0 density fluctuation and S=1 spin density fluctuation. The two modes have opposite effect: the former one is attractive, the latter is repulsive. From Eq. (1), we can write

$$\mathcal{V}_1 = \frac{1}{2}\Lambda(q)_0 G_0^{ph} G_0^{ph} - \frac{3}{2}\Lambda(q)_1 G_1^{ph} G_1^{ph}. \tag{6}$$

The factor 3 is due to the multiplicity of the spin mode. For the following discussion we should notice that Λ is negative. In turn, each G-matrix can be expressed as a superposition of G-matrices, projected onto two particle states (see Eq. (2)),

$$G_0^{ph} = \frac{1}{2}(-G_0 - 3G_1); \qquad G_1^{ph} = \tfrac{1}{2}(G_0 - G_1). \tag{7}$$

Since G_0 is attractive and G_1 is repulsive, assuming their magnitude to be comparable, we get $G_0 \approx G_1$[16] and the multiplicity plays the main role in establishing the dominance of the spin density mode over the density mode. In the latter calculation G_0 and G_1 are only roughly comparable, as it can bee seen in Fig. 2, nevertheless the conclusion is still valid. A quenching of $^1S_0$ pairing in neutron matter is to be expected, a result established long ago[17] and confirmed by many calculations in various approximations (see Ref. 3 and references therein).

In nuclear matter the situation could be quite different since the isospin fluctuations also come into play. The screening interaction now is split as follows

$$\mathcal{V}_1 = \frac{1}{4}(\Lambda(q)_{00}G_{00}^{ph}G_{00}^{ph} + \Lambda(q)_{01}G_{01}^{ph}G_{01}^{ph}) - \frac{3}{4}(\Lambda(q)_{10}G_{10}^{ph}G_{10}^{ph} + \Lambda(q)_{11}G_{11}^{ph}G_{11}^{ph}).$$

The various contributions are plotted in Fig. 2 in terms of pp states, the individual ph contributions are expressed as

$$G_{00}^{ph} = \frac{1}{4}(G_{00} + 3G_{10} + 3G_{01} + 9G_{11}); \; G_{10}^{ph} = \tfrac{1}{4}(-G_{00} - 3G_{10} + G_{01} + 3G_{11});$$
$$G_{01}^{ph} = \frac{1}{4}(-G_{00} - 3G_{10} + G_{01} + 3G_{11}); \quad G_{11}^{ph} = \tfrac{1}{4}(G_{00} - G_{10} - G_{01} + G_{11}). \tag{8}$$

In nuclear matter the pp G-matrix elements are dominated by the deuteron channel ($^3SD_1$ coupled pp channel), which is very attractive and therefore it reinforces the density mode and weakens the spin mode. In other words, the main isospin effect is to reverse the role of the medium, i.e. antiscreening instead of screening. In previous papers this effect has been discussed in terms of proton-proton ph screening against neutron-neutron ph screening in the neutron-neutron $^1S_0$ channel.[18] The latter gives repulsion the former attraction. At variance with Ref. 16 the proton-proton ph screening is stronger than neutron-neutron ph screening. This effect is to be traced back to stronger in medium renormalization of the force in the $T = 0$ channel than in the $T = 1$ one. Antiscreening is the overall effect.

In nuclear matter, as we discussed before, the screening effects in fact reinforce the attractive strength of the bare interaction. In neutron matter the situation is the other way round. The screening is repulsive, and small in the full RPA calculation, but still enough to produce a sizeable quenching of the pairing gap. These predictions confirm at least at qualitative level the corresponding results obtained with the Gogny force.[4]

3.2. *Pairing gap*

The present calculation is focussed on the $^1S_0$ neutron-neutron (or proton-proton) pairing. One can distinguish the bare interaction which is responsible for the pairing

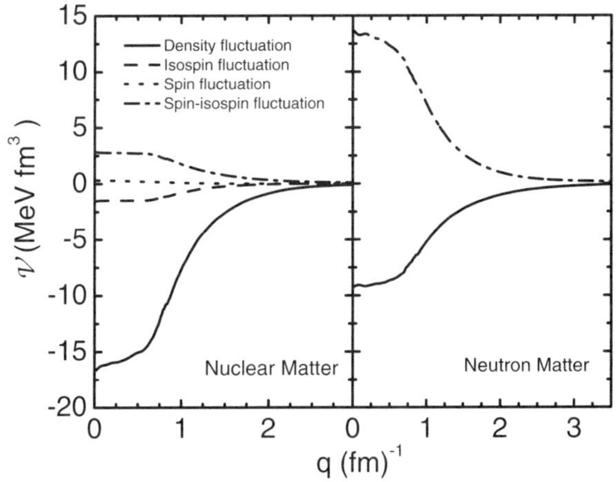

Fig. 2. Individual components of the ph residual interaction.

between the two particles in the $^1S_0$ state, from the screening interaction induced by the surrounding particles. Therefore the interaction, projected onto the $^1S_0$ channel, can be cast as follows

$$< k|\mathcal{V}|k' >= \int \frac{d\Omega}{4\pi}[\mathcal{V}_0(\vec{k},\vec{k}') + \mathcal{V}_1(|\vec{k}-\vec{k}'|)]. \quad (9)$$

As bare force we use Argonne $V18$, the same as for calculating the G-matrix and the selfenergy. We solved the general gap equation. The results are plotted in Fig. 3. In neutron matter the screening effect is small and just reduces the gap. At variance with previous calculations existing in the literature[3] the full RPA screening is much less effective than the one bubble approximation because of the stronger renormalization of the spin fluctuations vs the density fluctuations in the induced interaction. However this finding confirms the preceding predictions with Gogny force (see Fig. 8 of Ref.[4]).

In nuclear matter, due to the antiscreening effect we discussed earlier, the magnitude of the gap variation is the other way around and much more sizeable: the gap rises up from 3 MeV to 5 MeV for Fermi momentum $k_F = 0.8\ fm^{-1}$. This is displayed in Fig. 3.

There are two kinds of selfenergy effects: dispersive effect and Fermi surface depletion. The first one is a correction to the sp spectrum in the energy denominator. Usually it entails a reduction of the pairing gap since the effective mass, beyond BHF approximation, is less than the unity. But at very low density the effective mass is larger than unity[19] and it reduces the quenching rate of the gap due to the interaction. This effect can be seen in the low density side of the neutron gap with Σ_{total} (upper right panel). Additional strong reduction is due to the depletion of the Fermi surface which hinders transitions around the Fermi surface. The maximum gap in a complete calculation is $1.5 - 2\ MeV$ at $k_F \approx 0.8\ fm^{-1}$.

In nuclear matter the selfenergy effects are much stronger already at moderately

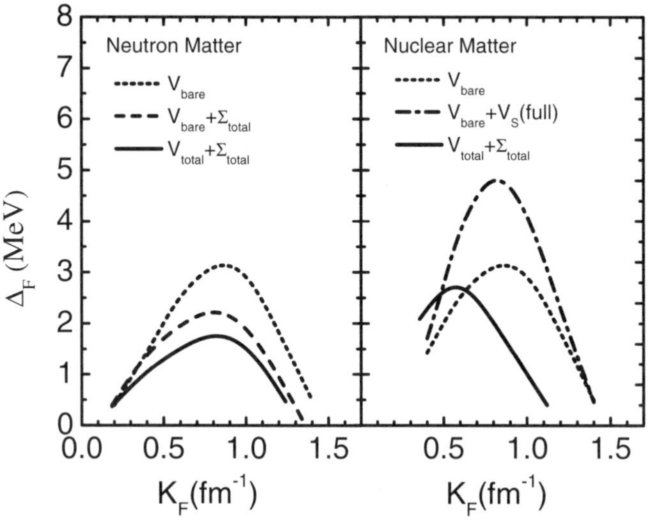

Fig. 3. Pairing gap in the $^1S_0$ channel for pure neutron matter and nuclear matter.

low density, as it has to be expected, and the peak value shifts down to very low density $k_F \approx 0.5–0.6\ fm^{-1}$. The Z-factor plays the major role: it quenches from 0.84 in neutron matter to 0.68 in nuclear matter at $k_F = 0.8\ fm^{-1}$. But the magnitude is about 0.5 MeV less than the value with only bare interaction. Therefore we can conclude that a strong cancelation occurs as soon as vertex corrections and self energy effects are simultaneously included in the gap equation. But this happens only in nuclear matter as an effect of antiscreening. We will come back to this point below.

4. Discussion and Conclusions

In this paper an exhaustive treatment of the $^1S_0$ pairing in nuclear and neutron matter has been reported. The medium polarization effects on the interaction and the selfenergy corrections to the mean field, both developed in the framework of the Brueckner theory, have been included in the solution of the gap equation.

Within the pure mean field approximation[20] the $^1S_0$ gap is not affected by the medium, either nuclear or neutron matter. So far the medium effects have not been considered in the case of nuclear matter except in Ref. 4. The vertex corrections due to neutron matter all give a reduction of the pairing, the magnitude depending on the adopted approximation.[3] The explanation relies on the competition of the attractive density excitations against the repulsive spin density excitations. The present calculation, based on G-matrix, also predicts a large quenching in agreement with almost all previous predictions, but only at the one bubble level. In the most complete calculation (full RPA) the quenching is largely reduced in apparent agreement with a recent Monte Carlo calculation.[21] But the inclusion of selfenergy effects definitely results in a large suppression as expected from basic properties of a strongly correlated many body system (see Introduction).

In the case of nuclear matter the most remarkable result is the antiscreening effect of the medium polarization. In fact in nuclear matter isospin modes arise that reverses the competition between the attractive density modes and the repulsive spin-density modes due to the presence of isospin modes. The argument addressed in Ref. 16 that the p-n (T=0) interaction is small compared to the n-n (T=1) is based on the vacuum scattering T-matrix and does not consider the strong medium renormalization of G-matrix, which inverts the strength of the two channels. However the enhancement of the gap to almost 5 MeV is almost completely suppressed by the strong correlation effects on the selfenergy. But, even a small variation of the force strength implies a large variation of the gap. These effects also push to lower density the peak value of the gap.

Acknowledgments

This work was partially supported by the grant appointed to the European Community project *Asia-Europe Link in Nuclear Physics and Astrophysics*, CN/ASIA-LINK/008(94791).

References

1. L. N. Cooper, R. L. Mills and A. M. Sessler, *Phys. Rev.* **114**, 1377 (1959).
2. D. Pines and A. M. Alpar, *Nature* **316**, 27 (1985).
3. U. Lombardo and H.-J. Schulze, *Lecture Notes in Physics* **578**, pp. 30-54, Eds. D. Blaschke, N. K. Glendenning and A. Sedrakian (Springer Verlag, 2001).
4. Caiwan Shen *et al.*, *Phys. Rev. C* **67**, 061302(R) (2003); Caiwan Shen, U. Lombardo and P. Schuck, *Phys. Rev. C* **71**, 054301 (2005).
5. F. Barranco *et al.*, *Eur. Phys. J. A* **21**, 57 (2004); F. Barranco *et al.*, *Phys. Rev. Lett.* **83**, 2147 (1999); F. Barranco *et al.*, *Phys. Rev. C* **72**, 054314 (2005).
6. J. P. Jeukenne, A. Lejeune and C. Mahaux, *Phys. Rep.* **25C**, 83 (1976).
7. M. Baldo and A. Grasso, *Phys. Lett. B* **485**, 115 (2000); U. Lombardo *et al.*, *Phys. Rev. C* **64**, 021301(R) (2001); H. Müther *et al.*, *Phys. Rev. C* **72**, 054313 (2005).
8. R. B. Wiringa, V. G. J. Stoks and R. Schiavilla, *Phys. Rev. C* **51**, 38 (1995).
9. M. Baldo, *Nuclear Methods and the Nuclear Equation of State*, International Review of Nuclear Physics, Vol. 8 (World Scientific, Singapore, 1999),
10. J. Dechargé and D. Gogny, *Phys. Rev. C* **21**, 1568 (1980).
11. S. Babu and G. E. Brown, *Ann. Phys. (N.Y.)* **78**, 1 (1973).
12. M. Baldo and L. S. Ferreira, *Phys. Rev. C* **50**, 1887 (1994).
13. L. G. Cao, U. Lombardo, C. W. Shen and N. V. Giai, *Phys. Rev. C* **73**, 014313 (2006).
14. H.-J. Schulze *et al.*, *Phys. Lett. B* **375**, 1 (1996).
15. P. Grangé, A. Lejeune, M. Martzolff and J.-F. Mathiot, *Phys. Rev. C* **40**, 1040 (1989).
16. H. Heiselberg *et al.*, *Phys. Rev. Lett.* **85**, 2418 (2000).
17. J. M. C. Chen *et al.*, *Nucl. Phys. A* **555**, 59 (1993).
18. U. Lombardo, P. Schuck and C. W. Shen, *Nucl. Phys. A* **731**, 392 (2004).
19. U. Lombardo, P. Schuck and W. Zuo, *Phys. Rev. C* **64**, 021301(R) (2001).
20. M. Baldo *et al.*, *Nucl. Phys. A* **515**, 409 (1990); *Nucl. Phys. A* **536**, 349 (1992).
21. A. Fabrocini *et al.*, *Phys. Rev. Lett.* **95**, 192501 (2005).

PINNING AND BINDING ENERGIES FOR VORTICES IN NEUTRON STARS: COMMENTS ON RECENT RESULTS

P. M. PIZZOCHERO

Dipartimento di Fisica, Università degli Studi di Milano,
and Istituto Nazionale di Fisica Nucleare, Sezione di Milano,
Via Celoria 16, 20133 Milano, Italy
E-mail: pierre.pizzochero@mi.infn.it

We investigate when the energy that pins a superfluid vortex to the lattice of nuclei in the inner crust of neutron stars can be approximated by the energy that binds the vortex to a single nucleus. Indeed, although the pinning energy is the quantity relevant to the theory of pulsar glitches, so far full quantum calculations have been possible only for the binding energy. Physically, the presence of nearby nuclei can be neglected if the lattice is dilute, namely with nuclei sufficiently distant from each other. We find that the dilute limit is reached only for quite large Wigner-Seitz cells, with radii $R_{\mathrm{WS}} \gtrsim 55$ fm; these are found only in the outermost low-density regions of the inner crust. We conclude that present quantum calculations do not correspond to the pinning energies in almost the entire inner crust and thus their results are not predictive for the theory of glitches.

Keywords: Neutron stars; Pulsar glitches; Vortex pinning; Superfluid neutron matter.

1. Pinning and Binding Energies

Pulsar glitches are sudden spin-ups in the otherwise steadily decreasing frequency of rotating magnetized neutron stars. According to the vortex-model, glitches may represent direct evidence for the existence of a macroscopic superfluid inside such stars.[1] This scenario involves the inner crust of the star, namely the density range $\rho_d \leq \rho \leq 0.6\rho_o$, where $\rho_d = 4 \times 10^{11}$ g/cm^3 is the neutron drip density and $\rho_o = 2.8 \times 10^{14}$ g/cm^3 is the nuclear saturation density. Here, the quantized vortices that form in the superfluid of unbound neutrons to carry its angular momentum can attach themselves to the lattice of neutron-rich nuclei co-existing with the neutrons in the inner crust, thus freezing part of the superfluid angular momentum. As a consequence of the star's spin-down, hydrodynamical forces then develop (Magnus force) which tend to detach the vortices from the lattice in order to let the angular momentum free to decrease. If a large number of vortices can be unpinned simultaneously and deliver their angular momentum to the star surface, its corresponding sudden spin-up is observed as a glitch. The process naturally repeats itself so that part of the superfluid angular momentum is released in discrete time steps, thus explaining the approximate periodicity of glitches.

A crucial microscopic input of the model is the structure of the nuclear lattice at

Table 1. Properties of five zones along the inner crust, as obtained in Ref. 2.

Zone	ρ_B (g/cm^3)	n_G (fm^{-3})	$k_{F,G}$ (fm^{-1})	R_{WS} (fm)	R_N (fm)
1	1.5×10^{12}	4.8×10^{-4}	0.24	44.0	6.0
2	9.6×10^{12}	4.7×10^{-3}	0.52	35.5	6.7
3	3.4×10^{13}	1.8×10^{-2}	0.81	27.0	7.3
4	7.8×10^{13}	4.4×10^{-2}	1.09	19.4	6.7
5	1.3×10^{14}	7.4×10^{-2}	1.30	13.8	5.2

the different densities found in the inner crust. Calculations predict a bcc structure for the coulomb lattice, namely each nucleus is at the center of a cubic cell of side $\ell_p = 2R_{WS}$ with nuclei at each vertex. The bcc lattice can also be seen as a superposition of layers of cubic cells; each layer, of thickness ℓ_p, is delimited by two planes of nuclei (those at the vertices of the cells) and contains in its middle a third plane of nuclei (those at the center of the cells). In Table 1 we list the results from the classical paper by Negele and Vautherin[2] for five zones along the inner crust. The parameters which characterize the lattice structure are the baryon mass density (ρ_B), the number density (n_G) and corresponding Fermi momentum ($k_{F,G}$) of the gas of unbound neutrons, the radius of the Wigner-Seitz (WS) cell (R_{WS}) and the radius of the nucleus (R_N) at the center of the cell.

The other crucial microscopic input is the vortex-lattice interaction potential as a function of the vortex position in the lattice, from which the pinning force could be derived. Due to the complicated spatial geometry of the general problem, however, a simpler approach has been followed so far which calculates only the *pinning energy*. This is defined as the difference in energy between two vortex-lattice configurations which are relevant to the pinning mechanism and possess a reasonable degree of symmetry. More precisely, one starts by taking the vortex axis oriented along a symmetry direction of the lattice and then considering only the interaction of the vortex with the nuclei lying in a plane perpendicular to the vortex axis (the middle plane of a layer). This allows to determine the pinning energy 'per site' (i.e. per layer). The pinning energy 'per unit length', which is the one relevant to the vortex-model for glitches, can then be obtained if one knows the number of pinning sites (layers) per unit length. Statistical estimates of this quantity allow to evaluate pinning also for random vortex-lattice mutual directions.

In Fig. 1 (left) the black disks represent the nuclei in the given plane, each positioned at the center of a cube. The nuclei at the vertices of the cubes are shown with white disks; they lie in the planes that delimit the layer, at a distance R_{WS} above and below the original plane. The vortex core position is schematically represented by the dashed line and the vortex axis, perpendicular to the plane, is shown by a star. The nuclear pinning (NP) configuration corresponds to the vortex pinned on a nucleus, e.g. in position (1). The interstitial pinning (IP) configuration corresponds to the vortex away from and equidistant between two nearby nuclei in

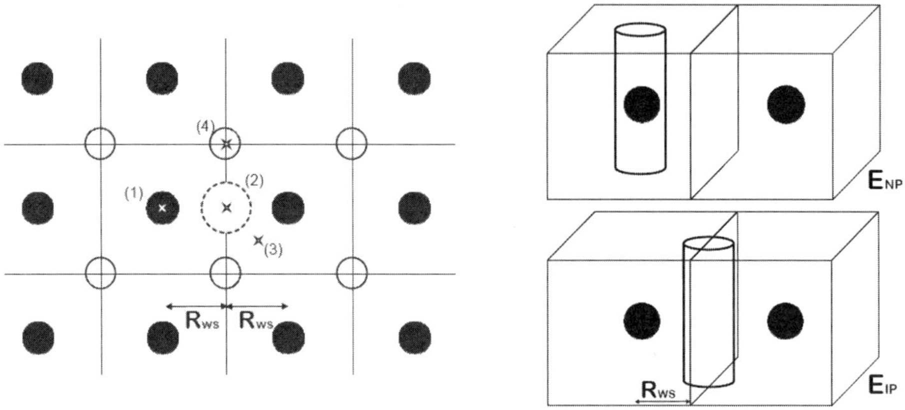

Fig. 1. Left: vortex positions in a bcc lattice. Right: configurations for the pinning energy.

the plane, e.g. in position (2). Notice that in the NP configuration the vortex also pins to the nuclei in the layers above and below, which are aligned along its axis with spacing $2R_{\rm WS}$. Conversely, in the IP configuration the vortex does not pin to any nucleus in the lattice. Position (4) is not allowed as an interstitial pinning site; actually, it is physically equivalent to nuclear pinning, since the vortex will pin to the nuclei in the layers above and below; in other words, positions (1) and (4) are equivalent once we consider the pinning energy per unit length.

The pinning energy is the energy necessary to move the vortex by a distance $R_{\rm WS}$, from the NP to the IP configuration. The *pinning energy per site* is then defined as $E_{\rm pin} = E_{\rm NP} - E_{\rm IP}$, i.e. the difference in energy between the two configurations. Physically, the vortex interacts significantly only with first neighbors; therefore, it is enough to restrict the attention to a system of two adjacent cells, as shown in Fig. 1 (right), since contributions from cells further away will cancel out. If $E_{\rm pin} < 0$ the vortex is attracted by the nucleus ('nuclear pinning') and the average pinning force per site is obtained as $F_{\rm pin} = |E_{\rm pin}|/R_{\rm WS}$. If $E_{\rm pin} > 0$ the vortex is repelled by the nucleus ('interstitial pinning'); is such a case the pinning force is several orders of magnitude smaller than $F_{\rm pin}$ and practically negligible. This because the vortex can be moved around the lattice completely avoiding the nuclei, e.g. from (2) to (3), which cost almost no energy.[3]

If one considers a *single* nucleus instead of the actual lattice of nuclei, a simpler quantity can be introduced: the *binding energy* of the vortex-nucleus system is the energy necessary to detach the vortex from the nucleus and take it to infinity, where the background matter is uniform (unaffected by the presence of the nucleus or the vortex). It can be defined as the difference in energy between two configurations, one with the vortex bound to the nucleus (i.e. with vortex-nucleus separation $d = 0$) and the other with the vortex far away and unbound from the nucleus (i.e. with vortex-nucleus separation $d \to \infty$). The short range nature of nuclear forces and the rapidly decreasing kinetic potential ($\propto 1/r^2$) associated to the vortex flow,

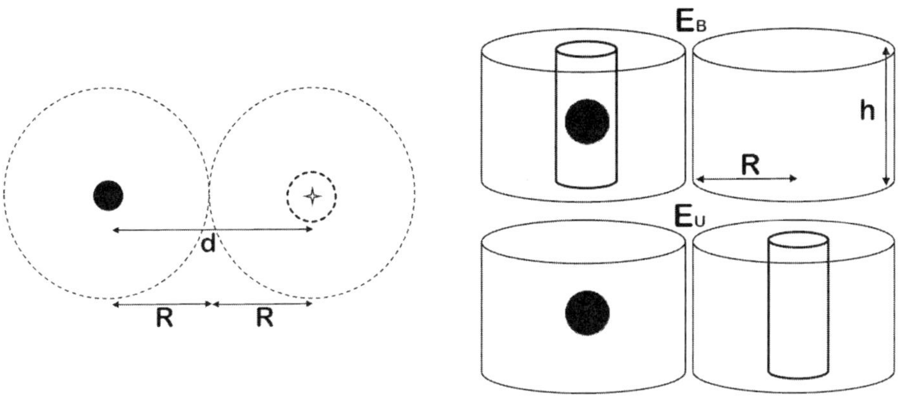

Fig. 2. Left: vortex-nucleus system with separation d and cylindrical quantization boxes of radius $R = \frac{1}{2}d$. Right: configurations for the binding energy.

however, imply that the vortex and the nucleus interact significantly only when they are relatively close to each other. Therefore, to a good approximation the binding energy is reached not for $d \to \infty$, but already for some finite separation $d = D_{\rm conv}$. The convergence distance $D_{\rm conv}$ corresponds to a configuration presenting a zone between the nucleus and the vortex where neutron matter has practically reached uniformity; taking $d > D_{\rm conv}$ only adds more uniform matter in between, whose contributions cancel out in the binding energy.

In Fig. 2 (left) the vortex and the nucleus are shown at a distance d apart. Such a system can be described by two cylinders of radius $R = \frac{1}{2}d$ and height h, which are also natural quantization boxes to evaluate the energy of this geometry. Indeed, the vortex-nucleus bound (B) and unbound (U) configurations are those represented in Fig. 2 (right) and the binding energy is defined as $E_{\rm bind} = E_{\rm B} - E_{\rm U}$. Of course, for this to make sense the result must converge; as explained, this will happen only when the separation is $d \geq D_{\rm conv}$ and thence when the cylinders have radius larger than $R_{\rm conv} = \frac{1}{2}D_{\rm conv}$. The physical interpretation of the convergence radius is the same as before: when $R = R_{\rm conv}$, matter has practically reached uniformity at the surface of both cylinders of Fig. 2. Taking larger boxes would only add more uniform matter at their boundaries, whose energy contributions cancel out in the binding energy. In this respect, we notice that the choice of cylindrical boxes, although natural, is by no means unique; for example one could as well take two cubic boxes of side $\ell = d$. As long as $\ell \geq 2R_{\rm conv}$, energy contributions from the additional uniform matter will again cancel out in the final binding energy.

The complicated geometry of the pinning configurations, with its interplay of different symmetries (spherical for the nucleus, axial for the vortex, periodic for the lattice), makes the calculation of pinning energies a very difficult problem at the quantum level. Conversely, binding energies are simpler to calculate since the convergence requirement allows to split both the U and B configurations into inde-

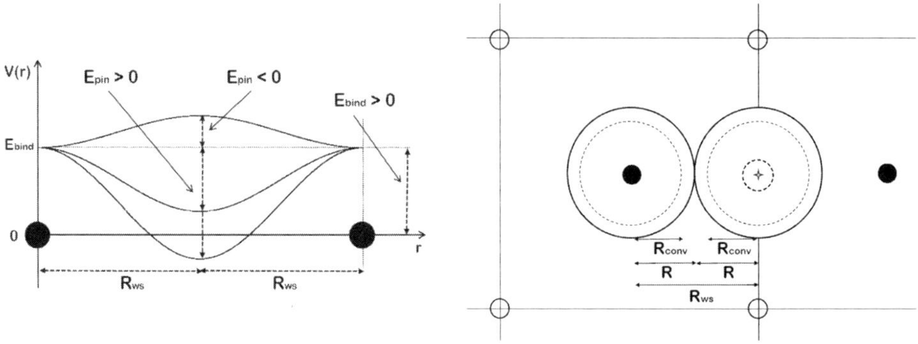

Fig. 3. Left: example of different possible pinning energies corresponding to the same binding energy. Right: lattice in the dilute limit $R_{\mathrm{WS}} > 2R_{\mathrm{conv}}$.

pendent axially symmetric problems (the pairs of cylinders). The question is under which conditions pinning and binding energies can be expected to be equivalent.

2. Dilute Limit for the Lattice

In general, pinning and binding energies are different and unrelated quantities. The binding energy represents the interaction between a vortex and a single nucleus while pinning involves the interaction of the vortex with a lattice of nuclei. Moreover, binding is defined as the value to which the difference $E_{\mathrm{B}} - E_{\mathrm{U}}$ converges for large enough vortex-nucleus separation and thus it does not depend on any distance parameter; conversely, pinning is defined with respect to a specific configuration, namely the difference $E_{\mathrm{NP}} - E_{\mathrm{IP}}$ must be evaluated at vortex-nucleus separation R_{WS}, and thus it depends crucially on the length parameter of the lattice. Figure 3 (left) illustrates the difference between E_{bind} and E_{pin}: binding is related only to the value of the interaction potential at the center of the WS cells, while pinning is determined by the local extremum in the interaction potential which, by symmetry reasons, must develop at the boundary between two WS cells. Whether this is a maximum or a minimum will depend on the particular radial dependence of the interaction potential; as long as this dependence is not known, as it is the case for the problem under study, no prediction on the value or *even the sign* of the pinning energy can be extracted from the knowledge of the binding energy, as obvious from the figure. We point out that the NP configuration of pinning is physically equivalent to the B configuration of binding, at least as long as the vortex core is smaller than the WS radius so that the presence of nearby nuclei can be neglected. Therefore, the difference between E_{pin} and E_{bind} follows essentially from the physical difference between the IP and U configurations.

If the spacing of the nuclei in the lattice is sufficiently large, however, one expects that in the IP configuration the vortex is sufficiently distant from the nuclei that it does not interact significantly with them anymore, so that the IP configuration becomes physically equivalent to the U configuration. If this is the case, the pinning

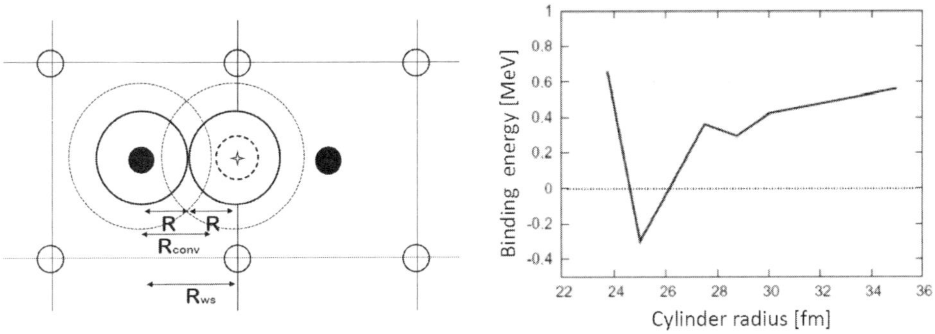

Fig. 4. Left: lattice in the dense limit $R_{\rm WS} < 2R_{\rm conv}$. Right: dependence of the binding energy on the radius of the quantization cylinder (figure taken from Ref. 7).

energy will coincide with the binding energy; we call this scenario, where $E_{\rm pin} = E_{\rm bind}$, the *dilute limit* for the lattice. Quantitatively, such a limit is reached when the vortex-nucleus distance in the IP configuration ($R_{\rm WS}$) is larger than the separation for which they do not appreciably interact anymore ($D_{\rm conv}$).

Figure 3 (right) shows the lattice in the dilute limit, namely with $R_{\rm WS} > D_{\rm conv} = 2R_{\rm conv}$. The IP configuration with vortex-nucleus separation $d = R_{\rm WS}$ can be represented by two cylinders (solid circles) of radius $R = \frac{1}{2}d = \frac{1}{2}R_{\rm WS} > R_{\rm conv}$. Since this radius is larger than the convergence radius (dotted circles), the IP configuration in this limit is indeed equivalent to the U configuration (cf. Fig. 2) and the pinning energy thus calculated is the same as the binding energy. Conversely, if $R_{\rm WS} < D_{\rm conv} = 2R_{\rm conv}$ the lattice is in the dense limit and $E_{\rm pin} \neq E_{\rm bind}$. Figure 4 (left) illustrates this scenario; the cylinders needed to represent the IP configuration have radii smaller than $R_{\rm conv}$ and therefore this geometry is not equivalent to the U configuration, which requires convergence to be properly defined. Consequently, the energy difference $E_{\rm pin}$ calculated from this geometry cannot be equal to $E_{\rm bind}$, the latter being reached only for larger radii of the cylinders.

In conclusion, the important parameter to be determined is the convergence distance $D_{\rm conv}$ for binding. A lower limit can easily be found by considering the kinetic energy contribution of a nucleus added to the vortex flow at distance d from the vortex axis.[4] This energy becomes negligible only for vortex-nucleus separations larger than $d \approx 30$ fm,[5] which implies that $D_{\rm conv}$ cannot be smaller than this value.

3. Comments on Recent Results

To date, the only consistent and realistic calculation of pinning energies corresponding to the parameters of the inner crust (Table 1) have been performed in the framework of the Local Density Approximation (LDA); details of the model are given in Ref. 5.

The purely additive nature of energy contributions from adjacent volumes of matter in this semi-classical approximation allows on the one hand to deal with the difficult geometry of pinning; but on the other hand, it completely neglects 'proximity' effects associated to the non-local quantum nature of pairing phenomena. In the absence of credible quantum calculations for the vortex-lattice system, however, which can confirm or modify the semi-classical predictions, the LDA results are the only physically reasonable input for the vortex theory of glitches.

Binding energies are relevant to glitch theory in two related aspects: they allow to determine the convergence distance $D_{conv} = 2R_{conv}$ and they are equal to the pinning energies for the density regions where the lattice is dilute, i.e. where $R_{WS} > D_{conv}$. Recently, a quantum calculation of binding energies has been performed by solving the mean-field Hartree-Fock-Bogoliubov (HFB) equations. Although improperly and quite misleadingly called pinning energy, the quantity calculated in Ref. 6 is obtained precisely as shown in Fig. 2 (right), and convergence is correctly required as a consistency condition for the model. In Fig. 4 (right) we show their study of the convergence of E_{bind} as a function of the cylinder radius R. At the value $R = 30$ fm used in Ref. 6, *reasonable* convergence is finally reached (the still increasing E_{bind} is included in the error bars in their figures). Being optimistic one may go down to $R_{conv} \approx 27 - 28$ fm, but lower values present wild oscillations (at $R \sim 25$ fm, E_{bind} even reverses its sign), obviously out of control and not related to any physics but only to the artificial boxes of the quantization procedure (the 'dripped' neutrons occupy positive-energy *continuum* states). One thus finds $D_{conv} \approx 55$ fm, and thence the dilute limit corresponds to $R_{WS} \gtrsim 55$ fm. The results of Ref. 2 show that this condition corresponds to only the outermost layers of the inner crust, with densities $\rho_B \lesssim 4.7 \times 10^{11}$ g/cm^3 and $n_G \lesssim 10^{-4}$ fm^{-3} ($k_{F,G} \lesssim 0.14$ fm^{-1}). This is very unwelcome, since it does not allow to use binding in place of pinning in most of the crust, particularly in the density regions which appears to be relevant to glitches in the LDA, i.e. around zone 3.[5]

In Fig. 5 (left) we report the comparison between LDA and HFB results used in Ref. 6 to support their main conclusion, namely that around zones 3 and 4 pinning is not nuclear, but interstitial. This conclusion is wrong and misleading: as shown by the vertical line at $k_{F,G} = 0.14$ fm^{-1}, the figure compares two different energies in a regime where they do not represent the same quantity. To state it more vividly, in order to describe pinning in zone 1 ($R_{WS} = 44$ fm) the HFB approach should use cylinders with $R = 22$ fm and the result should already converge at this small radius; to describe zone 3, convergence should be obtained at $R = 14$ fm!

Where the dilute limit is reached, however, the HFB results for E_{bind} can and should be compared to the LDA results for E_{pin}, in order to assess whether the semi-classical model is realistic or not, at least in this regime. One can actually exploit the 'locality' of the LDA to extend the analysis a bit further. Indeed, the absence of proximity effects is such that, when applied to the calculation of E_{bind}, the LDA converges to a definite final value already for $D_{conv} \approx 30$ fm. Therefore, at densities $n_G \lesssim 0.01$ fm^{-3} ($k_{F,G} \lesssim 0.6$ fm^{-1}), where $R_{WS} \gtrsim 30$ fm, the LDA values for pinning are the same as its results for binding and as such they could be compared to the corresponding HFB results. However, before claiming that at very

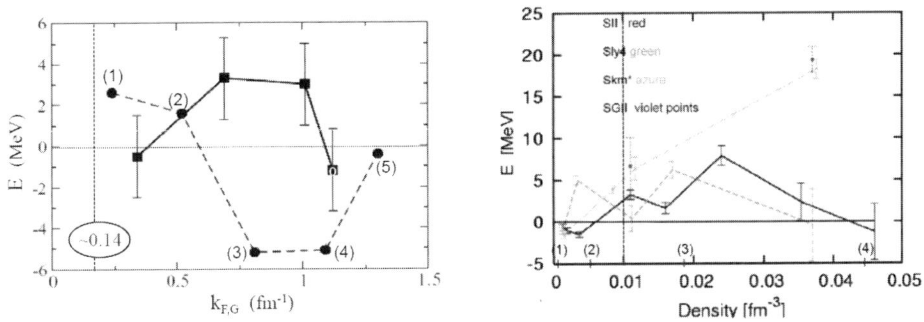

Fig. 5. Left: comparison of LDA results for pinning energies (dashed line) and HFB results for binding energies (solid line); the numbers indicate the zones of Table 1 (figure taken from Ref. 6). Right: comparison of HFB results for binding energies corresponding to different choices of the pairing interaction (figure taken from Ref. 7).

low densities (between zones 1 and 2) the HFB and LDA predictions for binding have opposite sign, like Fig. 5 (left) may suggest, some crucial issues should be kept in mind: *(i)* Figure 5 (right) shows the HFB results for different pairing interactions; the corresponding binding energies between zones 1 and 2 can differ even in the sign. Thence the comparison must be done with the 'same' pairing interaction, in the sense that the density dependence of the neutron gap in uniform matter must be the same in HFB and LDA. Instead, the interaction used in Ref. 6 is equivalent to Gogny D1S, which for $k_{F,G} > 1$ fm^{-1} yields larger gaps than the Argonne used in the LDA. *(ii)* As carefully explained in Ref. 5, the pinning energy should be evaluated at *fixed chemical potential* (the particle bath represented by the macroscopic inner crust which surrounds the widely spaced vortices), while the results of Ref. 6 correspond to a fixed number of particles. Since $E_{\rm bind}$ comes from the difference between very large numbers, great attention must be devoted to this kind of issue. *(iii)* Different zones present nuclei with different radii, while in Ref. 6 the same nucleus (with $R_N = 7$ fm) was taken at all densities.

With these issues under control, the comparison of quantum and semi-classical results for binding at low densities could become quite instructive.

References

1. P. W. Anderson and N. Itoh, *Nature* **256**, 25 (1975).
2. J. W. Negele and D. Vautherin, *Nucl. Phys. A* **207**, 298 (1973).
3. B. Link and R. Epstein, *Astrophys. J.* **373**, 592 (1991).
4. R. Epstein and G. Baym, *Astrophys. J.* **328**, 680 (1988).
5. P. Donati and P. M. Pizzochero, *Phys. Rev. Lett.* **90**, 211101 (2003); *Nucl. Phys. A* **742**, 363 (2004); *Phys. Lett. B* **640**, 74 (2006).
6. P. Avogadro *et al.*, *Phys. Rev. C* **75**, 012805 (2007).
7. P. Avogadro *et al.*, talk given at the meeting *Theoretical issues in nuclear astrophysics*, IPN Orsay (2007) (http://snns.in2p3.fr/meetings/0704meeting.html).

STRUCTURE OF A VORTEX IN THE INNER CRUST OF NEUTRON STARS

P. AVOGADRO

Dipartimento di Fisica, Università degli Studi di Milano, Milano, 20133, Via Celoria 16, Italy
E-mail: paolo.avogadro@mi.infn.it

F. BARRANCO

Departamento de Fisica Aplicada III, Universidad de Sevilla, Escuela Superior de Ingenieros Sevilla, 41092 Camino de los Descubrimientos s/n, Spain

R. A. BROGLIA

Dipartimento di Fisica, Università degli Studi di Milano, Milano, 20133, Via Celoria 16, Italy;
INFN, Sezione di Milano, Milano, 20133, Via Celoria 16, Italy;
The Niels Bohr Institute, University of Copenhagen, Copenhagen Ø, Blegdamsvej 17 2100, Denmark

E. VIGEZZI

INFN Sezione di Milano, Milano, 20133, Via Celoria 16, Italy

We study the vortex-nucleus interaction in the inner crust of neutron stars within the framework of quantum mean field theory. We use the SLy4 Skyrme interaction in the particle-hole channel, and a density dependent contact interaction in the particle-particle channel. We discuss the results obtained for the spatial dependence of the pairing gap and the density dependence of the pinning energy. A comparison with a semiclassical model is also presented.

Keywords: Inner Crust; Neutron star; Pinning energy; Glitches.

1. Introduction

A neutron star is the remnant of the explosion of a massive star. The typical radius of a neutron star is of the order of 10 km; its mass on the other hand is of the order of the solar mass ($M_\odot \approx 2 \cdot 10^{33} g$). The density of a neutron star is thus comparable to the density found in atomic nuclei. For this reason models of neutron star structure are largely based on the experimental data and on the theoretical tools used for atomic nuclei. The outer layer of a neutron star is called outer crust; in this region nuclei form a Coulomb lattice and are surrounded by relativistic electrons. Going towards the center of the star, the density increases and nuclei become more and more neutron rich in order to fulfill the β-stability condition. When the density reaches about 0.001 n_0 ($n_0 = 2.8 \cdot 10^{14}$g/cm^3 being the saturation density in nuclei), all the neutron levels are filled and the so called drip-line is reached, and neutrons start to form a gas that surrounds the nuclei. This represents the beginning of the inner crust, which extends in the density range between approximately $10^{-3}n_0$ and $0.5n_0$. As the density increases, the lattice step decreases, until nuclei start to deform and different configurations are energetically favoured in the so called pasta phase.[1] In the following we shall only deal with the inner crust at densities smaller than 0.3 n_0, where the assumption of spherical nuclei should be safe. The first microscopic study of this region was carried out by Negele and Vautherin,[2] who used density functional theory to calculate the lattice constant and the isotopic composition of the nuclei in this region. They did not take into account pairing correlations, which have been considered in recent calculations.[3] In fact, in all but the youngest stars the temperature is expected to be not higher than 0.1 MeV, much lower than the typical values of the neutron pairing gap ($\approx$ 1-3 MeV, depending on the different approximations adopted) obtained in uniform matter at the densities of the inner crust, and for this reason the free neutrons are expected to be superfluid. If the neutron star rotates, vortices are expected to form in the superfluid neutron gas. As explained in Section 2, vortex dynamics may be at the basis of the phenomenon of glitches. In Section 3 we present a microscopic theory for the calculation of a vortex, taking into account the inhomogenous character of the crust.[4] In Section 4 we discuss our results concerning the spatial dependence of the pairing gap and the density dependence of the pinning energy, and make a comparison with a semiclassical model.

2. Vortices and Glitches

Neutron stars are nowadays observed in other parts of the electromagnetic spectrum, but they were first discovered as pulsars in radio waves. The signal emitted is very stable and reflects the period of rotation of the star (typical periods range from few ms to few s). The pulsar period increases steadily, but there are some irregularities: particularly important are the so called glitches, which consist in a sudden decrease of the period, followed by a recovery phase. Different models have been proposed in order to explain the glitches; one of the most promising is due to Anderson and Itoh[5] and involves the motion of vortices in the inner crust. The dynamics of the vortices in the inner crust is affected by the presence of the nuclei which compose the Coulomb lattice. In a superfluid where currents are present a free vortex line moves according to the velocity field. However, if the interaction between vortex and nuclei is strong enough the vortices are pinned on the nuclei and cannot move freely. A difference then develops between the velocity of the star (which slows down constantly due to electromagnetic braking) and the velocity of the neutron sea (which is fixed, because it depends only on the number of vortices). This difference induces a hydrodynamic force, the so called Magnus force, which acts on vortices. The Magnus force per unit length is given by $\mathbf{F}_M = n_s \mathbf{\Gamma} \times (\mathbf{V}_v - \mathbf{V}_s)$, where n_s is the neutron density, $\mathbf{\Gamma} = \hat{z} \oint \mathbf{V} \cdot \mathbf{dl}$ ($\mathbf{V}$ being the velocity field around the vortex line), $\mathbf{V}_v$ is the velocity of the vortex line and $\mathbf{V}_s$ is the velocity field of the neutrons in absence of the vortex line at that point. When the difference between the two velocities is large enough the Magnus force overcomes the pinning force and the vortices are free to move outwards (and eventually annihilate at the edge of the inner crust). This reconfiguration leads to a modification of the velocity field and to a transfer of angular momentum from the neutron gas to the rest of the star, thus causing the glitches. In order to develop a quantitative model able to describe glitches, it is essential to know in detail the interaction between the vortices and nuclei in the inner crust.

3. Quantum Microscopic Description of the Inner Crust

We shall base our microscopic calculation of the properties of the inner crust on the Hartree-Fock-Bogoliubov (HFB) equations. These equations yield a self-consistent mean field including pairing correlations, and for zero-range forces they can be written as

$$\begin{cases} (H(\mathbf{x}) - \lambda)U_i(\mathbf{x}) + \Delta(\mathbf{x})V_i(\mathbf{x}) = E_i U_i(\mathbf{x}), \\ \Delta^*(\mathbf{x})U_i(\mathbf{x}) - (H(\mathbf{x}) - \lambda)V_i(\mathbf{x}) = E_i V_i(\mathbf{x}), \end{cases} \tag{1}$$

where λ denotes the chemical potential, $H(\mathbf{x})$ and $\Delta(\mathbf{x})$ are the Hartree-Fock Hamiltonian and the pairing field, while E_i is the quasi particle energy, U_i and V_i being the associated quasiparticle amplitudes. To solve the HFB equations we enclose the system in a cylindrical box. We assume perfectly reflecting walls, with single-particle wavefunctions vanishing at the edge of the box. We have checked that the box radius ($R_{box} \sim 30$ fm) is large enough to ensure the stability of the results. We shall assume that the system is axially symmetric with respect to the $z-$axis, and expand the quasiparticle amplitudes on a single-particle basis $\varphi_{n,m,k}(\rho, z, \phi)$ composed by the solutions of the Schrödinger equation in the cylinder:

$$U^{qm}(\rho, z, \phi) = \sum_{nk} U^{qm}_{nk} \varphi_{n,m,k}(\rho, z, \phi), \tag{2}$$

$$V^{qm}(\rho, z, \phi) = \sum_{nk} V^{qm}_{nk} \varphi_{n,m-\nu,k}(\rho, z, \phi). \tag{3}$$

The labels qm define a quasiparticle level q with associated projection of the angular momentum m on the $z-$axis, while the index n denotes the different number of nodes of the radial part of the basis functions and the index k labels the different wavenumbers of the plane waves along the z axis. In the particle-hole channel we shall use the Skyrme SLy4 interaction,[6] while in the particle-particle channel we use a density dependent contact interaction, parametrized similarly to Ref. 7 :

$$V_{pair}(\mathbf{x}, \mathbf{x}') = V_0 \left(1 - 0.7 \cdot \left(\frac{n(\mathbf{x})}{0.08}\right)^{0.45}\right) \delta(\mathbf{x} - \mathbf{x}'), \tag{4}$$

where $V_0 =$ -481 MeV fm^3 . The use of a contact interaction requires the introduction of a cutoff. We adopt $E_{cut} \approx 60$ MeV.

A vortex is an excited state of a superfluid system in which all Cooper pairs carry the same angular momentum, defined by the vortex number ν, which appears in Eq. (3). We use the standard ansatz for the pairing gap associated with a vortex:

$$\Delta(\rho, z, \phi) = \Delta(\rho, z)e^{i\nu\phi}, \tag{5}$$

taking the vortex number ν equal to 1. Choosing instead $\nu = 0$ one obtains a solution with no vortex, analogous to those of previous studies of the neutron pairing gap in the inner crust.[8,9] Considering only neutrons we can describe a piece of uniform neutron matter, while including protons in the calculations (we take $Z = 40$, following Ref. 2), we can describe a nucleus immersed in the neutron sea. We shall solve the HFB equations considering four different configurations (cf. the right part of Fig. 2): a cell with a nucleus and a vortex (a), a cell with a nucleus and no vortex (b), a cell with a vortex in a uniform sea of neutrons (c), a cell with a uniform sea of neutrons and no vortex (d). In each case we shall calculate, besides the total energy of the system, the neutron density:

$$n(\rho, z) = 2\sum_{qm} |V^{qm}(\rho, z, \phi)|^2, \tag{6}$$

the abnormal neutron density:

$$\tilde{n}(\rho, z, \phi) = \sum_{qm} U^{qm}(\rho, z, \phi)(V^{qm}(\rho, z, \phi))^*, \tag{7}$$

and the current density of the superfluid neutrons:

$$\Phi(\rho, z) = -\frac{i\hbar}{m\rho}\sum_{qm}(V^{qm}(\rho, z, \phi))^*\frac{\partial V^{qm}(\rho, z, \phi)}{\partial\phi}, \tag{8}$$

from which we obtain the velocity field, Φ/n. The pairing gap is obtained self-consistently from the abnormal density, $\Delta = V_{pair}\tilde{n}$. We note that Eqs. (2-7) imply that the single-particle levels that constitute the Cooper pair associated with a $\nu = 1$ vortex have opposite parity, because for the cases we consider $\Delta(\rho, z) = \Delta(\rho, -z)$.

4. Results

We could check our HFB calculations for a vortex in uniform neutron matter (configuration (c) above) with previous studies.[10] Instead, quantum calculations of a vortex in the presence of the nucleus (configuration (a)) had only been performed until now with the little realistic assumption of cylindrical symmetry.[11] In Fig. 1 we show our results for the spatial dependence of the neutron density, of the pairing gap and of the velocity field corresponding to the asymptotic value (i.e. far from the nucleus) of the neutron density $n \approx 0.012$ fm^{-3}. It is seen that the gap is strongly suppressed inside the nuclear volume. This is related to the fact that at this density, due to shell effects, only single-particle resonances of a given parity are

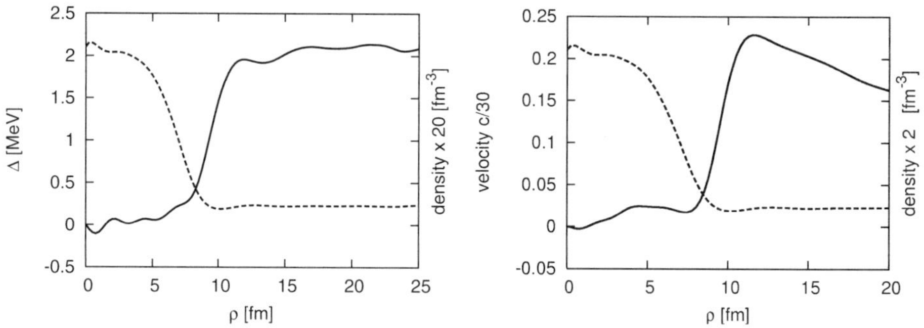

Fig. 1. Spatial dependence of a vortex passing through a nucleus (configuration (a) in the text, cf. Fig. 2a, right panel). The spherical nucleus is placed at the origin of the cell, while the vortex line is directed along the z-axis. The pairing gap (left figure) and the velocity field (right figure) are shown (solid curves) as a function of the distance from the center of the nucleus in the equatorial plane ($z = 0$). We also show the (scaled) the neutron density (dashed curve).

found close to the Fermi energy. Consequently it is difficult to form $\nu = 1$ Cooper pairs concentrated within the nuclear volume. At higher densities, shell effects in the single-particle continuum become much weaker and the gap can penetrate into the nuclear volume.

The interaction between a nucleus and a vortex is calculated in terms of the *pinning energy*. This quantity was studied by Epstein and Baym,[12] in a semiclassical study of the free energy profile of the nucleus-vortex system as a function of the distance between the vortex axis and the center of the nucleus. They defined the pinning energy as the difference between the minimum energy of the system and the energy when the vortex is far away from the nucleus. In our framework, due to the limitation of axial symmetry, we can only compare the two extreme configurations: a vortex placed on a nucleus, and a vortex far from it. We shall then compare the cost E^{vor}_{nuc} to pin a vortex on a nucleus with the cost E^{vor}_{unif} to create a vortex in uniform matter (cf. the right part of Fig. 2). The pinning energy is then obtained as $E_{pinning} = E^{vor}_{nuc} - E^{vor}_{unif}$, and negative values of $E_{pinning}$ indicate that the pinned configuration is favoured. To compute E^{vor}_{nuc} we need to solve the HFB equations for two configurations: (a) a cell containing a nucleus at its center, and a vortex with its axis passing through the nucleus; (b) a cell with only the nucleus and without the vortex. The two cells must have the same number of particles. To compute E^{vor}_{unif} we solve the HFB equations for the two following configurations: (c) a cell containing only the vortex and (d) a cell containing only neutrons, having the same number of particles. Furthermore, the neutron density far from the nucleus in cell (b) must be the same as in cell (d).

Our study of pinning involves only one vortex and a single nucleus. While the distance between vortices is estimated to be macroscopic, one should be concerned about the interaction of the vortex with other nuclei of the lattice. In order to address the validity of our approximation in a quantitative way, one would need to

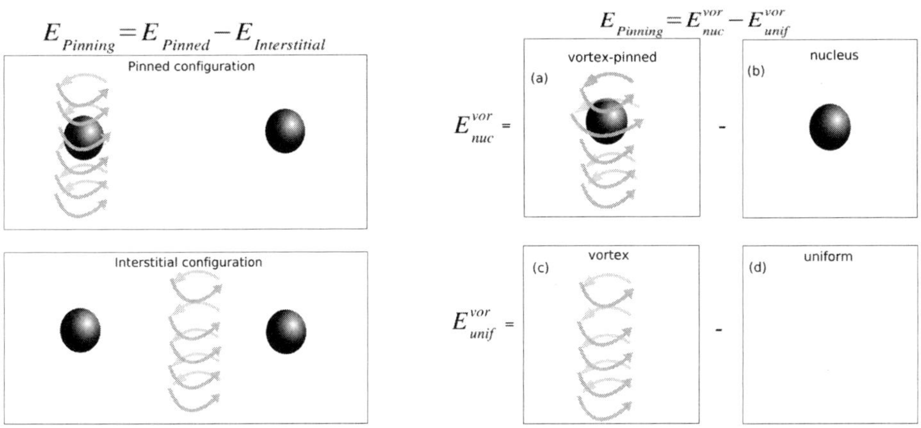

Fig. 2. Scheme used to calculate the pinning energy by Donati and Pizzochero[14] (left) and in the present paper (right).

perform systematic calculations taking into account the band structure associated with the lattice.[13] One expects, however, that our approximation should be reasonable if the vortex radius is smaller than the distance between nuclei. We have estimated that this is the case in the density range we consider (cf. Fig. 4d below).

In a semiclassical study of the problem under discussion[14] the pinning energy was defined as the difference of the energies associated with the pinned and the interstitial configurations (cf. the left part of Fig. 2), taking explicitly into account the presence of one of the neighbouring nuclei. This definition of pinning energy reduces to ours, when the interaction of the vortex with the second nucleus is negligible. For a comparison with the semiclassical model, and in order to answer some concerns raised about the reliability of our calculations,[15,16] we have tested the importance of the neighboring nucleus within the semiclassical model. We first computed the energy of each of the relevant configurations using the semiclassical theory, the two pairing interactions (Argonne and Gogny) and the parameters presented in Ref. 14 , instead of solving the HFB equations. We then calculated the pinning energy using the scheme proposed in Ref. 14 (cf. the left part of Fig. 2): the values are shown by the dashed curves in Fig. 3, and they reproduce to a good accuracy the values of Ref. 14 (cf. the dashed curve in Fig. 4(a) for the Argonne interaction, taken from Ref. 14). They are compared in Fig. 3 with the pinning energy obtained with our scheme (cf. the right part of Fig. 2), also using semiclassical energies. The two schemes yield very similar results, showing that the differences existing between the quantal and the semiclassical calculations, which we will now discuss, are not due to the geometrical procedure used to derive the pinning energy.

The pinning energy obtained from the quantum calculations is shown in Fig. 4, where it is compared to the semiclassical result with the Argonne potential (note that our pairing interaction (Eq. 4) closely reproduces the pairing gap obtained with the Argonne interaction in neutron matter). We find pinning in the regions at low

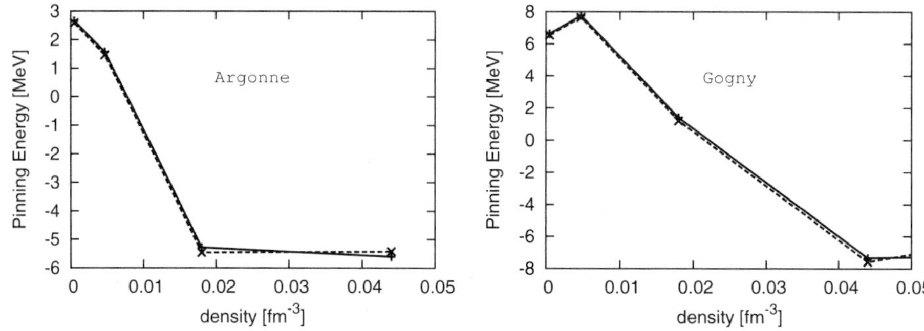

Fig. 3. Solid and dashed curves show respectively the pinning energies calculated using our scheme (Fig. 2, right part), or the scheme of Ref. 14 (Fig. 2, left part). In both cases, we have computed the energies of the different configurations using the semiclassical model, with the Argonne pairing interaction (left panel) or the Gogny pairing interaction (right panel).

density and antipinning at medium density, in contrast with the behaviour found in the semiclassical calculation. In order to get some insight on this different behaviour, it us useful to consider the semiclassical expressions for the kinetic energy associated to the vortex flow, E_{flow}, and the condensation energy E_{cond}. For a given configuration, E_{flow} and E_{cond} are obtained integrating respectively the associated kinetic

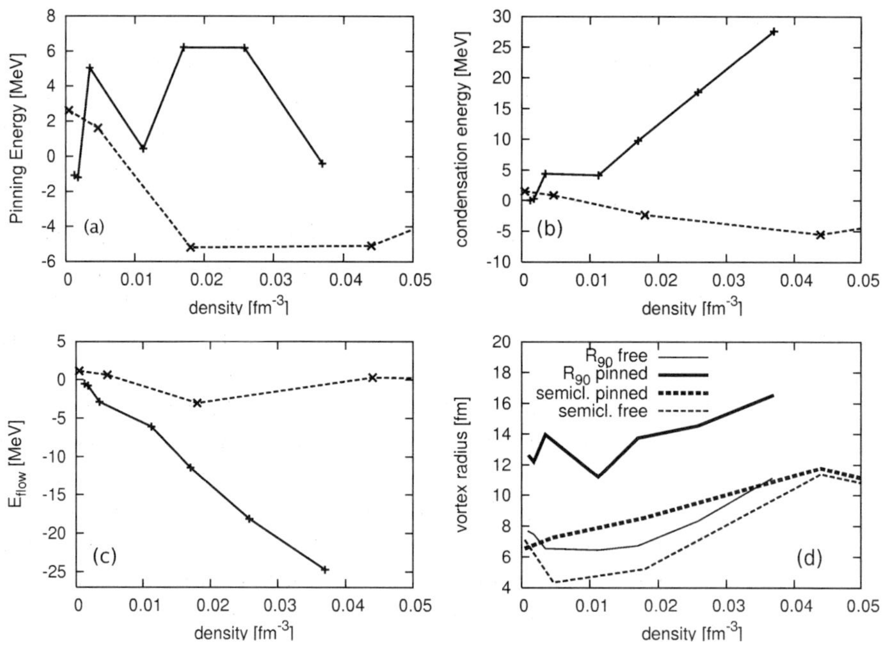

Fig. 4. Pinning energy (a), condensation energy (b), kinetic energy of the vortex flow (c) and vortex radius of the free and pinned vortex (d) as a function of the density, according to our quantal model (solid curves) and to the semiclassical model (values taken from Ref. 14, dashed curves).

energy density 1/2 Φ^2/n, and the density -3/8 $\left[n(\rho,z)\Delta^2(\rho,z)/(\lambda - V_{nuc}(\rho,z))\right]$, where V_{nuc} is the nuclear mean field potential. The contributions of these quantities to the pinning energy are shown by dashed lines in Fig. 4b and Fig. 4c for the semiclassical calculation,[14] and comparing with Fig. 4a it can be seen that in this case E_{cond} largely determines $E_{pinning}$. In Fig. 4b and Fig. 4c we also show the corresponding quantities, calculated using the densities, pairing field and potentials obtained in our quantum calculations. The two estimates of E_{cond} and E_{flow} show a very different dependence. This striking difference can be related to the strong suppression of the pairing gap around the nucleus in the pinned configuration found in the quantal calculation, which implies a large loss of condensation energy, as well as a suppression of the velocity field (cf. Fig. 1). This effect, and the consequent "expulsion" of the vortex from the nuclear volume does not take place in the semiclassical calculation, and reflects itself also in the vortex radius (cf. Fig. 4d).[17] The radii of the free vortex are rather similar, while the radius of the pinned vortex becomes much larger in the quantum calculation. Summarizing, in the pinned configuration the volume affected by the loss of pairing and flux velocity around the nucleus in the quantal model is, due to shell effects, much larger than in the semiclassical model and as a consequence the condensation and flux kinetic energies predicted by the semiclassical model must be strongly corrected.

References

1. C. J. Pethick and D. G. Ravenhall, *Annu. Rev. Nucl. Part. Sci.* **45**, 429 (1995).
2. J. W. Negele and D. Vautherin, *Nucl. Phys. A* **207**, 298 (1973).
3. M. Baldo, E. E. Saperstein and S. V. Tolokonnikov, *Nucl. Phys. A* **775**, 235 (2006).
4. P. Avogadro, F. Barranco, R. A. Broglia and E. Vigezzi, *Phys. Rev. C* **75**, 012805(R) (2007).
5. P. W. Anderson and N. Itoh, *Nature* **256**, 25 (1975).
6. E. Chabanat, P. Bonche, P. Haensel, J. Meyer and R. Schaeffer, *Nucl. Phys. A* **635**, 231 (1998).
7. E. Garrido, P. Sarriguren, E. Moya de Guerra and P. Schuck, *Phys. Rev. C* **60**, 064312 (1999).
8. N. Sandulescu, Nguyen Van Giai and R. J. Liotta, *Phys. Rev. C* **69**, 045802 (2004).
9. F. Montani, C. May and H. Müther, *Phys. Rev. C* **69**, 065801 (2004).
10. Y. Yu and A. Bulgac, *Phys. Rev. Lett.* **90**, 161101 (2003).
11. F. De Blasio and Ø. Elgaroy, *Astron. Astrophys.* **370**, 939 (2001).
12. R. I. Epstein and G. Baym, *Astrophys. J.* **328**, 680 (1988).
13. N. Chamel, S. Naimi, E. Khan and J. Margueron, *Phys. Rev. C* **75**, 55806 (2007).
14. P. M. Donati and P. M. Pizzochero, *Nucl. Phys. A* **742**, 363 (2004).
15. P. M. Pizzochero, email to the participants of the conference *Theoretical issues in nuclear astrophysics*, Orsay, 10-13 April 2007.
16. P. M. Pizzochero, these proceedings.
17. In the semiclassical work the radius has been obtained equating the pressures of the superfluid and normal phases of the vortex, yielding a value close to the coherence length. In the microscopic calculations we define the radius R_{90} as the distance from the vortex axis where the pairing gap recovers 90% of its asymptotic value (cf. the left part of Fig. 1).

THE DYNAMICS OF VORTEX PINNING IN THE NEUTRON STAR CRUST

B. LINK

Department of Physics, Montana State University,
Bozeman, Montana 59717, USA
E-mail: link@physics.montana.edu

The energetically-preferred rotational state of the inner-crust superfluid (SF) involves *pinning* of the neutron vortices to the lattice through a vortex-nucleus interaction. The question of whether or not vortices actually reach this state is crucial toward understanding the rotational modes of a neutron star (NS). I describe the physics of relaxation of unpinned vortices to the pinned state and show that vortices can pin only if the differential velocity between the SF and the solid is very low, $\lesssim 10$ cm s^{-1}. I argue that the pinned state is possibly *dynamically inaccessible* in a typical NS. I conclude with discussion of the implications for understanding NS spin jumps and long-period precession (nutation).

Keywords: Neutron stars; Pulsars; Dense matter; Rotating stars.

1. Introduction

Isolated neutron stars (NSs) exhibit a wealth of dynamical behavior. Over one hundred spin jumps ("glitches") have now been identified (see, e.g., Ref. 1). Evidence for precession (nutation) has been found in several NSs (e.g., Refs. 2–4). All isolated NSs show stochastic spin variations (timing noise) at some level (e.g., Ref. 5). Recently, quasi-periodic oscillations were seen in two magnetars (NSs with magnetic fields in excess of $\simeq 10^{14}$ G) following explosions in soft gamma-rays;[6,7] these oscillations have been interpreted as representing shear waves in the crust.[8,9]

To interpret these phenomena and obtain constraints on the ground state of matter in β-equilibrium and its dynamics, a theory of NS seismology needs to be developed. Particularly important in this connection is a full understanding of the dynamics of the rotating quantum liquids a NS is expected to contain and how these fluids couple to the crust, whose spin rate we directly observe. The coupling depends on how the *vortex lines* that thread a rotating superfluid (SF) interact with the crust. Initial estimates,[10] supported by current state-of-the-art calculations[11,12] presented at this meeting, indicate that the vortices of the $^1S_0$ neutron SF of the inner crust are attracted to the nuclei in some regions with energies of order an MeV per nucleus. Hence, the SF can lower the energy of the system if the vortices *pin* to the nuclei in regions where the interaction is attractive. The local SF angular momentum is determined by the spatial arrangement of the vortices; if the vortices were perfectly pinned, the SF would develop excess angular momentum as the crust spins down under electromagnetic torque. If a macroscopic number of vortices unpins as a consequence of some instability and moves dissipatively with respect to the crust, the crust will spin up, as in a glitch. The notion that pinning/unpinning transitions are responsible for glitches was proposed not long after pulsars were discovered.[13] Much of the physics, however, remains to be elucidated to bring these ideas into the form of a quantitative theory with which predictions can be made. While I focus here on the possibility of vortex pinning in the inner crust, vortices may also pin to magnetic flux tubes in the outer core[14] or to crystalline quark matter in the inner core[15] (if such a state exists).

If the inner-crust SF is to act as a reservoir from which glitches originate, vortices must pin in the first place. Exactly how this happens has not been explained. Here I present ongoing work on this question. I argue that the pinned state, while energetically favored, may be *dynamically inaccessible* in a spinning-down NS.

2. Vortex Properties and the Pinned State

A rotating neutron SF is threaded by vortices within which the fluid is normal and about which the SF circulates. In the inner crust the vortices are ~ 10 fm across, of macroscopic lengths and separated by $\sim 10^{-2}$ cm. As in an ordinary, incompressible fluid, the angular momentum of the fluid is determined by the distribution of vorticity. Unlike an ordinary fluid, however, the vortices of a SF are stable hydrodynamic structures about which each particle in the condensate (or each Cooper pair in a Fermi liquid) carries exactly $\hbar$ of angular momentum.

In the state of minimum free energy for a given angular momentum, the vortices corotate with the SF and its container. In a NS, differential rotation occurs if the SF and the rest of the star are coupled through a dissipative torque. To address the question of angular momentum exchange between the SF and the crust it is therefore crucial to understand dissipation in vortex motion.

The inner crust is expected to be amorphous, generally lacking long-range order,[16] rather than a regular lattice. Consider an initially straight vortex in the presence of randomly-positioned nuclei to which it is attracted. The transverse components of the vortex-nuclear interaction force sum to nearly zero along the vortex to very high precision because of the randomness of nuclear positions. The vortex has no preferred position in this situation and so is not pinned. For the vortex to pin it must bend. Sharp bends are energetically prohibited because a vortex has an enormous self-energy ("tension") of order ~ 1 MeV fm^{-1}. The origin of the tension is primarily the kinetic energy of the flow field about the vortex core. If the tension were infinite, the vortex could never pin. (*Some* pinning would be possible through the displacement of nuclei from their minimum-energy positions toward the vortex, but this is a very small effect).

The energy of a vortex segment has two contributions: 1) the potential energy in the presence of a nearby nucleus, and, 2) the bending energy associated with distortion of the vortex's velocity field. The pinned state is a configuration that minimizes the energy from these two competing contributions, and to reach that state energy must be dissipated. The primary energy loss process appears to be coupling of vortex excitations to lattice vibrations. The rate at which the vortex can damp into the pinned state will be determined by how quickly a vortex segment can excite a nearest-neighbor nucleus into oscillation, and how fast that oscillating nucleus can then share its energy with the rest of the lattice. Without the loss of vortex energy to the lattice, the vortex could never lose energy and would never pin.

These ideas can be conveniently described using the concept of vortex drag. Suppose there is an ambient superfluid flow, as for the general case in which the SF is not corotating with the solid. If there were no drag, the vortices would corotate with the superfluid and would never pin. As a result of dissipation, there is a drag force per unit length of vortex $\mathbf{f}_{\rm drag} = \eta \mathbf{v}$, where $\mathbf{v}$ is velocity of the vortex segment and η is the associated drag coefficient (all velocities are in the rest frame of the solid). There are two interesting limits to consider. If $\eta' \equiv \eta/\rho_s\kappa << 1$ (ρ_s is the SF mass density, κ is the vorticity quantum), the vortex moves in a state of low drag and nearly co-moves with the SF; in this case damping to the pinned state may never occur. In the opposite limit, $\eta' >> 1$, the vortex moves under high drag; it nearly co-moves with the solid, and so is effectively pinned. For a moving vortex to go from an unpinned state to a pinned one, it must make a transition from a state of low drag to one of high drag. I will argue that due to a mismatch of time-scales that determine the vortex dynamics, vortices may always be in a low-drag state, and so might never pin in a NS.

A relevant time-scale to consider is how long it takes for a vortex to pin. That is, if a vortex that is initially straight but co-moving with the solid, how long does it take for the vortex to assume a minimum energy, pinned configuration? Now if an ambient flow of velocity $\mathbf{v}_s$ is established, an initially unpinned vortex will move at $\simeq \mathbf{v}_s$ for $\eta' << 1$; a transition to pinning will occur if the vortex is moving sufficiently slowly past prospective pinning nuclei that each vortex segment interacts with each nucleus for a time longer than the pinning time $t_{\rm pin}$, i.e., if $v_s < R_{vn} t_{\rm pin}^{-1}$ where R_{vn} is the characteristic length scale of the vortex-nucleus interaction.

3. Vortex Pinning Dynamics and Dissipation

I now present a classical calculation of $t_{\rm pin}$, and will comment later on how this treatment is likely to *overestimate* the effectiveness of pinning. A classical hydrodynamic treatment of the vortex dynamics is valid, since the vortex excitation wave numbers during satisfy $k\xi << 1$ (ξ is the vortex core dimension). Treating the lattice dynamics classically is not so well justified, but gives a good starting point. I treat the vortex as a classical, massless "string" with tension (though it obeys different equations of motion than a string; see below). Over a vortex segment of length equal to the nuclear spacing a, it interacts with a single nearest nucleus through a parameterized potential. Each nucleus rests in a harmonic potential, and is coupled to its nearest neighbors by Hooke's Law. The equation of motion of a vortex displaced by $\mathbf{r}_v(z)$, interacting with a nucleus is

$$T_v \frac{d^2\mathbf{r}_v}{dz^2} + \rho_s \boldsymbol{\kappa} \times \frac{d\mathbf{r}_v}{dt} + \nabla V = 0, \tag{1}$$

where T_v is the vortex tension (self-energy per unit length), $\boldsymbol{\kappa}$ is the (conserved) vorticity vector of the vortex and V is the vortex-nucleus potential; the length scale R_{vn} of V is unknown; dimensionally, we expect it to be comparable to the SF coherence length for an isolated vortex near a single nucleus. But in a solid, the potential will be much more complicated, and it likely to change over a length scale comparable to the nuclear spacing a. I therefore assume $R_{vn} = a$ in the following. The first term represents the restoring force of tension as the vortex is bent. The second term is the Magnus force exerted on a unit length of vortex as it is forced by the nucleus to move with respect to the ambient superfluid; the origin of this force is that flow past a vortex creates a lift force orthogonal to the direction of the flow. In the absence of the nucleus, the vortex would remain straight and motionless with respect to the solid. Eq. (1) admits Kelvin modes, *kelvons*, circularly-polarized vortex waves that carry energy and angular momentum.

To account for losses to phonons, Eq. (1) is supplemented by

$$m_N \frac{d^2\boldsymbol{r}_N}{dt^2} + m_N \omega_N^2 \boldsymbol{r}_N - \nabla V + \mathbf{F}_d = 0, \tag{2}$$

where m_N is the mass of the nucleus, $\boldsymbol{r}_N$ is its displacement from equilibrium, ω_N is the lattice frequency and $\mathbf{F}_d$ is a damping force due to the loss of kinetic energy of the nucleus to phonons; $\mathbf{F}_d$ is proportional to $d\boldsymbol{r}_N/dt$ times the lattice frequency.

In this treatment, there are three relevant frequencies in the problem: the typical kelvon frequency, the lattice frequency and the damping rate of kinetic energy of the nucleus by the phonon field. The typical kelvon frequency is $\sim 10^{21}$ s^{-1}. The frequency of a single nucleus displaced from equilibrium will be somewhat higher than the ion plasma frequency, and is comparable to the kelvon frequency. The damping rate of nuclear motion due to coupling to the phonon field, calculated from a separate simulation of lattice dynamics in a Hooke's Law approximation, is $\gamma_N \simeq 0.03\omega_N$.

Eqs. (1) and (2) were solved over 50 zones with periodic boundary conditions on the ends of the vortex, requiring the solution of 300 coupled non-linear differential equations. The system exhibits some very interesting non-linear dynamics, shown on the next page. Fig. 1 shows the initial motion of a vortex segment and its nearest nucleus; the nucleus accelerates toward the nucleus, while the vortex takes evasive action and initiates an orbit. This figure depicts one zone in the simulation; motion of similar character is occurring everywhere else along the extended vortex. As the orbit continues, the vortex segment and nucleus damp towards new equilibrium positions (Fig. 2a) finally getting there in Fig. 2b. The detailed motion is

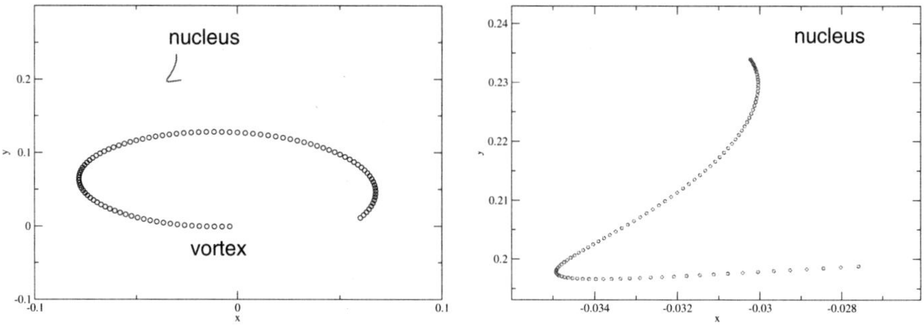

(a) (b)

Fig. 1. The initial dynamics of the vortex-nucleus system in one Wigner-Seitz cell, showing the motion of the cross section of the vortex and the nearest nucleus in one of 50 planes perpendicular to the unperturbed vortex. Points are separated by equal times. The vortex segment begins at the origin and the nucleus begins at the top end of the red curve. (a) The nucleus always accelerates towards the vortex segment, while the vortex segment generally moves in a transverse direction and initiates an orbit. (b) Detail of the nuclear motion.

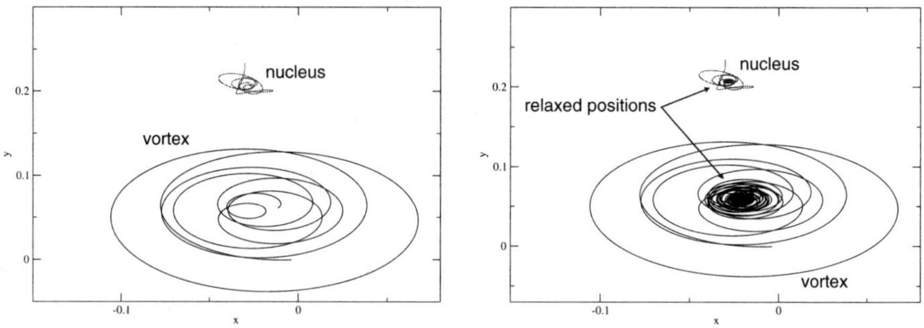

(a) (b)

Fig. 2. (a) Same as Fig. 1, later in the evolution. (b) Still later in the evolution, showing the relaxation of the vortex and nucleus to equilibrium positions. The two do not coincide because vortex tension prevents the vortex from threading each nucleus.

complicated by the fact that a vortex segment is coupled to all other parts of the vortex through tension. Hence, as one vortex segment is excited, kelvons propagate along the vortex, undergoing partial reflection and transmission at points along the vortex that are near a nucleus.

Though the motion is complicated, the pinning time scales simply:

$$t_{\rm pin} \propto \frac{T_v \omega_N^4}{E_{vn}^2 \gamma_N},$$

where E_{vn} is the vortex-nucleus interaction energy. If the vortex tension is infinite, or the nuclei cannot move (and dissipate energy), the vortex will never pin. Similarly, if either the vortex-nucleus interaction or the energy exchange rate of a moving nucleus with the phonon field is negligible, the vortex will never pin. Numerically, the pinning rate is $t_{\rm pin}^{-1} \sim 10^{12}$ s^{-1} for the fiducial values $\rho_s = 10^{14}$ gm s^{-1}, $R_{vn} = a = 50$ fm and $E_{vn} = 1$ MeV. This rate is much smaller than the ion plasma frequency and the typical kelvon frequency. The corresponding drag coefficient ($\propto t_{\rm pin}^{-1}$ is $\eta' \sim 10^{-8}$. Hence, the vortex damps to the pinned state in a regime of very low drag. This classical calculation, moreover, almost certainly overestimates the damping rate. The energy scale for exciting nuclei from their ground states is $\hbar\omega_N \sim 1$ MeV, which is comparable to the energy per vortex segment that must be dissipated for pinning to occur. Hence, quantum effects will impose phase space restrictions on the energy transfer rate that have been ignored here. The transfer of energy from one excited oscillator to the phonon field will also be restricted.

This calculation of the pinning time assumed there is no ambient SF flow, so the vortex was initially stationary. The vortex *will* pin in this situation, after $\sim 10^{-12}$ s. In general there will be a net flow $\mathbf{v}_s$ of the SF with respect to the solid that drives the vortex past potential pinning sites at very low drag ($\eta' \sim 10^{-8}$). In the absence of a full dynamical calculation of a *translating* vortex, we can estimate pinning to occur if $v_s \lesssim v_{\rm pin} \sim R_{vn} t_{\rm pin}^{-1}$. The implied critical velocity below which pinning can occur is then $v_{\rm pin} \sim 10$ cm s^{-1} for $R_{vn} = 50$ fm, a typical nuclear spacing in the inner crust. By contrast, the critical velocity to unpin an *already pinned* vortex is $\sim 10^5$ cm s^{-1}.[17] The difference can be traced back to the fact that $t_{\rm pin}^{-1}$ is $\sim 10^9$ times smaller than both the lattice and kelvon frequencies.

4. Superfluid Dynamics in a Spinning-Down Neutron Star

The NS crust (angular velocity Ω_c) is spun down by electromagnetic torque while the neutron SF (angular velocity Ω_s) spins down through the dissipative torque exerted on it by the spinning-down crust. If the drag were zero, the SF's angular velocity would remain constant, and the local velocity difference between the SF and the solid, $\Delta v = r(\Omega_s - \Omega_c)$, would increase indefinitely (r is the distance of the vortices from the rotation axis, approximately the stellar radius for the inner crust). For any finite drag, a steady state velocity difference develops; the higher the drag, the lower the equilibrium velocity difference. For a uniform vortex distribution, conservation of local vorticity gives the following expression for the spin-down rate of the SF:

$$\frac{d\Omega_s}{dt} = 2\frac{\Omega_s}{r} v_r ,$$

where v_r is the radial component of the the vortex drift velocity under drag. For vortex motion in the low-drag limit ($\eta' << 1$), as the above calculations indicate is the case, $v_r \simeq \Delta v \eta'^2$.

Suppose the vortices are unpinned and the SF angular velocity Ω_s differs locally from that of the crust. Eventually an equilibrium will be established in which the crust and SF are spinning down at the same rate, with a local velocity difference between them $\Delta v \propto \eta'^{\,-2}$. Unless Δv established by vortex drag is less than $v_{\rm pin}$, the SF and the crust will spin-down indefinitely without the vortices ever pinning.

Only if the drag is sufficiently strong in the first place to ensure $\Delta v < v_{\rm pin}$ can the vortices pin. The drag required for this is still weak ($\eta' << 1$), as can be seen through the following argument. The spin-down age of a NS is defined as $t_{\rm age} = \Omega_c/2|\dot{\Omega}_c|$. (The spin-down age is believed to be a reasonable estimate of the star's true age). Equating $\dot{\Omega}_c$ to $\dot{\Omega}_s$ above, and requiring $\Delta v < v_{\rm pin}$, leads to a lower limit on the drag below which pinning cannot occur

$$\eta'_{\rm crit} \sim 10^{-7} \left(\frac{t_{\rm age}}{10^4 \text{ yr}} \right)^{-1/2} ,$$

about an order of magnitude higher than the estimate for η' given above. It appears quite possible that vortices never pin in a typical NS because it is spinning down too fast.

Being more careful, we must consider that the vortex-nucleus interaction energy, the nuclear spacing, the lattice frequency and the vortex tension all depend on the local mass density. The state-of-the-art calculations of the vortex-nuclear interaction are those of Donati & Pizzochero (2006),[11] based on a local density approximation, and Avogadro et al. (2007),[12] based on a full quantum treatment of the vortex, the nucleus and the condensate. These two calculations are in substantial disagreement, and I leave discussion of the possible reasons for the discrepancies to the contributions of those authors to these proceedings. Pizzochero & Donati (2006) find an attractive vortex-nuclear interaction of ~ 4 MeV at densities around 0.018 fm^{-3} (calculation PD); elsewhere the interaction is repulsive and cannot lead to strong pinning. Avogadro et al. (2007), by contrast, find attractive interactions of 2-3 MeV at densities near 1.8×10^{-3} fm^{-3} (calculation ABBV, low density) and 0.05 fm^{-3} (calculation ABBV, high density). The estimates of η' and the corresponding critical value below which vortices cannot pin due to spin-down of the crust are as follows for these three calculations. PD: $\eta' \simeq \eta'_{\rm crit} \simeq 10^{-7}$. ABBV (low density): $\eta' \simeq 0.1\eta'_{\rm crit} \simeq 10^{-7}$. ABBV (high density): $\eta' \simeq \eta'_{\rm crit} \simeq 5 \times 10^{-8}$. The critical value is different for the three cases since $\eta' \equiv \eta(\rho_s)/\kappa\rho_s$ has density dependence.

The calculated η' and its critical value are very close to those from both calculations and at three very different densities, rather astonishing given the different frequencies, time scales (such as the spin-down rate of the NS crust) and densities that enter into this problem. But since the classical calculation overestimates η, these results suggest that pinning is marginal at best, and might not occur at all. It could be that the general state of vortex motion in the crust consists of low-drag motion past nuclei. The next problem to be solved is to consider a vortex being forced past nuclei with which it is interacting and to simulate the transition from moving to pinned to see if the estimates and arguments presented here are correct. This is work in progress.

5. Implications for Neutron Star Precession

Pinning of vortices anywhere in the star, such as to nuclei in the inner crust or flux tubes of the outer core, is at odds with the strong evidence that at least some NSs precess with periods of years;[2–4] pinning produces much faster precession with a period of order $\sim 1 - 100$ spin periods,[17–20] instead of years, even if the vortices are imperfectly pinned.[19,21] In a star showing long-period precession, the vortices cannot be pinned anywhere. In fact for the wobble angle of several degrees as inferred for PSR1828-11, the Magnus force in the precessing state would be so large as to prevent vortices from pinning in the crust,[17] thus allowing the long-period precession to persist so long as $\eta' << 1$. A very telling observation would be that of a giant glitch in a star that is showing clear, long-period precession. We would then know that glitches have nothing whatsoever to do with pinning/unpinning transitions (at least in precessing stars) and we would have to look for another physical mechanism for glitches, such as transitions out of states of SF turbulence.[22] Without such an observation, and given the complexity of NS physics, we cannot rule out the possibility that pinning really is marginal and only occurs in some stars.

Acknowledgments

I thank I. Wasserman for valuable discussions and the INFN, Sezione di Catania, for their hospitality where some of this work was performed. This work was supported under NSF grant AST-0406832.

References

1. S. L. Shemar and A. G. Lyne, *Mon. Not. Roy. Astr. Soc.* **282**, 677 (1996).
2. I. H. Stairs, A. G. Lyne and S. L. Shemar, *Nature* **406**, 484 (2000).
3. T. V. Shabanov, A. G. Lyne and J. O. Urama, *Astrophys. J.* **552**, 321 (2001).
4. F. Haberl, R. Turolla, C. P. D. Vries, S. Zane, J. Vink, M. Mendez and F. Verbunt, *Astron. Astrophys.* **415**, L17 (2006).
5. Z. Arzoumanian, D. J. Nice, J. H. Taylor and S. E. Thorsett, *Astrophys. J.* **422**, 671 (1994).
6. G. L. Israel *et al.*, *Astrophys. J.* **628**, L53 (2005).
7. T. E. Strohmayer and A. L. Watts, *Astrophys. J.* **637**, L117 (2006).
8. A. L. Piro, *Astrophys. J.* **634**, L153 (2005).
9. A. L. Watts and S. Reddy, *Mon. Not. Roy. Astr. Soc.* **379**, 63 (2007).
10. M. A. Alpar, *Astrophys. J.* **213**, 527 (1977).
11. P. Donati and P. M. Pizzochero, *Phys. Lett. B* **640**, 74 (2006).
12. P. Avogadro, F. Barranco, R. A. Broglia and E. Vigezzi, *Phys. Rev. C* **75**, 012805 (2007).
13. P. W. Anderson and N. Itoh, *Nature* **256** (1975).
14. J. A. Sauls, in *Timing Neutron Stars*, eds. H. Ögelman and E. P. J. van den Heuvel (Proc. NATO ASI, Dordrecht, 1989).
15. M. G. Alford, K. Rajagopal, T. Schaefer and A. Schmitt, *Rev. Mod. Phys.*, submitted.
16. P. B. Jones, *Mon. Not. Roy. Astr. Soc.* **321**, 167 (2001).
17. B. Link and C. Cutler, *Mon. Not. Roy. Astr. Soc.* **336**, 211 (2002).
18. J. Shaham, *Astrophys. J.* **214**, 251 (1977).
19. A. Sedrakian, I. Wasserman and J. M. Cordes, *Astrophys. J.* **524**, 341 (1999).
20. B. Link, *Phys. Rev. Lett.* **91**,101101 (2003).
21. B. Link, *Astron. Astrophys.* **458**, 881 (2006).
22. C. Peralta, A. Melatos, M. Giacobello and A. Ooi, *Astrophys. J.* **635**, 1224 (2005).

PART H
Poster Session

MICROSCOPIC DATA AND SUPERNOVAE EVOLUTION

P. BLOTTIAU* and PH. MELLOR

CEA/DIF, B.P.12,
Bruyères-Le-Châtel, 91680, France
**E-mail: patrick.blottiau@cea.fr*

J. MARGUERON

Institut de Physique Nucléaire,
Orsay, 91406, France
E-mail: jmargue@ipno.in2p3.fr

Due to the difficulty of hydrodynamic simulations to reproduce type II supernovae explosions, we investigate possible missing microscopic physics, such as neutrino trapping near the critical temperature of the nuclear liquid-gas phase transition, temperature dependant neutrino mean free paths or electron capture rates on nuclei to evaluate the impact on the improvement of the supernova outgoing shock propagation.

Keywords: Supernovae; Neutrino diffusion; Liquid-gas phase transition; Random phase approximation; Spinodal instability.

1. Progress in Type II Supernovae Physics and Simulations

The global scenario of type II supernovae has been understood, since several decades:[1] stars with masses above eight times the solar mass develop an onion skin structure, with an iron core, which collapses due to electron capture on free protons or nuclei and to photodissociation of iron-group nuclei. Neutrinos produced, by particle reactions, can escape freely at first, and are then trapped when the density of collapsing matter reaches about 5.10^{11} g.cm^{-3} : efforts have been done to build exact neutrino transport schemes.[2,3] When the inner homologous part of the core exceeds the saturation density $\rho_0 = 0.17$ nucl.fm^{-3}, the nuclear equation of state becomes stiff, infalling matter bounces and induces a strong shock wave, that might blow the envelop away. It has been now admitted that kinetic energy is not sufficient to produce such a 'prompt explosion' and neutrino heating of the outer layers is necessary to allow a successful shock. In spite of numerical improvements and physical refinements, brought by the 2D and 3D hydro codes generation, able to incorporate convection, Rayleigh-Taylor instabilities or rotation, it still remains difficult to produce a satisfying shock energy.[3] In that context and using our 1D hydro code tool,[4] we try to investigate new physical ingredients, as revisited electron capture rates, nuclear equations of state, and we will insist here on the impact of the Random Phase Approximation on neutrino mean free path calculations.

2. New Ingredients and Shock Efficiency Consequences

Contrary to previous assumptions, many-body models and recent data[5] have shown that electron capture on nuclei dominates for $A > 55$, and we will study important effects, as excited nuclear configurations, configuration mixing and allowed Gamow-Teller transitions in nuclei. The Gogny D1P interaction has been also employed to treat nuclear asymmetric matter in beta equilibrium,[6,7] and we introduced temperature dependant mean free path, related to RPA reponse functions performed in the Landau approximation, into the hydro code. In the nonrelativistic limit, the neutrino mean free path λ is given by the well-know expression :

$$\frac{1}{\lambda(k_\nu, T)} = \frac{G_F^2}{16\pi^2} \int d\boldsymbol{k}_3 [c_V^2 (1 + cos\ \theta) S^{(0)}(q, T) + c_A^2 (3 - cos\ \theta) S^{(1)}(q, T)], \quad (1)$$

where k_ν is the neutrino energy-momentum, T is the temperature, G_F is the Fermi constant, c_V and c_A are the vector and axial coupling constants, $\boldsymbol{k}_3$ is the final neutrino momenta, $q = k_\nu - k_3$ the transferred energy momentum, and $\cos\theta = \hat{\boldsymbol{k}}_\nu \cdot \hat{\boldsymbol{k}}_3$. $S^{(0)}$ and $S^{(1)}$ are the dynamical structure factors and describe the response of nuclear matter to excitations induced by neutrinos, for, respectively, density and spin-density fluctuations. Close to the liquid-gas transition, it has been found that RPA correlation induces a strong reduction of the neutrino mean free path.[7] This critical opalescence effect, for densities in the range between about $0.1\rho_0$ and $0.6\rho_0$, could increase the neutrino pressure inside a thin layer $\epsilon \sim 100$ m in the neutron star of radius $R \sim 10$ km, by a huge factor ~ 100, if we assume the extreme hypothesis that all neutrinos are trapped in this layer. The real factor should be less important since dynamical and Pauli effects are neglected in this estimation. Some recent calculations[8] incorporate the Coulomb interaction, and it has been found that long-range Coulomb interaction wash out the instability as expected, but it could be restored at finite transfered momentum. In conclusion, using our 1D code tool, we hope that neutrino trapping inside a thin layer, near the critical point of the liquid-gas phase transition, would add a strong pressure effect, able to grandly help the currently admitted 'delayed' mechanism.

References

1. S. A. Colgate and R. H. White, *Astrophys. J.* **143**, 626 (1966).
2. Ph. Mellor, J. P. Chièze and J. L. Basdevant, *Astron. Astrophys.* **197**, 123 (1988).
3. R. Buras, M. Rampp, H.-Th. Janka and K. Kifonidis, *Astron. Astrophys.* **447**, 1049 (2006).
4. P. Blottiau, PhD thesis, Paris VII University, (1989).
5. K. Langanke, G. Martínez-Pinedo *et al.*, *Phys. Rev. Lett.* **90**, 241102 (2003).
6. J. Margueron, J. Navarro and N. Van Giai, *Nucl. Phys. A* **719**, 169 (2003).
7. J. Margueron, J. Navarro and P. Blottiau, *Phys. Rev. C* **70**, 028801(2004).
8. C. Ducoin, J. Margueron and Ph. Chomaz, in preparation (2007).

PARITY DOUBLET MODEL APPLIED TO NEUTRON STARS

V. DEXHEIMER,* S. SCHRAMM and H. STOECKER
FIAS, Johann Wolfgang Goethe University,
Frankfurt am Main, Germany
** dexheimer@th.physik.uni-frankfurt.de*

The parity doublet model containing the SU(2) multiplets including the baryons identified as the chiral partners of the nucleons is applied for neutron star matter. The chiral restoration is analyzed and the maximum mass of the star is calculated.

Keywords: Neutron Star; Chiral symmetry; Parity model.

1. The Model

The doublet parity model assumes, besides the nucleons, the presence of baryons with opposite chirality named chiral partners. In a high density environment as the one present in the interior of a neutron star, there is the possibility of creating such heavy particles. The main difference between this model and the usual chiral models (Ref. 1) is that in this one there is a bare mass term in the lagrangean density. This mass called M_O mixes the original fermion fields and leads to mass terms for the baryons and their chiral partners after diagonalization of the theory.[2]

It is assumed that the star is in chemical equilibrium and the baryons interact through the mesons σ, ω and ρ (the latter one is taken into account in order to reproduce the high asymmetry between neutrons and protons). Electrons insure charge neutrality. The lagrangean density of the system contains besides the kinetic and the interaction terms an explicit symmetry breaking term in order to reproduce the masses of the pseudo-scalar mesons. The coupling constants of the baryons are adjusted to reproduce the baryonic vacuum masses. In the high-density limit the nucleon and its chiral partner have degenerate masses ($M_0 = 790 MeV$) as the sigma field goes to zero and chiral symmetry is restored:

$$M_{\pm}^{*} = \sqrt{\left[\frac{(M_{N_+} + M_{N_-})^2}{4} - M_0^2\right]\frac{\sigma^2}{\sigma_0^2} + M_0^2} \pm \frac{M_{N_+} - M_{N_-}}{2}\frac{\sigma}{\sigma_0}. \qquad (1)$$

A good candidate for the nucleon chiral parter is the N'(1535), but since the identification of the chiral partners is still not clear we study the case with N'(1200) and N'(1500) for comparison. This variation has drastic consequences on the results. Four different cases first studied in Ref. 2 are applied to neutron stars.

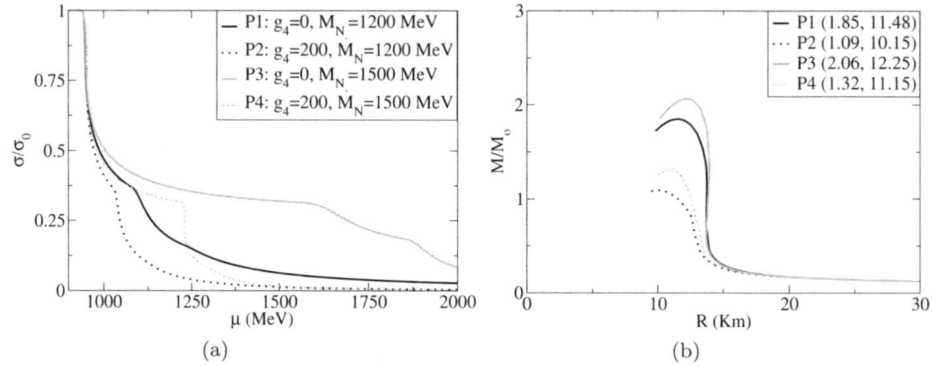

Fig. 1. (a) Scalar condensate versus chemical potential. (b) Star mass versus radius.

2. Results

When the density increases, the protons, the neutron and the proton chiral partners appear, respectively. Since the matter is not symmetric, there is no need for the chiral partners to appear at the same density. The transition to the chiral phase is abrupt for the cases with the fourth order autointeraction term of the vector mesons $g_4 = 0$ (P2 and P4) because of the sudden appearance of the chiral partners (Fig. 1(a)).

The calculation of the mass of the star shows that bigger maximum values are reached for the case that non-linear vector meson interactions are not included. Including those non-linear effects reduces the value of the vector field ω and consequently its repulsive effect is diminished such that the star will be unstable with respect to gravitational collapse at lower mass values (Fig. 1(b)).

3. Conclusion

With increasing density,i.e. towards the center of the star, chiral partners begin to appear, reaching a point where they exist at the same rate as the corresponding particles. The decrease in the scalar condensates signals the restoration of the chiral phase. Depending on the parameters, the phase transition turns out to be a continuous cross-over or of first order for P4.

The maximum mass of the star is higher without non-linear vector meson interactions (P1 and P3). These values are in agreement with the most massive star observed, that has $M = 2.1^{+0.4}_{-0.5} M_\odot$ (Ref. 3).

References

1. S. Schramm, *Phys. Lett. B* **560**, 164 (2003).
2. D. Zschiesche, L. Tolos, J. Schaffner-Bielich and R. Pisarski, *J. Phys. G* **31**, 935 (2005).
3. D. J. Nice, E. M. Splayer, I. H. Stairs, O. Loemer, A. Jessed, M. Kraemer and J. M Cordes, *Astrophys. J.* **634**, 1242 (2005).

STRUCTURE OF HYBRID STARS

D. JACCARINO,[1,*] U. LOMBARDO[1,2] and G. X. PENG[3]

[1] *INFN-LNS, Via S. Sofia 62, I-95123 Catania, Italy*
[2] *Physics and Astronomy Department, Via S. Sofia 64, I-95123 Catania, Italy*
[3] *Institute of High Energy Physics, P.O. Box 918-4, Beijing 100049, China*
**E-mail: danilo.jaccarino@alice.it*

The transition to the quark phase in the core of a neutron star is studied with both Maxwell and Glendenning construction. The hadron phase is described within the Brueckner theory with three body forces. For the confined quark phase we adopt the Density Dependent Quark Mass Model, which is consistent with chiral symmetry requirements.

Keywords: Hybrid stars; Density dependent quark mass; Phase transition.

1. Introduction

The aim of this work is to study the interior of hybrid stars. This can be done building an Equation of State (EoS) that contains a quark-hadron phase transition and to use it as input for the Tolman Oppenheimer Volkov equations. In this work the quark part of the hybrid EoS is described with a new semi-phenomenological model, as easy to use as the Bag Model but more realistic, because it satisfies the chiral symmetry requirements: the Density Dependent Quark Mass (DDQM) Model.[1]

2. The DDQM Model: an Overview

Quark matter is described as a free gas of u, d, s quarks and electrons in $\beta-$equilibrium. The model is based on the assumption that the quark effective masses do vary on the environment (temperature and density), satisfying a relation consistent with linear confinement:

$$m_q = m_{q0} + \frac{D}{n_b^z}\Big[1 - \frac{8T}{\lambda T_c}\exp\Big(-\lambda\frac{T_c}{T}\Big)\Big],$$

in which m_{q0} is the quark current mass, T is the temperature, n_b is the total baryon density and T_c, λ are constants; the remaining parameters D, z are guessed from stability arguments about strange quark matter. At the limit for $T \to 0$ the expression becomes simpler; the density dependence modifies the usual thermodynamical

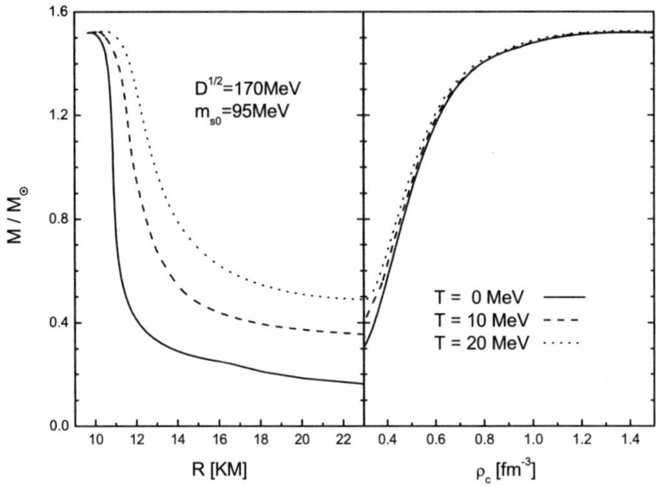

Fig. 1. Mass-Radius plots for the hybrid EoS in the present model (Glendenning phase transition).

equations introducing some derivatives of the flavour masses in the relations between the chemical potentials and the densities n_i (or Fermi momenta ν_i):

$$\mu_i = n_i\sqrt{\nu_i^2 + m_i^2} + \sum_j n_j\left(\frac{\partial m_j}{\partial n_i}\right) f\left(\frac{v_j}{m_j}\right) \qquad \text{with } i,j = \{u,d,s,e\}.$$

One can finally calculate the EoS from the thermodynamical potential as usual. With such a treatment the artificious bag constant can be dropped, as confinement is reached automatically by means of the density dependence of quark masses.

3. Results

In this work nuclear matter was described in the framework of the Brueckner theory with microscopic three body forces to better reproduce the saturation properties. In first approximation a first order phase transition is built in the Maxwell scheme; the results however suffer of the typical plateau in the mixed phase EoS. For this reason the Glendenning construction[2] is adopted, and the resulting hybrid EoS chosen as input for calculations. Integrating the TOV equations one finally finds stable configurations up to $1.6M_{\odot}$ (see Fig. 1). Stable hybrid configurations (pure quark core, hadron matter shell) are found, as the transition sets on a few times the nuclear saturation point. Unfortunately the results strongly depend on the D constant: one can see that for $\sqrt{D} < 160\,\text{MeV}$ no phase transition is allowed. A deeper investigation in SQM stability arguments can give stronger constraints and therefore more reliability to the model.

References

1. G. X. Peng, *Nucl. Phys. A* **747**, 75 (2005).
2. N. K. Glendenning, *Phys. Lett. B* **114**, 392 (1982).

NUCLEAR THREE-BODY FORCE FROM THE NIJMEGEN POTENTIAL

Z. H. LI,[1,*] U. LOMBARDO,[1,2] H.-J. SCHULZE,[3] AND W. ZUO[4]

[1] *INFN-LNS, Via S. Sofia 62, I-95123 Catania, Italy, *E-mail: zhli@lns.infn.it*

[2] *Physics Department, Catania University, Via S. Sofia 64, I-95123 Catania, Italy*

[3] *INFN Sezione di Catania, Via S. Sofia 64, I-95123 Catania, Italy*

[4] *Institute of Modern Physics, Chinese Academy of Sciences, Lanzhou 730000, China*

A nuclear three-body force based on the meson-exchange approach is constructed using the same meson parameters and the exponential form factors as in the Nijmegen potential, involving four kinds of important mesons, π, ρ, ω, and σ [$f_0(975)$ and $\epsilon(760)$]. For the 2π-exchange three-nucleon component, we adopt the new expansion strength constants a, b, c consistent with the contemporary πN-scattering data base and the corresponding dipole form factor. An effective two-body interaction is derived by averaging out the third nucleon, and is self-consistently used together with the Nijmegen potential in the Brueckner-Hartree-Fock approximation. The empirical nuclear matter saturation properties are reproduced very well. At higher density the equation of state becomes rather stiff due to the strong repulsion from the (σ,ω)-N three-body contribution.

Keywords: Three body force; Nuclear matter; Equation of state; Incompressibility.

1. Introduction

It is well known that three-body forces (TBF) are necessary to reproduce the empirical saturation properties for symmetric nuclear matter in the non-relativistic Brueckner-Hartree-Fock (BHF) approach.[1–4] In order to treat two-body force and TBF on the same footing, we start with the Nijmegen93 potential, which is a meson-exchange two-body interaction, and construct a new accordant TBF. For the detailed description of the different meson masses, meson-nucleon coupling constants, and cutoffs of the Nijmegen potential we refer to Ref. 5.

The microscopic TBF is established from the meson-exchange current approach,[6] which contains the contribution of the two-meson exchange part of the NN interaction medium modified by the intermediate virtual excitation of nucleon resonances, the term associated to the non-linear meson-nucleon coupling required by chiral symmetry, and the two-meson exchange diagram with the virtual excitations of nucleon-antinucleon pairs. These components are depicted in Fig. 1(a), taken from Ref. 6. By averaging out the third nucleon, an effective two-body interaction is derived. Keeping the same coupling constants and the form factor as used in the Nijmegen potential, the corresponding TBF contributions are constructed in terms of the four important meson(π,ρ,σ,ω)-exchange properties. In the Nijmegen po-

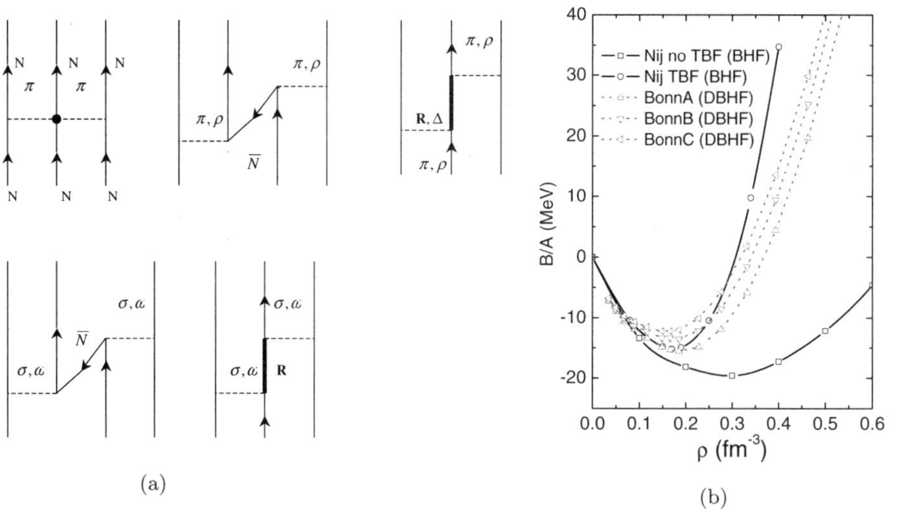

Fig. 1. (a) The main diagrams of the microscopic three-body force. (b) Equation of state from Nijmegen potential without (squares) and with (circles) TBF, compared to other calculations.

tential the isoscalar meson σ corresponds to two mesons f_0, ϵ. Now we present the results with this new constructed TBF correction.

2. Results

In the framework of the BHF approach, complemented by the new TBF contribution when the Nijmegen potential is chosen as the two-body interaction, the equation of state is derived as a function of the density, as shown in Fig.1(b). From this figure, we can see that the empirical nuclear matter saturation properties are reproduced very well. However, at higher density the equation of state becomes rather stiff due to the strong repulsion from the (σ,ω)-$N\overline{N}$ three-body contribution. As a result, this leads to a too large incompressibility ($K = 308$ MeV) at normal density.

Acknowledgments

This work was supported by the EU grant CN/ASIA-LINK/008(94791).

References

1. R. Machleidt, *Adv. Nucl. Phys.* **19**, 189 (1989).
2. M. Baldo, I. Bombaci and G. F. Burgio, *Astron. Astrophys.* **328**, 274 (1997).
3. A. Lejeune, U. Lombardo and W. Zuo, *Phys. Lett. B* **477**, 45 (2000).
4. W. Zuo, A. Lejeune, U. Lombardo and J.-F. Mathiot, *Nucl. Phys. A* **706**, 418 (2002).
5. V. G. J. Stoks, R. A. M. Klomp, C. P. F. Terheggen and J. J. de Swart, *Phys. Rev. C* **49**, 2950 (1994).
6. P. Grangé, A. Lejeune, M. Martzolff and J.-F. Mathiot, *Phys. Rev. C* **40**, 1040 (1989).

MONOPOLE EXCITATIONS IN QRPA ON TOP OF HFB

JUN LI,[1,2,*] GIANLUCA COLÒ[1] and JIE MENG[2]

[1] *Dipartimento di Fisica, Università degli Studi and INFN Sez. di Milano, 20133, Milano, Italy*
[2] *School of Physics, Peking University, Beijing, 100871, China*
**E-mail: jun.li@mi.infn.it*

We present results for the monopole excitation strength function in tin isotopes obtained by means of a self-consistent Quasiparticle-Random-Phase-Approximation (QRPA) which employs the canonical Skyrme-Hartree-Fock-Bogoliubov (HFB) basis. The effect of pairing correlations on the monopole excitation strength function, and centroid energy, is studied by comparing with the results of RPA.

Keywords: Skyrme-Hartree-Fock-Bogoliubov; Monopole excitation.

1. Introduction

The isoscalar giant monopole resonance (ISGMR) is a compressional mode, and its energy is related to the nuclear incompressibility K_∞.[1] For this reason it has been studied both in experimental and theoretical aspects for many years.

Recently, many works have studied the nuclear collective excitations in non-relativistic RPA based on Skyrme-Hartree-Fock (HF) mean field[2,3] (or QRPA based on Skyrme-HF mean field plus Bardeen-Cooper-Schrieffer[4] (BCS) pairing field[5–7] or based on Skyrme-HFB)[8] or relativistic RPA based on relativistic mean field[9,10](RMF) (or relativistic QRPA based on RMF plus BCS).[11]

If the density functionals are characterized by a nuclear incompressibility K_∞ around 230$\sim$240 MeV (or 250$\sim$270 MeV in the case of RMF), they give the right ISGMR centroid energies compared with the experimental data in ^{208}Pb.[7,9] However, most of them also overestimate the centroid energies in the tin isotopes.[9,10] The questions arises whether there is need of more detailed considerations on nuclear pairing in the case of tin. To understand this question, we analyze the problem of the ISGMR by means of a self-consistent QRPA calculation on top of Skyrme-HFB.

2. Numerical Method and Results

The HFB equations in coordinate representation are solved in a box with radius fixed at 20 fm and a small mesh (0.5 fm). In the particle-hole channel, we use the Skyrme force SLy5,[13] and in the particle-particle channel, we use a zero-range volume pairing force $v(\mathbf{r}_1, \mathbf{r}_2) = v_0\delta(\mathbf{r}_1 - \mathbf{r}_2)$, where v_0 is fixed by fitting the experimental neutron gap of ^{120}Sn (Δ_n = 1.32 MeV, v_0= -170.92 MeV fm^3). Due to the zero range of the

Table 1. Systematics of the moment ratios m_1/m_0 for the ISGMR with the Skyrme force SLy5 and the volume pairing force in HFB in the tin isotopes. They are evaluated in the energy interval between 10.5 and 20.5 MeV. The experimental data are extracted from Ref. 14.

	^{112}Sn	^{114}Sn	^{116}Sn	^{118}Sn	^{120}Sn	^{122}Sn	^{124}Sn
HFB-QRPA	16.77	16.76	16.73	16.69	16.62	16.54	16.47
HF-BCS-QRPA	17.33	17.24	17.13	17.02	16.90	16.80	16.72
HF-RPA	17.28	17.14	17.04	16.90	16.74	16.67	16.61
Exp.	16.2±0.1	16.1±0.1	15.8±0.1	15.8±0.1	15.7±0.1	15.4±0.1	15.3±0.1

Skyrme force, ee cutoff the quasiparticle states up to the quasiparticle energy being 200 MeV and a maximum angular momentum $j_{max} = 15/2$. After solving the HFB equations, we obtain the canonical basis by diagonalizing the density matrices.

The operator related to the isoscalar monopole excitation is $F^{IS}_{monopole} = \sum_{i=1}^{A} r_i^2$, and the strength function is $S(E) = \sum_n |\langle n|F^{IS}_{monopole}|0\rangle|^2 \delta(E - E_n)$. The moments of the strength function are $m_k = \int E^k S(E) dE$, and the centroid energy can be defined as m_0: $E_0 = m_1/m_0$.

Table 1 shows the centroid energy in the tin isotopes with the Skyrme force SLy5 and the volume pairing force. For the comparison, we also give the results of HF-RPA and HF-BCS-QRPA. The moment ratios are evaluated in the energy interval between 10.5 and 20.5 MeV, and the experimental data come from Ref. 14. HF-RPA and HF-BCS-QRPA overestimate the centroid energy about 1 MeV in tin isotopes. As shown in Tab. 1, HFB-QRPA makes the theoretical results significantly closer to experiment, especially in ^{112,114,116}Sn. However, the trend of the energies is not as in experiment. More work along this line is in preparation.[15]

This work was supported by the EU grant CN/ASIA-LINK/008(94791).

References

1. S. Stringari, *Phys. Lett. B* **108**, 232 (1982); J. P. Blaizot, *Phys. Rep.* **64**, 171 (1980).
2. B. K. Agrawal, S. Shlomo and A. I. Sanzhur, *Phys. Rev. C* **67**, 034314 (2003); B. K. Agrawal and S. Shlomo, *Phys. Rev. C* **70**, 014308 (2004).
3. T. Sil *et al.*, *Phys. Rev. C* **73**, 034316 (2006).
4. L. Cooper, J. Bardeen and J. Schrieffer, *Phys. Rev.* **108**. 1175 (1957).
5. G. Colò *et al.*, *Phys. Rev. C* **70**, 024307 (2004); *Nucl. Phys. A* **722**, 111c (2006).
6. S. Fracasso and G. Colò, *Phys. Rev. C* **72**, 064310 (2005).
7. N. Paar *et al.*, *Rep. Prog. Phys.* **70**, 691 (2007) and references therein.
8. J. Terasaki *et al.*, *Phys. Rev. C* **71**, 034310 (2005).
9. J. Piekarewicz, *Phys. Rev. C* **76**, 031301(R) (2007).
10. H. Sagawa *et al.*, nucl-th/07060966 (2007).
11. Cao Li-Gang and Ma Zhong-Yu, *Chin. Phys. Lett.* **21**, 810 (2004).
12. W. Nazarewicz, T. R. Werner and J. Dobaczewski, *Phys. Rev. C* **50**, 2860 (1994); J. Dobaczewski *et al.*, *Phys. Rev. C* **53**, 2809 (1996).
13. E. Chabanat *et al.*, *Nucl. Phys. A* **643**, 441 (1998).
14. T. Li, U. Garg *et al.*, to be published.
15. Jun Li, Gianluca Colò and Jie Meng, to be published.

THE INFLUENCE OF THE δ-FIELD ON NEUTRON STARS

A. J. MI
School of Nuclear Science and Technology, Lanzhou University, 730000 Lanzhou, China
E-mail: miaijun@gmail.com

W. ZUO
Institute of Modern Physics, Chinese Academy of Sciences, 730000 Lanzhou, China
E-mail: zuowei@impcas.ac.cn

A. LI
School of Physics, Science and Technology, Lanzhou University, 730000 Lanzhou, China

The effects of scalar-isovector meson δ field on the neutron star matter is investigated in the framework of relativistic mean field (RMF) theory. We find that the δ-field reduces the binding energy per baryon and enhances the strangeness contents of the neutron star. The moment of inertia of neutron stars is enhanced by including the δ-field.

Keywords: Relativistic mean field theory; Scalar-isovector meson field; Neutron stars.

1. Introduction

It is generally believed that the interior of neutron star which contains not only nucleons but also hyperons is a highly isospin-asymmetry system. The neglect of the δ meson in the investigation of neutron star matter may miss an important contribution to the isospin degree of freedom which must be considered to correctly describe the properties of strongly isospin-asymmetric matter.[1,2] In fact, the classical RMF model without δ meson field is no fully selfconsistent, as for isospin asymmetric matter the scalar density $< \overline{\psi}\tau_3\overline{\psi} > \neq 0$ whereas the mean field with the same quantum numbers vanishes.[3] Recently, some works give more attention to the scalar-isovector meson δ field.[1-9] In present work, our motivation is to investigate the effect of the δ meson field on the hyperon-rich neutron star matter.

2. Results and Conclusion

To investigate the effects of the δ meson field, we keep the same symmetry energy coefficient of $a_4 = 36.8$ MeV at saturation density and generate the different C_δ^2 and C_ρ^2 parameters combinations. $C_\delta^2 = 0$ is the case of absence of the δ-field. From the Fig. 1(1)(a) we can see that the binding energy per baryon is pinned down with the presence of hyperons at baryonic density of $\rho_B \approx 0.35 fm^{-3}$. From this density to

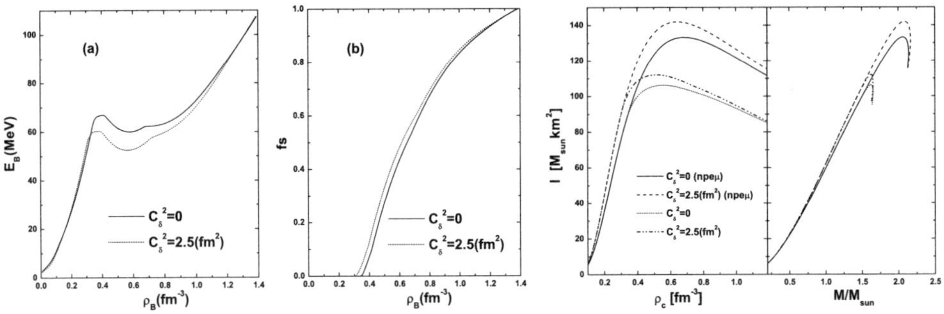

(1) (2)

Fig. 1. (1)(a) The binding energy per baryon of neutron star matter. (1)(b) The strangeness contents in neutron star. (2) The moments of inertia of neutron star and nucleonic star (npeμ).

the high density about 8 times of saturation density, the binding energy per baryon becomes smaller with including the δ-field. While at higher density, the difference of the binding energy per baryon with or without the δ-field is slight because that the neutron start tends to isospin symmetric matter. From the Fig. 1(1)(b) we can find that the inclusion of the δ-field slightly enhances the strangeness contents of neutron star. The moments of inertia curves for the neutron star and nucleonic star are shown in Fig. 1(2). The inclusion of hyperons softens the EOS of neutron star and predicts the lower maximum mass of neutron star, so the corresponding values of the moment of inertia of the neutron star with hyperons are lower. From the figure we can see that the moment of inertia of the neutron star were markedly enhanced by the presence of the δ-field.

Acknowledgments

This work was supported by the EU grant CN/ASIA-LINK/008(94791). One of the authors (A. J. Mi) would like to thank Prof. Umberto Lombardo for his kindly help on my study and living in Catania. Special thanks are also due to Dr. Z. H. Li and Dr. H.-J. Schulze.

References

1. H. Huber, F. Weber and M. K. Weigel, *Nucl. Phys. A* **596**, 684 (1996).
2. H. Huber, F. Weber, M. K. Weigel and C. Schaab, *Int. J. Mod. Phys. E* **7**, 301 (1998).
3. S. Kubis, M. Kutscheraa and S. Stachniewicza, *Acta Phys. Polonic. B* **29**, 809 (1998).
4. S. Kubis and M. Kutschera, *Phys. Lett. B* **339**, 191 (1997).
5. B. Liu *et al.*, *Phys. Rev. C* **65**, 045201 (2001).
6. M. Di Toro *et al.*, *Nucl. Phys. A* **775**, 102 (2006).
7. B. Liu, H. Guo, M. Di Toro and V. Greco, *Eur. Phys. J. A* **25**, 293 (2005).
8. D. P. Menezes and C. Providência, *Phys. Rev. C* **70**, 058801 (2004).
9. A. Sulaksono, P. T. P. Hutauruk and T. Mart, *Phys. Rev. C* **72**, 065801 (2005).

MAGNETIZATION OF COLOR-FLAVOR LOCKED MATTER

J. NORONHA

Frankfurt Institute for Advanced Studies, J.W. Goethe–Universität, Frankfurt am Main, D-60438, Germany
E-mail: noronha@fias.uni-frankfurt.de

I. A. SHOVKOVY

Department of Physics, Western Illinois University, Macomb, IL 61455, USA
E-mail: I-Shovkovy@wiu.edu

We show that the magnetization in color-flavor locked superconductors can be so strong that homogeneous quark matter becomes metastable for a wide range of magnetic field values. This indicates that magnetic domains or other type of magnetic inhomogeneities can be present in the quark cores of magnetars.

Keywords: Color superconductivity; Magnetization; Magnetic instabilities.

1. Introduction and Conclusions

The cold and superdense inner cores of compact stars are likely to consist of color superconducting quark matter. Assuming that the very strong surface magnetic fields found in magnetars ($B \sim 10^{14} - 10^{16}$ G) can be transmitted to the their inner cores, it is worth studying whether the effects of strong magnetic fields on color superconductors are important for understanding the physics of magnetars. In Ref. 1 the effects of moderately strong magnetic fields on the properties of color-flavor locked (CFL) superconductors (Ref. 2) were studied numerically in the framework of a Nambu-Jona-Lasinio model. It was shown that the ground state of 3-flavor quark matter undergoes a continuous crossover from the CFL phase into the magnetic CFL (mCFL) phase, which was initially introduced in Ref. 3.

The free parameters of the model (the diquark coupling constant and the ultraviolet cutoff) were set to yield a CFL gap of either $\phi_0 = 10$ MeV or $\phi_0 = 25$ MeV when the quark chemical $\mu = 500$ MeV and $B = 0$. It was shown in Ref. 1 that the mCFL gaps display de Haas-van Alphen oscillations with respect to eB/μ^2 (see also Ref. 4). The magnetization of the system M is given by $M = (\partial\Gamma/\partial B)$, where Γ is the one-loop quark contribution to the free energy evaluated at the stationary point. Our numerical results for the magnetization are presented in Fig. 1.

It was shown in Ref. 5 that the magnetization of hadronic matter is negligible even for magnetar conditions, i.e., $4\pi M/B \ll 1$ for $B \lesssim 10^{19} G$. In the case of quark

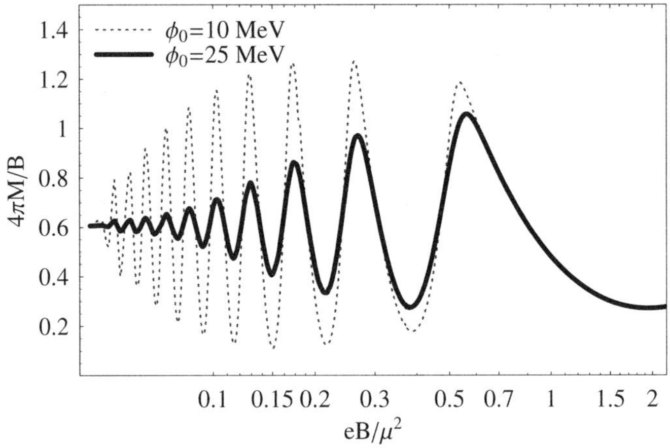

Fig. 1. Ratio $4\pi M/B$ versus eB/μ^2 for two sets of parameters that yield $\phi_0 = 10$ MeV and $\phi_0 = 25$ MeV.

matter, however, the situation is very different. In Fig. 1 we plotted $4\pi M/B$ versus eB/μ^2 for a mCFL superconductor. The magnetization is significantly larger in this case and it displays de Haas-van Alphen oscillations with very large amplitude. The large magnitude of these magnetic oscillations create regions in which $(\partial H/\partial B)_\mu < 0$, where $H = B - 4\pi M$ is the magnetic field present in the outer layers of the star. These regions correspond to unstable or metastable states where, depending on the geometry of the system, a transition into a magnetic domain configuration may occur. This could lead to a wealth of different physical phenomena.

For instance, successive phase transitions coming from discontinuous changes of B during the star's evolution can release a vast amount of energy that would heat up the star, which would then cool down by for example the emission of neutrinos. Therefore, sudden bursts of neutrinos coming from magnetars with color superconducting cores even after the deleptonization period could be expected. On the other hand, if a mixed phase with microscopic domains of nonequal magnetizations is formed, the relative size of domains with different magnetizations would change with H in order to keep the average induced magnetic field B continuous. In either case, since the magnitude of the fields involved is enormous the system could potentially release an immense amount of energy. If observed, this type of phenomena could help to distinguish magnetars with quark cores from their purely hadronic counterparts.

References

1. J. L. Noronha and I. A. Shovkovy, arXiv:0708.0307 [hep-ph].
2. M. G. Alford, K. Rajagopal and F. Wilczek, *Nucl. Phys. B* **537**, 443 (1999).
3. E. J. Ferrer, V. de la Incera and C. Manuel, *Phys. Rev. Lett.* **95**, 152002 (2005).
4. K. Fukushima and H. J. Warringa, arXiv:0707.3785 [hep-ph].
5. A. Broderick, M. Prakash and J. M. Lattimer, *Astrophys. J.* **537**, 351 (2000).

AB INITIO PAIRING GAP CALCULATION FOR A SLAB OF NUCLEAR MATTER WITH PARIS AND ARGONNE V18 BARE NN-POTENTIALS

S. S. PANKRATOV,[1,*] M. BALDO,[2] U. LOMBARDO,[2] E. E. SAPERSTEIN[1]
and M. V. ZVEREV[1]

[1] *Kurchatov Institute, 123182 Moscow, Russia*
[2] *INFN, Sezione di Catania, Via S. Sofia 64, I-95123 Catania, Italy*
[*] *E-mail: pankratov@mbslab.kiae.ru*

The gap equation for $^1S_0$ NN-pairing is solved for a nuclear slab with two realistic bare NN-interactions, the separable form of the Paris potential and the Argonne potential v_{18}. For solving the gap equation the renormalization method based on the effective pairing interaction concept is used. Calculations are performed for various values of the chemical potential μ in the range from -8 MeV, which corresponds to stable nuclei, to -0.1 MeV which corresponds to nuclei in the vicinity of the drip line. A detailed comparison of the results for two nuclear potentials is performed.

Keywords: Nucleon pairing; Effective pairing interaction.

1. Introduction

We solved the gap equation $\Delta = -\mathcal{V}\varkappa$, where V is a bare NN interaction and $\varkappa$ is the anomalous density, for $^1S_0$ pairing in a nuclear slab system. A slow convergence of these equation with increase of the cutoff momentum[1,2] is overcome with the help of the effective interaction concept.[3] The complete Hilbert space is split on the model subspace S_0 and the complementary one S'. The gap equation in the S_0 subspace,

$$\Delta = -V^p_{\text{eff}}\, \varkappa_0\,, \tag{1}$$

is solved in terms of the effective pairing interaction V^p_{eff} which is found in the S' subspace:

$$V^p_{\text{eff}} = \mathcal{V} + \mathcal{V}(GG)'\, V^p_{\text{eff}}\,. \tag{2}$$

The pairing effects can be neglected in the latter provided the model space is sufficiently wide.

2. Results

The effective interactions obtained for the two NN-potentials are very close to each other. The faster convergence of V^p_{eff} with respect to the cutoff momentum (S'

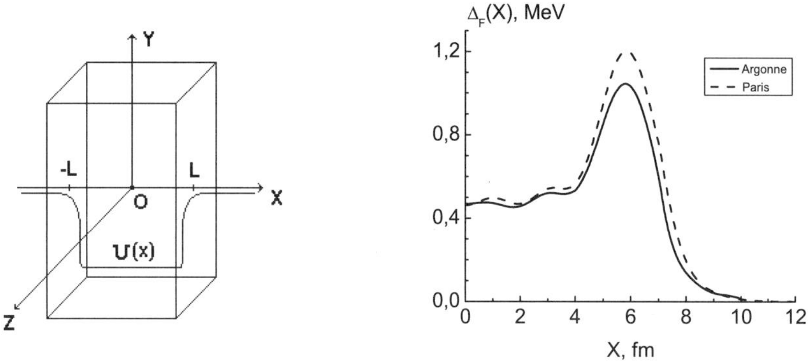

Fig. 1. Left side: Slab of nuclear matter. Right side: Fermi averaged pairing gap for Paris and Argonne v_{18} NN-interactions.

subspace size) for Argonne potential is confirmed.[4] The pairing gap difference for Argonne and Paris NN-potentials is about 15%. This result is in agreement with calculations for infinite nuclear matter. Values of the gap matrix elements are close to the experimental ones for heavy atomic nuclei. The dependence of the gap on the chemical potential μ was examined within two models. In the first model, the potential well depth $U_0 = -50$ MeV is fixed. In the second one, the local Fermi momentum $k_F(x = 0) = \sqrt{2m\,(\mu - U_0)} = 1.42$ fm^{-1} is constant. The results turns out to be model dependent.

Acknowledgments

This research was partially supported by the Russian Ministry for Education and Science (grant no. NS-8756.2006.2), the Russian Foundation for Basic Research (projects nos. 06-02-17171-a and 07-02-00553-a). Three of us (S.P. , E.S. and M.Z.) thank INFN and Catania University for hospitality rendered during EXOCT 2007 conference.

References

1. F. Barranco, R. A. Broglia, G. Colo, G. Gori, E. Vigezzi and P. F. Bortignon, *Eur. Phys. J. A* **21**, 57 (2004).
2. S. S. Pankratov, M. Baldo, U. Lombardo, E. E. Saperstein and M. V. Zverev, *Nucl. Phys. A* **765**, 61 (2006).
3. M. Baldo, U. Lombardo, E. E. Saperstein and M. V. Zverev, *Phys. Rep.* **391**, 261 (2004).
4. S. S. Pankratov, M. Baldo, U. Lombardo, E. E. Saperstein and M. V. Zverev, *Physics of Atomic Nuclei* **70**, 658 (2007).

HYBRID NEUTRON STARS WITHIN THE NAMBU-JONA-LASINIO MODEL AND CONFINEMENT

M. BALDO, G. F. BURGIO, P. CASTORINA, S. PLUMARI and D. ZAPPALÀ

Dipartimento di Fisica e Astronomia, Università di Catania and INFN, Sezione di Catania, Via S. Sofia 64, I-95123 Catania, Italy

Recently, it has been shown that the standard Nambu-Jona-Lasinio (NJL) model is not able to reproduce the correct QCD behavior of the gap equation at large density, and therefore a different cutoff procedure at large momenta has ben proposed. We found that, even with this density dependent cutoff procedure, the pure quark phase in neutron stars (NS) interiors is unstable, and we argue that this could be related to the lack of confinement in the original NJL model.

Keywords: Dense matter; Neutron stars.

1. NJL at Large Density

In the interior of astrophysical compact objects, like NS, nuclear matter is expected to reach a density which is several times nuclear saturation density. In these conditions, calculations only based on nucleonic degrees of freedom becomes highly questionable and a phase transition to quark matter becomes possible. The quark matter equation of state (EoS) derived from standard NJL model is soft enough to render NS unstable at the onset of the deconfined phase, and no pure quark matter can be present in its interior. Though the NJL model reproduces correctly the phenomenological low energy data on hadron properties, it is not able to reproduce the correct behavior of the solution of QCD gap equation at large density, as pointed out in Ref. 2. In order to clarify this point, we have studied a modified NJL model with a density dependent momentum cutoff, which preserves the low energy properties of the theory. According to this procedure, a μ dependent cut-off $\Lambda(\mu)$ is introduced, which implies a μ dependent coupling constant $g(\mu)$. Since there is no compelling restriction to a specific functional form of $\Lambda(\mu)$, we choose some smooth monotonically increasing functions of μ. For details, the reader is referred to Ref. 1. Considering different slopes in the μ dependence of the cut-off is an important issue, because this implies different growth behaviors of the pressure, as a function of the baryon chemical potential μ_B, which is crucial to determine the transition point to quark matter. By assuming a first order phase transition, as suggested by the indications coming from lattice calculations,[3] we have adopted for the hadronic phase a nucleonic equation of state obtained within the Brueckner-Bethe-Goldstone (BBG)

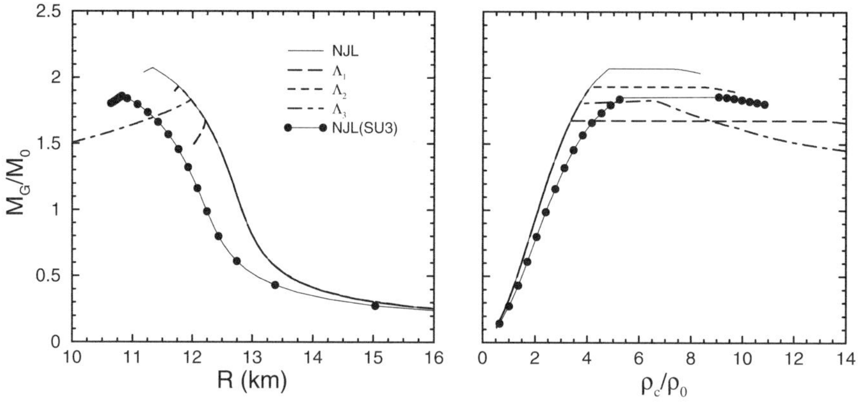

Fig. 1. The gravitational mass (in units of the solar mass $M_\odot = 2 \times 10^{33} g$) is plotted as function of the radius (left panel) and the central density (right panel), for different choices of the cut-off behavior.

approach,[4] while for the quark phase, we have used the standard parametrization of the NJL model[5] and the above prescriptions.

In Fig. 1 we report the neutron star masses as a function of the radius (left panel), and of the central density (right panel) for the different choices of the density dependent cut-off discussed above. For comparison also the results for the standard NJL model (three flavor) of Ref. 6 are reported (full circles). The plateau in the mass-central density plane is a consequence of the Maxwell construction. In all cases one can see that at the maximum mass the plot is characterized by a cusp, which corresponds to the instability mentioned above. As discussed in Ref. 1 the origin of the NS instability could be related to the missing quark confinement in the model. In fact, when chiral symmetry is restored, the NJL model behaves like the MIT model with a bag constant $B_{NJL} \simeq 140\ MeV/fm^3$. Therefore, if one adds by hand a confining potential which is switched off at the chiral phase transition,[7] the instability could be removed since the effective bag constant would be correspondingly reduced.

References

1. M. Baldo, G. F. Burgio, P. Castorina, S. Plumari and D. Zappalà, *Phys. Rev. C* **75**, 035804 (2007).
2. R. Casalbuoni, R. Gatto, G. Nardulli and M. Ruggieri, *Phys. Rev. D* **68**, 034024 (2003).
3. Z. Fodor and S. D. Katz, *JHEP* **0404**, 050 (2004).
4. C. Maieron, M. Baldo, G. F. Burgio and H.-J. Schulze, *Phys. Rev. D* **70**, 043010 (2004), and references therein.
5. M. Buballa, *Phys. Rep.* **407**, 205 (2005).
6. M. Baldo, M. Buballa, G. F. Burgio, F. Neumann, M. Oertel and H.-J. Schulze, *Phys. Lett. B* **562**, 153 (2003), and references therein.
7. S. Lawley, W. Bentz and A. W. Thomas, *J. Phys. G* **32**, 667 (2006).

A STUDY OF PAIRING INTERACTION IN A SEPARABLE FORM

YUAN TIAN[1] and ZHONGYU MA[2]
China Institute of Atom Energy, Beijing 102413, China
E-mail: [1]tiany@ciae.ac.cn, [2]mazy12@ciae.ac.cn

P. RING
Physik Department, Technische Universität München, D-85747 Garching, Germany
E-mail: ring@ph.tum.de

Recently Duguet proposed a microscopic effective interaction to treat pairing correlations in the $^1S_0$ channel.[1] He starts in nuclear matter and introduces a separable pairing force of a product of two Gaussians adjusting the parameters to the gap at the Fermi surface $\Delta_F(\rho)$ as a function of the density, calculated with the bare nucleon-nucleon interaction. This effective interaction is tailored in such a way that reproduces the pairing properties of the bare force. Starting from this simple ansatz and summing up *pp*- and *hh*-ladders he derives after using several approximations an effective pairing force in the nuclear medium for Hartree+BCS calculations. This effective force is non-local and of finite range. It depends on the momentum and on the density, but it contains no additional phenomenological parameters. In a further approximation he derives from this effective force a density dependent zero range force with a smooth cut-off parameter proportional to the BCS occupation numbers v_k^2. This is a very ambitious concept because it not only starts with the nucleon-nucleon interaction, which is well known, but it takes into account in addition correlations within the medium. On the other side it contains a number of approximations and assumptions which may be valid in nuclear matter, but which are hard to control in finite nuclei. In addition this force is tailored for Hartree+BCS calculations and it is not clear to what extend it can be used in case where one needs Hartree-Bogoliubov theory for the description of pairing.

Our new concept is much more modest. We start from the fact, that the phenomenological Gogny force[2] provides an excellent description of pairing properties in finite nuclei. This has been shown over the years in numerous applications of this force by the group of Gogny but also by the Madrid group.[3] In addition the pairing part of this force has been used with great success in the framework Relativistic Hartree-Bogoliubov (RHB) theory (for a review see Ref. 4.) On the other side this force is of finite range and not separable. Therefore the numerical effort is considerable and this is certainly the reason why this pairing force is not applied more frequently by other groups.

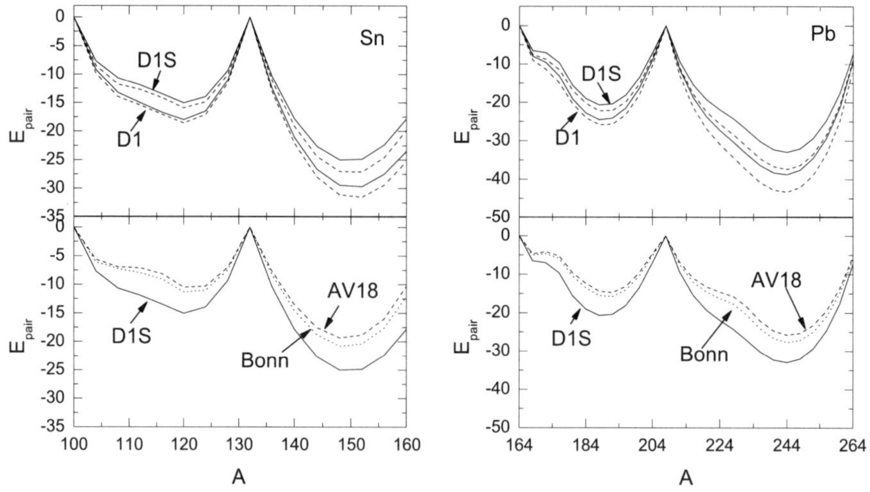

Fig. 1. Pairing energies in Sn-isotopes ^{100}Sn$\sim^{160}$Sn and Pb-isotopes ^{164}Pb$\sim^{264}$Pb. Upper figure shows the result of RHB calculation with Gogny(D1 and D1S) pairing interactions(solid curves) and their corresponding separable forms pairing interaction(dotted curves). The lower figure shows the result of RHB calculation with the Gogny D1S pairing interaction (solid curves), and the separable pairing interactions corresponding to the potentials Bonn A (dotted curves) and AV18 (dashed curves).

We start with the same separable ansatz in nuclear matter as Duguet. However, we do not consider this as a bare force from one has to derive an effective pairing interaction by summing up *pp*- and *hh*-ladders. On the contrary we determine this simple separable force from the pairing properties of the Gogny force in nuclear matter and we use it for extensive calculations in finite nuclei. First we transform it to *r*-space and to oscillator space. We show that is has a relatively simple form in both cases too. In particular we find that the matrix elements of this force can be represented in oscillator space by a sum of separable terms, which converges very quickly. Therefore this force can be without any difficulties in full Hartree-Bogoliubov calculations.

Acknowledgments

This work was supported by the EU grant CN/ASIA-LINK/008(94791).

References

1. T. Duguet, *Phys. Rev. C* **69**, 054317 (2004).
2. J. Dechargé and D. Gogny, *Phys. Rev. C* **21**, 1568 (1980).
3. J. L. Egido and L. M. Robledo, in *Lecture Notes in Physics*, edited by G. Lalazissis, P. Ring and D. Vretenar (Springer-Verlag, Heidelberg, 2004), Vol. 641, p. 269.
4. D. Vretenar, A. V. Afanasjev, G. A. Lalazissis and P. Ring, *Phys. Rep.* **409**, 101 (2005).
5. M. Serra and P. Ring, *Phys. Rev. C* **65**, 064324 (2002).

ISOSPIN DEPENDENCE OF NUCLEAR MATTER

E. N. E. VAN DALEN

Departament d'Estructura i Constituents de la Matèria, Universitat de Barcelona, E-08028 Barcelona, Spain
E-mail: vandalen@ecm.ub.es

C. FUCHS, P. GÖGELEIN and H. MÜTHER

Institut für Theoretische Physik, Universität Tübingen, D-72076 Tübingen, Germany

Exploring the isospin dependence of the nuclear matter is one of the main challenges of modern nuclear physics. The *ab initio* calculations are the proper tool for these investigations. Results of the Dirac-Brueckner-Hartree-Fock calculations for asymmetric nuclear matter, which are based on improved approximation schemes, are presented. Furthermore, the application to finite nuclei is discussed.

Keywords: Nuclear equation of state; Isospin dependence; Relativistic Brueckner approach; Density dependent relativistic mean field theory; Finite nuclei.

The investigation of isospin asymmetric matter is of importance for astrophysical and nuclear structure studies. In the former field it is important for the physics of supernovae and of neutron stars, whereas in the latter field it is of interest in neutron-rich nuclei. The *ab initio* calculations based on realistic interactions, as for instance the Dirac-Brueckner-Hartree-Fock (DBHF) approach, are the proper tool for these studies.

We describe isospin asymmetric nuclear matter in the framework of the relativistic DBHF approach based on projection techniques using the Bonn A potential. In this approach one schematically has the following system of coupled equations,

- $T = V + i \int VQGGT$ Bethe-Salpeter(BS) equation,
- $G = G_0 + G_0 \Sigma G$ Dyson equation,
- $\Sigma = -i \int_F (Tr[GT] - GT)$ Self-energy Σ in Hartree-Fock approximation,

with T the T matrix, V the potential, G the Green's function of an in-medium nucleon, and G_0 that of a free nucleon. This system represents a self-consistency problem, which has to be iterated until convergence is reached. Furthermore, the optimal representation scheme, the subtracted T matrix representation scheme, is applied.[1,2] In addition, in the np channel we abandon the approximation of an

Table 1. Charge radius and binding energy per nucleon.[6]

nuclei	DDRMF		Experiment	
	r_{ch} [fm]	E_b [MeV]	r_{ch} [fm]	E_b [MeV]
^{16}O	2.78	-8.35	2.74	-7.98
^{40}Ca	3.44	-8.73	3.48	-8.55
^{48}Ca	3.45	-8.73	3.47	-8.67
^{90}Zr	4.17	-8.74	4.27	-8.71
^{208}Pb	5.31	-7.87	5.50	-7.87

averaged *np* mass in the solution of the BS equation and distinguish explicitly between the different isospin dependent matrix elements. As a consequence, the potential and T matrix are evaluated with a sixth independent helicity or covariant amplitude,[2,3] instead of five as in the *nn* and *pp* channel.

At the saturation density $n_{sat} = 0.181$ fm^{-3} the value of the binding energy is $E_{b,sat} = -16.15$ MeV. Furthermore, the binding energy shows a nearly quadratic dependence on the asymmetry parameter β as expected. In addition, our DBHF calculations predict a mass splitting of $m^*_{D,n} < m^*_{D,p}$ in neutron-rich matter. However, the nonrelativistic mass derived from our DBHF approach in general shows the opposite behavior, i.e. $m^*_{NR,n} > m^*_{NR,p}$, which is in agreement with the results from nonrelativistic BHF calculations.[4] This opposite mass splitting is not surprising, since these masses are based on completely different physical concepts.

Within the framework of density dependent relativistic mean field (DDRMF) theory, effective density dependent coupling functions can be obtained from the Brueckner self-energy components[5] using a renormalization scheme to account for the differences between DBHF and DDRMF concerning the structure of the self-energy. Next these renormalized density dependent coupling functions are fitted and a local modification of the omega coupling function is added ($n_{sat} = 0.178$ fm^{-3};$E_{b,sat} = -16.25$ MeV) in order to move the saturation point closer to the empirical value. Although the lighter nuclei are a little too much bound, the binding energies for heavy nuclei are well reproduced in table 1. This parameterization is also used for neutron star crust calculations.[6]

Our DBHF approach for asymmetric nuclear matter gives quite good saturation properties. It predicts a Dirac mass splitting of $m^*_{D,n} < m^*_{D,p}$ in neutron-rich matter. Furthermore, the DDRMF results are quite satisfactory for finite nuclei.

References

1. E. N. E. van Dalen, C. Fuchs and A. Faessler, *Nucl. Phys. A* **744**, 227 (2004); *Phys. Rev. Lett.* **95**, 022302 (2005); *Phys. Rev. C* **72**, 065803 (2005).
2. E. N. E. van Dalen, C. Fuchs and Amand Faessler, *Eur. Phys. J. A* **31**, 29 (2007).
3. F. de Jong and H. Lenske, *Phys. Rev. C* **58**, 890 (1998).
4. T. Frick, Kh. Gad, H. Müther and P. Czerski, *Phys. Rev. C* **65**, 034321 (2002).
5. C. Fuchs, H. Lenske and H. Wolter, *Phys. Rev. C* **52**, 3043 (1995).
6. P. Gögelein, E. N. E. van Dalen, C. Fuchs and H. Müther, arXiv:0708.2867 [nucl-th].

EJECTED ELEMENTS FROM THE ENVELOPE OF COMPACT STARS BY QCD PHASE TRANSITION

N. YASUTAKE* and S. YAMADA

Science and Engineering, Waseda University,
Tokyo 169-8555, Japan
**E-mail: yasutake@heap.phys.waseda.ac.jp*
www.heap.phys.waseda.ac.jp/yasutake/opening.html

T. NODA and M. HASHIMOTO

Department of Physics, Kyushu University,
Fukuoka 810-8560, Japan
E-mail: tsune@gemini.rc.kyushu-u.ac.jp

K. KOTAKE

Division of Theoretical Astronomy, National Astronomical Observatory of Japan,
Tokyo 181-8588, Japan
E-mail: kkotake@yso.mtk.nao.ac.jp

We perform two-dimensional, magneto-hydrodynamical simulations of "petit" collapse and bounce of neutron stars by QCD phase transition. We study how the phase transition affects the elements of the ejecta near the epoch of core-bounce.We find that the elements of envelope can eject without nucleosynthesis for such a "petit" collapse and bounce by the transition. In addition, we estimate the gravitational wave using the quadrupole formula.

Keywords: QCD phase transition; Nucleosynthesis; Gravitational waves.

1. Introduction

There has been extensive work devoted to studying the evolution of compact stars in the context of cooling processes and X-ray bursts.[1] The envelopes of compact stars might be the sites to produce p-process or rp-process elements.

In this paper, we use magneto-hydrodynamics (MHD) code, and estimate the possibility of the ejection of such elements from compact stars with QCD phase transition.

2. Input Physics

The numerical method in this paper is based on the ZEUS-2D code.[2] We take into account the general relativistic correction[3] to the Newtonian gravity. Since there

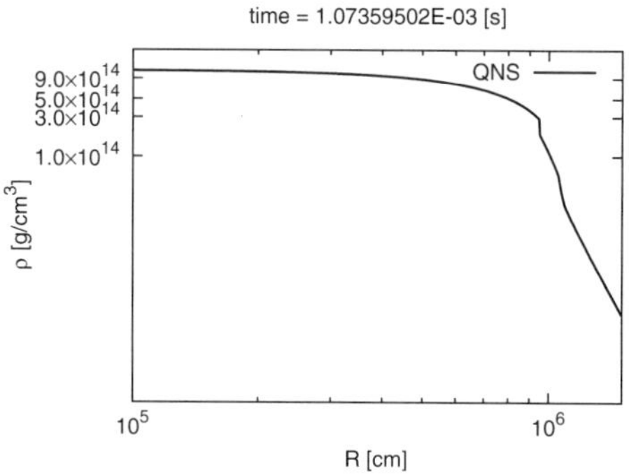

(a)

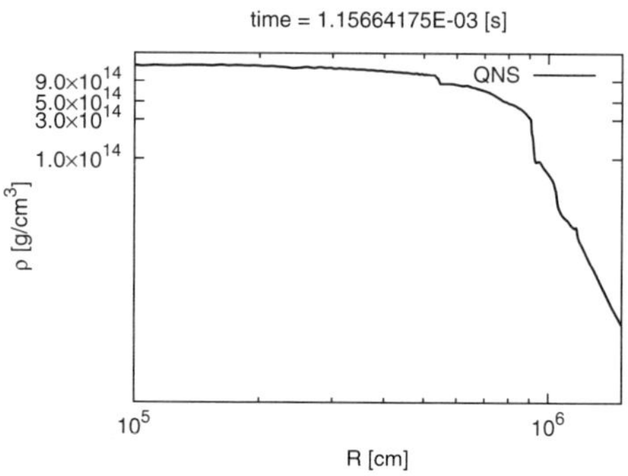

(b)

Fig. 1. The shock propagation. (a) Initial stable star. (b) After the transition. We can see the 'petit' shock at the outside of the core.

does not exist any reliable EOS that describes QCD phase transition, we follow the method adopted in Gentile et al.[4] We assume that the first order phase transition occurs during the collapse beyond some critical density. Our initial model is a realistic spherical neutron star including the cooling effects and/or the nucleosynthesis of envelope such as X-ray burster.[1] This initial structure of the neutron star is perfectly stable during ten times of our calculational time scale without QCD phase transition (see Fig. 1(a)). We add the rotation and magnetic field to the initial models, parametrically. We do not consider neutrino effects, since the time scale of the collapse and bounce is very short, ~ 0.1 ms.

3. Results and Discussion

We find that the elements of envelope can eject by "petit" collapse and bounce by the transition. The temperature range is wide ($\sim 10^{6-9}$K), since it depends on the rotation, magnetic field, initial model structure, coupling constant, vacuum energy. Figure. 1 is one result of our test calculations, and shows 'petit' shock by QCD phase transition. Here, this is calculated on spherical model, bag constant $B = 94$ MeV/fm^3, QCD coupling constant $\alpha = 0$, and baryon critical density $\rho_c = 7.0 \times 10^{14}$g/cm^3. We will show the result of similar estimates under the more realistic baryonic EOS in nearly future.

References

1. O. Koike, M. Hashimoto, R. Kuromizu and S. Fujimoto, *Astrophys. J.* **603**, 242 (2004).
2. J. M. Stone and M. L. Norman, *Astron. & Astrophys. Suppl. Ser.* **80**, 753 (1992).
3. A. Marek, H. Dimmelmeier, H.-T. Janka, E. Muller and R. Buras, *Astron. & Astrophys. Suppl. Ser.* **445**, 273 (2006).
4. N. A. Gentile, M. B. Aufderheide and G. J. Mathews, *Astrophys. J.* **414**, 701 (1993)

MICROSCOPIC THREE-BODY FORCE EFFECT ON NUCLEON-NUCLEON CROSS SECTIONS

H. F. ZHANG,[1] U. LOMBARDO,[2,3] Z. H. LI,[2] P. Y. LUO[4] and W. ZUO[4]

[1] *School of Nuclear Science and Technology, Lanzhou University, Lanzhou 730000, China*
[2] *INFN-LNS, Via S. Sofia 62, I-95123 Catania, Italy, *E-mail: zhli@lns.infn.it*
[3] *Physics Department, Catania University, Via S. Sofia 64, I-95123 Catania, Italy*
[4] *Institute of Modern Physics, Chinese Academy of Sciences, Lanzhou 730000, China*

We provide a microscopic calculation of neutron-proton and proton-proton cross sections in symmetric nuclear matter at various densities, using the Brueckner-Hartree-Fock approximation scheme with the Argonne V_{14} potential including the contribution of microscopic three body force. In the present calculation, the rearrangement contribution of three body force is considered, which will reduces the neutron and proton effective mass, and suppresses the amplitude of cross section. The effect of three body force is shown to be repulsive, especially in high densities and large momenta, which will suppress the cross section markedly.

Keywords: Three body force; Effective mass; Nucleon-nucleon cross section.

1. Introduction

During the impact phase between two heavy ions from high-energy central collisions, the nucleon density can reach values 2-3 times of the saturation density. Therefore one expects that, even if the mean field and the Pauli blocking still play a role in the medium modification of the nucleon-nucleon (NN) cross section, the effect of the modification of the interaction strength has the main role indeed. In the high-density domain the NN interaction is in fact deeply affected by the presence of the three-body force (3BF).[1,2] In the Brueckner theory the G-matrix plays the role of the in-medium scattering amplitude, the medium effects being the mean field and the Pauli blocking. In the high-density limit the additional medium effects are vertex corrections due to $N\bar{N}$ excitations and nucleon resonances (isobar Δ or Roper $N^*(1440)$). Their effect can be described in terms of a 3BF.[2] Recently it has been proved that the rearrangement contribution of 3BF to the self-energy is quite large,[3] and since it gives rise to a large reduction of the effective mass, one has to expect that it might have a strong influence on the in-medium cross section.[4–6]

2. Results

One can see in Fig. 1 that the 3BF induces a further suppression to the total cross section of identical nucleons as well as non identical nucleons. But here the

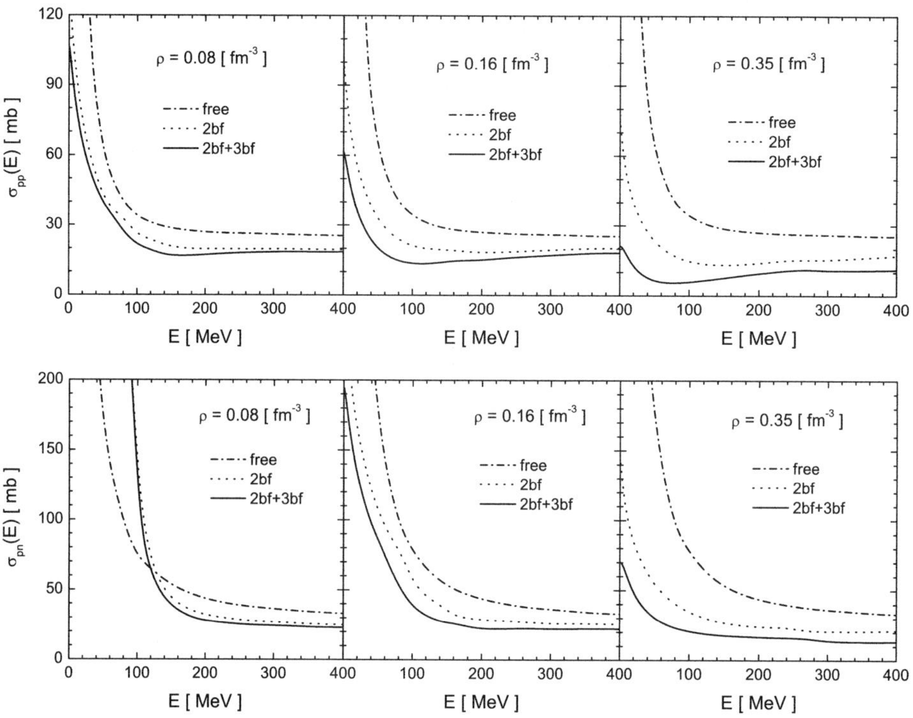

Fig. 1. Total nucleon-nucleon cross section with (solid lines) and without (dotted lines) the effect of TBF. The free space cross section is also shown for comparison.

main effect is the strong reduction of the density of states in the entrance and exit channels due to the rearrangement term of the self energy, which however is also traced to the 3BF. The effect of 3BF is shown to be repulsive, especially in high densities and large momenta, which suppresses the cross section markedly.

Acknowledgments

This work was supported by the EU grant CN/ASIA-LINK/008(94791) and the Natural Science Foundation of China (NSFC) under Grants 10775061, 10575119.

References

1. P. Grange, A. Lejeune, M. Martzolff and J.-F. Mathiot, *Phys. Rev. C* **40**, 1040 (1989).
2. W. Zuo, A. Lejeune, U. Lombardo and J.-F. Mathiot, *Eur. Phys. J. A* **14**, 469 (2002); *Nucl. Phys. A* **706**, 418 (2002).
3. W. Zuo, U. Lombardo, H.-J. Schulze and Z. H. Li, *Phys. Rev. C* **74**, 014317 (2006).
4. A. Bohnet *et al.*, *Nucl. Phys. A* **494**, 349 (1989).
5. G. Giansiracusa, U. Lombardo and N. Sandulescu, *Phys. Rev. C* **53**, R1478 (1996).
6. G. Q. Li and R. Machleidt, *Phys. Rev. C* **48**, 1702 (1993); **49**, 566 (1994)

TENSOR CORRELATIONS AND SINGLE-PARTICLE STATES IN MEDIUM-MASS NUCLEI

WEI ZOU, GIANLUCA COLÒ and PIER FRANCESCO BORTIGNON

Dipartimento di Fisica, Università degli Studi and INFN, Sezione di Milano, 20133 Milano, Italy

E-mail: weizou2048@gmail.com, gianluca.colo@mi.infn.it, pierfrancesco.bortignon@mi.infn.it

ZHONGYU MA

China Institute of Atomic Energy (CIAE), Beijing 102413, China

E-mail: mazy12@ciae.ac.cn

HIROYUKI SAGAWA

Center for Mathematical Sciences, University of Aizu, Aizu-Wakamatsu, Fukushima 965-8560, Japan

E-mail: sagawa@u-aizu.ac.jp

In the attempt to reach a global view of the effects on single-particle states caused by the tensor interaction, we analyze the evolution of the spin-orbit splittings in the Ca isotopes and in the N=28 isotones.

Keywords: Tensor force; Spin-orbit splitting; Hartree-Fock.

The role of the tensor effective interaction in the evolution of the single-particle states was first discussed, within the Skyrme Hartree-Fock (SHF) framework, by Fl. Stancu and co-workers,[1] almost thirty years ago. Until very recently, not so much attention has been devoted to the analysis of the effects of the tensor force on the evolution of the shell structure in the atomic nuclei. The issue has been addressed recently in Ref. 2. In Ref. 3, it has been shown that the effects of the tensor force can be understood in a very transparent way by inspecting the SHF formulas; in this work, the detailed study of the Sn isotopes and N=82 isotones has performed, by taking into account the effects of the triplet-even and triplet-odd tensor contributions separately. In the present work, which is a follow-up of Ref. 3, we extend our study of the tensor correlations to lighter systems. The goal is to see whether the same conclusions reached before, can be confirmed by looking at empirical data in another mass region.

We have performed a study of the single-particle states in the Ca isotopes by using the Skyrme interaction SLy5, and two different pairing interactions: a volume and a surface one. By comparing the pairing gap near the Fermi surface, and

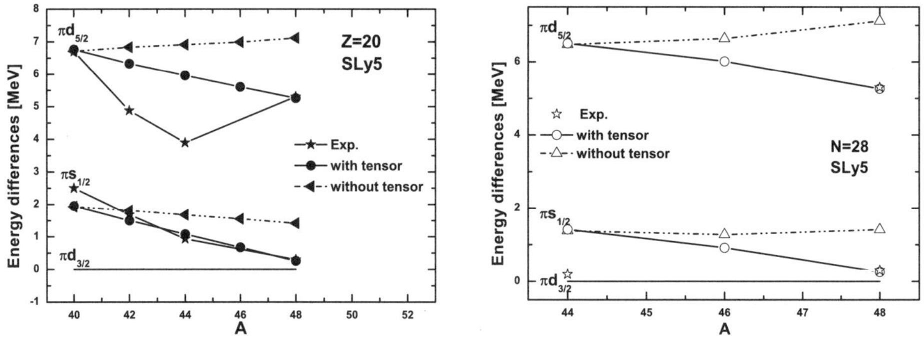

Fig. 1. Energy differences between the $1d_{3/2}$ single-proton state and, respectively, the $2s1/2$ and $1d5/2$ states, plotted for the Ca isotopes (on the left) and $N = 28$ isotones (on the right). The calculations are performed within HF-BCS, using the force SLy5 and the a density-dependent pairing force in the pairing channel. Results without and with the tensor terms are displayed; the experimental data are taken from Ref. 5.

the average pairing gap, with the results of the three-point and five-point formulas respectively, we can fit suitable parameters for both pairing forces. We have found (cf. the figure) that without the inclusion of the tensor terms, one cannot reproduce the trend of the experimental data. One can notice instead a substantial improvement due to the inclusion of the tensor interaction with the same parameters already employed in Ref. 3. In particular, the experimental results in the magic isotopes ^{40}Ca and ^{48}Ca are very well reproduced.

In order to get more understanding on the effects of the tensor force on single-particle states in nuclei, we also consider the evolution of the spin-orbit splitting along the N=28 isotones. Again, the figure shows that after considering the tensor force the results are much better.

In summary, our work shows that the inclusion of the tensor terms can fairly well explain the isospin dependence of spin-orbit splittings, not only in heavy nuclei but also in medium-mass nuclei, reinforcing consequently the conclusion reached in Ref. 3. More work along this line is in progress.

Acknowledgments

We acknowledge the financial support from the Asia-Link project [CN/ASIA-LINK/008(94791)] of the European Commission.

References

1. F. Stancu, D. M. Brink and H. Flocard, *Phys. Lett. B* **68**, 108 (1977).
2. T. Otsuka, T. Suzuki, R. Fujimoto, H. Grawe and Y. Akaishi, *Phys. Rev. Lett.* **95**, 232502 (2005); T. Otsuka, T. Matsuo and D. Abe, *Phys. Rev. Lett.* **97**, 162501 (2006).
3. G. Colò, H. Sagawa, S. Fracasso and P. F. Bortignon, *Phys. Lett. B* **646**, 227 (2007).
4. D. M. Brink and F. Stancu, *Phys. Rev. C* **75**, 064311 (2007).
5. P. Doll, G. J. Wagner and K. T. Knopfle, *Nucl. Phys. A* **263**, 270 (1976).

Participants

PARTICIPANTS

Deborah Aguilera – University of Alicante, Spain
deborah.aguilera@ua.es

Mark Alford – Washington University, St. Louis, USA
alford@wuphys.wustl.edu

Paolo Avogadro – Università di Milano, Italy
paolo.avogadro@mi.infn.it

Marcello Baldo – INFN Sezione di Catania, Italy
marcello.baldo@ct.infn.it

Francisco Barranco – University of Sevilla, Spain
barranco@us.es

Michał Bejger – Copernicus Astronomical Center, Warsaw, Poland
bejger@camk.edu.pl

Omar Benhar – INFN, Università "La Sapienza," Roma, Italy
benhar@roma1.infn.it

David Blaschke – University of Wroclaw, Poland
blaschke@ift.uni.wroc.pl

Patrick Blottiau – CEA/DIF, Bruyères-le-Châtel, France
patrick.blottiau@cea.fr

Ignazio Bombaci – Università di Pisa, Italy
bombaci@df.unipi.it

Alfio Bonanno – INAF - OACT, Catania, Italy
abo@oact.inaf.it

Luca Bonanno – Università di Ferrara, Italy
bonanno@fe.infn.it

Fiorella Burgio – INFN Sezione di Catania, Italy
fiorella.burgio@ct.infn.it

Li-Gang Cao – China Institute of Atomic Energy, Beijing, China
caolg@impcas.ac.cn

Brandon Carter – LUTH, CNRS Observatoire de Paris, France
brandon.carter@obspm.fr

Paolo Castorina – Università di Catania, Italy
paolo.castorina@ct.infn.it

Nicolas Chamel – Université Libre de Bruxelles, Belgium
nchamel@ulb.ac.be

John Clark – Washington University, St. Louis, USA
jwc@wuphys.wustl.edu

Gianluca Colò – Università di Milano, Italy
colo@mi.infn.it

Maurizio Consoli – INFN Sezione di Catania, Italy
maurizio.consoli@ct.infn.it

Verônica Dexheimer – FIAS, Goethe University, Frankfurt, Germany
dexheimer@th.physik.uni-frankfurt.de

Massimo Di Toro – Università di Catania and LNS-INFN, Italy
ditoro@lns.infn.it

Alex Dieperink – KVI, Groningen, Netherlands
dieperink@kvi.nl

Alessandro Drago – Università di Ferrara, Italy
drago@fe.infn.it

Camille Ducoin – LPC/ENSI, Caen, France
ducoin@lpccaen.in2p3.fr

Nicola Farina – Università di Roma "La Sapienza," Italy
nicola.farina@roma1.infn.it

Valeria Ferrari – Università di Roma "La Sapienza," Italy
valeria.ferrari@roma1.infn.it

Christian Fuchs – University of Tübingen, Germany
christian.fuchs@uni-tuebingen.de

Pawel Haensel – Copernicus Astronomical Center, Warsaw, Poland
haensel@camk.edu.pl

Matthias Hempel – Goethe University, Frankfurt, Germany
hempel@th.physik.uni-frankfurt.de

Hiroyuki Sagawa – University of Aizu, Japan
sagawa@u-aizu.ac.jp

Antonio Insolia – University of Catania, Italy
antonio.insolia@ct.infn.it

Danilo Jaccarino – Dipartimento di Fisica e Astronomia, Catania, Italy
danilo.jaccarino@libero.it

Elias Khan – Institut de Physique Nucléaire, Orsay, France
khan@ipno.in2p3.fr

Sachie Kimura – LNS-INFN, Catania, Italy
kimura@lns.infn.it

Kei Kotake – National Astronomical Observatory, Mitaka, Japan
kkotake@th.nao.ac.jp

James Lattimer – Stony Brook University, USA
lattimer@astro.sunysb.edu

Salvo Leotta – Università di Catania, Italy
salvo.leotta@ct.infn.it

Ang Li	– Lanzhou University, China `easylion@gmail.com`
Jun Li	– Università di Milano, Italy `jun.li@mi.infn.it`
Zenghua Li	– INFN-LNS, Catania, Italy `zhli@lns.infn.it`
Bennett Link	– Montana State University, USA `link@physics.montana.edu`
Umberto Lombardo	– Università di Catania, Italy `lombardo@lns.infn.it`
Zhongyu Ma	– China Institute of Atomic Energy, Beijing, China `mazy12@ciae.ac.cn`
Ruprecht Machleidt	– University of Idaho, Moscow, USA `machleid@uidaho.edu`
Chiara Maieron	– INFN Sezione di Catania, Italy `chiara.maieron@ct.infn.it`
Jérôme Margueron	– IPN Orsay, France `jmargue@ipno.in2p3.fr`
Toshiki Maruyama	– Japan Atomic Energy Agency, Tokai, Japan `maruyama.toshiki@jaea.go.jp`
Ai Jun Mi	– Lanzhou University, China `miaj05@lzu.cn`
Emilio Migneco	– Università di Catania and INFN-LNS, Italy `migneco@lns.infn.it`
Hamid Reza Moshfegh	– University of Tehran, Iran `hmoshfegh@ut.ac.ir`
Takumi Muto	– Chiba Institute of Technology, Japan `takumi.muto@it-chiba.ac.jp`
Tsuneo Noda	– Kyushu University, Fukuoka-shi, Japan `tsune@gemini.rc.kyushu-u.ac.jp`
Jorge Noronha	– FIAS, Goethe University, Frankfurt, Germany `noronha@fias.uni-frankfurt.de`
Sergey Pankratov	– Kurchatov Institute, Moscow, Russia `pankratov@mbslab.kiae.ru`
Guangxiong Peng	– Institute of High Energy Physics, Beijing, China `gxpeng@ihep.ac.cn`
M. Ángeles Pérez-García	– University of Salamanca, Spain `mperezga@usal.es`
Catia Petta	– Università di Catania, Italy `catia.petta@ct.infn.it`
Pierre Pizzochero	– Università degli Studi di Milano, Italy `pierre.pizzochero@mi.infn.it`

Salvatore Plumari – Università di Catania, Italy
salvatore.plumari@ct.infn.it

Artur Polls – Universitat de Barcelona, Spain
artur@ecm.ub.es

Jose Pons – Universitat de Alicante, Spain
jose.pons@ua.es

Constança Providência – Universidade de Coimbra, Portugal
cp@teor.fis.uc.pt

Nunzio Randazzo – INFN Sezione di Catania, Italy
nunzio.randazzo@ct.infn.it

Gerd Röpke – University of Rostock, Germany
gerd.roepke@uni-rostock.de

Valerio Russo – Università di Catania, Italy
valerio.russo@ct.infn.it

Basil Sa'd – FIAS, University of Frankfurt am Main, Germany
sad@th.physik.uni-frankfurt.de

Irina Sagert – Goethe University, Frankfurt am Main, Germany
sagert@th.physik.uni-frankfurt.de

Hideyuki Sakai – University of Tokyo, Japan
sakai@phys.s.u-tokyo.ac.jp

Michelangelo Sambataro – INFN Sezione di Catania, Italy
samba@ct.infn.it

Francesca Sammarruca – University of Idaho, Moscow, USA
fsammarr@uidaho.edu

Nicolae Sandulescu – NIPNE, Bucharest, Romania
sandules@ipno.in2p3.fr

Alexandre Santos – Universidade de Coimbra, Portugal
magno@teor.fis.uc.pt

Ananda Santra – Bhabha Atomic Research Centre, Mumbai, India
santra@barc.gov.in

Eduard Saperstein – Kurchatov Research Centre, Moscow, Russia
saper@mbslab.kiae.ru

Jürgen Schaffner-Bielich – Goethe University, Frankfurt, Germany
schaffner@astro.uni-frankfurt.de

Peter Schuck – Institut de Physique Nucléaire, Orsay, France
schuck@ipno.in2p3.fr

Hans-Josef Schulze – INFN Sezione di Catania, Italy
schulze@ct.infn.it

Andrew Steiner – NSCL/Michigan State University, USA
steinera@nscl.msu.edu

Baoyuan Sun – Peking University, Beijing, China
sunbaoyuan@gmail.com

Toshitaka Tatsumi	– Kyoto University, Japan tatsumi@ruby.scphys.kyoto-u.ac.jp
Pasquale Tomasello	– Università di Catania, Italy pasquale.tomasello@ct.infn.it
Roberto Turolla	– University of Padova, Italy turolla@pd.infn.it
Marco Valli	– Università di Roma “La Sapienza,” Italy marco.valli@roma1.infn.it
Eric Van Dalen	– DECM, Universitat de Barcelona, Spain vandalen@ecm.ub.es
Isaac Vidaña	– University of Barcelona, Spain vidana@ecm.ub.es
Xavier Viñas	– University of Barcelona, Spain xavier@ecm.ub.es
Hermann Wolter	– University of Munich, Germany hermann.wolter@lmu.de
Kentaro Yako	– University of Tokyo, Japan yakou@phys.s.u-tokyo.ac.jp
Dmitry Yakovlev	– Ioffe Institute, Saint-Petersburg, Russia yak@astro.ioffe.ru
Nobutoshi Yasutake	– Waseda University, Shinjuku, Japan yasutake@heap.phys.waseda.ac.jp
Tian Yuan	– China Institute of Atomic Energy, Beijing, China ytian@ph.tum.de
Dario Zappalà	– INFN Sezione di Catania, Italy dario.zappala@ct.infn.it
Julian Leszek Zdunik	– Copernicus Astronomical Center, Warsaw, Poland jlz@camk.edu.pl
Hongfei Zhang	– Lanzhou University, China zhanghongfei@impcas.ac.cn
En-guang Zhao	– Institute of Theoretical Physics, Beijing, China egzhao@mail.itp.ac.cn
Wei Zou	– Università di Milano, Italy zouwei2048@gmail.com
Wei Zuo	– Institute of Modern Physics, Lanzhou, China zuowei@impcas.ac.cn
Mikhail Zverev	– Kurchatov Institute, Moscow, Russia zverev@mbslab.kiae.ru

Participants of the Symposium I

Participants of the Symposium II

Members and Students of the ASIA-LINK Network during the Poster Session

AUTHOR INDEX